D1000265

The theories of continental drift and plate tectonics are underpinned by paleomagnetic studies. Paleomagnetism also connects with many other related geosciences, such as paleontology, structural geology and stratigraphy. This book explains the use and techniques of paleomagnetism to map the movement of major portions of the Earth's surface through time. Written for a geological audience, the initial chapters provide the basic essentials for understanding the value and significance of paleomagnetic results. The later chapters are unique in bringing together the vast amounts of available paleomagnetic data and analyses. This information is integrated with the paleogeography and tectonic movements of the blocks and placed in context with current tectonic hypotheses. A considerable proportion of the present continents are considered – that is the land areas that are now found around the Atlantic and Indian oceans and in Asia. This vast region of land was built, destroyed, deformed, incorporated and consolidated on the shores of the ancient Atlantic, Tethys and Iapetus oceans.

Also presented is an extensive catalogue of paleomagnetic results for all major continents and the displaced elements that they contain.

The evolving Phanerozoic worldscape opens up through the book, revealing the contribution that paleomagnetic results have made to the understanding of Phanerozoic tectonics.

Paleomagnetism

of the Atlantic, Tethys and Iapetus Oceans

Paleomagnetism
of the Atlantic, Tethys and Iapetus Oceans

Rob Van der Voo

Department of Geological Sciences
University of Michigan, Ann Arbor

Published by the Press Syndicate of the University of Cambridge
The Pitt Building, Trumpington Street, Cambridge CB2 1RP
40 West 20th Street, New York, NY 10011-4211, USA
10 Stamford Road, Oakleigh, Melbourne 3166, Australia

First published 1993

Printed in Great Britain at the University Press, Cambridge

A catalogue record for this book is available from the British Library

Library of Congress cataloguing in publication data

Van der Voo, R. (Rob)
Paleomagnetism of the Atlantic, Tethys, and Iapetus oceans/Rob Van der Voo.
p. cm.
Includes bibliographical references and index.
ISBN 0-521-41941-7
1. Paleomagnetism. 2. Plate tectonics. I. Title
QE501.4.P35V36 1993
538'.72–dc20 92-575 CIP

ISBN 0 521 41941 7 hardback

RL

Contents

Preface and acknowledgements

In the 1960s the study of paleomagnetism produced a revolution in the Earth Sciences, with the development of the theory of plate tectonics crowning many decades of debate about continental drift. Michael W. McElhinny published a book in 1973 that described these then-new developments. Although since 1973 several books have appeared on paleomagnetism, none have attempted to present and update the paleomagnetic contributions to Phanerozoic tectonics in the last two decades in a systematic way, that is with an emphasis on the paleomagnetic results available for a wide range of continental elements, rather than on paleomagnetic theory. This book attempts to remedy this situation in ways more fully introduced in the Prologue (Chapter 1).

It is a tremendous pleasure to acknowledge the invaluable help of many colleagues. As experts in tectonics, David Rowley and Shangyou Nie of the University of Chicago and Chris Scotese of the University of Texas at Arlington collectively produced some thirty single-spaced pages of scientific comments and suggestions for improvement. I am equally indebted to Eric Essene of the University of Michigan and John Geissman of the University of New Mexico, who also read through the entire text; they corrected many stylistic problems and helped clarify conceptually difficult issues. Paleomagnetists such as Michael McElhinny, Ron Merrill, and Valerian Bachtadse also read the entire book and provided many helpful comments. Many others (in alphabetical order) read parts or chapters: Enrique Banda, Bob Butler, Rob Hargraves, Dennis Kent, Cathérine Kissel, Carlo Laj, Hamza Lotfy, Bill Lowrie, Chad McCabe, Joe Meert, Neil Opdyke, Josep Maria Parès, Hervé Perroud, Steve Potts, Mustafa Saribudak, Trond Torsvik, Ben van der Pluijm, Zhongmin Wang and Hans Wensink. To all these colleagues I am very grateful for their efforts. Assistance with the figure drafting by Dale Austin, Joe Meert and Zhongmin Wang is greatly appreciated.

This book would not have come about without financial help from many sources. The data compilation occupied me for about four years and was funded by Shell Development Company in Houston, through Scott Cameron and Larry Garmezy of the Global Geology Section, and Peter Ziegler and Berend van Hoorn of the Hague. Ongoing support from the Division of Earth Sciences, the National Science Foundation, grants EAR 86-12469, EAR 88-03613 and EAR 89-05811, further facilitated my work. Financial support for the figure drafting has been provided by Dean John D'Arms of the Horace H. Rackham Graduate School of the University of Michigan. The actual writing took place mostly during a sabbatical leave, funded by the University of Michigan and the Ministerio de Educación y Ciencia, Dirección General de Investigación Cientifica y Téchnica (DGICYT) in Madrid. The warm and sustained welcome provided by colleagues and staff at the Institut Jaume Almera in Barcelona, especially Enrique Banda and Josep Maria Parès, the staff of the up-to-date library of the University of Barcelona, and the wonderful people of Catalunya deserves special mention.

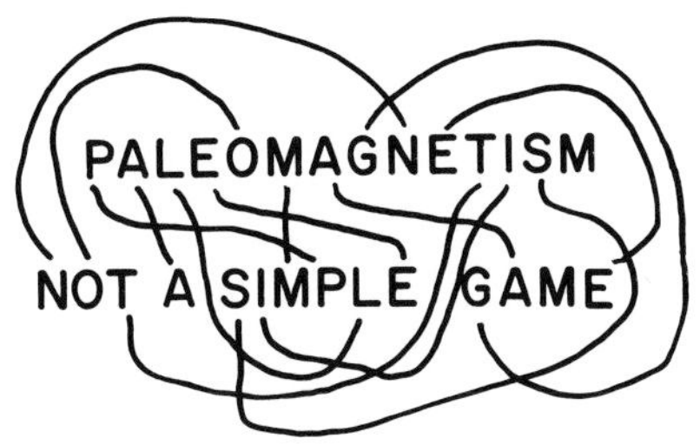

This drawing is an anagram, displaying the same letters on the top and bottom lines, but in different order. It is reproduced with permission of Chris Scotese, of the University of Texas at Arlington, who constructed this. It is an appropriate first 'quote' for this book, because it is true. After each of the chapters of this book, pages such as this one will come with other quotes or thoughts, nothing terribly profound, but just a chance to look back on the previous text and glance forward to the next set of issues; sometimes these quotes give me a chance to give a personal reaction to the material.

1 Prologue

Most people prefer to read of the exciting, unexpected and unusual. Scientists are no exception and the statement applies equally to newspaper items and scientific journal articles. Before starting this chapter, I was reading a brief article in a Dutch newspaper on 'Man Bites Sea Snake' (*De Telegraaf*, March 13, 1991, front page) – much more of an eye-catching story than the more numerous but less highlighted reports on 'Dog Bites Man'. For the 'man' involved, however, the encounter with the dog may have been just as exciting.

It is no different in paleomagnetism. We are awash in a sea of paleomagnetic results, and encounter many snakes in the grass as well as in the sea.

Paleomagnetic researchers these days must try to avoid the snakes by doing everything possible in terms of field sampling and re-sampling, rock magnetism and laboratory experiments, data analysis and fancy plotting to convince themselves and their colleagues that the result is indeed the useless, uninteresting, secondary and post-tectonic overprint that they thought it was when they did the first few measurements. Without the elaborate work, they would not succeed in getting the results published. A man-biting dog does not make the front page, for sure.

There are, however, also very gratifying moments in the life of a paleomagician, when she or he can pull a neat result out of the hatful of possible sampling targets; a result, moreover, that settles, or at least appears to settle, a long-standing controversy. The problem is then, that the audience – not skilled in the intricacies of paleomagic – does not have the faintest clue how to evaluate the ambitious and biting claim of our graduate student who is presenting her own first and supposedly significant results to a national or international public, either in person or in writing. Because I do not share the belief that paleomagnetism is wizardry, as I have heard some people proclaim, the subject matter must be made more accessible. This is one fairly obvious reason why I wrote this book.

Thus, this book aims to help the non-paleomagnetist to learn a few of the tricks of the trade and gradually to acquire useful skills in the evaluation of the results presented. It may not always be possible to distinguish dogs from snakes, but at least we can learn to tell whether the man's bite was successful!

But there is more to this book than just that. I have also tried to bring together the data (not so raw any longer, but still in their elementary and individual form), the regional geology in so far as relevant to the topic, the previous opinions, my own analysis, and a perspective on what further work is needed for a variety of regions and global problems. This is the second, and principal, reason for writing this volume: to bring together, inside one cover, the multitude of data and analyses now scattered throughout scores of journal articles. I believe that a need exists for such a book covering a majority of the continental paleomagnetic data.

Paleomagnetic results are now so abundant that no one can possibly be aware of all results, for the whole world, for all of geological time. To write a globally comprehensive monograph would not only be a herculean task, it would also be impossible because new data appear in the literature as fast as one can write a section of a chapter in a book like this. Because a new result usually changes the mean pole for the period and the continent concerned, this means that the writer must redo his averaging, which in turn will cause a change – usually a little, sometimes a lot – in the reference values used for terrane analysis. When the terrane displacements inferred from this analysis change as well, our writer, this Hercules, turns into a Sisyphus who has to redo his chapter. By that time, again a new result appears and the mini-(Wilson?-)cycle starts over. It is more or less like doing a census in a country the size of the USA without computers: by the time the human manpower (say, 1300 clerks) is able to come up with results, the population and demographics have changed dramatically. This present book, then, is limited in scope to post-Precambrian time and it also excludes the Pacific Ocean and most of its rims. This still leaves a fair amount of terrain and terranes to cover, nevertheless!

What *is* described is the evolving Phanerozoic worldscape, but exclusive of Panthalassa, that large hemispherical precursor of the Pacific Ocean that existed when almost all the continents were joined together in Pangea. Dutch sailors in the golden century of Holland called the Pacific the 'Silent Ocean', so I will be in good pirate company in keeping quiet about circum-Pacific issues. On the other hand, the area dealt with does include the continental bits and pieces that today are found around the Atlantic and Indian Oceans and in Asia (Figure 1.1).

The Greeks and Romans had it easy, although they may not have appreciated it. Their oceans were 'Our Sea' (*Mare nostrum*), which we call the Mediterranean, and the far-away Atlantic. People say knowledge about geography is declining universally, and perhaps this is because we have so many more oceans and seas to worry about. To top it all, the self-respecting geoscientist is expected to know also about

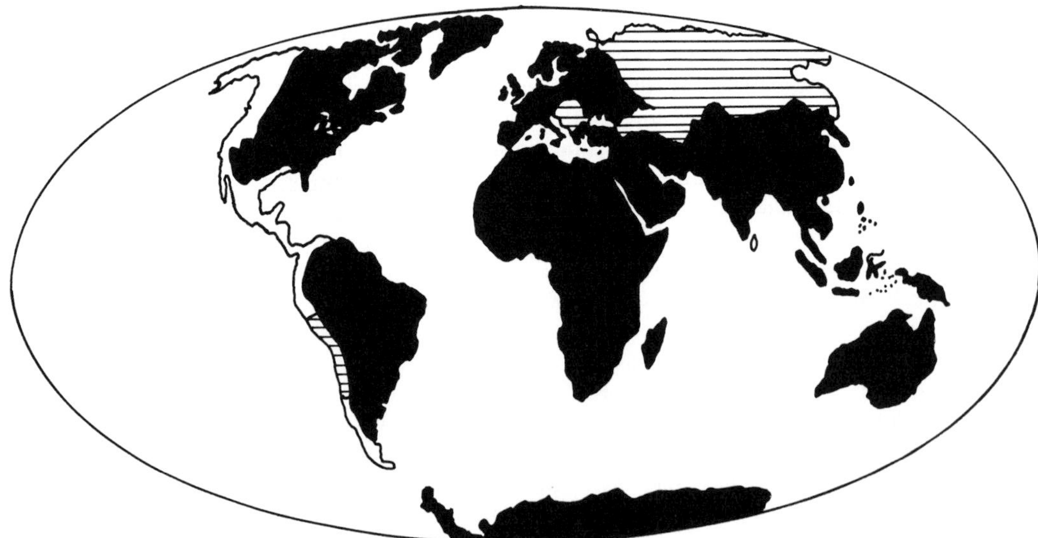

Figure 1.1. Map of the present-day world indicating the areas (black) for which the available Phanerozoic paleomagnetic data are discussed in this book; these data have been compiled in the appendices. Shaded areas are discussed in this book, but the data have not been included in the appendices. Continental areas left unshaded are not being discussed.

ancient oceans, long gone, which translates into *Kula* in some Native American tongue. There are not many snakes left in the Mediterranean, but there probably were lots of snakes in the Tethys.

Besides forming a coherent geographical package, the part of the world that I hope to cover was also built, destroyed, deformed, incorporated and consolidated on the shores of three remarkable oceans that have resounding names: Atlantis, Tethys, Iapetus. These three, moreover, were family.

Tethys and Iapetus were brother and sister, two of the 12 Titan siblings who as a group enjoyed – in those mythological days – a rather unpleasant reputation, because of incest, infanticide and cannibalism. Atlas, after whom the Atlantic was named (I guess), was the son of Iapetus. His grandparents were Heaven and Earth (Uranus and Gea). Atlas held the heavens on his shoulder and is usually the center statuette on large antique pendulum wallclocks. He was – literally – petrified by Perseus in the end. His father Iapetus was 'married' to Atlas' aunt Themis, also one of the Titan sisters, and was considered the father of all mankind. Tethys, another aunt of Atlas, also had an incestuous relationship with her brother Oceanus by whom she had an offspring of some 3000 oceanids as well as all of the Earth's rivers. Tethys, besides being the mother of all oceans, was also known as the lovely queen of the sea. Lastly, one of the more famous deities of Greek mythology was Zeus, cousin of Atlas, who had many wives after he was saved by mother Rhea from being eaten alive by his father Cronus; some of these spouses were also members of the Titan clan.

Şengör (1987) reports that the first use of the idea of Tethys as an oceanic domain that separated Africa, India and Australia on the one hand, from Europe and Siberia on the other, was by Melchior Neumayr and his father-in-law, Eduard Suess. Suess (1893, 1901) recognized that the present Mediterranean and the surrounding Alpine–Himalayan mountain belt contained the relicts of an influential ancient oceanic domain. This concept received wide recognition early on, because of the prestige of the Viennese school of geological thought, and it has been a central feature of paleogeography, paleontology, sedimentary facies analysis and tectonics ever since.

With the nowadays general acceptance of continental drift and plate tectonics, the Tethys concept has gained enormously in relevance. The drift of the major continents and the mobility of smaller continental elements now incorporated into the spectacular Alpine–Himalayan mountain belt provide a framework for understanding the changing face of this part of the Earth. As we will see in this book, it is very unlikely that there are any remnants of Tethys left as modern ocean floor. The present-day oceanic depressions in the Mediterranean are all successor basins to Tethys.

The name Iapetus for an ancient ocean that disappeared in the Caledonian mountain building episode was coined by Harland & Gayer (1972). The outline, extent and Paleozoic longevity of this Iapetus Ocean are poorly known, but from continental geology in Greenland, Norway, the British Isles, the Canadian Maritime Provinces and the Atlantic Seaboard of the USA it is clear that the once-continuous Caledonian–Appalachian mountain chain formed in a continent–continent collision after the elimination of the ancient intervening Iapetus Ocean basin. The Iapetus concept, then, is also firmly rooted in continental drift and plate tectonics.

Tethys and Iapetus were Paleozoic oceans, with the former continuing to exist into the Mesozoic and Early Tertiary. The birth of their nephew, the Atlantic Ocean, took place about 175 million years ago (Ma) in the Mesozoic.

This book will examine the drift of the major continents and the movements of the smaller flotsam, that is, the continental pieces and relicts of the ocean floor, in, around, and along the shores of the Atlantic, Tethys and Iapetus oceans. We will see how these peripatetic continental elements have continuously changed their positions during the last 600 million years.

My younger son, who is in transition from high school to freshman year in college, has complained to me that he found it hard to understand the text of the manuscript pages that lie scattered in the Barcelona appartment, which we rented while I spent my sabbatical there. He is correct. This book is not directly written for the layman. To avoid all misunderstandings, let me emphasize for whom this book is, and is not, conceived.

Who should probably not read this book. The unsuspecting scholar of Greek classics and mythology, perhaps attracted by the rootword παλαιος in the title, will probably be disappointed – unless he or she has a strong side interest in geoscience and has had about a year of college geology, so that the jargon is not totally unfamiliar. While it is not a requirement that readers of this book know the meaning of words like 'ophiolite' or 'granodiorite', it is recommended that such persons know where to look up their definition.

A beginning student of geology or related fields would probably do well to wait until the material of a few courses of introductory geology has been digested. It is very likely that high school students will find the material far too esoteric, judging by my son's reaction.

Who might benefit from reading this book. Researchers and graduate students in paleomagnetism seem to be the logical audience, but they are not the only ones I had in mind. Rather, this volume is intended for geologists with no prerequisite knowledge of the field of paleomagnetism. While there exists a considerable body of background knowledge in rock and paleomagnetism, this book includes only that material that I believe to be essential for grasping the value and significance of paleomagnetic results. The first three chapters after this Prologue provide these essentials.

Although paleomagnetism is generally grouped together with geodesy, seismology and potential fields in the 'Solid-Earth Geophysics' category, this book is not a geophysics book and it is certainly not mathematical. One of the nicer aspects of paleomagnetism is that it connects and bridges easily with many of the other geoscience disciplines, such as paleontology, structural geology and tectonics, oceanography, or sedimentology and stratigraphy, through the applications of paleomagnetism such as magnetostratigraphy, the determination of rotations, terrane displacement analysis and paleogeography.

Two last caveats. Although this book deals with a great variety of interesting geological problems, it is not a comprehensive treatment of regional or global tectonics. The topics included are specifically those where paleomagnetism has had, or can have, a chance to contribute. The second and important notice I wish to give is that this book is not intended to transform an earth-science student into a paleomagnetist, 'licensed' to practice the 'magic'. There are many better and more extensive texts available for that purpose; the latest one I might refer to is Butler (1991). While the following three chapters provide background material, necessary to *understand* paleomagnetic data, they do not constitute a manual for *executing* paleomagnetic analysis. Chapters 5–8, on the other hand, contain the real 'meat' of this book and provide, in my opinion, the sort of integrated paleomagnetic analysis that cannot be found elsewhere within one cover.

Well, on with the subject. Let's get our flutes out, snake charmers, and go blue.

We didn't get beat. We were just behind when the clock ran out.

Coach H. Schnellenberger after his Louisville team lost to that of Ohio State University, September 15, 1991.

The culture differences between campuses in America and elsewhere are quite large. The influence of collegiate sports in America, in particular, is noteworthy. Europeans are likely to wonder what 'Go Blue' means (see the last lines of chapter 1), whereas US-trained scientists just know; it is a yell spurring the team on, of course.

What does this, or the above quote, have to do with paleomagnetism? Well, in American traditions many things are described in terms of a game. And most games end when the clock runs out. Not so in the game of the Paleomagnetists versus Mother Nature. Fortunately for the paleomagnetists the clock does not run out, at least not soon. It remains to be seen which team has the longest breath, but – as you will see in this book – the paleomagnetists appear to be less behind now than they were a decade ago.

2 Paleopoles and paleomagnetic directions

The measurement of a rock sample magnetization is expressed by three values: the two angles of declination (D) and inclination (I) and the scalar magnetic intensity value (J). The latter is, of course, of limited value for tectonic purposes. The declination is an angle in the locally defined horizontal plane, measured clockwise from 0 to 360°, with reference to the geographic meridian (i.e., true, present north). In terms of the present-day geomagnetic field, it would be the angle between the direction in which a compass needle is pointing and the reference line of the meridian. Modern day declinations, except in very high latitudes, are typically less than 15°; because the magnetic poles are close to the geographic ones, the compass needle pointing towards the magnetic (north) pole, points also more or less in the direction of the geographic (north) pole.

The inclination is the angle, in the vertical plane, between the magnetic direction and the horizontal. Thus a horizontal magnetic direction has an inclination of 0°, and for a vertical direction I = 90°. Convention dictates positive I for downward and negative I for upward directions.

While the declination tells us the direction in which to seek the pole, the inclination provides the distance between the site and the pole. Together, then, D and I define a magnetic direction and its corresponding pole. However, some assumptions are made along the way, so it may not be as simple as that.

The magnetic field today is that of a geocentric dipole field only for about 80% of its total value. 'Geocentric' means that this dipole field can be thought of as originating at the center of the Earth. The remaining 20% of the field are contributions from dipole fields that are not geocentric (e.g., remanent magnetizations in the crust), or non-dipole fields, or fields of external origin that relate to the solar and interplanetary magnetic fields. This 'restfield' is difficult to understand for the typical geologist not trained in geophysical theory. For those wishing to acquire a better understanding of it, I refer to Merrill & McElhinny (1983) or Butler (1991). However, to appreciate paleomagnetic results, such detailed knowledge is unnecessary. The following is merely an elementary introduction to the issue. Terms such as dipole (with two poles), quadrupole, octupole (with eight poles), or geocentric fields are representative of the mathematical/geometrical descriptions that we can use to define the components that make up the total field. The effects of non-geocentric dipole fields are equivalent in part to those of geocentric non-dipole fields. Thus, the bothersome 20% restfield can be described well mathematically, but the physical basis is non-unique and hence, poorly understood as to its origin.

A further complexity of the geomagnetic field, however, is a blessing in disguise. The restfield described above, partly of internal and partly of external origin, displays rapid variations. The external field, for instance, can vary enormously in just a few hours during magnetic storms. The internal field varies rapidly, that is on a geological time scale, such that every five years new world magnetic maps of declination, inclination and field strength need to be made for precise navigation purposes. In a hundred years or so, the declination may change some 10° in a quasi periodic pattern that is called secular variation (secular is a word derived from Latin, implying 'on a century-scale'). A fuller discussion of secular variation can be found in Butler (1991).

Because of the rapid temporal variations of the restfield, we need only to sample the field for several centuries and the variations begin to be averaged out. Even the non-coaxial components of the dipole field, when averaged over a long time, cancel themselves this way, resulting in a long-term average field nearly exclusively of geocentric, dipolar origin aligned with the rotation axis (coaxial field). The long-term non-dipole field, sampled and averaged over millions of years, may nevertheless not be completely zero (Merrill & McElhinny, 1983; Schneider & Kent, 1990 a,b), but the effects are small, amounting to an uncertainty in paleomagnetic pole positions of up to 5° or so. These 5° (556 km) may not be considered small on a human scale, but we will see below that the typical paleomagnetic uncertainty is generally of the same order or greater.

Thus, paleomagnetic studies are all based on a fundamental assumption – in early papers this was always explicitly stated as 'granted the assumption of a geocentric dipole field . . .'. If the assumption is indeed granted, then we can use the dipole model to convert an inclination into (paleo)latitude (λ) using the dipole formula

$$\tan I = 2 \tan \lambda$$

The dipole formula is graphically illustrated for both normal and reversed polarities in Figure 2.1, whereas Figure 2.2 shows cross sections through the Earth to illustrate representative inclinations for normal and reversed fields. I re-emphasize the fact that for the inclination–latitude relationship to be valid, it is implied that effects of secular variation and non-dipole fields have been 'cancelled' (averaged) by sampling over at least many thousands of years. At first glance, the two plots of Figure 2.2 seem very similar, but the vectors point in opposite directions. Thus, after a reversal the declinations have become antipodal and point to what used to be the south pole (i.e., they have changed by 180°) and the inclinations changed from downward to upward or vice versa (i.e., they have changed sign, but not absolute value).

When we have the paleolatitude (= distance to the equator), we automatically

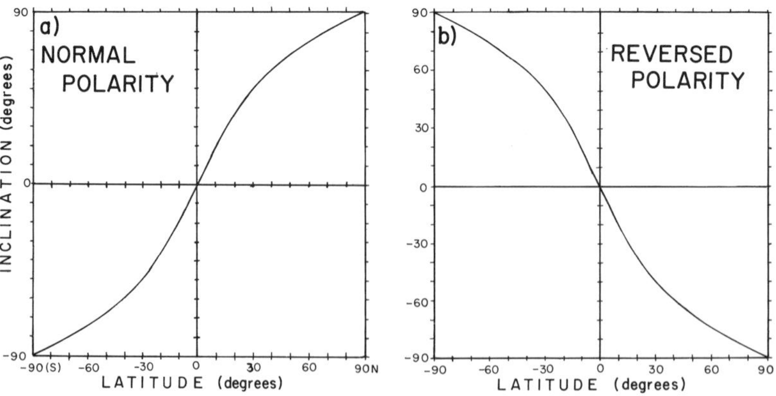

Figure 2.1. Inclination as a function of latitude for normal and reversed polarity, according to the dipole formula.

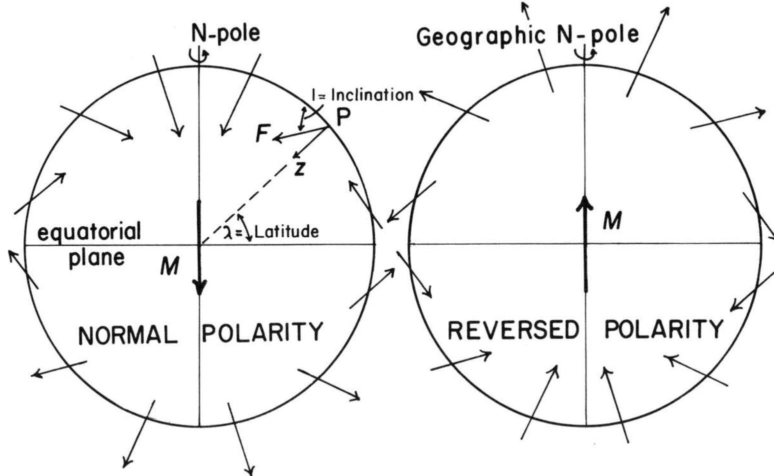

Figure 2.2. Magnetic field vectors as a function of latitude for normal and reversed polarity plotted in a cross section through the Earth. For normal polarity, the declinations (which cannot be seen directly in the figure) all are pointing to the north pole, whereas for reversed polarity the declinations point to the south pole. The angle between a local horizontal plane (tangential to the Earth's surface) and the plotted vectors is the inclination, positive when downward; for reversed polarity, for instance, the inclinations are downward (positive) in the southern hemisphere.

know the supplementary distance to the pole (p), and recalling that the declination tells us which way to travel from our site (at modern geographic latitude λ^* and longitude ϕ^*), we arrive easily at our pole position. In practice, the calculation of a pole position (at modern latitude λ' and longitude ϕ') is done with spherical trigonometric formulae by computer, according to the following steps (see Figure 2.3 for visualization of the parameters; after McElhinny, 1973, his figure 18 and text of p. 25)

$$\sin \lambda' = \sin \lambda^* \cos p + \cos \lambda^* \sin p \cos D \qquad (-90° \leq \lambda' \leq +90°)$$

$$\phi' = \phi^* + \beta \text{ when } \cos p \geq \sin \lambda^* \sin \lambda'$$

or

$$\phi' = \phi^* + 180° - \beta \text{ when } \cos p < \sin \lambda^* \sin \lambda'$$

where

$$\sin \beta = \sin p \sin D / \cos \lambda' \qquad (-90° \leq \beta \leq +90°)$$

Latitudes are positive (negative) for the northern (southern) hemisphere, whereas longitudes are measured positive eastward from the Greenwich meridian.

The fact that the geomagnetic dipole field is known to have reversed itself many times in the geological past (more than one hundred times since the Cretaceous alone), means that we do not automatically know whether to call our paleopole a north pole or a south pole. Not knowing whether a pole is a north or south pole means that we do not know whether the paleolocation of our sampling areas was in the northern or southern hemisphere; this hemispheric ambiguity can play a serious role in the determination of terrane displacements, as illustrated in the following example.

For most major continents, a sufficient number of pole positions is available to see the logical progression from today's north pole to older and older north pole positions, so that the hemispherical ambiguity is only a problem in the Precambrian. But for small displaced terranes, the luxury of having many paleopoles at our disposal usually does not exist. One example is the displaced terrane of Wrangellia, now found dispersed in locations ranging from Alaska's Wrangell Mountains through Vancouver Island to perhaps a small location in Idaho (Jones, Silberling & Hillhouse, 1977). Triassic basalts, very diagnostic of each of these dispersed locations, have yielded excellent paleomagnetic data (e.g., Hillhouse, 1977; Yole & Irving, 1980; Panuska & Stone, 1985). But the Triassic field is also known to have reversed itself many times, so that the polarity is not immediately known. Therefore, the inclination of 28° observed at one of the localities in Alaska can mean a paleolatitude of provenance of either 15° N (according to Figure 2.1a) or 15° S (according to the reversed polarity diagram of Figure 2.1b), as illustrated in Figure 2.4. Thus, a choice of hemisphere of provenance for the Triassic of Wrangellia either depends on other, non-paleomagnetic evidence, or it must await an abundance of younger paleopoles, more or less continuous in time from the Triassic to the present, so that we can tie in the Triassic pole with either the present-day north or south pole.

Imperceptibly, we have already waded into one swampy issue that often confuses beginning students of paleomagnetism: why does everyone show a moving paleopole, when it is the continent or terrane that moved? Guilty, as charged. Although it is possible that the polar axis has moved with respect to the whole Earth, a process called true polar wandering (TPW), the magnitude of TPW has by all standards been small to negligible in comparison to the drift velocities of the plates since 150 Ma (Jurdy & Van der Voo, 1975; Jurdy, 1981; Harrison & Lindh, 1982a;

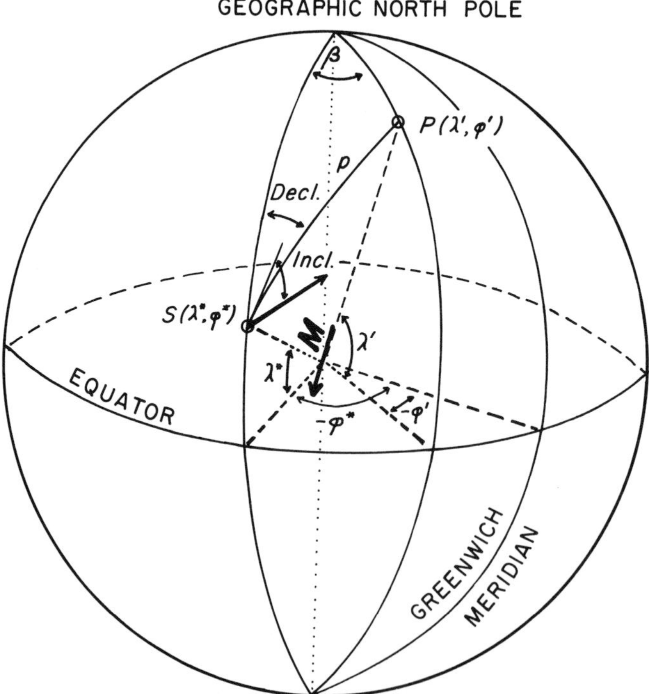

Figure 2.3. Plot showing the variables in the calculation of a paleopole P (with coordinates λ', ϕ') from the declination (*Decl*), inclination (*Incl*) and field site coordinates S (λ^*, ϕ^*). Latitude is represented by λ (positive, when northerly), and longitude by ϕ (positive eastward, negative westward). Figure redrawn and modified after McElhinny (1973).

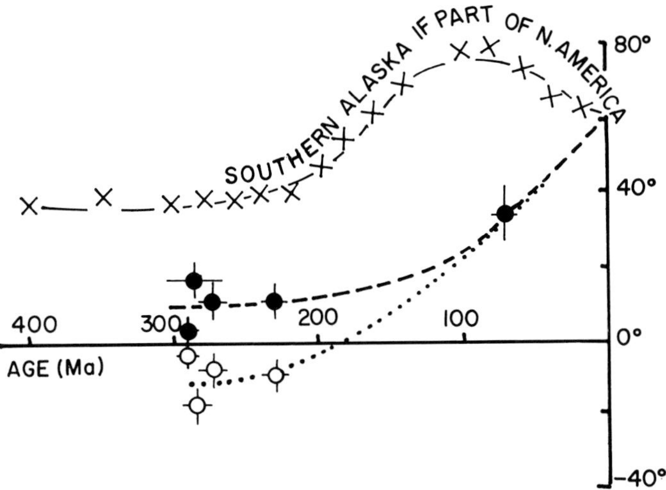

Figure 2.4. A plot of (paleo)latitude versus age for selected paleomagnetic results from the Wrangellia terrane in southern Alaska. Both northern (closed circles) and southern paleolatitude options (open circles) are possible, unless the polarity of the results is assumed. The predicted paleolatitudes for southern Alaska, if it had remained attached in its present location with respect to the North American craton, are shown by crosses. The magnitude of the northward drift of Wrangellia with respect to the craton is therefore at least some 20° (northern hemisphere original location) and possibly as large as 40° or more (southern hemisphere choice). Modified and redrawn after Howell (1989) with data from Hillhouse & Grommé (1984) and Panuska & Stone (1985), and published with permission of the author (Howell, D., 1989) and the publisher (Chapman and Hall).

Andrews, 1985; Gordon, 1988). On the other hand, there is little doubt that the continents moved and still move; modern satellite-based distance measurements (Chapter 6) show this convincingly, as does an entire array of plate tectonic evidence. Thus, ideally, paleomagnetists should hold the pole fixed and show a moving continent. But here, many problems arise. First, it is easier to plot a pole (a point), than it is to draft a continental outline of, say, Europe as it moves successively in steps over the surface of the Earth. With computers, the latter can be done today with some ease, but even so the plotting of a point is much simpler, and the same applies to a series of sequential points (an apparent polar wander path, or APWP) as a curve. Second, the paleopole is a unique point, well determined from the input of site location, D and I. However, the inverse problem is accompanied by introduction of a degree of freedom, such that when the site location, as well as D and I are used to position the continent holding the pole fixed, the longitudinal position of the continent is indeterminate. This is illustrated in Figure 2.5, showing that, when a continent is placed at a certain latitude, then all longitudinal positions give the same D and I with respect to the polar axis held fixed. This results, of course, from the axial symmetry built into the dipole model.

Third, a paleopole position is uniquely and simply described by its latitudinal and longitudinal coordinates in a table, but listing the coordinates of an entire continent would be a waste of space. Printing D and I for this continent requires that the site be specified, as these parameters (but not the paleopole) change from location to location. Fourth and last, it might be possible and certainly defensible to show a paleolatitudinal position of the continent, with respect to a fixed polar axis, but what about the illustration of a paleomagnetic direction from an accreted sliver or rotated thrust sheet? Thus, as paleomagnetism developed as a subdiscipline over the years, the custom took hold to plot paleopoles for a fixed (modern day) position of the continent. Paleopoles, arranged in a temporal order, together form an APWP. The path is called 'apparent' to remind us that it is only an imaginary movement of the pole with respect to the continent held fixed. Figure 2.6 illustrates the APWP concept side-by-side with a plot of the continent moving while holding the pole fixed; in the plot of the paleogeographical locations of the continent, no more than about half a dozen positions can be given until the figure becomes too crowded. Once again it must be noted that in such a figure the longitudes of the ancient continental positions are arbitrary.

The secular variation of the geomagnetic field implies that a spot reading of the field, such as that obtained from a rapidly cooled lava flow, is very unlikely to represent a pure geocentric dipole field. Such a result is called a virtual geomagnetic pole (VGP). A sequence of many lava flows, needed to average the secular variation, ideally would cover tens of thousands of years. Poles obtained from sections of sediments, spanning some tens of meters, or from slowly cooled igneous bodies, on the other hand, probably sampled the ancient field sufficiently to average to a geocentric dipole direction; they can then be called 'paleopoles', signifying that secular variation is thought to have been averaged.

Typical paleomagnetic results should, therefore, display some scatter in their sample directions (e.g., Figure 2.7a); if this scatter is too low, the mean direction may be only a VGP. In the case of many paleomagnetically studied Paleozoic limestones from the North American interior, the scatter is low (Figure 2.7b) for a different reason: these rocks are thought to have been remagnetized during the Late Carboniferous–Permian period (McCabe & Elmore, 1989), by new growth of magnetite during (very) late diagenesis (Suk, Van der Voo & Peacor, 1990a; Suk, Peacor & Van der Voo, 1990b). This growth may have taken such a long time, that secular variation is averaged on a sample scale by the slightly different ages of the

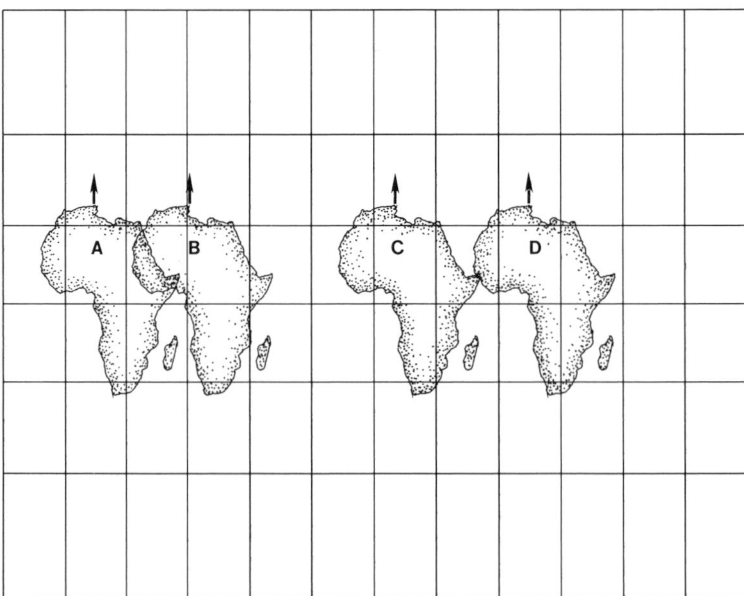

Figure 2.5. Plot showing that the magnetic vector remains the same for a given site location if the continent does change longitude (positions A–D), that is, if it is rotated only about the polar rotation axis, keeping latitude and meridional orientation unchanged.

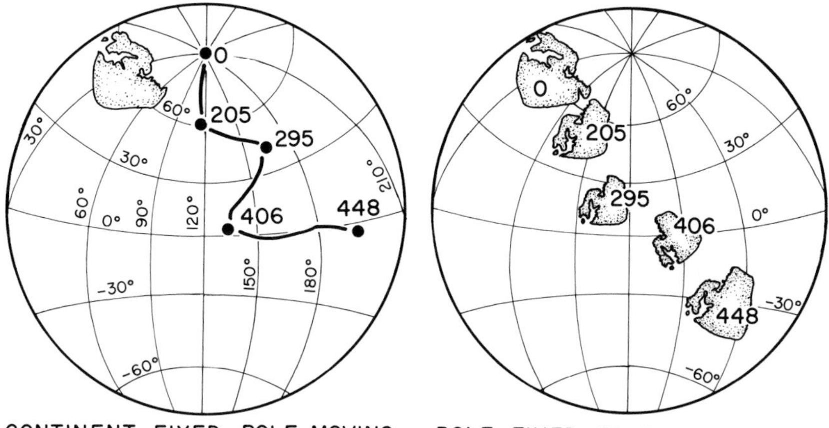

Figure 2.6. Paleopoles, displayed for a continent held fixed (left), are connected to form an apparent polar wander path. On the right, the same continent (the Baltic Shield/Russian Platform of Europe) is shown in its paleolatitudinal positions for five different times, while holding the pole fixed. These paleogeographic positions correspond to the five paleopoles on the left labeled with their ages (Ma). Note that the paleogeographic locations have been assigned arbitrary paleolongitudes, that is, the continent can be moved east- or westward with impunity.

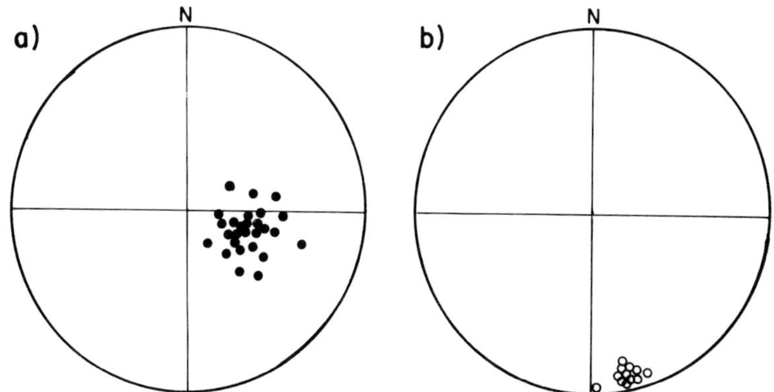

Figure 2.7. Two examples of typical well-clustered paleomagnetic directions. a) Scatter that is normal for a primary magnetization acquired over some length of time, and b) scatter that is not large enough (results are too clustered) for the magnetization to be primary. In this case the magnetization was acquired over such a long time that it averaged typical secular variation in each sample.

millions of tiny, submicron magnetite grains (e.g., Figure 2.8) that are found in a typical sample of ten cubic centimeters.

'Scatter' in paleomagnetic results is a vague term. It can be quantified by the use of statistical parameters devised by Fisher (1953), who – as the anecdote has it – produced the treatment ready for publication from a drawer in his desk when his colleague Keith Runcorn approached him about the best way to deal with typical paleomagnetic errors. A full treatment of these parameters is given in books covering the basis of paleomagnetic measurement and analysis (e.g., Irving, 1964; McElhinny, 1973; Tarling, 1983; Collinson, 1983; Piper, 1987), so only the essentials will be presented here and the formulae given below will not be derived or justified. Paleomagnetic results can be thought of as directions penetrating a sphere of unit radius. For circularly distributed directional statistics, in which vector length is uniform (unity), two parameters are commonly used: one is α_{95}, the semi-angle of the cone of confidence around an observed mean, such that the true mean of a

Figure 2.8. Scanning electron microscope images of small magnetite grains in thin sections of carbonate rocks from New York State. Bars in μm for scale. a) Spheroidal aggregate of mostly magnetite, but with occasional bright cores (see arrow) where relict pyrite is found. The interpretation is that the magnetite is replacing the pyrite. b–d) Similar spheroidal aggregates in voids or cracks. Note the octahedral crystal shape. The matrix is calcite. From Suk, Peacor & Van der Voo (1990b), reprinted by permission from Nature. Copyright © 1990 Macmillan Magazines Ltd.

population is thought to lie with 95% probability within this cone. Occasionally α_{63} is used (63% probability), which approaches (but is not identical to) the standard deviation about the mean. α_{95} is calculated from the formula

$$\cos \alpha_{95} = 1 + [(N - R)/R]\,[1 - (20)^{**}(1/(N - 1))]$$

where N is the number of entries (e.g., samples or site-mean directions), R is the resultant sum of the N vectorially added individual unit vectors, such that $R \leq N$, and ** signifies to the power of the term that follows.

The other commonly used measure of scatter is the precision parameter (estimate) *k*, which is an approximation of the parameter $\varkappa$, which itself governs the probability density distribution (P) of an ideal population of directions on a sphere, such that

$$P = [\varkappa/(4\,\pi \sinh \varkappa)]\,\exp\,(\varkappa \cos \psi)$$

where ψ is the angle between the direction of a sample and the true mean direction at which $\psi = 0$ and density is a maximum. Because this distribution P can only be determined by an infinite number of sample measurements, $\varkappa$ must be estimated when a limited number of samples is used. This precision estimate is represented by k, such that

$$k = (N-1)/(N-R)$$

where N and R are as defined above.

Convention prescribes the use of the symbols k and α_{95} when the statistical parameters are calculated for the mean of N directions, each consisting of declination and inclination pairs. K and A_{95} are used, when the mean of N VGPs is calculated. Alternatively, an oval area of confidence about the paleopole can be calculated from α_{95}. This oval has principal axes aligned parallel with (dp) and perpendicular to (dm) the great circle connecting paleopole and site, such that

$$dp = \alpha_{95}\,[1 + 3\,(\cos p)^{**}2]/2$$
$$dm = \alpha_{95} \sin p/\cos I$$

where p is the distance between site and paleopole (the colatitude).

In the use of A_{95} or the oval of confidence described by dp and dm, lies a fifth reason why paleomagnetists hold the continent fixed, while plotting the 'apparent' movement of the pole (see earlier discussion). Paleomagnetism has its roots in the principles of physics; physicists, when measuring, are accustomed to describing the error limits (e.g., α_{95}) of their results. This is easily done and illustrated when the continent is held fixed and the pole is allowed to move. On the other hand, plotting a continental paleoposition with respect to a fixed pole does not allow the associated uncertainties to be displayed.

The parameter k can range from one (for large N when $R = 0$, i.e., completely randomized results) to infinity (when $R = N$, i.e., all directions are parallel). Typically $k < 3$ means that the result is essentially random such that its mean direction and α_{95} are meaningless, whereas $k > 150$ (for entries representing samples) is often taken to mean that the results are too well clustered and may represent only a VGP or a remagnetized result.

It should also be noted that k is more or less independent of N for typical paleomagnetic collections, where N is usually greater than 10 and in the last few decades often includes 60 samples or more. On the other hand, α_{95} depends directly on the number of samples; the larger N, the smaller α_{95} for the same distribution of directions (same k). By simply collecting and measuring more and more samples, one can reduce α_{95} to values around a few degrees for typical paleomagnetic results. This is inherent in Fisherian statistics, but one has to keep in mind that in a typical paleomagnetic study other sources of error are propagated that are not necessarily averaged by taking more and more samples; such errors may include those related to non-dipole fields, mentioned earlier, incomplete demagnetization (not infrequent, see Chapter 4), errors in the determination of structural corrections for tilt (frequently several degrees), systematic bias introduced by laboratory measurements such as the remanence of a sample holder (rare), or systematic but small orienting errors in the field (not so rare). My own rule of thumb is to take the typical 'area of confidence' around a result to have a radius of 5° or more, regardless of the value listed in the original publication for its α_{95}, unless I am convinced that extreme care has been taken in eliminating systematic biases.

This previously discussed minimum of the 'radius of confidence' becomes impor-

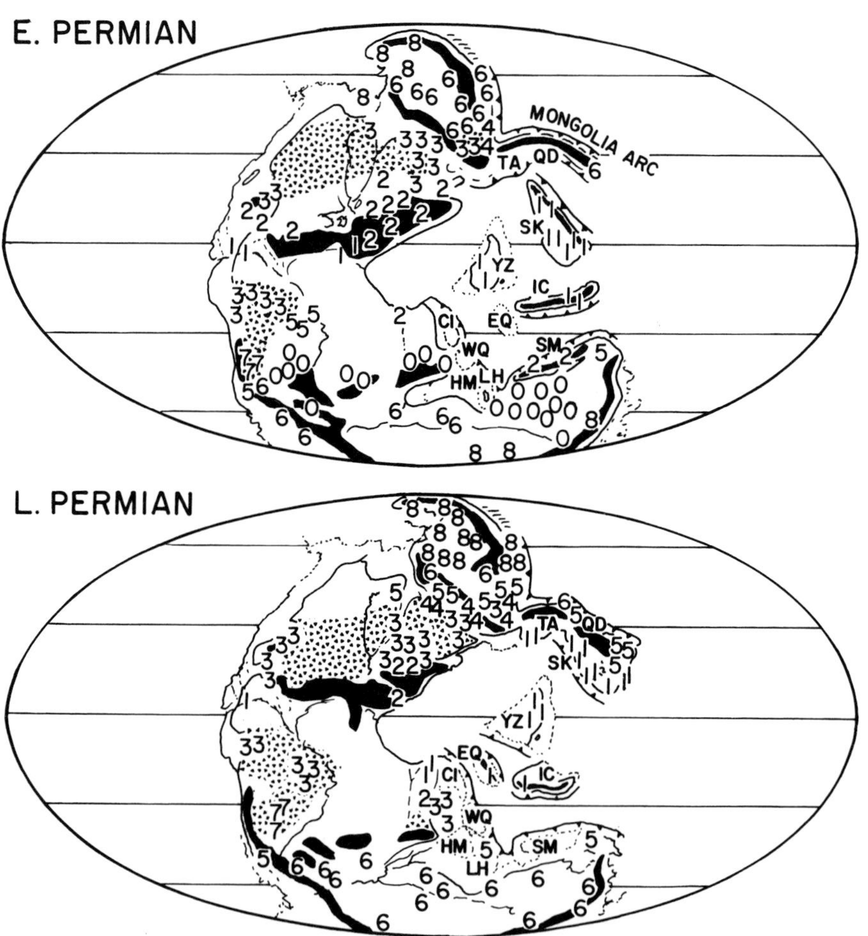

Figure 2.9. Early (E) and Late (L) Permian paleogeography with quasi-zonal vegetation designations (called biomes; from Ziegler, 1990), ranging from tropical 'everwet' (biome 1) to polar 'glacial' (biome 10, indicated by 0). Dotted areas indicate the presence of evaporites, black areas are mountains. Abbreviations: CI = Central Iran, EQ = East Qiangtang, HM = Helmand (Afghanistan), IC = Indochina, LH = Lhasa, QD = Qaidam, SK = North China and Korea, SM = Sibumasu, TA = Tarim, WQ = West Qiangtang, YZ = Yangtze (i.e., South China). The reconstruction is from Lottes & Rowley (1990), and the figure is redrafted after Nie, Rowley & Ziegler (1990), and published with permission from the Geological Society Publishing House.

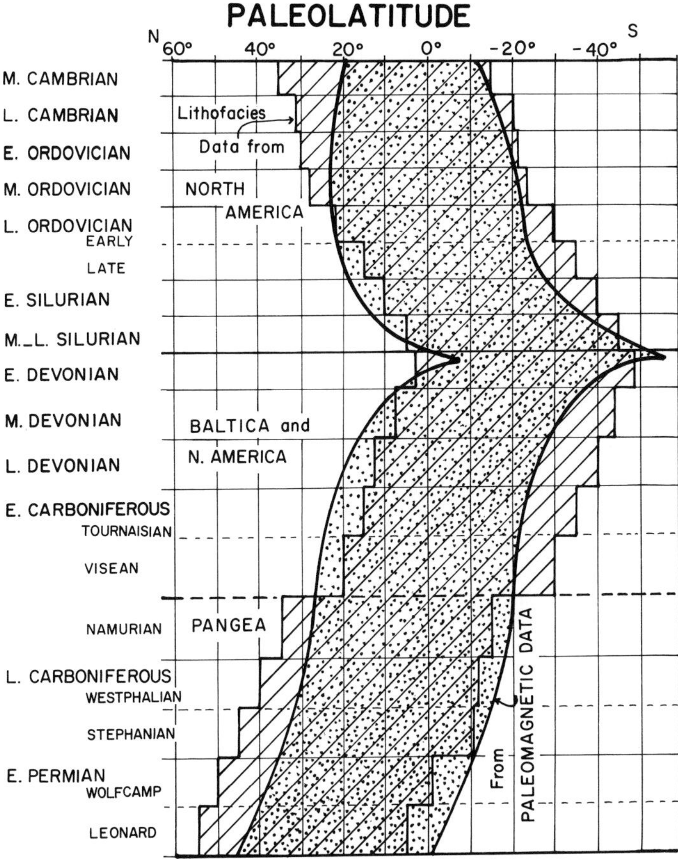

Figure 2.10. A comparison between two independently determined paleolatitude patterns for Paleozoic time for the North American craton. Oblique ruling contained by thin solid lines represents the paleolatitude range derived from sedimentary facies in North America (after Witzke, 1990), whereas the stippled area contained by heavy solid lines represents this range as based on paleomagnetic poles (from Table 5.1). The lithofacies data for post-Silurian time are derived from Laurussia (= North America, Greenland, the Baltic Shield and Russian Platform combined), whereas for Namurian and later time the data are derived from all of Pangea. Abbreviations: E = Early; M = Middle; L = Late. Published with permission from the author and the Geological Society Publishing House.

tant when one compares two results in a statistical fashion. Suppose that a mean direction is obtained from displaced terrane Utopia with a mean inclination of 28° and α_{95} = 2.5°. The nearby craton of Walhalla yields, for the same site and age interval, a reference inclination of 35° and α_{95} of 4°. It is clear that the two cones of confidence do not overlap, and that, *statistically*, the two results are indeed significantly different. Can one then say that Utopia is definitely displaced with respect to Walhalla? My own cautious reaction is that that would be unwise.

As there is, in my opinion, a minimum radius of confidence for practical purposes, there is also a need to specify that if α_{95} exceeds a certain value, then the result is less reliable. I will return to this issue in Chapter 4.

In summary, paleomagnetic directions and corresponding paleopoles can be valuable pieces of information, provided that certain assumptions are granted and that statistical parameters associated with the means are treated with proper circumspection. The basic assumption is that the direction obtained is representative of a geocentric dipole field. This assumption is untestable for any but the last 100 Ma, as so eloquently explained by McElhinny (1973, p. 27; see also Irving, 1964). For the last 100 Ma, the assumption is granted to within an allowable deviation of no more than some 5° (Merrill & McElhinny, 1983).

The average geomagnetic field has been ideally, but not necessarily, aligned with the rotation axis in the geological past. This is not a necessary condition for tectonic purposes, because a non-coaxial, yet geocentric, field still yields a global model that can be used to calculate a pole (or positions of many different continents with respect to that same pole). The only consequence of a non-coaxial field is that the geographic and geomagnetic poles did not coincide; however, many studies show good agreement between paleomagnetically and paleogeographically determined poles. Considerable success has been achieved by testing the alignment of the paleopoles and the geographic rotation axis in the geological past through the use of paleoclimatic indicators. Factors controlling climate and temperature gradients from pole to equator are on average zonal, yet independent of the magnetic field. Nevertheless, local deviations from a purely zonal (latitude controlled) pattern

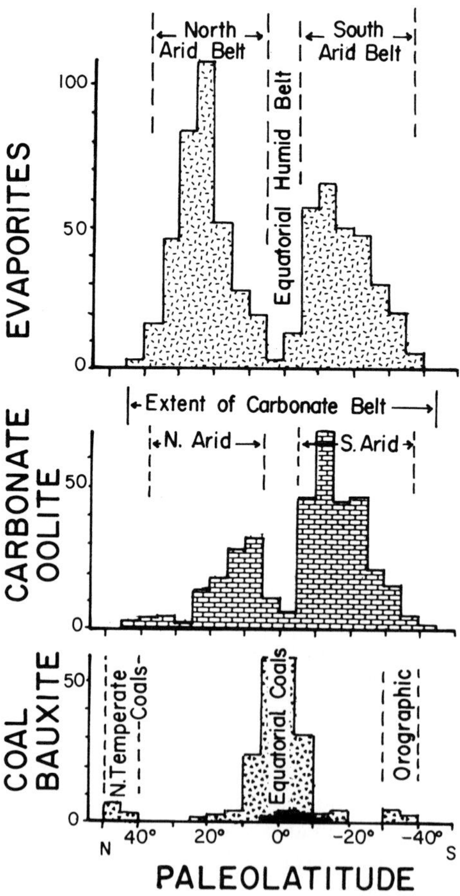

Figure 2.11. The distribution of Paleozoic lithic paleoclimatic indicators for North America and Laurussia (after the Silurian) as a function of inferred paleolatitude. Redrawn after Witzke (1990) and published with permission from the author and the Geological Society Publishing House.

may arise from elevation and continent–ocean contrasts. An example of global maps for the Early and Late Permian (Figure 2.9) shows the excellent correlation between the paleomagnetically determined latitude grid and the paleoclimate-related biome designations which are grosso modo paleolatitude indicators (Nie, Rowley & Ziegler, 1990; Ziegler, 1990). On a more regional scale, the Paleozoic lithofacies analyses of Witzke (1990) show remarkable agreement with paleomagnetically determined paleolatitudes for North America, suggesting that even on this smaller scale good resolution can be had. I may add, moreover, that Witzke and his colleagues have been ahead of the paleomagnetic community for many years in insisting that the Devonian equator ran through Arctic Canada (e.g., Heckel & Witzke, 1979), instead of through the Great Lakes area as mistakenly inferred by us paleomagneticians in the late 1970s (Kent & Opdyke, 1978; Irving, 1979; Van der Voo, French & French, 1979). We now know the Devonian reference field for the North American craton (a little) better (Miller & Kent, 1986a, b, 1988; Stearns & Van der Voo, 1988; Stearns, Van der Voo & Abrahamsen, 1989) and it turns out that Heckel and Witzke were correct (Figure 2.10).

The equatorial to temperate facies, used as paleolatitude indicators by Witzke (1990) are shown in Figure 2.11, and it is only one of the more recent examples of the use of such lithofacies. Earlier and important papers comparing paleopoles and paleogeographic latitude indicators started with the work of Irving (1956) and Opdyke & Runcorn (1960) and were followed by many others (e.g., Blackett, 1961; Runcorn, 1961; Irving & Briden, 1962; Briden & Irving, 1964; Briden, 1968a; Ziegler *et al.*, 1981, 1987; Parrish, Ziegler & Scotese, 1982). With increasing precision, these studies have demonstrated that in general the ancient magnetic field was indeed co-axial on average.

Everything should be made as simple as possible, but not simpler.

A. Einstein

This quote from Albert Einstein, is clearly an appropriate motto for the first few chapters.

The treatment of Chapters 2 and 3 is made as simple as possible, but indeed (I hope) not simpler. These chapters could even be skipped (what a horrible suggestion and that from the author himself!), but that might leave readers rather uncomprehending at some later stage, unless they had already been exposed to much of the background material of paleomagnetism and tectonics.

I re-emphasize that there is a lot more to paleomagnetism and rock magnetism than is covered here, but inclusion of such material would duplicate other books and would make the length of this volume too long and the price too high!

3 Megaplates, microplates, blocks, terranes, accreted slivers, thrusts and olistostromes

In the 40 years or so since the first paleomagnetic directions were obtained from rocks older than the last few million years, some 7000 paleopoles have been published for rocks from every imaginable location on Earth, with ages ranging from the very recent to the Archean. To bring some order to this wealth of data, it is necessary to divide the Earth's surface into geologic/geographic elements; the nature and definition of these elements form the subject of this chapter. If a paleopole is representative of the ancient geomagnetic dipole field at the time of rock formation, it is a valuable datum (Chapter 2) for the ancient geographic position of the rocks and its surroundings. What are these 'surroundings'?

For a discussion of this, we need to look briefly at the paradigm of plate tectonics and its consequences for the structure of the crust. The present-day surface of the Earth is made up of about ten major lithospheric plates, many with continental as well as oceanic crust (e.g., the North American, South American, African, Indo-Australian, Eurasian, Caribbean and Antarctic plates), but a few with predominantly oceanic crust only (e.g., the Pacific, Cocos and Nazca plates). The plate boundaries are the relatively narrow zones where relative motions occur between the plates: these motions can be divergent, convergent or strike-slip (along transform faults). The interiors of the plates remain – in theory – undeformed and rigid. If this is true, then a paleopole is valid for the entire plate interior, its continental as well as its oceanic parts.

Modern-day continents, however, are long-lived products of an ever-changing plate tectonic regime, and an ancient plate boundary in the middle of a continent may have become inactive, say 50 Ma. A paleopole for rocks of 100 Ma is thus valid only for that part of this continent that since that time retained rigidity and internal coherence with the site. An example is found in the continental area of the Iberian Peninsula (Spain and Portugal), which is today part of the European continent and the Eurasian plate. However, the Iberian Peninsula was welded onto the main part of Europe during the Pyrenean orogeny in the Early Tertiary; before that time, in the later part of the Mesozoic, it behaved as an independent small plate, a 'microplate'.

All major continents have such fossil plate boundaries of various ages criss-crossing their interiors. Now-extinct plate boundaries existed during the Mesozoic, for instance, within the present-day North American plate between the Colorado Plateau and areas of Mexico and all along the Canadian Rocky Mountains on their west side. Some of the plate boundaries are scarcely recognized and motion at these boundaries may have been relatively minor or short-lived. An example of relatively minor movements with respect to the US Mid-Continent is found in the Colorado Plateau: a clockwise rotation of less than 6° about a vertical axis in, or close to, the Plateau describes this motion of Laramide (Early Tertiary) age (Bryan & Gordon, 1986, 1988). Some scientists would even argue that the Rocky Mountains between the Colorado Plateau and the Mid-Continent should not be called a true fossil plate boundary; instead, they would argue that this zone is merely a broad and fuzzy belt of intra-continental deformation.

It is clear, then, that a paleopole must be assessed within the framework of geological knowledge about the continental crustal evolution since its time of magnetization acquisition. For North America in the Phanerozoic, a major part retained internal coherence and rigidity and this is called the North American craton (Figure 3.1). Surrounding the craton is a 'disturbed' margin, which in Paleozoic times (primarily in the east and south, less importantly in the west and north) and in Mesozoic to Early Tertiary times (in the west) was affected by orogenesis. However, this disturbed zone did not undergo large displacements with respect to the craton; it remained more or less in place, while undergoing rotations or regional thrusting. Paleolatitudes determined for this disturbed zone should therefore be representative of the craton, whereas declinations may have been perturbed.

For Precambrian times, the craton concept gradually loses its validity, even though at this time we are not certain whether the craton was amalgamated in the Proterozoic by coalescing elements that underwent large-scale relative displacements in a modern-type plate tectonic scenario (e.g., Hoffman, 1988) or whether these elements were never far separated. However, the Precambrian, as interesting as it is, falls outside the scope of this book.

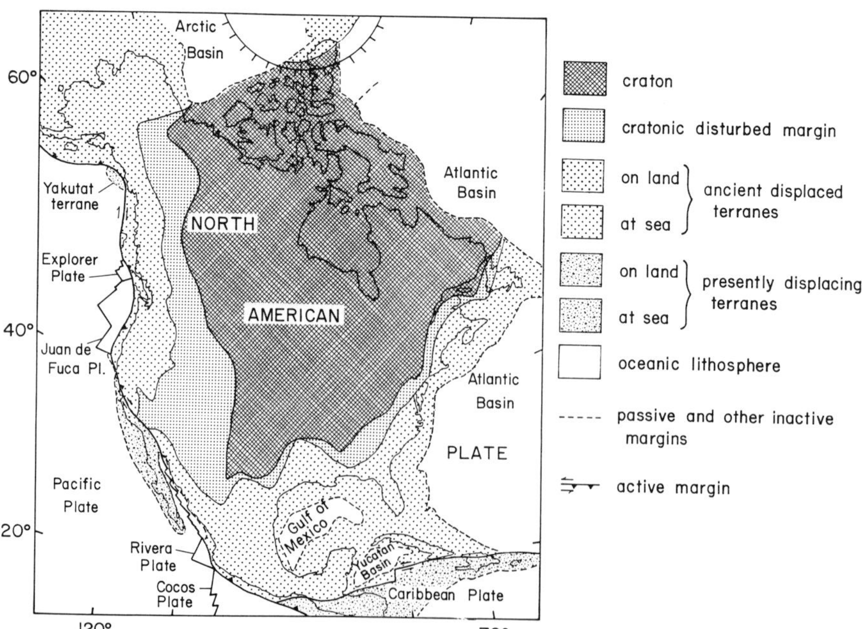

Figure 3.1. Tectonic framework of North America, outlining the craton, the cratonic disturbed margin, displaced (or displacing) terranes and oceanic lithosphere. Figure from Van der Voo (1990a), as adapted from the Geological Society of America index map of North American continent–ocean transects.

So far we have covered (mega-)plates (e.g., modern Eurasia, North America, including the adjacent ocean crust of these plates), microplates (e.g., the Iberian Peninsula) and blocks, such as the Colorado Plateau. The latter might also be called a microplate, although the definition of its fossil plate boundaries is very vague. When is a microplate too small to be called as such?

Plates consist of crust and a portion of the upper mantle, together called the lithosphere. The depth of detachment from the underlying mantle, the asthenosphere, is generally poorly defined, but this depth is well below the Moho, that is, well below the crust/mantle interface. However, for very small elements, and certainly for thrust sheets, the depth of detachment has clearly been within the crust. Such elements cannot be called plates. An example of one of the smallest ancient (micro-)plates is probably formed by the islands of Corsica and Sardinia in the western Mediterranean, measuring horizontally at most a few 100 km in diameter. These islands rotated, independently, away from the Mediterranean coasts of France and Spain during the Tertiary (Alvarez, Cocozza & Wezel, 1974; Westphal, Orsini & Vellutini, 1976), by formation of new oceanic crust in their wake. It is likely that the level of detachment of the Corsica–Sardinian crust from the underlying (non-rotating) material was below the Moho, although there is no precise information on this. I will return to the western Mediterranean microplates in Chapter 7.

Elements that cannot be called microplates may be denoted as blocks or terranes. The former show some sort of internal rigidity, while the latter may be internally deformed during orogenesis. Examples of blocks and terranes abound in the western part of North America (Figure 3.1), and for many of these, sizeable displacements have been demonstrated by paleomagnetic, faunal, lithofacies, and/or structural analyses (Beck, 1976, 1980; Hillhouse, 1977; Jones *et al.*, 1977, 1987; Coney, Jones & Monger, 1980; Howell, 1985, 1989), with respect to the North American craton. Likewise, but with less detailed evidence, ancient displaced terranes are found along the Atlantic Coast of North America (Figure 3.1). I will return to these Paleozoic displaced terranes in Chapter 8.

Some terranes are labeled 'suspect'. This is a convenient way to express doubts about the otherwise presumed coherence with the craton, for areas which may have been displaced but where compelling evidence is lacking. An example from southern South America is illustrated in Figure 3.2, showing the Patagonian area of Argentina as a Paleozoic suspect terrane (e.g., Dalziel & Forsythe, 1985). Separating this terrane from the craton is the front of a mobile belt that early on was called the western part of the 'Samfrau' geosyncline, a late Paleozoic to early Mesozoic mobile belt that fringes the cratonic portions of the Gondwana Supercontinent (see Chapter 5) from eastern Australia, through Antarctica and southern South Africa, into northern Argentina to end up joining the Andean mobile belt. There is at this time no conclusive proof that the Patagonian terrane has been displaced with respect to the South American craton farther north, although Rapalini & Vilas (1991) recently reported preliminary data suggesting this.

Present-day plate boundaries, of course, define modern, 'displacing' areas that may become the terranes of tomorrow (e.g., Figure 3.1). Displacing continental elements can be found in the Caribbean and in Baja California, besides oceanic elements. When in the future some plate boundaries once again fossilize as the plate tectonic regime adjusts to global changes, such displacing elements may become incorporated into the North American plate. Similarly, future displacing terranes may form through the inception of new boundaries within the North American craton, slicing off a corner, so to speak.

Most ancient displaced terranes display continental affinities; truly oceanic crustal elements now incorporated into the continental crust are rather rare. This is because of the recycling of oceanic lithosphere in subduction zones, continental crust being generally too buoyant to sink. Those ancient oceanic elements that have been preserved generally owe their survival to some ancient structural elevation (e.g., ocean islands, seamounts, ancient fracture zones). However, at converging plate margins the top layers of the ocean floor, much of it sedimentary in nature, may be scraped off (Figure 3.3) above the subduction zone or trench (e.g., Cadet *et al.*, 1985). When such accreted margins become exposed, they show strong imbrication and may consist of a mixture of rock types such as clayey pelagic sediments, blocks of limestone, pillow lavas, gabbro and diabase, or even slices of upper mantle ultramafic rocks. Such a mixture is called a mélange, and it may contain large, internally coherent blocks tens of kilometers in size (accreted slivers). Olistostromes are sedimentary deposits often associated with displaced blocks in a setting not necessarily related to subduction.

Paleomagnetic data obtained from elements in a mélange or olistostrome are, of necessity, valid for areas of small extent only. Nevertheless, some interesting conclusions may be obtained from such blocks. In the Franciscan Group of California, large blocks of Cretaceous limestone show good internal stratification and the younging direction could be deduced. Their upward inclinations (Figure 3.4) yield a paleolatitude of about 15° (Alvarez *et al.*, 1980a; Tarduno *et al.*, 1986; Tarduno, McWilliams & Sleep, 1990). Paleomagnetic directions of the blocks are older than the tilting of the strata and were originally assumed to be of the same age as the rocks. Thus, they could be inferred as being of normal polarity, acquired during the long normal Cretaceous interval, which results in a southern hemisphere provenance. However, this has recently been disputed by Hagstrum (1990), who argued for a pre-deformation, but secondary, age of magnetization (and implicitly for a reversed polarity and northern hemisphere provenance). Be that as it may, the original

Figure 3.2. The location of the Patagonia suspect terrane (possibly displaced in Paleozoic times) in Argentina, South America.

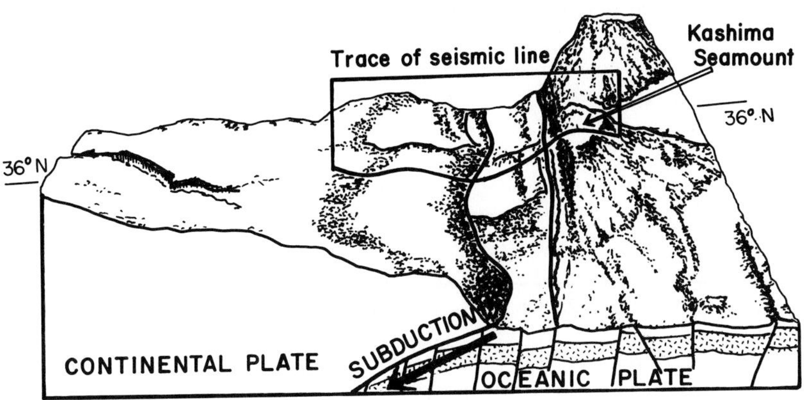

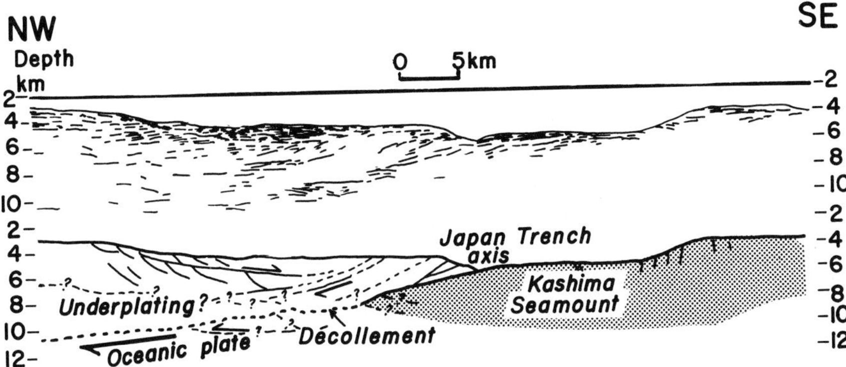

Figure 3.3. A block diagram and interpretations of a seismic section across the Japan Trench, where the Pacific plate (with the Kashima Seamount riding on it) is being subducted. From Howell (1989), based on data and interpretations from Cadet *et al.* (1987) and Lallemand, Cullota & Von Huene (1989). Published with permission from the authors, Chapman and Hall Ltd., (Howell, D., 1989), and Elsevier Science Publishers B. V.

hypothesis of Alvarez and colleagues forms an illustrative example, and is discussed here under the (possibly false) assumption that the magnetization is primary. The North American craton was located at intermediate northerly paleolatitudes in the Cretaceous, and given a 15° S paleolatitude for the limestone blocks in the Franciscan Group, the conclusion was inescapable that the blocks traveled northward for thousands of kilometers before accreting to North America, presumably riding on one of the Pacific oceanic plates. On the other hand, if Hagstrum's arguments about a pre-deformation but secondary age of magnetization are correct, the displacements are less but still quite significant. In the latter interpretation, the magnetization dates back to an interval during accretion but before post-accretion northward displacement of the limestone blocks now found in the Franciscan Group.

In general, though, it is difficult to arrive at such spectacular conclusions: often, the blocks are not stratified and, without information about the paleohorizontal, paleomagnetic directions are meaningless for paleolatitude determinations. Alternatively the direction of younging may be poorly known, so that a paleolatitude determination may be either northerly or southerly, diminishing its value. Furthermore, as Hagstrum's comments show, the magnetization may not be primary and the polarity may therefore be indeterminate. Even if the paleohorizontal is known,

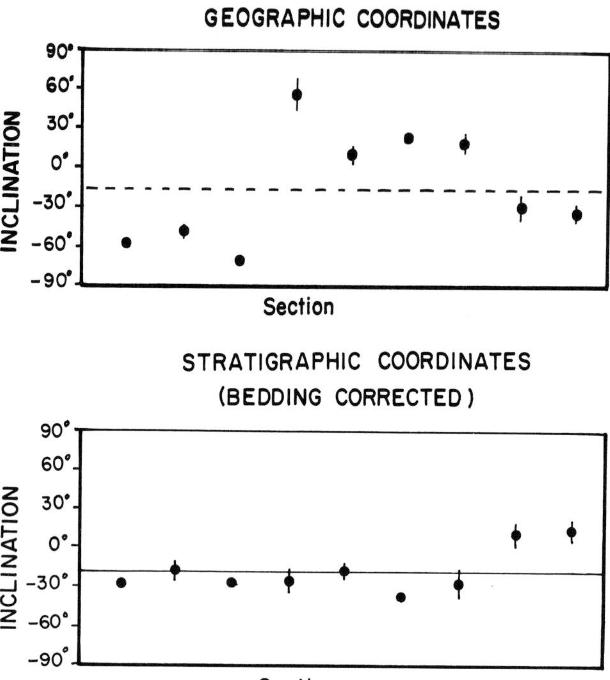

Figure 3.4. Paleomagnetic inclination results before (above) and after correction for the tilt of the strata (below) from the Laytonville Limestone blocks in the Franciscan mélange of California. Error bars represent the cones of 95% confidence limits. Note the better grouping after correction, indicating magnetization acquisition before the tilting (i.e., the final emplacement). Redrawn after Tarduno, McWilliams & Sleep (1990).

a far-traveled block may escape paleomagnetic detection if its displacement has been in an east–west sense; because of the axial symmetry of the geomagnetic dipole field model, paleolongitudes cannot be determined.

In the chapters that form the core of this book (5–8) I will use the concepts described above to group the available paleopoles into subsets for different geographic elements, dealing first with the major cratons (Chapter 5). These are North America, Stable Europe, the Gondwana Continents, Siberia, North and South China. Then, sequentially, I will discuss the histories of the Atlantic, Tethyan and Iapetus oceans as revealed by paleopoles for the peripatetic continental elements that belonged to their domains.

I tried Geology and found it curious, but, finally, non-sustaining.

(C. Bukowski in his 1979/1980 preface to John Fante's book Ask the Dust*)*

I too find Geology curious and so do almost all of my colleagues and most of the students I come in contact with. But non-sustaining? No, I disagree, but it *is* a matter of personal taste, isn't it? We will see later how extraordinary and interesting the drift and plate tectonic histories of the continents and blocks around the Tethys and Iapetus Oceans have been. I find that very sustaining, even for an entire lifetime. I invite Charles Bukowski to read some books on plate tectonics; for instance, John McPhee's books are very accessible and entertaining.

In the next chapter, though, we need to go into the details of paleomagnetic analysis, before we can appreciate paleomagnetic data and what they can and cannot tell about drift histories. Some of this material is intricate and – at first glance – complex and confusing. But it is necessary background material or we will not be able to know what to believe and what to suspect.

4 Paleomagnetic information – what makes a paleopole valuable?

As already mentioned, some 7000 paleopoles have been published in the last four decades. As with any discipline, there are 'good' and 'bad' data in this extensive data base. The problem is to decide which is which, and to make this decision with some degree of objectivity. It simply will not do to state that the paleopoles produced in one's own laboratory are, by definition, superior to those produced elsewhere!

In the preceding chapters, I have treated the main assumptions that underlie the calculation of a paleopole, as well as the basis for dividing the Earth's surface into smaller tectonic elements. In this chapter, I will examine the evidence and information that can (and oftentimes must) be provided to make a pole valuable or reliable, including that related to field and laboratory aspects of paleomagnetic studies.

One of the more important considerations in any geological endeavor is the determination of a geological or radiometric age. Unfortunately, paleopoles cannot be dated directly, nor have any radiogenic isotope age determination techniques been fully applied yet to the dating of magnetic minerals, such as (titano-) magnetite, (titano-) hematite, goethite or pyrrhotite. The discussion begins, therefore, with the age constraints for the rocks from which a paleopole has been derived, but at the end of this chapter I will return to the still more important matter of constraints on the age of a paleopole, in the meantime having dealt with the important experiments and tests provided by demagnetizations, folds, conglomerates, baked contacts and reversals.

The age of the rocks (= the age of the magnetization?)

Let us assume that a sequence of rocks has yielded a paleopole, which can be considered as 'primary'. By primary, paleomagnetists generally mean that the magnetization was acquired when the rocks formed (when they were deposited, or when they solidified and cooled) or within a short time thereafter. 'Short' is a relative term, and its definition very much depends on the purposes of a study; for the time being, we define short as within 1% of the total age of the rock. There are many ways in which rocks can acquire a magnetization: thermal blocking of the magnetization, chemical growth of minerals to a size where they are able to carry a stable remanence,

Table 4.1. *Different types of remanent magnetizations*

Abbr.	Brief Description
NRM	Natural Remanent Magnetization, or the total remanence in a rock before any demagnetization has been carried out. NRM may consist of any combination of naturally occurring types of remanence.
ARM	Anhysteretic Remanent Magnetization, which is generally laboratory induced, and occurs when a sample is exposed to a combination of alternating and direct magnetic fields.
CRM	Chemical Remanent Magnetization, acquired when new grains grow (at temperatures well below the Curie temperature) in a magnetic field.
DRM	Depositional Remanent Magnetization, acquired when previously magnetized grains settle in a water column under the influence of a magnetic field to form part of a sediment.
pDRM	Post-Depositional Remanent Magnetization, acquired when grains already in a sediment column are allowed to re-orient themselves in an existing field. This re-orientation may be caused, for instance, by burrowing.
IRM	Isothermal Remanent Magnetization, acquired in a strong magnetic field, such as that produced by a lightning strike or by an electromagnet in the laboratory.
SIRM	Saturation IRM, produced in the laboratory by a high magnetic field, such that a sample acquires the maximum remanence it can possess.
TRM	Thermo-Remanent Magnetization, acquired when minerals cool through the Curie temperature.
pTRM	Partial Thermo-Remanent Magnetization, acquired when minerals cool through a specific temperature interval.
TCRM	Thermo-Chemical Remanent Magnetization, a CRM acquired at elevated temperatures.
VRM	Viscous remanence, acquired when grains have such low relaxation times that their magnetizations realign themselves with the ambient field in a matter of seconds to a few million years.
VpTRM	Viscous partial Thermo-Remanent Magnetization, acquired when minerals acquire a viscous remanence while cooling from elevated temperatures.

For further explanation of these remanences, see text.

depositional or post-depositional alignment of magnetic grains in a sediment, or exposure to a high magnetic field (i.e., during passage of a bolt of lightning), to name a few (Table 4.1). Apart from lightning strikes, many of these mechanisms produce a magnetic direction in the rocks which is parallel to that of the geomagnetic field. Having eliminated the effects of secular variation by sampling over sufficient time, let us say we have obtained a 'good' paleopole.

This paleopole tells us the orientation and paleolatitude of the rocks and their surroundings for the time the magnetization was acquired. However, for accurate comparisons with similar-aged paleopoles from other continents or terranes, a relatively precise age designation is necessary for our paleopole as well as for the other results in the comparison. Once again, one of these vague terms: 'relatively precise'! And, once again, its definition depends on the purposes of the study. In order to show how fast (±10%) India drifted northward with respect to Asia between the beginning and the end of the Paleocene (66–55 Ma), we need rock ages to be correct within one million years, whereas to illustrate the validity of the Pangea

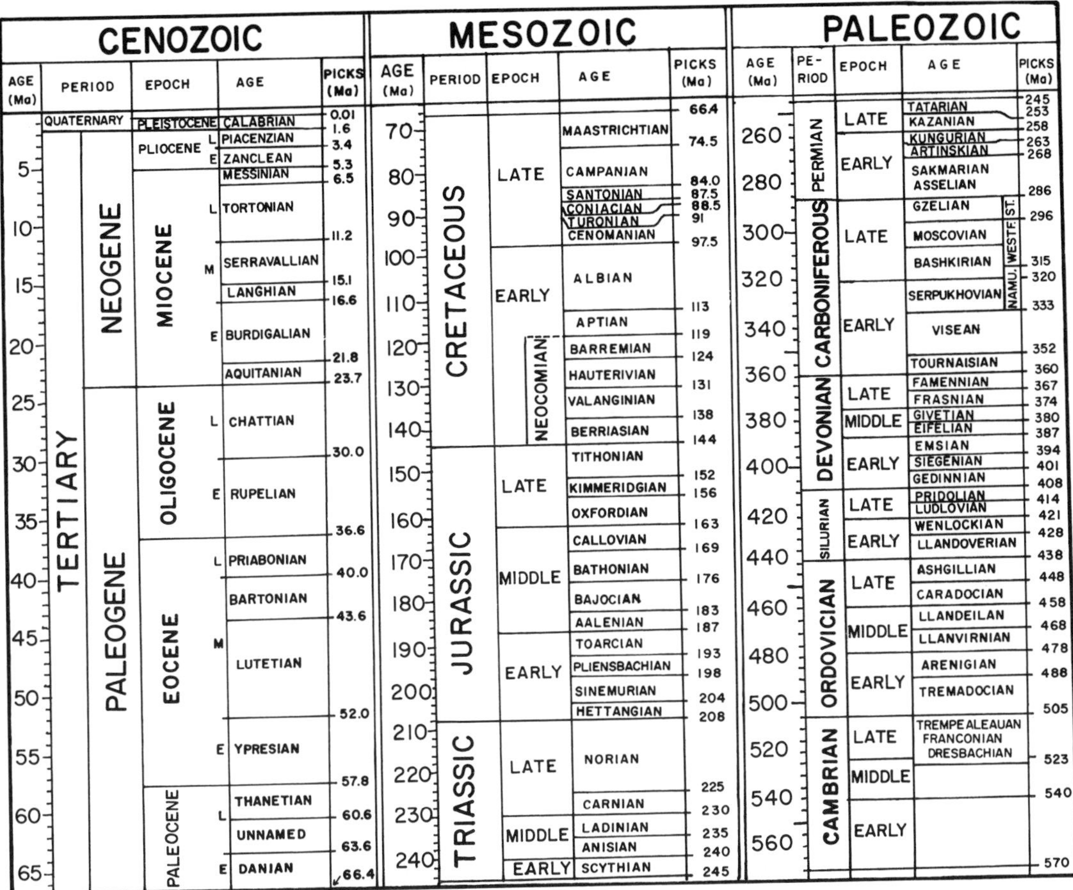

Figure 4.1. The DNAG time scale, as used throughout this book (after Palmer, 1983). Published with permission from the author.

supercontinent configuration between Europe and North America with Permo-Triassic paleopoles, a precision of 35 million years is more than adequate.

The majority of paleomagnetically studied rocks for the Phanerozoic are sediments, and their ages are determined stratigraphically. Conversion of stratigraphic to radiometric ages must be done using a time scale. In this book, I have based all ages on the time scale of Palmer (1983), also referred to as the DNAG time scale (Figure 4.1). It is very similar to that of Harland *et al.* (1982). It is quite probable that the future will bring further refinements in such a time scale, because for some parts of Phanerozoic time there are still important differences of opinion (e.g., for the Ordovician–Devonian; see Tucker *et al.*, 1990, or McKerrow, St Lambert & Chamberlain, 1980, vs. Gale, Beckinsale & Wadge, 1980).

In the last few decades, the demands for relatively precise ages of paleomagnetically studied rocks have become more stringent. Compare the statement in McElhinny (1973, p. 106) that ages must be known to within half an era, with one of the reliability criteria of Van der Voo (1990a,b) that ages must be known to within half a period: early Paleozoic vs. Late Silurian (?!).

A 'good' paleopole for Gondwana dated no better than early Paleozoic may be very valuable, because after all it must still lie on the APWP, regardless of whether its age is Cambrian, Ordovician or Silurian. In 1973, this was important because the Paleozoic APWP of Gondwana was so poorly known that any result would be helpful, just to constrain the shape of the path. On the other hand, if we now wish to calculate the amount of displacement of Wrangellia with respect to North America on the basis of Mesozoic paleomagnetic data, we must have age assignments to within half a period (e.g., Late Triassic). The arguments for this quantitative assessment can be found in an examination of the typical rate of change in paleopole position with time, that is, the rate of apparent polar wander. In about 25 Ma (i.e., about half a period, on average) the typical APWP changes, again on average, by about 8°. This is of the order of the combined uncertainties in the Wrangellian and the North American mean paleopoles. Any precision better than 25 Ma would be exceeded anyway by the typical confidence limits about the paleopoles and would, therefore, be unnecessary.

A precision as poor as 'Early Paleozoic' (570–408 Ma), in contrast, would allow comparisons with an accuracy of only some 50°; this value is much too large to be called 'accurate' for tectonic purposes, although the paleopole may be useful and supportive material to construct the form, if not the temporal calibration, of the APWP.

In some cases, age determinations of well-dated fossiliferous late Mesozoic or Cenozoic sediments have begun to approach precisions to within a million years or so, not in the least because of magnetostratigraphic information in addition to that provided by the microfossils or ammonites (e.g., Lowrie & Heller, 1982).

For the Precambrian, a radiometric age of the rocks is paramount. With modern radiogenic isotope techniques (e.g., U–Pb on zircons or badelleyite) extraordinary precision can be achieved, whereas earlier Rb–Sr or K–Ar age dating sometimes had large associated uncertainties of tens or even hundreds of millions of years. Again keeping in mind the typical APWP rate, a reliability criterion for the age uncertainties of Precambrian rocks may be of the order of ±4% or ±40 Ma, whichever is smaller (Van der Voo & Meert, 1991).

Number of samples and statistical uncertainties

The number of samples determines, generally, whether secular variation and random field orientation and laboratory measurement errors have been averaged. In addition, a sizeable number (≥25) of samples analyzed makes it plausible that no observable aspect of the magnetization has gone unrecognized. This criterion, as well, has become more stringent over the years: McElhinny's minimum in 1973 was eight samples. I should note that a result based on only ten samples may be perfectly adequate, but chances are that it is not. The minimum number of 25 for this criterion to be met, is of course a bit arbitrary; why not specify a minimum of, say, 47? Practical experience, however, has shown that a mean direction can still change significantly when the number of samples is increased from 15 to 25, whereas an increment from 25 to 45 samples generally makes no further significant difference, all else being satisfactory (such as demagnetization behavior; see below).

My preference for minimum k (or K) is 10, and for maximum α_{95} (or A_{95}) is 16°. As discussed earlier, k is a parameter intrinsically independent of the number of samples. A k somewhat less than 10 is not random (directions still mostly fall within one quadrant), but precision is poor (e.g., Figure 4.2). A typical α_{95} for N = 25 samples with k = 10 is of the order of 9° whereas α_{95} is about 18° for N = 8 samples and a k = 10 (Figure 4.2b). It may seem, therefore, that the maximum α_{95} is too generous, when compared to the conditions for k and N. However, many publi-

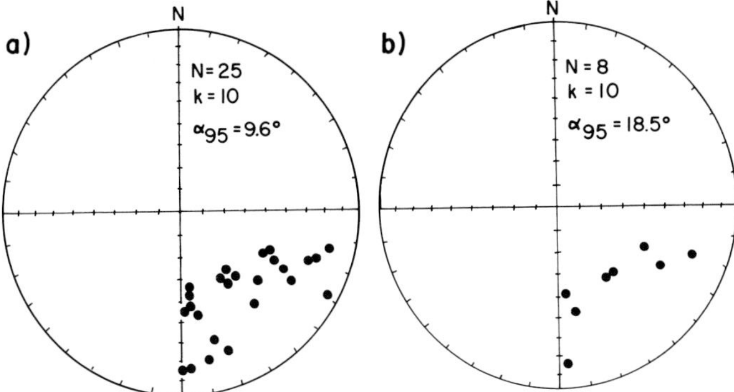

Figure 4.2. Typical, not-so-well-clustered paleomagnetic directions where the precision parameter (k) is 10, and where the number of samples is a) 25 and b) 8. The corresponding α_{95} of these populations are listed.

cations do not publish α_{95} values based on samples but on site means. This is done, for instance, when each site yields a VGP based on a single lava flow. Hence, my combined condition that N $\geq$ 25 samples and α_{95} (or A_{95}) $\leq$ 16° and k (or K) $\geq$ 10.

One may well ask what happens when this condition is not met. The result may still be perfectly valid, even if some such conditions are not met. However, the chance is diminished that the result is reliable. At the end of this chapter, when I have finished the discussion of many similar criteria, I will return to the issue of data selection and rejection.

Demagnetization – background

Laboratory demagnetization procedures constitute one of the more intricate and obscure aspects of our field for non-paleomagnetists, and it deserves considerable attention and explanation. The fundamentals of these treatments are rooted in solid state physics and in the subdiscipline of rock magnetism. As it is not the purpose of this book to provide a complete theoretical basis for all aspects of paleomagnetism, only the essentials will be described. Full treatments are provided in classic texts or summary discussions dealing with rock and paleomagnetism (e.g., Nagata, 1961; Irving, 1964; Collinson, Creer & Runcorn, 1967; McElhinny, 1973; Stacey & Banerjee, 1974; Valencio, 1980; Collinson, 1983; Tarling, 1983; O'Reilly, 1984; Piper, 1987). Much has been contributed by rock magnetic studies, but many aspects of this accumulated knowledge are irrelevant to a geologist who wishes to achieve an understanding of the tectonic applications of paleomagnetism.

Demagnetization is the process that – gradually – eliminates a natural remanent magnetization in a sample, without introducing new magnetizations; the latter are called spurious or laboratory-induced magnetizations. By repeating the measurements after each demagnetization step, one obtains by vector subtraction both the direction and the intensity of the magnetization component removed during that step.

Why is demagnetization necessary? During the history of a rock sequence, several different magnetizations may have been introduced: first, the 'primary' magnetization, but later also magnetizations acquired by chemically grown (diagenetic or weathering-related) iron-oxides or iron-sulfides. In addition, some magnetized grains may not be stably magnetized for long periods of time. 'Unstable' here means that after a geologically short time, the magnetization direction adjusts itself to the new ambient field and it keeps on adjusting itself after each similar successive interval. For an assembly of such unstable grains, the typical period in which this realignment takes place is related to what is called the relaxation time.

For stably magnetized grains, the relaxation time may exceed several billions of years. When the relaxation time is short, in the order of seconds to millions of years, the magnetization is called a Viscous Remanent Magnetization (VRM). VRMs are extremely common in almost all rock types and their effects are superimposed on the primary magnetization, if any exist. Because of the repeated realignments, most VRMs are found in directions parallel to the recent local geomagnetic field (called present-day field or 'pdf').

One more complexity enters into the discussion, involving the effects of temperature. Elevated temperatures reduce the relaxation time, generally, for an assemblage of grains. Hence, elevated temperatures enhance the acquisition of VRM, and the product is appropriately but cumbersomely called a Viscous partial Thermal Remanent Magnetization (VpTRM). Such VpTRMs can be the result of a brief ancient heating episode experienced by the rocks during burial; upon cooling to low temperatures, the relaxation time increases again significantly (exponentially), so that the VpTRM may survive subsequently for geologically long intervals.

Demagnetization, then, is necessary to separate – if at all possible – primary magnetizations from those that are ancient but secondary (e.g., VpTRMs), from those that are viscous at room temperature (VRMs). Ancient, but secondary, magnetizations may also relate to diagenesis affecting the magnetic carriers; see for instance the magnetites in Siluro-Devonian carbonates from New York State displayed in Figure 2.8, which are thought to have crystallized during the Permian from framboidal parent material consisting of pyrite. Because the magnetic carriers formed by chemical reactions, the magnetization is known as a Chemical Remanent Magnetization (CRM). In other case histories, secondary goethite (FeOOH) has been shown to carry ancient but secondary CRMs of late Miocene to early Pleistocene age in carbonates from southern Germany and northwestern Switzerland (Heller, 1978; Johnson, Van der Voo & Lowrie, 1984). Many other examples of such partial or complete remagnetizations exist; indeed this is the rule, not the exception, for most rocks. I will return to this issue in the last part of this chapter.

Separation of different magnetization components can be achieved if each shows different resistance to the demagnetization treatments applied. Demagnetization can be done by chemical leaching, heating in an oven (thermal demagnetization) or by subjecting the sample to alternating fields. In each method, care has to be taken that the sample does not acquire new, spurious magnetizations. Therefore, such treatments are done in a zero magnetic field. Chemical leaching removes first the smaller and more soluble grains (e.g., pigmentary hematite) and only during longer treatment (or with higher concentration of acid) are specular hematite and magnetite removed. The technique cannot be applied to all rock types; obviously, limestone samples will completely disintegrate in hydrochloric acid (HCl). The method works best with detrital sediments and has been very successful in the analysis of the remanence of red beds (e.g., Roy & Park, 1974; Henry, 1979).

Thermal demagnetization in a first incremental step of say, 200 °C, may 'unblock' the magnetizations of some of the titanomagnetite or (titano-) hematite grains, depending on their size, titanium content, or relaxation time (as a function of temperature). All magnetic materials must lose their remanence at a certain temperature (called the Curie temperature (Tc) for magnetite), but small grains may lose

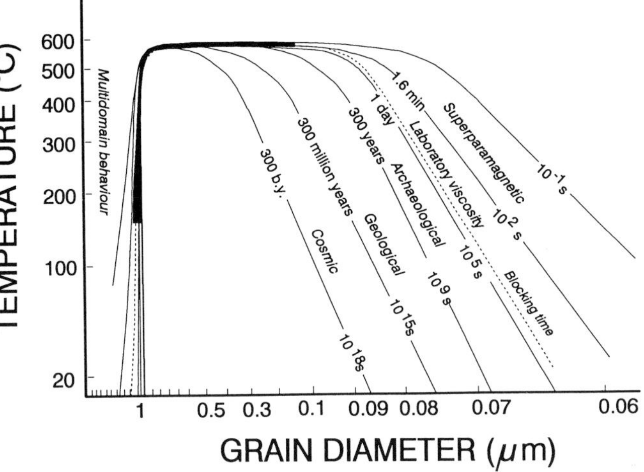

Figure 4.3. The relationship between temperature (vertical axis), grain diameter (horizontal axis) and relaxation time (labels along curves) for ordinary (titano-) magnetite grains. The (dashed) boundaries between superparamagnetic, single-domain and multidomain behavior are approximate. Figure redrawn after Tarling (1983). Published with permission from the author and publishers (Chapman and Hall Ltd).

their magnetization at temperatures well below Tc, because at that point the thermal energy is able to overcome the magnetic energy barrier that keeps the magnetization from re-aligning itself. For very small equidimensional magnetite grains (e.g., with diameter < 0.05 μm), the magnetic energy is relatively much lower than for grains of 0.1 to 1.0 μm (see Figure 4.3; after Tarling, 1983, his figure 2.6). This lower energy level can also be reflected in a shorter relaxation time, discussed earlier, from which it follows that VRMs should be removed at lower temperatures than other types of remanence. This, fortunately, is generally indeed the case, even for hematite (Dunlop & Stirling, 1977). When grains are so small that their relaxation time is much less than a second, they are incapable of carrying a measureable remanence. Although intrinsically ferromagnetic, they behave as paramagnetic material and because of this are called 'superparamagnetic' (Figure 4.3).

An important principle that governs thermal demagnetization is that a partial Thermo-Remanent Magnetization (pTRM) rapidly acquired by a rock in a temperature interval of, say, 500 to 570 °C, will also be eliminated in thermal demagnetization in that same temperature interval (Figure 4.4). This reciprocity means that TRMs in volcanic rocks are removed in temperatures just up to the Tc of the predominant magnetic minerals, generally (titanium-poor) magnetite, but occasionally hematite when the flows were extruded subaerially under oxidizing conditions.

Theory in rock magnetism predicts that different generations of remanent magnetizations (VRM, VpTRM, CRM, TRM, etc.; see Table 4.1) could well be removed in different intervals of a stepwise thermal demagnetization treatment. Typical increments in the heating are about 50 to 100 °C up to 450 °C, whereas these steps are usually 10 to 20 °C between 450 °C and the Tc. Tcs are about 575 °C for pure, titanium-free magnetite, and about 680 °C for pure hematite, whereas goethite has a Tc of only 100 to 150 °C. The presence of titanium in solid solution in the magnetite or hematite crystal structure (without exsolution of the mineral into separate phases) means that Tc is progressively lower with increasing 'dilution' of the pure magnetite endmember. The titanium-rich end members in the solid-

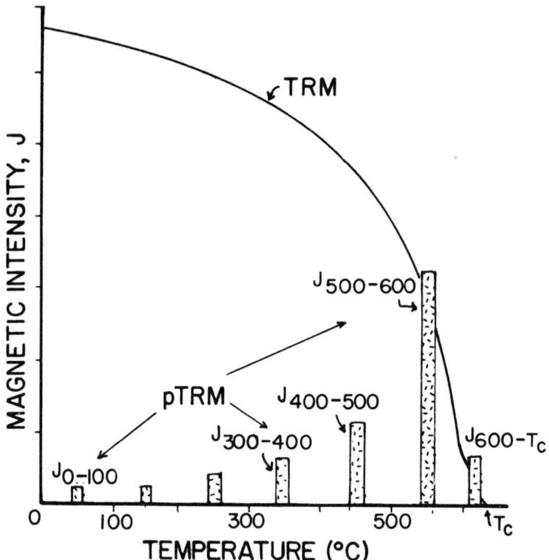

Figure 4.4. Intensity acquisition or decay, as a reversible process, of a typical thermo-remanent magnetization (TRM) as a function of temperature (upper curve). The partial TRMs acquired over discrete intervals are shown as vertical bars, and are the derivatives of (hence, add up to) the total curve. Tc is the Curie temperature. Figure redrawn after McElhinny (1973).

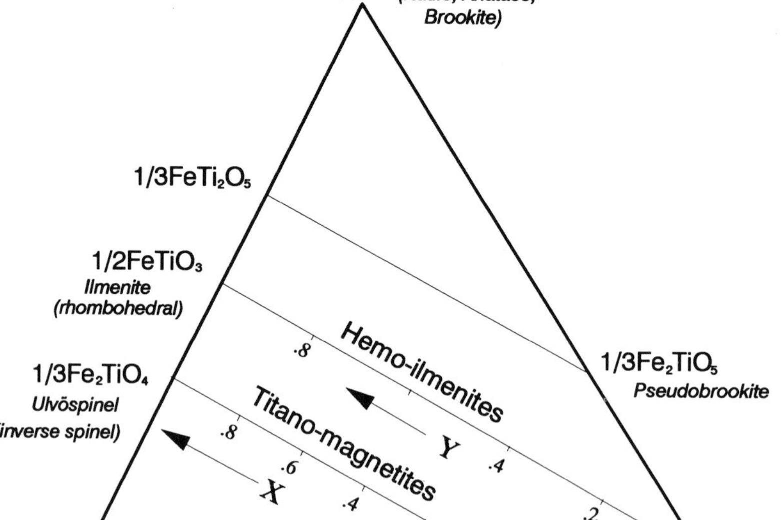

Figure 4.5. The ternary diagram of the iron-titanium oxides. Solid-solution series between magnetite and ulvöspinel and between hematite and ilmenite show increasing titanium content with increasing X and Y. For the titano-hematites (also called hemo-ilmenites for the titanium-rich part) the valency of iron simultaneously changes gradually from +3 to +2.

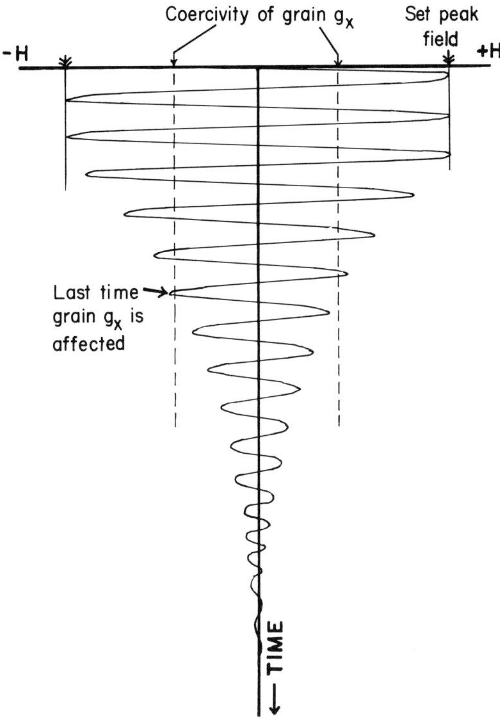

Figure 4.6. Plot of alternating field (AF) strength (H) as a function of time (vertical axis) during a single AF demagnetization step. The dashed lines indicate the coercivity of a given grain (g_x) which is initially exceeded by the field H. However, during decay of H to values below the coercivity value of g_x, this grain stops being affected by the AF and freezes its magnetization in the plus (or minus) direction of the field H.

solution series (ilmenite and ulvöspinel in Figure 4.5) have Tcs well below room temperature; they are paramagnetic at room temperature.

The third technique employed in demagnetization utilizes magnetic alternating fields (AF). With this technique the magnetizations of individual grains (of which there are up to millions in a standard rock specimen) are forced into realignment in different, opposite directions so that they cancel each other out. The magnetization of a grain may be re-aligned by a magnetic field, if that field is strong enough to overcome the magnetic energy barrier of the grain. The critical field strength just able to do this, is called the microscopic coercive force of a grain or also, its 'coercivity'. Different grains have, of course, different coercivities, depending again on size, chemical composition, imperfections in the crystal structure, and, indirectly, on parameters related to relaxation time. In AF treatment, each step consists of higher peak fields than the previous one; however, the step itself begins with the chosen peak field, which then gradually decays to zero (Figure 4.6) for reasons explained below. After each step, the sample is measured again. Each successive stepwise treatment will realign the magnetizations of more and more grains, as their coercivities are exceeded by the higher and higher fields. Because the applied magnetic fields are alternating, the grains with coercivities less than the applied field will realign their magnetizations back and forth. But since in a given step the field itself gradually decays to zero, their alignment will freeze in (in either a plus or a minus direction, so to speak) when the field has decayed to the level of the grain's coercivity (Figure 4.6). After the step, when the AF has decayed to zero, all magnetic directions of the affected grains are thus frozen into the two opposing directions, and generally half of the magnetizations of the grains will be aligned one way and the other half in the opposite direction. Thus with AF, the magnetization is not destroyed but simply made to progressively cancel itself.

As with the other demagnetization techniques, care must be taken not to introduce a spurious magnetization during treatment. A small direct magnetic field, for instance, superimposed on the alternating field of Figure 4.6 has the effect of displacing the zero-field axis, thereby destroying the symmetry of the treatment and causing preferential alignment of grain magnetizations in one direction. The undesirable remanent magnetization so introduced is called an Anhysteretic Remanent Magnetization (ARM) and is to be avoided unless one wishes to carry out special rock magnetic experiments. This is the reason, referred to earlier, that the AF must decay completely to zero before the sample can be removed for re-measurement, because otherwise the sample would suddenly acquire an ARM.

AF techniques also rely on the theoretical principle (and the investigator's hope and good faith) that magnetization components of different ages and origins will have different coercivity spectra. Generally, VRMs have lower spectra than TRMs

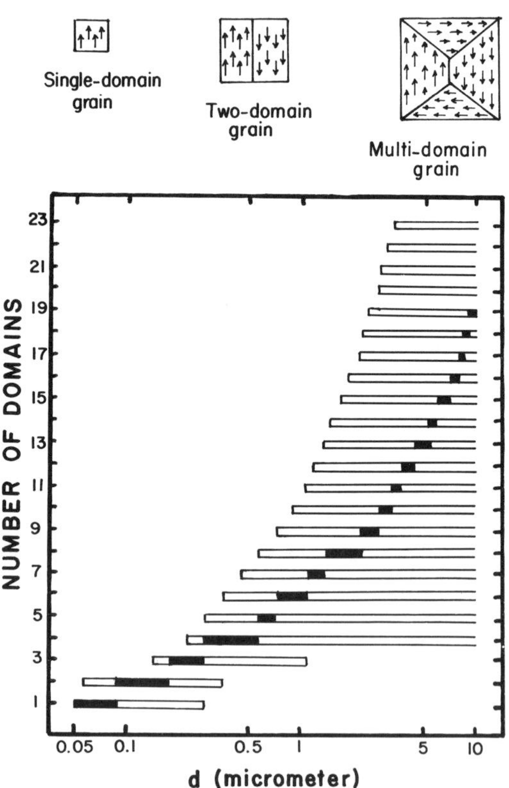

Figure 4.7. Single-domain, two-domain and multidomain grain configurations (top); in the bottom diagram, the likely number of domains is plotted vs. grain diameter for magnetite. The range of grain sizes for which that particular number of domains is the *lowest* energy domain state of a grain is represented by the filled portions of the bars, whereas the open bar segments represent additional but local energy minima. It can be seen that the likely single-domain size range overlaps with multidomain size ranges. Redrawn from Moon & Merrill (1985) and published with permission from the authors.

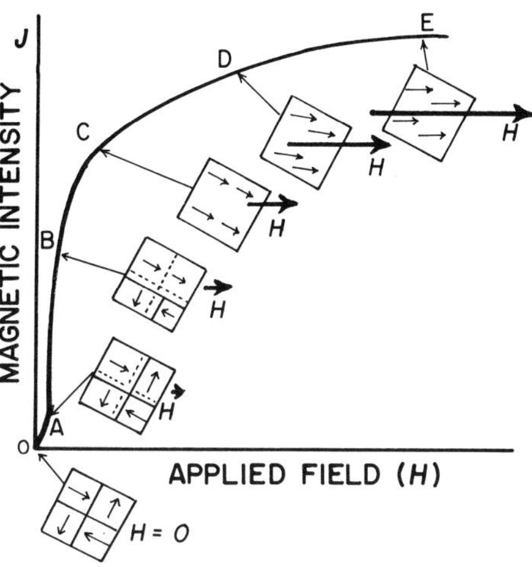

Figure 4.8. Magnetization acquisition as a function of applied field H along a typical partial hysteresis curve for a multidomain grain. The relative volumes of the magnetic domains change in the lower fields (between points A–C), followed by changes in the magnetization direction of the grain at higher fields (C–E) until the grain's magnetization is fully parallel with the direction of the applied field. Redrawn after Irving (1964) and Brailsford (1951), and published with permission of the publishers John Wiley and Sons, and Methuen Company.

or CRMs. However, Biquand & Prévot (1971) have shown that some VRMs of exceptionally high coercivity can exist.

In this discussion, I have thus far ignored the subjects of magnetic domain structures and magnetic hysteresis. These are extremely important, but (with one exception) an understanding of demagnetization analysis and paleomagnetic validity does not depend on them. Interested readers are referred to the books listed at the beginning of this section. The one exception relates to the larger grains, e.g., grains > 1 μm for magnetite, or $\gg 1$ μm for hematite. The reason that these grains are important lies in their 'visibility' when a thin-section is microscopically examined, whereas submicrometer-sized grains escape detection with optical microscopes. Microscopic observations are often reported in paleomagnetic papers, but it is important to realize that these observations do not generally represent the important smaller grains which carry the more stable magnetizations. Scanning or Scanning Transmission Electron Microscope (SEM/STEM) observations must be used to obtain images of these smaller grains (e.g., Figure 2.8).

The magnetic stability of the submicrometer grains is generally much higher than that of those large grains one may observe with a microscope. This is because of the natural tendency (based on energy minimization considerations) for larger grains, on average, to contain a larger number of magnetic domains (Figure 4.7). The magnetization directions in these domains are different from each other, with domain walls separating the domains. The domain walls are narrow, but have a finite thickness estimated to be between 0.02 and 0.1 μm. The walls, however, can move around with some ease when external energy is applied in the form of a magnetic field (e.g., during AF treatment). The volume of domains with a magnetization in the direction of the applied field will increase at the expense of other domains (Figure 4.8). The coercivity of such a multidomain grain is therefore generally lower than that of a single-domain grain, all other factors (such as defect density) held constant. Hence, AF demagnetization removes the bulk of the multidomain magnetizations before those magnetizations are eliminated that are carried by the single-domain grains; the latter, being more stable, are usually also the primary magnetizations if any exist, assuming that all magnetizations in the sample are carried by the same mineral (e.g., titanium-free magnetite). This 'primary' magnetization resides, therefore, preferentially in grains that one cannot observe microscopically!

In rocks with different co-existing magnetic mineralogies (e.g., hematite and magnetite), the resistance to demagnetization of each phase may vary, but greater resistance (often called 'hardness' in AF) of a magnetization does not automatically imply a greater age or better chance of being primary (Dunlop, 1979). Buchan (1978) has shown examples of older magnetizations removed before the younger (overprint) components are eliminated.

In this section, I have attempted to make the techniques and underlying principles of demagnetization understandable. It is clear that demagnetization *must* be performed for a paleopole to have any validity. My own experience has been that 99% of all the rocks that yield primary magnetizations also contain secondary magnetizations (overprints) in the same samples. Needless to say, measurements of undemagnetized samples reflect a vector sum of all these components and are virtually meaningless. Thus we must separate the magnetizations in terms of their directions; the principles of this are discussed in the following section.

Demagnetization – analysis of directions

The simplest situation a paleomagnetist may encounter is when samples contain only one magnetization. This means that during stepwise demagnetization, (s)he will observe that the intensity of magnetization gradually diminishes, while the direction (D and I) remains constant. This 'univectorial' decay can be depicted by showing the decaying intensity versus increasing level of treatment, and by reporting the direction, or plotting the D and I on a projection (Figure 4.9). The directions do not change appreciably during treatment, and the mean of these directions is called a 'stable endpoint'; it is 'stable' in the sense that there is almost no variation in D and I (stable, as used here, should be distinguished from stable in the sense of not changing with time).

Alternatively, the vectorial behavior during demagnetization may be visualized using a diagram based on geometry in cartesian coordinates, and combines intensity and directional information. It is based on classical geometry as developed some centuries ago in France and was not invented by paleomagnetists; nevertheless, such diagrams are now commonly called Zijderveld or As–Zijderveld diagrams, named after the scientists who first introduced their use in paleomagnetism (As & Zijderveld, 1958; As, 1960; Zijderveld, 1967a; see also Dunlop, 1979). A more neutral name for these diagrams is 'orthogonal demagnetization diagram', whereas Miller & Kent (1989a) have proffered the questionable term 'demagnetogram'.

The Zijderveld diagram projects the endpoint of the magnetization vector onto two planes simultaneously, in a manner to be discussed more fully below. Two planes are necessary because the vector is oriented in three-dimensional space. Any two orthogonal planes would do, but commonly the (geographic) horizontal and vertical planes are chosen, such that the horizontal plane shows a component corresponding to the declination. The vertical plane shows the vertical (upward or downward) component, which gives an indication of the inclination. The horizontal and vertical projections have one horizontal axis in common, optimally the one to which the declination is closest. The successive endpoints obtained during demagnetization reflect the intensity of magnetization in their distance from the origin. Figure 4.9b

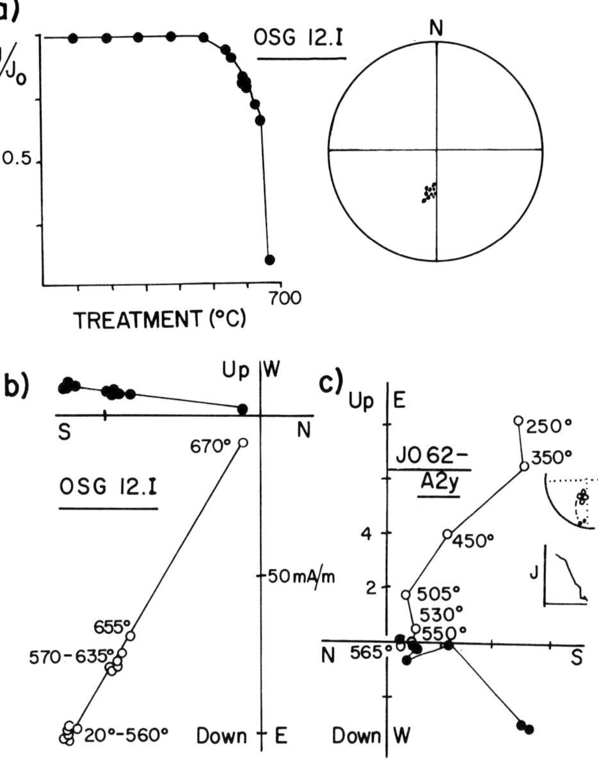

Figure 4.9. Illustration of the common visualization techniques to show the change in the direction of magnetization and its intensity during progressive demagnetization. a) Change in intensity and directions of magnetization as a function of treatment. In the case shown, these directions do not change appreciably, while the intensity decays; this indicates that the magnetization is univectorial. b) Orthogonal demagnetization diagram (also called Zijderveld plot) of the same sample as shown in a), illustrating that the trajectories through endpoints of the magnetization vector decay linearly to the origin. Solid circles represent projections of the magnetization vector onto the horizontal plane, open circles those onto the north–south vertical plane. For explanation of the projection method of this diagram see text or Figure 4.10. Intensity scale is labeled along the axes. The sample OSG 12.I is from Ordovician basalts from Newfoundland (from Van der Voo *et al.*, 1991). c) Sample showing 'noisy' behavior during thermal demagnetization with decay towards the origin that combines small spurious components of magnetization acquired during the treatment and a roughly univectorial decay of the remanence. The sample is from Devonian rocks in Scotland (Figure from Van der Voo & Scotese, 1981).

also shows a Zijderveld diagram for the simple univectorial decay, discussed earlier in terms of its directions in the stereonet. The labels next to the endpoint projections on the vertical plane (open symbols) give the level of treatment of the successive demagnetization steps, and the points are connected by a demagnetization trajectory, in this simple case two lines trending linearly to the origin. The 'direction' in which the magnetization is being removed in Figure 4.9b (i.e., towards the origin) is, of course, opposite to the direction of magnetization that exists in the sample. This magnetization direction is shown in Figure 4.9a as southerly and downward.

For the simple univectorial case, there is no advantage of one visualization technique of Figure 4.9 over the other; in fact, we might simply give the D and I for the sample in a table, since they have not changed within error limits. However, such univectorial behavior is extremely rare, especially in magnetite-bearing samples, although hematitic samples can occasionally be univectorial. Moreover, sometimes the linear decay to the origin is perturbed (called 'noisy') by small spurious magnetizations introduced during the treatment, resulting in a 'ragged' not-so-linear decay to the origin. In such a case, D and I stay roughly the same, but not necessarily within error limits. An example is shown in Figure 4.9c. Here the Zijderveld diagram has an advantage over the combined plots of stereonet with directions and the one with intensity decay (sometimes not even included in publications). The points plotted in the stereonet (inset of Figure 4.9c) could be mistaken as showing two directions, because at the higher temperature steps (above 505 °C) the direction seems to have moved from southwesterly and intermediate upward to southerly and nearly horizontal. This is probably illusory and due to mere noise of low intensity, superimposed on the true univectorial (?) direction, which above 505 °C has decayed to low intensities as well.

Most paleomagnetic samples contain two or more magnetizations, as illustrated in Figure 4.10a. With some practice, we now begin to see the real advantage of the Zijderveld plot, which shows two trajectories in each of the horizontal and vertical projections, whereas the stereonet simply shows a progressive change in direction. The lower-temperature trajectories of the Zijderveld plots in Figure 4.10a are linear, do not decay to the origin, and make a sharp angle with the higher-temperature trajectories, which do decay to the origin. The latter represent a stable endpoint, but the lower-temperature trajectories do not. The declinations of the lower-temperature trajectories can be interpreted directly as northerly in the horizontal projection (full circles), whereas the inclinations are steeply downward (below 430 °C) as can be seen in the vertical plot. The higher-temperature trajectory is south-southeasterly with steeply upward inclination (i.e., again opposite to the 'direction' in which the magnetization decays). The method of projection of the three-dimensional vector endpoints onto mutually orthogonal planes is illustrated in Figures 4.10 b, c. If these directions are not immediately seen by the reader, a bit of one's own plotting of what a southerly, easterly or northerly and upward (or downward) direction would look like will give sufficient practice to begin to be comfortable with this visualization technique. And the latter is just what it is, a mere tool to show others what one's measurements (sets of D, I and intensity values) reveal.

Early paleomagnetic studies in the 1960s usually did not completely demagnetize the magnetization all the way to near-zero (the origin of the Zijderveld diagram), but treated the samples with a field deemed sufficient to remove unwanted overprints, such as the lower-temperature trajectories of Figure 4.10a. This is called the 'blanket step' demagnetization technique and the directions after this step were called 'cleaned'. For these studies the question remained unanswered, though, whether the overprint had been completely removed; only one or just a few measurements (at worst a single point, at best only a few of the first seven points in the Zijderveld plot of Figure 4.10a) cannot reveal that. The investigators of those days relied on good clustering of the selected (cleaned) directions to gain confidence in their results. This led to an objectionable method, whereby the best clustered directions (with maximum precision parameter k) were selected from a series of blanket steps. My own experience has shown that this 'best-clustered grouping' is often still seriously contaminated by overprints.

It is not uncommon to find three or more magnetization trajectories coexisting in a Zijderveld plot (Figure 4.11), but often other problems begin to appear in such cases. Thus far, the examples shown in Figures 4.10 and 4.11 have sharply defined trajectories, with clear breaks between segments. Albeit ideal, it is, however, not common. Consider that each one of the straight-lined trajectories for a sample

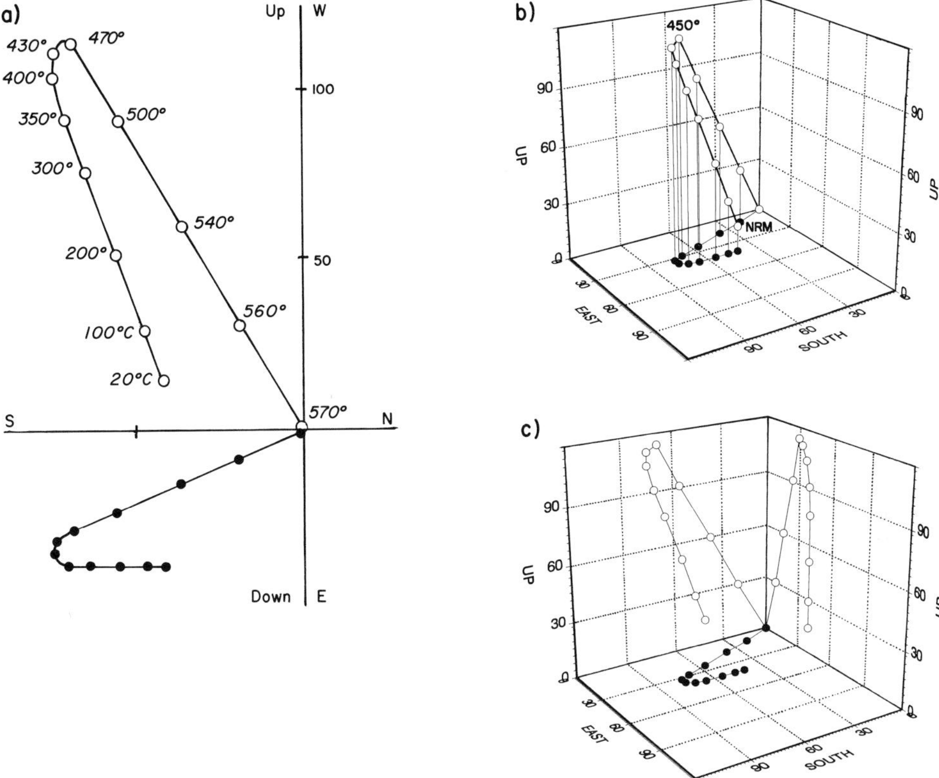

Figure 4.10. a) Zijderveld plot of a sample that contains two well-defined components of magnetization. The first removed component (between 20° and about 450°C) has a direction that is northerly and steeply downward; the second direction (450° to the origin at 570°C) has a south-southeasterly direction that is steeply upward. b) Illustration of the method to construct a Zijderveld plot for the same sample as shown in a). The open circles are 'floating' in three-dimensional space and represent the endpoints of the magnetization vectors as measured after each step of treatment; these endpoints are south-southwesterly and shallowly (at first, at point labeled NRM) to steeply upward (above 450°C). The three-dimensional endpoints are projected (with tie-lines) to the horizontal plane defined by the axes labeled south and east; these projected points are the full symbols. c) Further illustration of the construction of a Zijderveld plot. Here the 'floating' vector endpoints of b) have been omitted, whereas the solid circles representing the projections onto the horizontal plane are repeated. The open circles in *this* diagram are the projections onto the east–west and north–south vertical planes, all shown in perspective viewed from the southeast. The horizontal and the north–south vertical projections correspond to those combined in the two-dimensional plot in a).

simply means that only one component is being removed over a temperature (or AF or leaching time) interval. The unblocking temperatures (or coercivities or solubilities in acid) show what is called a spectrum for each magnetic component. In the samples of Figure 4.10 the two co-existing components each have their own discrete spectra, which do not overlap much. Completely non-overlapping spectra are illustrated in the hypothetical example of Figure 4.12a.

In most samples, on the other hand, the spectra overlap partially (Figure 4.12b). The corresponding synthetic Zijderveld plot is shown in Figure 4.12c. In the interval where the spectra overlap, the plot shows a 'rounding off' between the lowest and the highest straight-lined trajectories. This can be carried to greater extreme (Figure 4.12d) where much of the spectra overlap and the Zijderveld plot does not show any appreciably linear trajectory; instead it is continuously curved. The worst case is if the spectra overlap completely and have the same shape: both components are now removed at the same time and the trajectory is as linear as those of Figure 4.9, yet the direction of the 'isolated (?)' magnetization is meaningless!

In my experience complete overlap of the spectra is rare (there are tests, to be discussed later, that can detect this phenomenon). Especially when thermal, as well

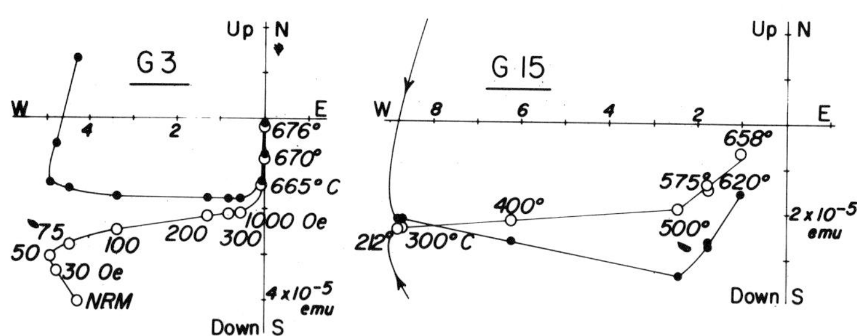

Figure 4.11. Zijderveld plot of two samples from red Mid-Proterozoic Beltian argillites in Montana (from Vitorello & Van der Voo, 1977) that show three components of magnetization; the first is likely to be of recent origin, the second, westerly and shallow component is thought to be latest Precambrian in age and the third (southerly to southwesterly and intermediate down) relates to about the time of deposition of these red beds. Published with permission from the Research Journals Division of the National Research Council of Canada.

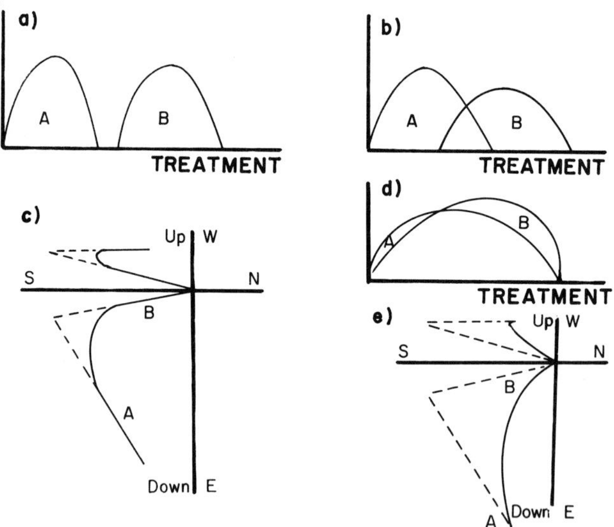

Figure 4.12. Spectra of unblocking temperatures (or coercivity or solubility in acid) for three hypothetical cases. a) The spectra A and B do not overlap; this means that magnetization components A and B are being removed in quite different temperature intervals and are 'discrete'. b) The spectra A and B overlap partially, such that in the overlap-interval both components of magnetization are being removed simultaneously. This is illustrated also in the synthetic Zijderveld diagram of c), where the rounding off between the two straight-line trajectories in each of the horizontal and vertical projections represents the interval of overlap. d) In this case, overlap of A and B is rather complete (although the spectra have different shapes), such that the two magnetizations are removed simultaneously but not at the same rate. This results in the accompanying Zijderveld diagram with continuously curved trajectories.

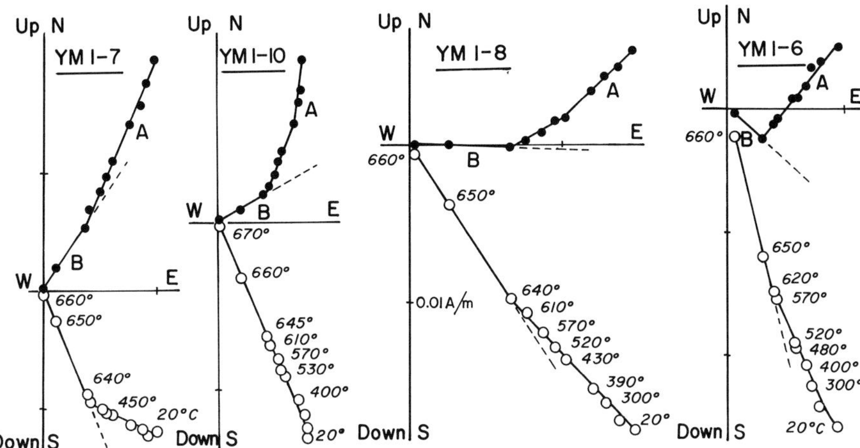

Figure 4.13. Zijderveld diagrams of four representative samples from a single outcrop of Carboniferous limestones of Yunnan, China (courtesy of Dr Fang Wu). The samples contain two components, A (northerly to northeasterly and down) and B. Component B appears to decay linearly to the origin, but its directions in the four samples can be seen to vary from northeasterly (left) to southeasterly (right). Despite the apparently linear trajectories of B, the suspicion is that component B in these samples is composite, that is, a mixture of component A and a component B′ which remains unidentified, but lies somewhere in the southern quadrants.

as alternating field or chemical demagnetization all give the same trajectory directions for the magnetic components, one can have quite a bit of confidence in the directions. Nevertheless, I have seen some apparently linear trajectories in Zijderveld plots that are obviously composite: the examples of Figure 4.13 show high-temperature components ranging from northeasterly to southeasterly in a progression of samples all from the same site. Given the lower-temperature overprint trajectory as northerly to northeasterly, the directions trend towards the southeastern quadrant and the farthest removed one is the last example of Figure 4.13. It may still be contaminated, but is a better approximation than the first sample shown in the figure.

Zijderveld diagrams (or stereonets) are, as already mentioned, mere visualization tools to help others see at a glance what the demagnetization measurements revealed. Directions can be estimated from them, but there is a computer algorithm (Kirschvink, 1980) which does it for us, based on principal component analysis (PCA). In this technique, lines and planes are fitted objectively to the successive endpoints (measurements) of the magnetization vector during progressive demagnetization. The advantage is that the program also gives a measure of maximum angular deviation (MAD angle) and that the measurements, usually stored in some memory format, can be processed directly.

When two components co-exist in a sample and are removed almost simultaneously, but not at the same rate, the direction in the stereonet is seen to change progressively, e.g., Figure 4.14. Two components, i.e., two vectors, define a plane which in the stereonet is a great circle (since the plane must pass through the origin). The great-circle trajectories of Figure 4.14 illustrate this. Halls (1976, 1978) has really advanced this technique enormously, although in earlier paleomagnetic papers (especially from authors in the USSR) it was already recognized that a 'streaking' along a great-circle trend must represent two co-existing magnetizations in the sample population.

The beauty of the great-circle analysis of Halls has been that under certain conditions these great circles intersect each other in a well-defined area of the stereonet (as they do in Figure 4.14). This intersection represents the line in common to all planes, which by definition is one of the two magnetizations. The conditions are that one of the two components co-existing in the collection of samples must be well clustered, whereas the other component must be much more dispersed. If both components are well clustered, all great circles pass through these clusters and they will be more or less parallel, thereby eliminating any chance of finding well-defined intersections. The PCA algorithm of Kirschvink (1980) also contains this planar analysis option.

The great-circle intersections can be divined from the stereonets of Figure 4.14, but many different intersections obtained from pairs need to be combined. This is done by taking the normals (somewhat awkwardly called 'poles') to the planes; each great-circle trajectory yields such a pole (triangles in Figure 4.14). When the sample great-circles have different trajectories in the stereonet, their poles will be similarly distributed in a combined plot (Figure 4.15) and it can be seen that a new (site-mean) great circle can be fitted through the triangles (poles) of the collection of individual samples. Now the pole to this site-mean great-circle is the desired mean intersection. Complicated? Not really: the sample poles (A1, A2, A3 . . .), although distributed, are all more or less perpendicular to the desired direction (B) and they define the A-series plane. By taking the normal (C) to the A-series plane, we have of course then C as perpendicular to this plane, whereas B is also perpendicular to the A-series plane. Thus B = C! By using best-fitting techniques, one also obtains estimates of error limits. More recently, Bailey & Halls (1984) have devised an

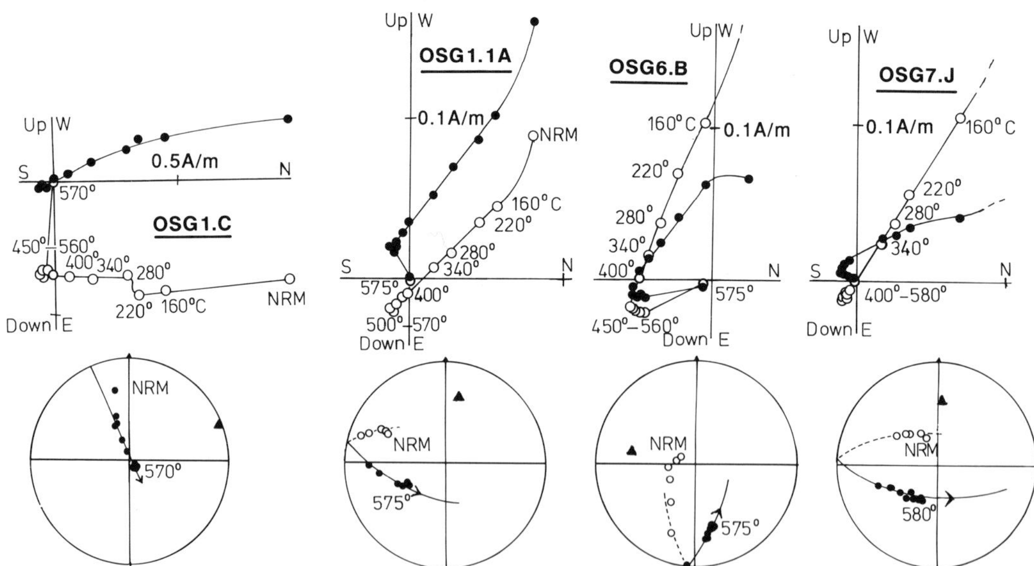

Figure 4.14. Zijderveld diagrams and stereonets showing changes in direction during thermal demagnetization of specimens from four samples from different sites with different structural tilts in Newfoundland Ordovician basalts (from Van der Voo *et al.*, 1991, in Geological Society of America Bulletin, v. 103, p. 1564–75). The trends in the stereonets converge towards a southeast and downward direction (characteristic for these basalts, after structural correction), indicating that the great circles represent planes that all have one component in common. The Zijderveld diagrams all show multivectorial behavior, but with overlapping spectra to such an extent that no stable endpoint is reached (the higher-temperature trajectories are composite and/or curved). The normals to the great-circle planes are shown as triangles.

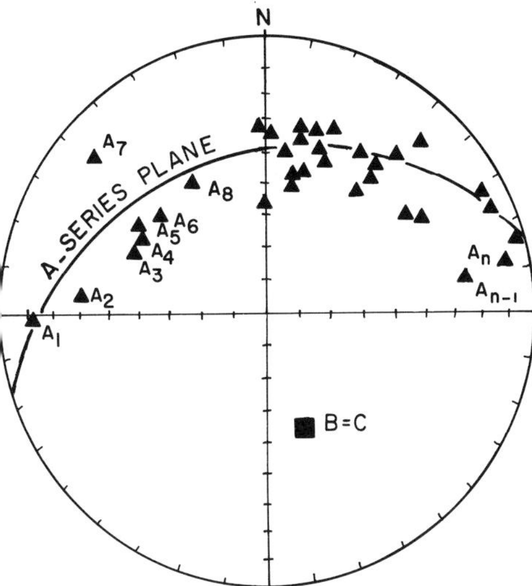

Figure 4.15. Stereonet showing all normals (Al, A2 . . . An) to great-circle planes for a large collection of samples (four of these were shown in Figure 4.14). The normals themselves define an overall great-circle or A-series plane as shown. The normal to *this* plane (square, B=C) is the direction in common to all samples. For further explanation see text.

algorithm that combines stable endpoint directions and great circle poles, so one can now obtain an overall mean (D, I) with k and α_{95} for such a set of results. The examples of Figures 4.14 and 4.15 are from a paper (Van der Voo *et al.*, 1991) which could use 80% of all demagnetization results this way, the remaining 20% representing only overprints. Without the Bailey & Halls technique, only those samples with stable endpoints (35% of the total) would have been usable.

When three components with overlapping spectra co-exist in a sample, they no longer define a single plane and the corresponding trend in the stereonet cannot be fitted by a great circle (e.g., Figure 4.16a). Let us call these components A, B and C, and assume that A and B overlap, and that B and C overlap, but A and C do not (Figure 4.16b). For such a case, Hoffman & Day (1978) proposed a technique that allows the determination of the direction of B. They argued that, if one takes the difference vectors between successive measurements, then plotting the directions of these vectors in a stereonet will show two great circle trends. The first will be the plane defined by A and B, the second will pass through the directions of B and C. The intersection of the two great circles, then, is the direction of B (Figure 4.16c). This method leaves, of course, the directions of components A and C yet to be estimated, but once B is known, perhaps other sample demagnetizations and great circle planes can be found that allow their determination.

In all these methods, the measurements must be relatively free of spurious magnetizations. In reality, this is often not the case, as we have already shown for one example of Figure 4.9. The greater the noise, the more difficult it will be to define individual planes and to obtain intersections with any precision. Thus there are practical limits to the great-circle techniques: the samples must show well-behaved demagnetization trajectories.

From the foregoing, it is clear that for a reliable paleopole, demagnetization is a condition *sine qua non*. I would also insist that the demagnetization analysis be published in the archival literature; it is all too frequent that unpublished results are included in a synthesis but, upon peer review, have undergone significant changes.

A further condition for reliability is that demagnetization must have been accompanied by directional analysis of magnetization components removed and isolated, either through the use of combined stereonet and intensity plots, the use of Zijderveld diagrams, or by PCA and/or great-circle analysis. Without such directional analysis, there is no guarantee that the reported directions represent uncontaminated magnetizations; a direction remaining in a sample after a given demagnetization treatment may still be multivectorial. This is the potential flaw in paleopoles obtained by the blanket-treatment or minimum-dispersion methods which were common until the mid-1970s. Since about 1976, almost all published paleomagnetic studies have used the Zijderveld plots pioneered by As & Zijderveld of the University of Utrecht. Some paleomagneticians have questioned this condition, saying that a single published Zijderveld plot is still no guarantee that the entire sample collection has been analyzed appropriately. This is, of course, true, but one must have faith in the integrity of the authors and hope that what they publish is a full, or at least representative, accounting of their results. Shangyou Nie of the University of Chicago has commented to me that representative should not mean 'i.e., the best'! He is correct; a good paleomagnetic publication should show the full range of demagnetization behavior in a series of Zijderveld plots.

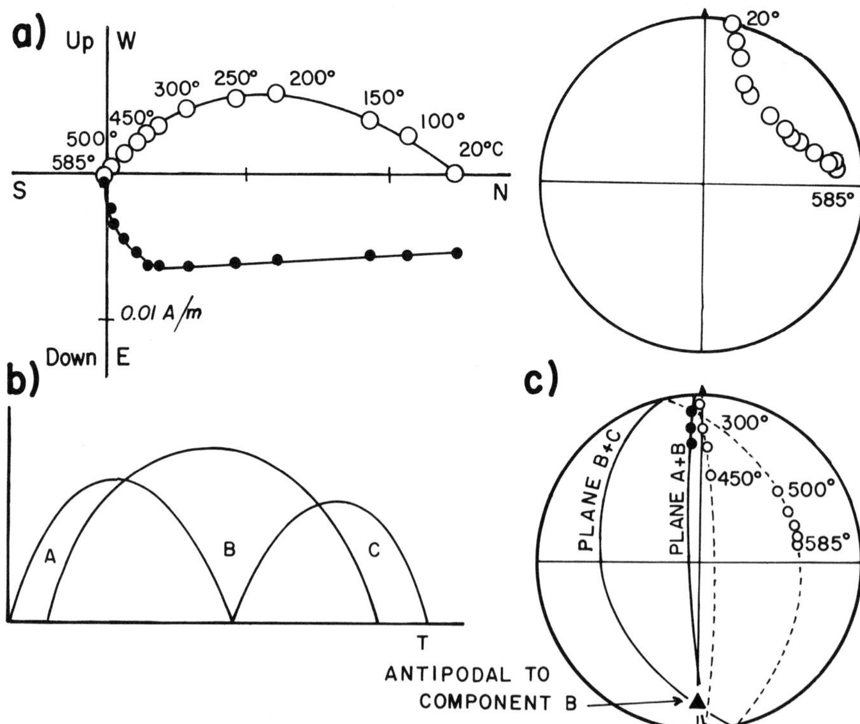

Figure 4.16. a) Continuously curved Zijderveld diagram and corresponding stereonet for a demagnetization in which no great-circle plane can be identified because at least three components of magnetization coexist that partially to completely overlap, as shown schematically in b). The appearance of a *small-circle* trajectory on the stereonet is coincidental. c) The Hoffman & Day (1978) method applied to the sample of Figure 4.16a, b) appears to show an intersection of the plane containing components A and B with the plane defined by components B and C, in a direction that is northerly and shallowly up (antipodal to triangle, southerly and shallow down). Plotted points represent vector differences between the vectors of successive measurements.

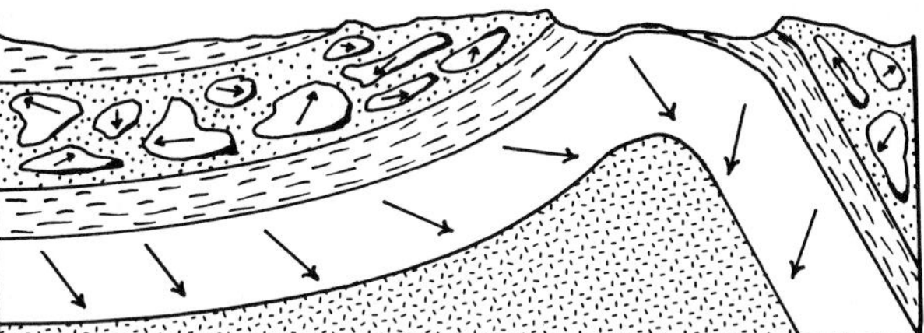

Figure 4.17. Field relationships in Graham's classical fold test and conglomerate test, showing magnetization directions in the folded layer and in the conglomerate cobbles that can be inferred to be older than the folding and redeposition, respectively. Redrawn and modified after McElhinny (1973) and Cox & Doell (1960).

Field tests to constrain the age of the magnetization

Favorable conditions may exist in the field setting of paleomagnetically studied rock sequences that can provide important constraints on the age of a magnetization. The best known of these conditions is the fold test, introduced by Graham (1949) and later refined on a statistical basis (McElhinny, 1964; McFadden & Jones, 1981; McFadden, 1990).

The fold test is based on the principle that magnetization directions older than the folding phase must have been clustered, before their bedding attitudes were altered by tilting, accompanied by dispersal of the magnetization directions in modern geographic coordinates (also called *in situ* coordinates). This is illustrated in the well-published diagram of Figure 4.17. When the bedding is restored to (paleo)horizontal upon correction for tilt, the *in situ* scattered directions should become more clustered in a statistically significant fashion. The statistical improvement is measured by the change in the precision parameter from k1 (before tilt correction, TC) to k2 (after TC); if the ratio k2/k1 exceeds a critical value (Figure 4.18), then the fold test is statistically significant (McElhinny, 1964).

The entries in the fold test analysis are usually the site-mean directions, but sometimes they are the individual sample directions. McFadden & Jones (1981) and McFadden (1990) have argued that this may bias the test towards the negative side. In the typical setting of folded rocks, synclines and anticlines may have roughly parallel fold axes, say, striking east–west. The limbs of the folds then have either northward or southward dips. If the two populations from the north and south dipping limbs have largely different numbers of sites, say nine in limb A and three in limb B, then the large number of sites on limb A that all have similar (but not identical) dips will overwhelm the outcome of the conventional fold test of McElhinny (1964). However, if the populations of directions from the two limbs have statistically similar clustering, the (previously 12) fold test entries can be reduced to a limited number (now two), representing groups of sites with similar dips. This modified test is less stringent than the McElhinny version.

The fold test is based on certain assumptions that generally can be verified. The first is, of course, that the beds must originally have been horizontal; in some continental sediments as well as for volcanic edifices, strata can be deposited on an incline of up to some 20°. The second assumption is that penetrative deformation of the beds has not occurred, and that the folded beds are simply tilted. For some special situations where internal strain of known magnitude and orientation has occurred, correction methods for its 'removal' have been proposed by Cogné & Perroud (1985). Plunging fold settings (with inclined fold axes) may have to be 'unplunged' before restoring the beds and their corrected (unplunged) directions to paleohorizontal about the equally corrected (unplunged) strikes (see MacDonald, 1980).

A negative fold test means that the directions cluster better *in situ* than after TC, so that the magnetization must be younger than the folding phase. In some cases, rocks which have undergone two phases of folding show magnetizations that are older than the later phase, but younger than the earlier one. The age of these secondary magnetizations is thereby bracketed by the ages of the folding phases, if known (e.g., Johnson & Van der Voo, 1989).

In the last decade, it has been discovered that some magnetizations neither pass nor fail the fold test, but that a statistically significantly better grouping of directions is obtained when the beds are corrected for only 40 to 70% or so of their full tilt. When the precision parameter k is plotted as a function of unfolding, performed stepwise in small increments, the best clustering occurs at the peak value of k (Figure 4.19). One interpretation of such results is that the magnetizations are contemporaneous with the folding, also called 'synfolding'. Another interpretation might be that there has been internal, penetrative deformation of the beds, although in several examples this could be regarded as unlikely; Kodama (1988), for instance, concluded that the inferred magnitude of magnetization rotation due to strain in the Allentown Dolomite of Pennsylvania could not explain the pattern of k vs. unfolding, whereas a synfolding acquisition of the magnetization could. In essence, the results of Johnson & Van der Voo (1989) are also synfolding, except that there

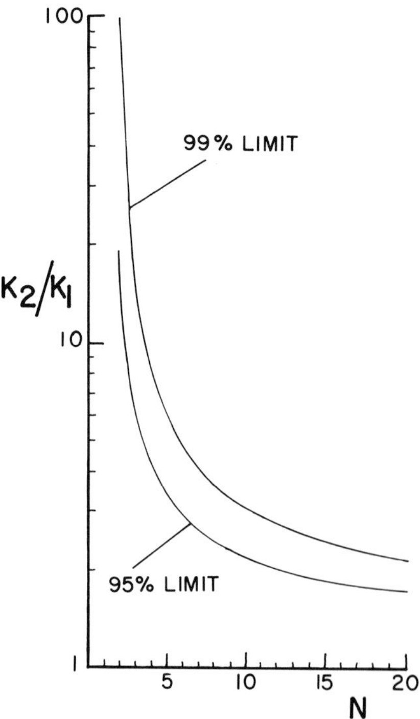

Figure 4.18. Semi-logarithmic plot of the k-ratio versus the number of entries in the statistical analysis (N), in which the two curves represent the critical values above which the fold test is positive at the significance level indicated. The k-ratio, k2/k1, is the ratio of the precision parameter after and before correction for the tilt of the strata. Figure redrawn after French (1978).

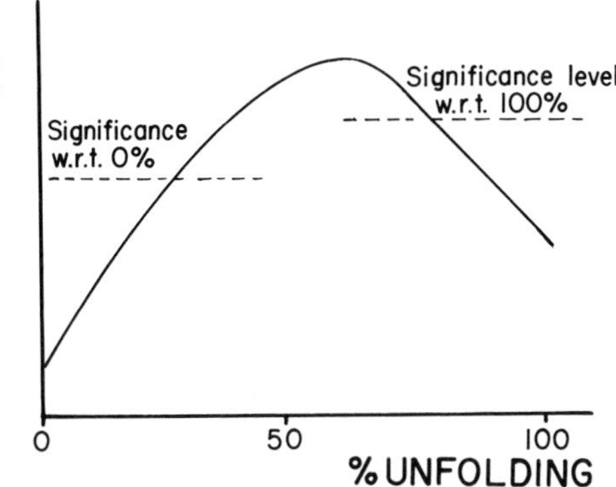

Figure 4.19. Plot of k (see Figure 4.18) for a case in which the highest value is obtained for a partially unfolded structural setting, suggesting acquisition of the magnetization *during* folding. The horizontal axis shows incremental corrections for structural tilt of the various limbs of folds from 0% (present-day, *in situ* coordinates) to 100% (all beds restored to paleohorizontal). Just as the significance of a conclusion that the magnetization is pre-folding is determined by testing whether k2/k1 exceeds a critical value (Figure 4.18), the significance of a syn-folding conclusion depends on the ratios of precision parameters, i.e., kmax/k. The significance is indicated with respect to 0 as well as 100% unfolding at the 95% confidence level.

the 'earlier' and 'later' episodes of folding could be separately identified and corrected for.

An inconclusive fold test, whereby neither k1/k2 nor k2/k1 exceeds the critical value for the number of entries N (Figure 4.18), obviously tells us nothing about the age of magnetization with respect to the age of folding. Even a positive fold test may reveal very little, if, for instance, the rocks are Paleozoic and the age of folding is late Mesozoic or Cenozoic as is the case in some settings of westernmost South China or western North America. Some magnetizations from the western part of the South China Block may be ancient but secondary, even though they pass the fold test. On the other hand, examples exist of magnetizations shown to be near-primary by a fold test because the folding occurred only a few million years after deposition: my own experience (Figure 4.20) includes a positive fold test for Late Carboniferous to Early Permian (Stephano-Autunian) red beds from Portugal which were folded in the Early to Middle Permian (Van der Voo, 1969).

A variant of the fold test is provided by syn-sedimentary slumps. A few examples (e.g., Smith, Stearn & Piper, 1983) are known that have proven the magnetization to be older than the slumping, that is, primary.

A second and powerful test to constrain the age of magnetization is provided by baked contacts (Everitt & Clegg, 1962). Intrusions cool after injection of molten material, and in the process they impart heat to the country rock. Within a critical distance from the intrusion (Figure 4.21), the country rock temperatures exceed, briefly, the unblocking temperatures so that the magnetization in this contact is reset and now dates back to the age of the cooling of the intrusion. The intrusion and its baked contact should therefore have the same magnetization direction, while the unaffected country rock at considerable distance from the intrusion should show an earlier unrelated direction of magnetization. If so, the contact test is positive for the magnetization of the intrusion and this magnetization has been shown to be primary. The age of the magnetization of the unaffected country rock is also constrained, because it can be inferred to be older than the age of the intrusion; this is called an 'inverse' contact test.

In many cases, the baked contact and intrusion give good directions, whereas the unaffected country rock turns out to be unsuitable for paleomagnetism. This could perhaps be called a partial contact test, but it may be of limited value as shown by the example of dikes and their contacts in New Mexico, both of which are thought to have been remagnetized in the same (younger) event (Jackson, Van der Voo & Geissman, 1988).

At a distance from the intrusion, but within range of its heating effects, the country rock may have retained some of its original magnetization and may have acquired a partial overprint in the ambient field during the intrusion (Figure 4.22). This takes place during intermediate temperatures less than the maximum unblocking temperatures of the samples. McClelland-Brown (1981) used the unblocking temperatures of such overprints to estimate and model the thermal regime at the time of the intrusion.

The third major test makes use of the remanence directions in the cobbles of a conglomerate (Graham, 1949; see Figure 4.17). Graham reasoned that if the cobbles, emplaced after considerable transport, would have retained their original magnetization, their directions would essentially be random. The test, if positive, constrains the age of the magnetization for the formation from which the cobbles have been derived. Buchan & Hodych (1989) showed that some cobbles in a Silurian conglomerate in Newfoundland have, in fact, the same demagnetization behavior as samples from the underlying flows from which they were derived. If the cobbles retained original stratification, one can even restore the magnetic directions to paleohorizontal: the inclinations thus corrected should all become similar (e.g., Figure 4.23). A variant of the conglomerate test can be used for breccias (Figure 4.24) and rip-up clasts, although here the transport (if not the rotations) of the fragments may have been minimal (e.g., Wisniowiecki *et al.*, 1983; Bachtadse, Van der Voo *et al.*, 1987a).

A microconglomerate test with small fragments is also possible (Geissman, 1980), and a further application was conceived by Elston & Purucker (1979) for detrital sediments in which grains are deposited on sloping foreset beds. They reasoned that some grains deposited in the foreset beds, which dip in different directions, should undergo rotations by rolling downslope for a short distance. The magnetic direction

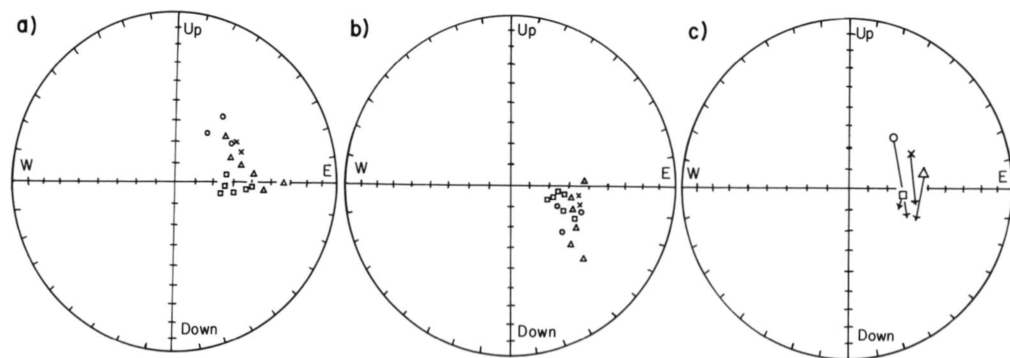

Figure 4.20. Positive fold test for earliest Permian rocks from Portugal that were folded during the later Permian (from Van der Voo, 1969). Note that the stereonets are vertical projections (to illustrate better the change from upper to lower hemisphere in the plots without resulting in crowding). a) Characteristic directions for all samples obtained during demagnetization, without tectonic correction, b) same, after tectonic correction, c) site-mean directions showing change from before to after tectonic correction by the arrows, indicating significantly better clustering after tectonic correction. Published with permission from Elsevier Science Publishers, BV

Figure 4.21. Photograph of a contact-test site in northern Greenland, where a Middle Proterozoic dike intrudes Archean gneisses. The schematized results of Figure 4.22 are based on several sites such as this one. Sample orienter (with 10 cm sample consisting of dike and baked contact material) for scale.

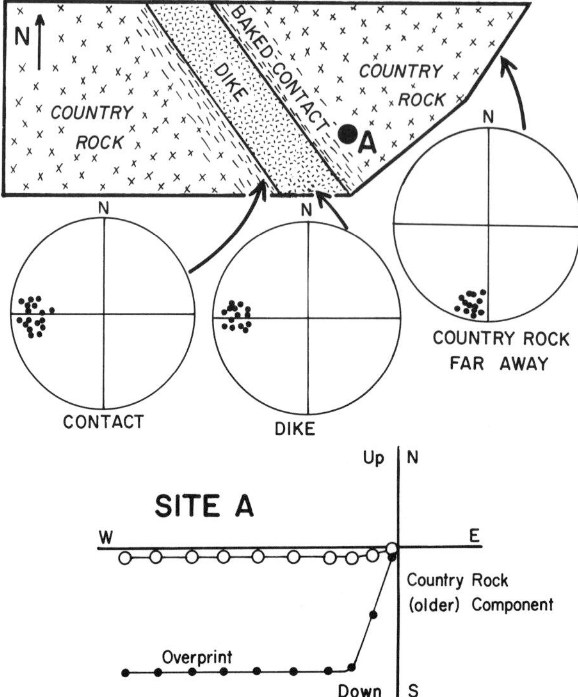

Figure 4.22. Illustration of a positive contact test between an intruding dike and its host (country rock). The schematic stereonets illustrate typical directions that would be found in such a case (based on the contact test described by Abrahamsen & Van der Voo, 1987; see Figure 4.21). Bottom diagram shows a Zijderveld plot for Site A (see map for location) where the thermal effect of the dikes is represented by a lower-temperature overprint that did not completely erase the older magnetization of the country rock (southwesterly and shallow down).

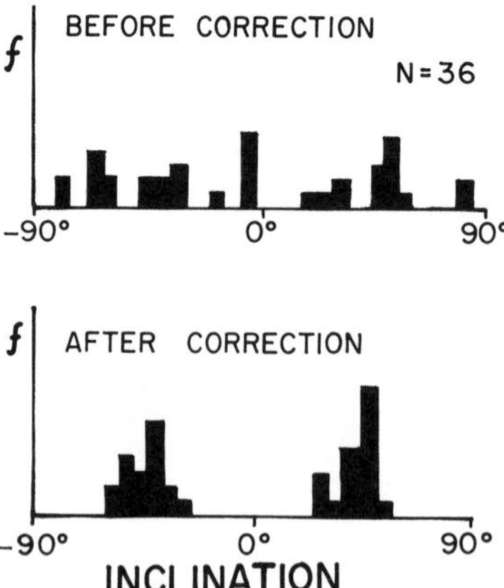

Figure 4.23. Histogram of inclinations that show a positive conglomerate test (not based on real data), wherein the individual pebbles or boulders are hypothesized to have retained their original stratification such that they could be restored to paleohorizontal. Note that the tops or bottoms of the bedding in the pebbles are likely to be unknown, so that inclinations after correction may be positive or negative, without necessarily indicating reversals.

Figure 4.24. Photograph of an Ordovician breccia sampled for a conglomerate test in southeast Alaska (the test turned out to be negative).

of the magnetic fraction of these grains has thereby changed somewhat, provided that the remanence was a Detrital Remanent Magnetization (DRM). A DRM implies that the grains are already magnetic when they are deposited and align themselves with the geomagnetic field if the sedimentation conditions are quiet. Elston & Purucker (1979) and later also Elmore & Van der Voo (1982) observed the expected effects as a systematic function of foreset dips and concluded that this indicated a primary DRM. As pleasing as this conclusion is, the direction of this magnetization is by definition perturbed and cannot be used to calculate a paleopole!

In summary, positive fold, conglomerate and baked contact tests can provide important constraints for the age of the magnetization and can prove its primary nature under certain conditions. One should not stipulate that a paleopole is only reliable when such tests are positive, because not all field settings lend themselves to these tests; on the other hand, these tests enhance the reliability of a result by some incremental value.

Structural control

In Chapter 3, it was shown that a paleopole is valid for a given craton (or block) only when the continuity between the site and this craton is assured for the time since the acquisition of the magnetization. In addition, the magnetic direction must be referred to the paleohorizontal before a paleopole can be called reliable. For stratified rocks, this is generally possible provided that the magnetization pre-dates later tilting. MacDonald (1980) discussed apparent rotations which may occur in mobile belts and perturbed cratonic margins during orogenesis, including rotations introduced by plunging folds. Only the declinations (and not the inclinations) are affected by such processes. The reason is simply that the inclination is the angle between the magnetic vector and the (paleo)horizontal, which is therefore invariant to any rigid-body rotations. If the reader is not convinced, a pencil at a fixed angle to a flat sheet of paper should be moved around together in whatever contorted ways one wishes: the invariance will become clear. Thus stratified rocks may provide ancient inclinations, and, hence, paleolatitudes even in the most disturbed orogenic settings.

The declinations, on the other hand, may generally be deflected from their original direction during orogenesis or thrusting. Thus, paleopoles from rocks in mobile belts that pre-date orogenesis may also be deflected, which makes such results potentially less reliable for APWP construction.

Needless to say, unstratified bodies such as plutonic and metamorphic rocks may lack complete knowledge about tilts subsequent to their formation and acquisition of one or more magnetizations. Only anorogenic intrusions in areas not further affected by deformation can be considered to have sufficient structural control. Sometimes, constraints on tilts of pre-orogenic rocks may be provided by overlying strata, but only rarely are these constraints tight enough.

Rotations about vertical axes are sometimes indicated by strike trends on a regional scale; following Carey (1958), this has been called 'oroclinal bending'. It is possible to correct for strike deviation (e.g., Perroud, 1983), but this does not provide adequate structural control because typically the reference direction is unknown and only arbitrarily chosen. A strike correction, however, may improve the clustering of paleomagnetic directions significantly and is an acceptable practice. If there is independent evidence for oroclinal bending, moreover, one may even consider a strike test in order to provide evidence that the magnetizations are older than the deformation that caused the rotations. This has not yet become common

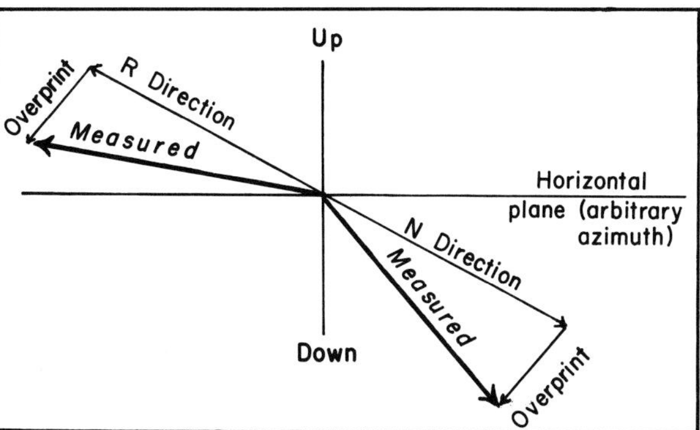

Figure 4.25. The effects of a systematic bias (overprint) on normal as well as reversed directions are shown in a vertical plane of arbitrary orientation. Superimposed on both polarity directions of primary origin (N and R) is a downward overprint directed to the left, resulting in composite measured directions.

practice, probably because the independent evidence is generally lacking, and without such evidence a 'strike test' would be based on circular reasoning.

Reversals

The presence of antipodal normal and reversed directions is an indication that (1) sufficient time has been sampled to average secular variation, and (2) the directions are not biased by a systematic overprint (Figure 4.25). Reversals are observed also in magnetization directions that are demonstrably based on remagnetizations (e.g., Johnson *et al.*, 1984; Johnson & Van der Voo, 1989), so they are no guarantee that a direction is primary. Normal and reversed polarity directions of secondary origin have been observed in red beds, carried by hematite as a CRM, and in goethite, also as a CRM. Thus far, normal and reversed directions in magnetite have to my knowledge not been conclusively demonstrated to have been secondary or of a chemical nature, although magnetite in metamorphic rocks may reveal reversals as TRMs which obviously were acquired only after cooling from the elevated temperatures during metamorphism.

A formal reversal test has been in existence (e.g., Cox & Doell, 1960; McElhinny, 1973, p. 85), but as recently shown by McFadden & McElhinny (1990), it is flawed. If the normal and reversed directions are less than 180° apart, the test decides whether the deviation from antipodality is statistically significant. This is determined by the cones of confidence (α_{95}) about the normal (N) and reversed (R) mean directions; if, after inversion of one of the two, the cones of confidence do not overlap, then the 'reversal test' fails. However, poorly clustered groupings (with large α_{95}) create an advantage, making it more likely that the test will pass. This seems at odds with what common sense dictates for a good test. McFadden & McElhinny's new test includes proper considerations of the size of cones of confidence, thereby remedying the problem. Published results have not yet been evaluated by this new method because it is so recent. My preference has been, therefore, to use the presence of reversals as a reliability criterion, but not to require a positive reversal test.

The presence of reversals allows us to determine whether magnetization directions are biased by systematic overprints (Figure 4.25), such as would be the case when the demagnetization spectra of two magnetic components overlap completely (see earlier discussion and Figure 4.13). When the systematic overprints are sizeable, the N and R directions deviate strongly (e.g., by more than 10°) from antipodality. On the assumption that the overprints are superimposed in equal measures on both groups, taking the average of the N and inverted R directions should eliminate the bias (Figure 4.25). Although this assumption is far from being automatically granted, the average will at least be a better approximation of the true direction than either N or R alone.

In the early days of paleomagnetism, a great deal of debate took place whether reversals observed in natural (igneous) rock samples were indeed due to a reversal of the geomagnetic field or whether there existed an intrinsic property of some minerals that allowed them to acquire a magnetization opposite to the direction of the applied field. The latter is called the 'self-reversal' mechanism. However, as demonstrated by many contact tests in subsequent decades, it appears that self-reversal in nature is rare; where it has been shown to exist, the magnetic mineralogy turned out to be of a very peculiar titaniferous magnetite composition as demonstrated by Curie point or electron microprobe determinations. Thus, self-reversals have ceased to be a worry, although rock magnetic experiments and positive contact tests remain essential elements in any study before it can be assumed that the reversals show a true record of the geomagnetic field.

Suspicion of remagnetization

It has been known since the early days of paleomagnetism that rocks can be completely remagnetized; in such a case, the results may fail the fold test (e.g., Irving, 1964). What was not realized, however, until the last decade, is that all rock types can be equally affected by such remagnetizations and these may easily go unrecognized. Nowadays one gets the impression that remagnetization is the rule rather than the exception.

Ancient, yet secondary, magnetizations are typically VpTRMs or CRMs, ignoring the acquisitions of TRMs by metamorphic rocks and baked contacts for the time being, because the latter are quite obvious from petrological and field observations. In igneous rocks, new magnetite may form by hydrothermal alteration, by oxidation/exsolution of iron-bearing phases (e.g., titanomagnetites), or by the breakdown of ferromagnesian silicates (e.g., amphiboles), whereas new hematite may form during oxidation. An example is provided by remagnetized granites in Aruba, where older Cretaceous country rocks and Cretaceous post-granitic dikes have the same easterly (presumably Cretaceous) directions, whereas the granite itself has a more northerly direction, inferred to be of secondary, Tertiary age (Stearns, Mauk & Van der Voo, 1982). The magnetization in the granite is characteristic of magnetite, and in thin section one observes very fine needles of magnetite in amphibole, as well as strings of equidimensional magnetite crystals along grain boundaries (Figure 4.26). These grains have all appearances of being secondary in nature; nevertheless, large, primary (titano-)magnetite grains are also observed, but presumably these do not carry a stable remanence.

In sediments, new magnetite may grow after precursor iron-bearing phases such as pyrite (Figure 2.8). Until the late 1970s, conventional wisdom among paleomagneticians was that all magnetite in carbonates must be detrital (e.g., Martin, 1975). We now know this to be an erroneous inference (e.g., McCabe & Elmore, 1989). Hematite in red detrital (or carbonate) sediments may have formed very early or may even be detrital (e.g., Purucker, Elston & Shoemaker, 1980), but it can also

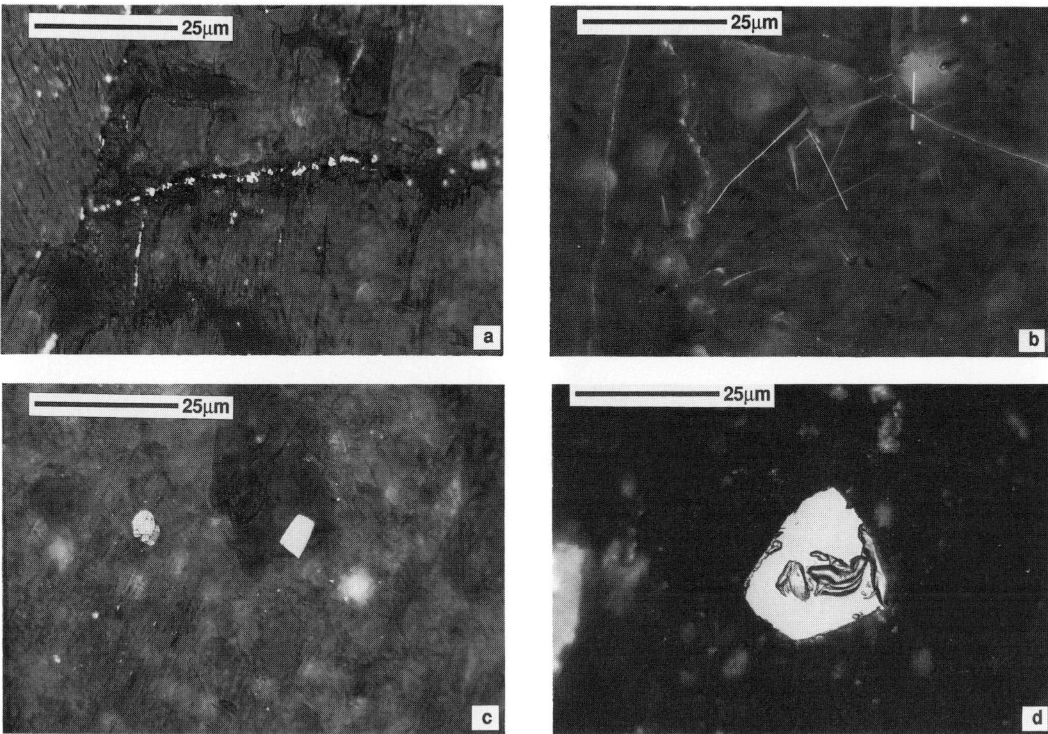

Figure 4.26. Photomicrographs of a) fine secondary magnetite crystals, b) secondary magnetite needles and c), d) larger, primary equi-dimensional magnetite grains observed in a granite from Aruba. The larger grains probably do not carry any stable remanence, whereas the fine-grained magnetites are thought to have formed by hydrothermal alteration and exsolution from mafic host minerals; they are inferred to carry a secondary magnetization (from the work of Stearns, Mauk & Van der Voo, 1982; photograph courtesy of J. Wm. Geissman).

carry ancient magnetizations acquired hundreds of millions of years after deposition, as shown by the studies of Appalachian red beds (Irving & Opdyke, 1965; Roy, Opdyke & Irving, 1967; French & Van der Voo, 1977; Irving & Strong, 1984; Kent & Opdyke, 1985; Miller & Kent, 1986a, 1986b, 1989a). Larson & Walker and their colleagues at the University of Colorado have argued for many years that the typical red bed magnetizations for Permian and Triassic rocks from the western United States are CRMs acquired during a long (how long?) period after deposition (e.g., Walker, Larson & Hoblitt, 1981; Larson *et al.*, 1982). In contrast, others have found that reversal zonations in these red beds are typically layer parallel, which would be fortuitous in the case of very late remagnetizations, and that some tests show evidence for primary detrital magnetizations (e.g., Elston & Purucker, 1979; Molina-Garza *et al.*, 1991). This 'red bed controversy' created an extensive debate that has now subsided, with the main issues of the controversies left unresolved. Most paleomagneticians now assert, on the basis of consistency between igneous and red bed paleopoles, that the magnetizations of these Permo-Triassic red beds were acquired within a few million years after deposition, i.e., within 1% of the age of the rocks. Thus, these magnetizations can be regarded as 'primary' using my earlier definition.

Goethite and pyrrhotite are also capable of carrying secondary magnetizations, usually acquired as stable and ancient CRMs. Thus, any magnetic mineral potentially can be secondary unless proven to be detrital or of primary origin in igneous rocks. Microscopic observations can be very helpful in illustrating the primary nature of magnetic minerals, but such observations are typically made on larger grains, whereas it may be the submicrometer grains that usually carry the stable remanence.

How does one recognize a remagnetization, if not through the tests discussed earlier? In essence, there is no unique and foolproof way (Figure 4.27). This is why the fold, conglomerate and baked-contact tests are extremely important. To test for

Figure 4.27. Photograph of a sequence of well-exposed Cambrian carbonate rocks in northern Greenland. Nothing in the appearance of the rocks indicates that they are completely remagnetized: however, all samples reveal only remanent magnetizations of recent origin.

remagnetization, one can compare a new paleopole with an existing APWP; if a result resembles paleopoles for ages younger (by a period) than that of the studied rocks, a strong suspicion exists that it is based on a remagnetization. I have used this resemblance (or rather the lack thereof) as a reliability criterion as well (Van der Voo, 1989, 1990b). It is, however, an assessment of the result in terms of 'guilty until proven innocent'!

Other considerations

Thus far, I have discussed seven separate areas of 'information' that may make a paleopole more reliable, if and when provided. These dealt with age precision, number of samples and statistical precision, demagnetization, field tests, structural control, reversals and a lack of suspicion about remagnetization. In addition, one may hope for good microscopic observations and adequate rock magnetic experiments. The latter two areas of information should not be specified in terms of reliability criteria, for the following reasons.

As already discussed, microscopic observations typically elucidate only the larger grains which may not be the carriers of a stable, ancient-remanence. Electron microscope techniques, which are able to image the submicrometer grains, have not yet been incorporated sufficiently in paleomagnetic studies to constitute a reliability requirement (e.g., Geissman, Harlan & Brearley, 1988; Lu, Marshak & Kent, 1990; Suk *et al.*, 1990a, b).

Rock magnetic experiments may provide information about the magnetic mineralogy (magnetite, maghemite, hematite, goethite or iron-sulphides), the titanium content of the iron oxides, the domain structures and sizes of the remanence-carrying grains, and information about coercivities or unblocking temperatures, in addition to that provided by demagnetization. This information is, obviously, of importance, but in general it does not contribute to our knowledge about the reliability of a paleopole. As discussed, any magnetic mineral can be secondary or primary and remagnetizations can be carried by all grain sizes. Although unexsolved titaniferous iron oxides are usually an indication of a primary igneous mineralogy, they rarely carry a Mesozoic or Paleozoic remanence in continental rocks.

Lastly, it should be mentioned that problems may occur with inclinations that are too shallow because of sediment compaction, as demonstrated by sedimentation experiments in a known laboratory field (King, 1955). Such an inclination error has also been observed in natural sedimentary rocks: Late Miocene to Pliocene marls from Sicily and Calabria in Italy show well-determined normal and reversed directions (e.g., Figure 4.28), which are too shallow by about 20° (Linssen, 1991; Langereis, 1984), with respect to the expected field direction. This means that the magnetizations must have been acquired very early, before dewatering and compaction. Simultaneously it illustrates that a paleopole calculated for these rocks would be erroneous. On the other hand, an analysis of magnetostratigraphic records from 52 deep-sea cores containing the Brunhes-Matuyama (0.7 Ma) boundary (Opdyke & Henry, 1969) and many comparisons of igneous and sedimentary rock directions have not revealed appreciable inclination errors (see Tarling, 1983, pp. 59–62, for a valuable discussion). Concerns about possible inclination errors should remain at the front-burner for paleomagneticians working with sedimentary rocks. At the same time, however, the phenomenon appears to be rare in pre-Quaternary rocks, probably because of the predominance of CRM in ancient sediments. Because of this, there does not seem to be a need to make a separate reliability requirement for the inclination-error problem.

Paleopole rejection and selection

When compiling paleopoles for a given continent or block, all authors nowadays use selection criteria; the most commonly agreed upon rejection criterion is that undemagnetized results should not be included. Most also disregard older papers that are superseded by newer and significantly different (!) results. From there on, almost every author uses different criteria (compare, for instance, Irving, Tanczyk & Hastie, 1976a, their categories B, A, A* and A**, with May & Butler, 1986; Westphal, 1986; Piper, 1987, 1988a; Van der Voo 1990b). The problem for the reader is often that the criteria are not spelled out in detail and that one is left not knowing whether a paleopole is excluded because it was rejected or because it was overlooked. With some 7000 paleopoles available for the whole world, it is inevitable that some results will be missed.

A consensus on selection criteria is unlikely to be achieved in the near future, not in the least because every compilation has its own specific goals which may require greater or lesser precision. I believe that a better approach is to establish 'information' criteria in a compilation, which allows readers and users to accept or reject a paleopole according to their needs. In the data base used for this book (see appendices) and the chapters that follow, the seven criteria enumerated in this chapter are evaluated for all available results. Table 4.2 lists these criteria in abbreviated form. Paleopoles not based on demagnetization of all samples, or superseded and essentially

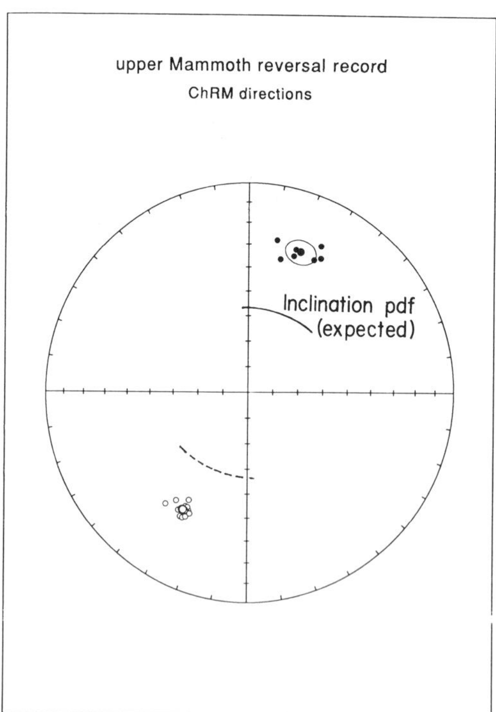

Figure 4.28. Example of normal and reversed directions observed in Pliocene marls (about 3.1 Ma) from southern Italy (after Linssen, 1991), which reveal inclinations that are too shallow by about 20° (inclination error). Many nearby and similar sections of different ages reveal similar inclination errors (Langereis, 1984; Linssen, 1991). The declinations are offset from expected values by a recent tectonic rotation of the area. Published with permission from the author.

Table 4.2. *Reliability criteria*

Number	Brief description
1	Well-determined rock age and a presumption that magnetization is the same age.
2	Sufficient number of samples (N>24), k (or K) $\geq$ 10 and α_{95} (A_{95}) $\leq$ 16.0°.
3	Adequate demagnetization that demonstrably includes vector subtraction.
4	Field tests that constrain the age of magnetization.
5	Structural control, and tectonic coherence with craton or block involved.
6	The presence of reversals.
7	No resemblance to paleopoles of younger age (by more than a period).

For detailed explanation of these criteria, see text.

unpublished results (abstract only) have been excluded. Also excluded are paleopoles (at least for the major continents) based on secondary magnetizations, unless independent evidence (e.g., synfolding age of the remanence) exists for their temporal calibration. It is now a common and acceptable practice to 'date' a secondary paleopole by comparing it to an established APWP. However, such a result should not be included for purposes of constructing a new APWP, because that would lead to circular reasoning.

My method requires that almost all paleopoles be listed, and that only subsequently rejection criteria are applied. Scientist A may decide that results without field tests do not get included, whereas scientist B may conclude that only results with reversals and structural control should be considered as useful for his or her purpose. My own preference (or should I say ambivalence?) has been to allow all results to be included, and then to make a selection based on the stipulation that they pass at least three reliability criteria, that is, any three of the total of seven. This means that all criteria satisfied are so marked, and that the total is listed as an 'information' or 'quality' factor, Q (see tables in appendices). A Q $\geq$ 3 means that at least three criteria are satisfied. In the future, when the data base may become very abundant, one could decide to raise this minimum.

The justification for the condition of Q $\geq$ 3 can be found in an analysis of the typical paleopole distributions for the well-studied continents of North America and Europe (Van der Voo, 1990b). After compiling all paleopoles with Q $\geq$ 3, their means were computed for periods of, on average, about 25 Ma. These mean paleopoles are, again on average, some 10° apart, and together form APWPs to be discussed in Chapter 5. If one defines a distance, psi (in degrees), between an individual paleopole and the mean for its period, a histogram of all psis should show a typical probability density function (P in Chapter 2), and as shown in Figure 4.29, this is indeed the case. The precision parameter of the entire distribution, k, is about 100. Since the mean paleopoles are about 10° apart, even a 'perfectly' fitting result may be located up to some 5° (i.e., midway) from the mean poles nearby. One can, furthermore, compare psi as a function of Q. The average psi for each group of poles with Q = 2, 3, 4, 5, 6 or 7 is shown in Figure 4.30. It can be observed that the spectrum is fairly flat for Q $\geq$ 3, whereas the average psi is higher for Q = 2 (and presumably for Q $\leq$ 2 in general, although there are too few low-Q pole entries to be included). Thus, paleopoles with Q $\leq$ 2 will scatter more than those with Q $\geq$ 3; this appears to be true regardless of the age of the rocks (Van der Voo, 1990b).

It is worth repeating once again that a low-Q paleopole may be perfectly correct in its representation of the ancient geocentric dipole field, and it is also possible that a high-Q paleopole is flawed. Readers can make their own selection criteria using the tables in the appendices. I would be surprised, however, if an alternative selection/rejection method would turn out to be superior to the one selected here, without reducing the available entries to unacceptably low numbers. Inspection of the appendices shows that paleopoles with $Q \geq 6$ are very rare, and that an APWP based on such a high minimum would be starkly underdetermined.

A reviewer of a paper using this scheme of evaluation with Q-values once remarked that the method is rather sophomoric. This comment is certainly appropriate! Never-

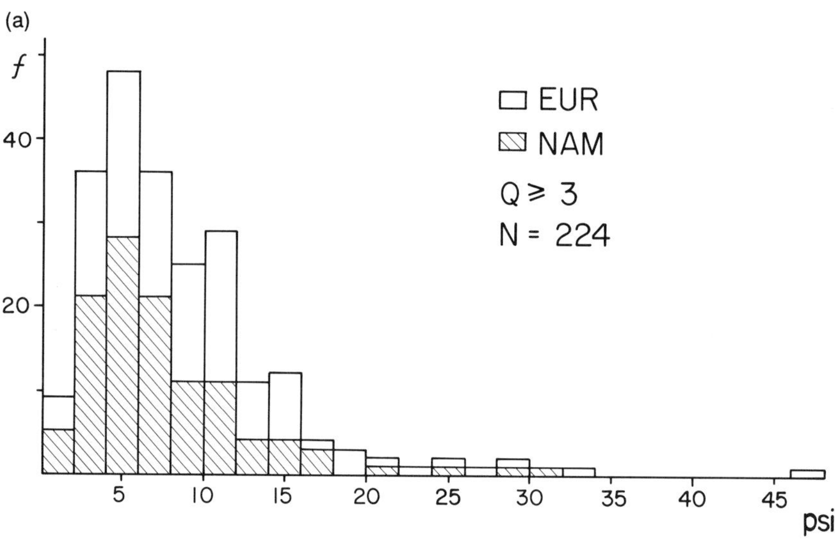

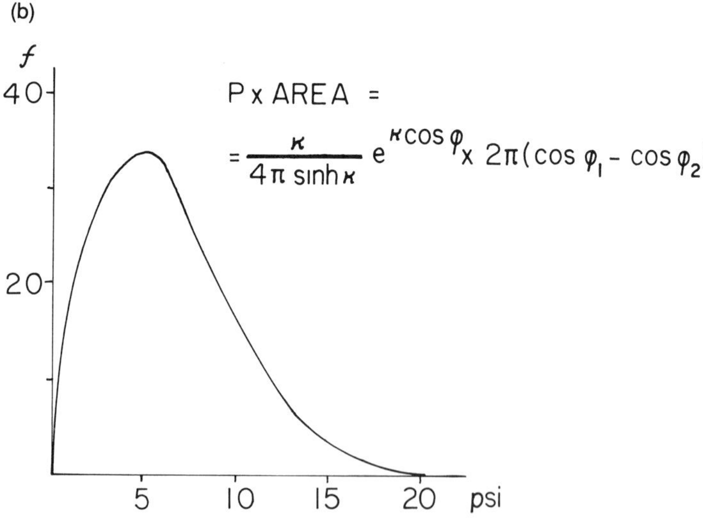

Figure 4.29. a) Histogram of the distance (psi) between the half-period mean poles and individual pole locations (as explained in the text), plotted for 252 paleopoles from Europe (EUR) and North America (NAM). Psi is measured as a great-circle distance (in degrees). b) A theoretical distribution of psi for a Fisherian population of directions with precision parameter (κ) of about 100. Note that b) is based on the probability that a given paleopole will be x degrees apart from the true mean for that period, which explains why the distribution goes through the origin (it is unlikely that a study will yield a paleopole that coincides exactly with the true mean). Figure from Van der Voo (1990b). Published with permission from Elsevier Science Publishers, BV.

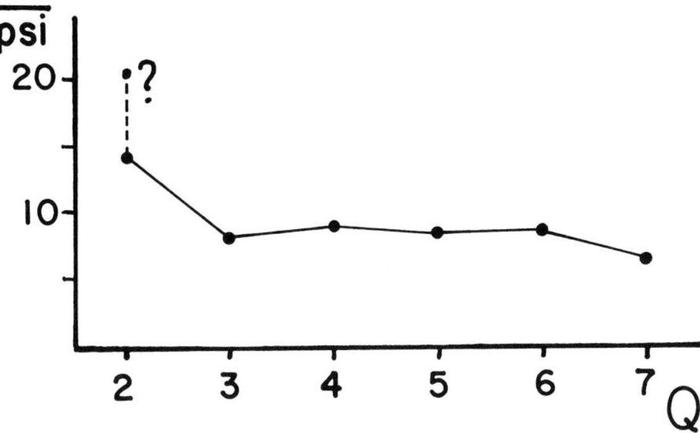

Figure 4.30. Plot of psi (as in Figure 4.29) averaged by groups of paleopoles with a quality factor (Q) as indicated along the horizontal axis. The Q factor as used throughout this book, is explained in the text. The plot shows that for Q ≥ 3, the average psi does not change much. However, for low Q the scatter increases dramatically. The dashed line represents an ambiguity in one data point which biases the average psi-value for Q = 2 significantly. Figure from Van der Voo (1990b). Published with permission from Elsevier Science Publishers, BV.

theless, the need exists to select 'good' paleopoles, and exclude, as best as one can, the 'bad' and poorly documented results, and to do so in an objective fashion. As with democracy, the technique can be terribly flawed but may yet be the least objectionable.

It is a natural tendency for experimentalists (and paleomagnetists are no exception) to want to make their own results appear to be as reliable as possible. The first requisite is to publish all significant information in appropriate detail. When making a compilation, one often discovers how many paleopoles are published with information missing about reversals, structural tilt corrections, certain statistical parameters, or even without information about fold tests which are seemingly positive, not to mention the typos in D, I and paleopole coordinates! If anything, this chapter in the book may induce my 'paleomagician' colleagues to achieve higher standards.

Quality is a characteristic of thought and statement that is recognized by a nonthinking process. Because definitions are a product of rigid, formal thinking, quality cannot be defined . . . But even though Quality cannot be defined, you know what quality is!

(Robert M. Pirsig (1974), Zen and the Art of Motorcycle Maintenance*).*

I really liked this book and couldn't agree more with Pirsig about quality. Yet I feel compelled to catalogue and catagorize and to attach quality labels; the manual for this has been presented in the previous chapter, the tables of data in the appendices contain these labels and the next chapters make use of the selections based on quality designations. So what am I doing?

What Pirsig implies, and the ways in which his implication is applicable to our field, is that we (paleomagnetists) almost always have an 'intuition' about a datum (paleopole): ask a colleague at national conventions about a result and there will be an opinion on its quality without quantification being possible.

The problem is that 'intuition' does not make good, communicable science, let alone that it could lead to a lot of accusations and law-suits. So better to use a flawed quality factor, than to make intuitive judgments or no quality selections at all, OK?

5 The major continents and Pangea

The major continents, geologically speaking, are North America, Greenland, Europe, Siberia, North and South China, and the Gondwana elements of South America, Africa, Madagascar, India, Australia and Antarctica. All consist of older, Precambrian nuclei, many of them true cratons, on which Paleozoic and younger sediments were deposited. This contrasts with the blocks and terranes, as defined in Chapter 3 and discussed in Chapters 7 and 8, which in general have little Precambrian basement exposed.

Paleomagnetic paleopoles have been compiled for all these continents, with the exception of Siberia, and are given in the Appendices. While this compilation was being made, an internationally funded effort started to establish a global paleopole database (McElhinny & Lock, 1990; Lock & McElhinny, 1991); this was deemed necessary because since 1980 (with poles up through 1978) the very helpful catalogues of the Earth Physics Branch in Ottawa and the lists in the Geophysical Journal of the Royal Astronomical Society had been discontinued or interrupted. My compilation has been cross-checked with that of McElhinny & Lock for completeness and accuracy as best as possible. The entries of this compilation, as discussed in the preceding chapter, have been evaluated with the reliability criteria of Table 4.2.

The entries of the data base of McElhinny & Lock have been compiled from all major national and international journals; to these I have added recently published results and paleopoles from the 'gray' literature, that is, publications not generally accessible in major libraries around the world, wherever I have been able to locate them. The nature of such publications, however, implies that some paleopoles may easily have been overlooked!

Wherever possible, references in the tables of the appendices are made to the original author(s). In some cases, a combined paleopole seems to fit the needs of the data base better; for these paleopoles reference is made to the previous compiler (e.g., Irving, Tanczyk & Hastie, 1976b, pole 8.189, or Khramov, 1975, pole 6.46).

Results have been included for the cratons and disturbed cratonic margins of each continent for the entire Phanerozoic. For ancient displaced or suspect terranes, results have been included for the periods that are thought to postdate accretion of these elements to the craton. In the following sections, maps that illustrate these tectonic elements and descriptions of the temporal definitions of the displaced terranes will be presented. For the major continents, results have not been compiled for post-Eocene times (<40 Ma) because the Neogene paleopoles generally fall very close to the present-day rotation axis. After discussing the available paleopoles and APWPs for each of the major continents, I will at the end of this chapter examine the evidence for the late Paleozoic to early Mesozoic supercontinent, called Pangea, in which almost all of them were assembled.

North America, Greenland and Europe

Undoubtedly, North America and Europe are paleomagnetically the best studied continents of the world. The compilation of their paleopoles in the Appendix, Tables A1 and A2, includes some 400 individual paleopoles for the Cambrian through Eocene periods. It may therefore be unwise to begin with these continents because it sets such high expectations. Yet, at the same time their data base illustrates the tremendous progress made since the paleomagnetics group at Imperial College in London (e.g., Clegg, Almond & Stubbs, 1954; Clegg *et al.*, 1957) and the group at Cambridge (later Newcastle-upon-Tyne) of Keith Runcorn (e.g., 1955, 1956), Edward Irving (1956) and Ken Creer (1957a,b) in England, started their now famous research programs in Europe and North America (see also Creer, Irving & Runcorn, 1954, 1957). This program led to the discovery that the two continents had to have moved with respect to each other. For detailed and charming accounts of these early days of paleomagnetism, I refer to Glen (1982), Irving (1988), and LeGrand (1988, 1990).

The British paleomagnetists were, however, not the first to undertake paleomagnetic studies for this purpose; in North America, John W. Graham had already measured many different ancient formations for paleomagnetic studies in the post-World War II years before 1954. However, due to conflicting results and interpretations, and certainly without the benefit of the demagnetization techniques that later proved essential, Graham became discouraged and skeptical about the possibility of using paleomagnetism to test continental drift (LeGrand, 1990). In contrast, the early British paleomagnetists seem to have had an uncanny sense of the rock types that might yield valuable paleomagnetic results; they also realized that well-grouped directions for a pole study were essential. With good intuition and without too much ceremony, poorly clustered results were ignored (LeGrand, 1990). Within a few years, APWPs for Europe and North America were produced that were located, grosso modo, 35° apart in longitude with respect to the continents held fixed (Figure 5.1). When the opening of the Atlantic is undone by fitting Europe and North America (and Greenland) together, the two APWPs of Figure 5.1 coincide.

The APWPs of Figure 5.1 were based on a handful of paleopoles only, so it is surprising to discover how well the general shapes of the paths have survived over the decades, as we will see below. This is, once again, testimony to the good insights and intuition of the early paleomagneticians. In the more than 35 years since the first APWPs were published, many publications – too numerous to discuss in detail – have synthesized new developments, added new data and considerable detail, or provided paleogeographic maps (e.g., Irving, 1964, 1977; Hospers, 1967; Wells & Verhoogen, 1967; Hospers & Van Andel, 1968; Krs, 1968; Deutsch, 1969; Larson & LaFountain, 1970; Phillips & Forsyth, 1972; McElhinny, 1973; Briden, Drewry & Smith, 1974; Van der Voo & French, 1974; Van Alstine, 1979; Harrison & Lindh, 1982b; Irving & Irving, 1982; Tarling, 1983; Westphal *et al.*, 1986; Frei & Cox, 1987; Piper, 1987; Lottes & Rowley, 1990; Torsvik *et al.*, 1990a; Van der Voo, 1989, 1990a). To trace the development of the two paths during these decades may be enlightening, but it would make for a lengthy enumeration with occasional digressions because of pitfalls barely avoided.

The latest synthesis at the time of writing, to my knowledge, has been my own paper (Van der Voo, 1990a), based on the data base available in 1989. Although the newest poles, published in the last year, have been added in the appendices, these have not perceptibly changed the APWPs, so the 1989 versions are reproduced here. Figure 5.2 illustrates individual paleopole entries ($Q \geq 3$) for Europe and North America, in present-day coordinates, whereas the mean paleopoles (from Van

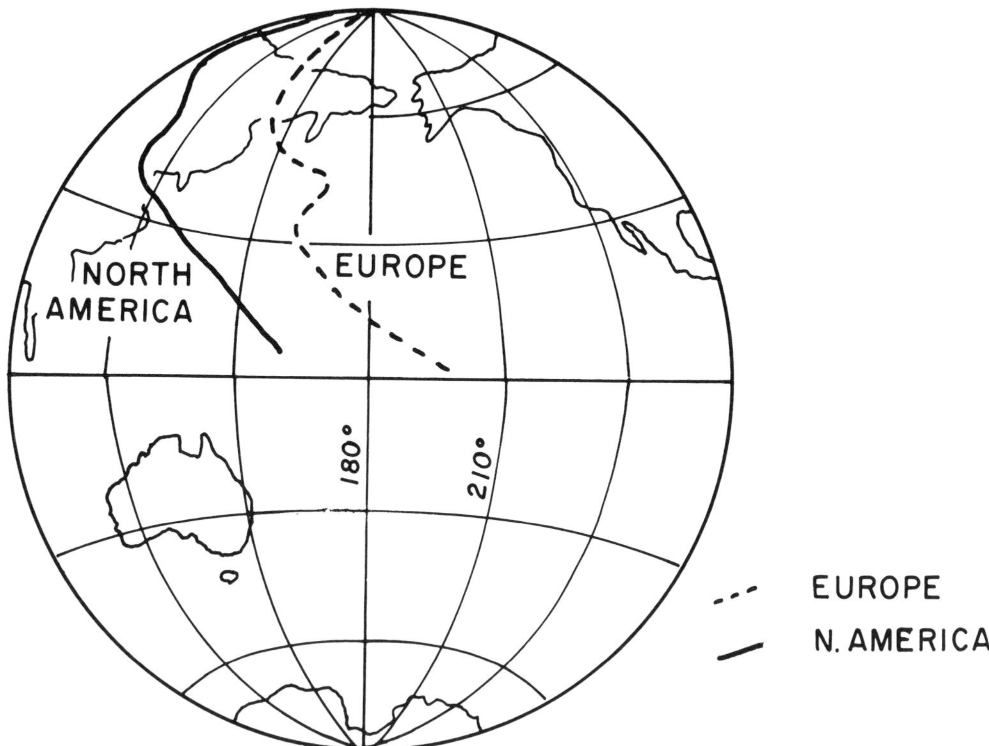

Figure 5.1. Apparent polar wander paths for Europe and North America in present-day coordinates, as determined from the first paleomagnetic studies of older Phanerozoic (Cambrian through Tertiary) rocks by Runcorn (1956) and his colleagues.

der Voo, 1990a) are shown in Figure 5.3 in present-day coordinates as well as after closure of the Atlantic Ocean according to the parameters of Bullard, Everett & Smith (1965). Updated mean paleopoles are presented in Table 5.1, with slight modifications from the earlier ones (Van der Voo, 1990a) used for the figures. The APWPs of Figure 5.3 cover the interval from Late Ordovician through Middle Jurassic. For Cambrian and Early Ordovician, there are insufficient paleopoles for a good comparison, whereas for Late Jurassic and younger times a comparison should take the slow opening of the Atlantic Ocean into account. The latter will be done separately in Chapter 6.

Throughout this book, I have used the same time intervals for grouping paleopoles as were used by Van der Voo (1990a). Generally, these were approximately half-periods, but not precisely. One of these time periods corresponds to the latest Silurian to earliest Devonian (398–414 Ma), comprising the Pridolian, Gedinnian and earlier part of the Siegenian (Figure 4.1), and may serve as an example. The reasons for selecting the limits of this interval are that: (1) there are several paleopoles for this interval with radiometric or stratigraphic ages that straddle the Siluro-Devonian boundary and thus cannot be included in either a Late Silurian or an Early Devonian interval without making arbitrary choices, (2) there may be some uncertainty which precise radiometric age (e.g., 408 Ma?) corresponds to this boundary anyway, and (3) the paleopoles for the selected interval appear to be significantly different from earlier and later intervals, e.g., the Early Silurian and Ludlovian

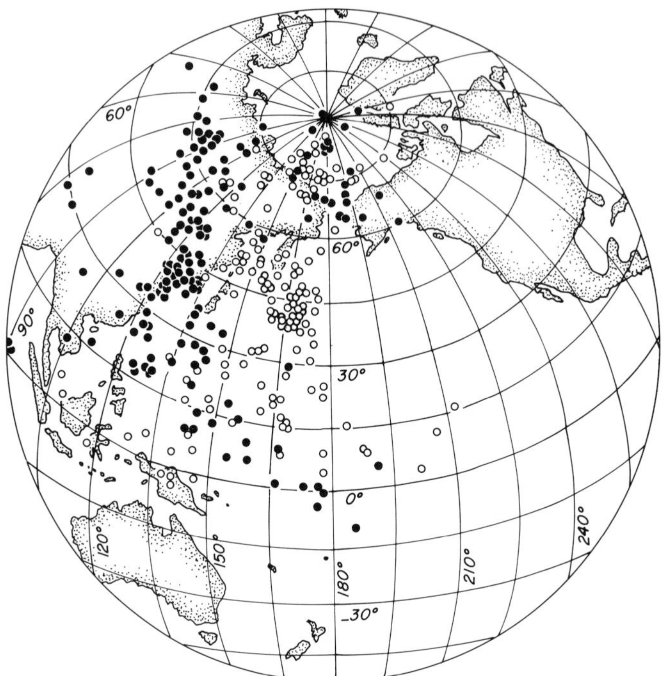

Figure 5.2. European (open circles) and North American paleopoles (solid circles) for Phanerozoic time and with $Q \geq 3$ in present-day position. For pre-Middle Silurian times, European paleopoles from the Laurentian terrane in northern Britain as well as from the Baltic Shield and Russian Platform have been included. Figure from Van der Voo (1990a).

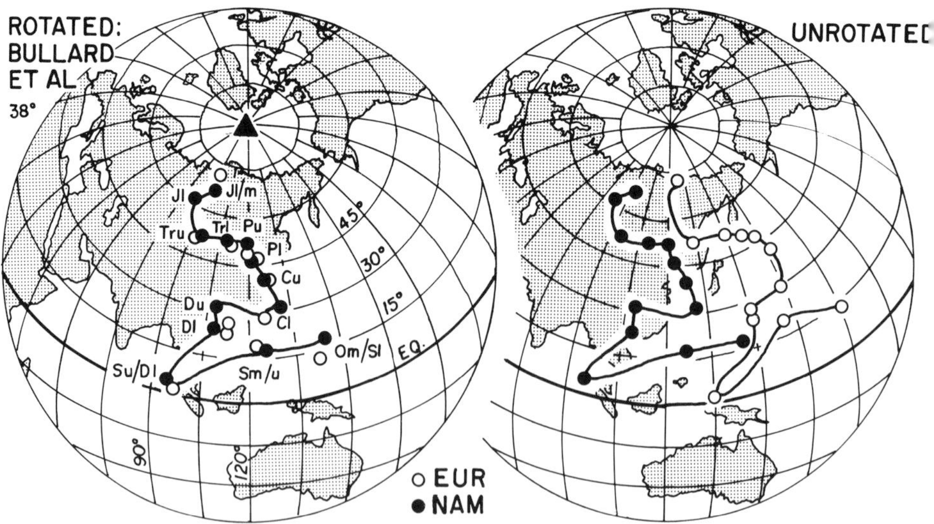

Figure 5.3. Apparent polar wander paths for Europe (EUR) and North America (NAM), for the Middle Ordovician through Early Jurassic interval, with mean poles based on $Q \geq 3$. The plot on the right is in present-day coordinates, whereas for the one on the left the Atlantic Ocean is closed with the parameters of Bullard, Everett & Smith (1965). The large triangle is the Euler pole of this rotation; the rotation angle is 38°. For an explanation of age abbreviations, see Table 5.1; for pre-Middle Silurian times, only European paleopoles from northern Britain have been used. From Van der Voo (1990a).

and the Emsian and Middle Devonian. A similar justification applies, for instance, also to the latest Triassic to earliest Jurassic interval (196–215 Ma) which is a time of rapid apparent polar wander resulting in a cusp in the North American APWP, as will be discussed later. By making this a separate interval, the paleopoles at the cusp are grouped together, and mean poles for the earlier and later intervals are not 'contaminated' by the significantly different Triassic–Jurassic boundary paleopoles.

Each paleopole in the data tables of the appendices has been assigned a numerical age range, based on a radiometric age for the rock unit and its uncertainties, or on stratigraphic ages converted into numerical ones using Figure 4.1. Thus, a Silurian rock unit with an age based on Upper Wenlockian, Ludlovian *and* Upper Pridolian fossils has been assigned an age range of 408–424 Ma (see Figure 4.1), yielding a 'median' or most representative age of 416 Ma. This result is then grouped with other poles for that interval in which it optimally belongs (e.g., 415–429 Ma). Results are included only once in mean pole calculations. Thus, even though the rock unit discussed above contains some latest Silurian fossils, it is placed in the late Early–early Late Silurian interval and not in the latest Silurian to earliest Devonian interval.

The method of construction of the APWPs has been conventional, grouping paleopoles by interval and calculating their Fisherian average (with associated K and A_{95}). However, other methods have been used in the past with greater or lesser success. These are either based on a moving-window analysis (e.g., Irving, 1977) or on the assumption that the path can be fitted by a smoothly varying function, such as those based on cubic splines (Thompson & Clark, 1981). For a well determined polepath based on many continuously distributed entries, the differences will be negligible. However, for segments of the path where few poles are available, the smoothing implied in both techniques will let the path go through most of the poles, but the time calibration may no longer correspond to the actual ages of the individual paleopoles, as we will see when the Paleozoic APWP of Gondwana is examined.

Because smoothing implies averaging of groups of individual paleopole entries, another problem is introduced by this technique when available paleopole locations are disparate for reasons that may be related to unrecognized remagnetizations. This is not a problem, of course, of the cubic spline technique *per se*, but of averaging in general. An example is illustrated in Figure 5.4, where a proposed Precambrian to Triassic APWP (from Larson *et al.*, 1985) is shown for North America together with the Early to Late Cambrian individual paleopoles of Table A1 (including all entries listed, whether deemed reliable or not). The APWP is highly smoothed, such that the Early to Middle Cambrian segment represents the average of all entries, which themselves are distributed all along the path segment. Undoubtedly the more northwesterly paleopoles, however, represent either a remagnetization in post-Cambrian times (e.g., Lynnes & Van der Voo, 1984) or a complex Cambrian APWP loop (Watts, Van der Voo & French, 1980a; Watts, Van der Voo & Reeve, 1980b). In either case the smoothed APWP now shows some of the Late Cambrian paleopoles (triangles in Figure 5.4) falling on the segment assigned to the Early Cambrian. The Late Cambrian rocks are precisely dated by fossils; since it is preposterous to assume that Late Cambrian rocks can carry Early Cambrian magnetizations, the simplified APWP must be erroneous.

The geographical basis for the North American path is shown in Figure 3.1. The craton and its disturbed margin have been used for the entire Phanerozoic, whereas paleopoles for the Appalachian displaced terranes north of New York City (41° N) have been included only for Devonian and younger times. Displaced terrane results from the Appalachians south of New York City have been included for Late Carboniferous and younger times. Results from terranes in Mexico, in the western North

Table 5.1. Mean paleopoles for North America and Europe ($Q \geq 3$)

Time interval	North America					Stable Europe				
	Mean	Pole	N	K	A_{95}	Mean	Pole	N	K	A_{95}
Tl (37–66)	82,	168	17	63	5	78,	177	20	75	4
@ Tl (37–66)	@77,	173	26	43	4					
Ku (67–97)	68,	192	5	250	5	72,	154	5	146	6
@ Ku (67–97)	@68,	192	7	227	4					
Kl (98–144)	69,	194	8	149	5	70,	193	4	41	15
Ju, uJm (145–176)	67,	133	8	44	9	66,	191	3	73	15
@,** Ju, uJm (145–176)	@68,	137	10	44	7	**72,	162	7	37	10
lJm, Jl (177–195)	68,	93	10	193	4	70,	126	4	86	10
Tru/Jl (196–215)	61,	81	15	68	5					
Tru, uTrm (216–232)	52,	96	7	180	5	52,	133	5	29	14
Trl/m, Trl (233–245)	52,	110	11	190	3	52,	150	13	103	4
Pu (246–266)	52,	120	5	371	4	50,	160	6	85	7
Pl (267–281)	45,	123	14	218	3	47,	164	21	134	3
Cu/Pl, Cu (282–308)	40,	128	15	314	2	41,	169	21	165	3
Cm, Cl, Du/Cl (309–365)	29,	131	20	49	5	25,	159	5	45	12
Du, Dm/Du (366–378)	30,	110	8	20	13	27,	151	6	38	10
Dm, Dl (379–397)	24,	108	5	23	16	24,	151	5	82	8
Su/Dl (398–414)	4,	95	4	75	11	3,	135	14	24	8
@ Su/Dl (398–414)	@4,	97	5	78	9					
Su, Sm (415–429)	18,	127	2	—	—	20,	161	11	44	6
Ou/Sl, Ou, Om (430–467)	18,	146	9	16	13					
*Same, N. Britain only						*13,	181	9	24	10
Ol/m, Ol (468–505)	17,	152	3	31	23	−24,	230	2	—	—
*Same, N. Britain only						*16,	212	1	—	—
Ꞓu, Ꞓm, Ꞓ (506–542)	9,	158	8	28	11					
Ꞓl (543–575)	5,	170	6	22	15	11,	231	2	—	—

Mean paleopooles are calculated for each of the two continents in their own coordinates from the entries, with $Q \geq 3$, in Tables A1 and A2 for North America and Europe. Abbreviations are: T = Tertiary, K = Cretaceous, J = Jurassic, Tr = Triassic, P = Permian, C = Carboniferous, D = Devonian, S = Silurian, O = Ordovician, Ꞓ = Cambrian, l = early, m = middle, u = late, with time in Ma given in parentheses, following the DNAG time scale (Palmer, 1983). N, K and A_{95} are the number of paleopole entries, the precision parameter and the cone of 95% confidence about the mean pole, respectively (Fisher, 1953).

* Great Britain north of the Iapetus suture only. For pre-Middle Silurian times, the Baltic Shield and Russian Platform, exclusive of Caledonian Europe are listed separately without (*).

@ Includes also the results from Greenland (in North American coordinates as in Table A3).

** Results for the late Jurassic of Europe are strongly bimodal. The second mean given is an option which includes four limestone poles (labeled by a Q value plus * in Table A2) excluded from the first mean.

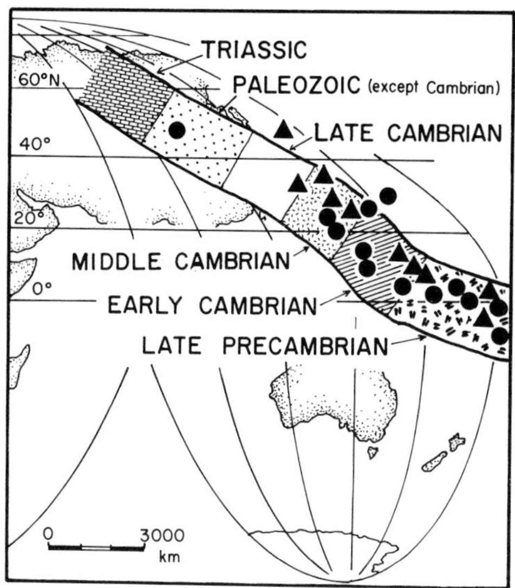

Figure 5.4. Simplified apparent polar wander path for the late Precambrian through Triassic of North America. The Cambrian paleopoles are strung out along this path; Late Cambrian paleopoles from Table A1 are shown as triangles (see text), Early and Middle Cambrian paleopoles are shown as solid circles. Figure redrawn and modified after Larson *et al.* (1985).

American cordilleras, and in Alaska have not been included as movements typically continued well into the Tertiary.

If in the disturbed margin some areas have rotated, their paleopoles will deviate significantly from those of the craton and they should either be excluded or restored if evidence is available about the amount of rotation. For two areas, the Colorado Plateau (including its surroundings in Colorado, Arizona and New Mexico; see e.g., Molina-Garza *et al.*, 1991) and the northern part of the Pennsylvania Salient in the Appalachian fold and thrust belt, such a correction for an inferred amount of rotation has been made, as indicated in Table A1. A few paleopoles that are included in the compilation have nevertheless not been used in the construction for the APWP; they are marked by a * in the Q-column and the reasons for exclusion are given in the Table's explanations.

The geographical basis for Europe is shown in Figure 5.5. The European craton is that part of the continent now found in Scandinavia, as well as the Russian Platform between the Urals and the Paleozoic or younger mobile belts at the surface or buried in Denmark, Germany, Poland, western Podolia (Ukraine), Romania and Bulgaria. Its disturbed margin is non-existent or mostly buried in the latter countries; in Scandinavia, it includes most of the Caledonide belt of Norway. The European cratonic part during Paleozoic times is called Baltica. There is a second ancient element (cratonic disturbed margin) in Europe in the northern part of the British Isles. This area of Scotland and northern Ireland throughout the Paleozoic is thought to have formed part of the combined North American–Greenland continent (called Laurentia) and includes what is called the Lewisian basement that used to be continuous with Greenland; when the Atlantic Ocean opened in the late Mesozoic to Tertiary, the break occurred to the west of this Lewisian basement area. The Paleozoic drift history between the European and the North American cratons, that is, between Baltica and Laurentia, is one of convergence and collision during Ordovician–Silurian times. Thus, pre-Late Silurian results from the northern

British Isles are not necessarily representative of Baltica ('Stable' Europe), but instead, after appropriate reconstruction against Greenland, are entirely representative of Laurentia. From the European perspective, the Lewisian basement and adjacent areas form an early Paleozoic 'displaced terrane'! Their pre-latest Silurian results have been listed separately at the end of Table A2 to highlight this situation.

Other displaced terranes of various ages occur to the south and west of Baltica and to the south of the northern British Isles. They can be divided into three groups. The first group, which was last affected by the Caledonian orogeny of Silurian to Early Devonian age, is found in the southern British Isles (almost all of England south of the Scottish border, Wales, and southern central Ireland). Its late Precambrian basement is often called the 'Midland craton', now mostly shallowly buried under younger Paleozoic and Mesozoic to Tertiary sediments in central-southern England, which with its disturbed margins had an independent drift history from Baltica and certainly Laurentia until possibly as late as the Late Silurian. Thus, only latest Silurian and younger results from the Midland craton and its margins have been included in the data base for the European APWP. The second group of displaced terranes were last affected by the Hercynian orogeny of Carboniferous age. The Hercynian front runs from southern Ireland through southernmost England to Belgium, Germany and southern Poland. Results from Hercynian Europe have been included only for latest Carboniferous and younger times (≤299 Ma). Earlier results will be discussed in Chapter 8, when I will examine the Hercynian displaced terrane(s), collectively called Armorica.

To the south of Hercynian Europe, the third phalanx is formed by Alpine Europe. Note that Alpine Europe contains large Hercynian basement massifs (e.g., Iberia, Corsica, Moesia) that have not been affected internally by Alpine deformation, but because these blocks have been displaced with respect to the rest of Europe, they

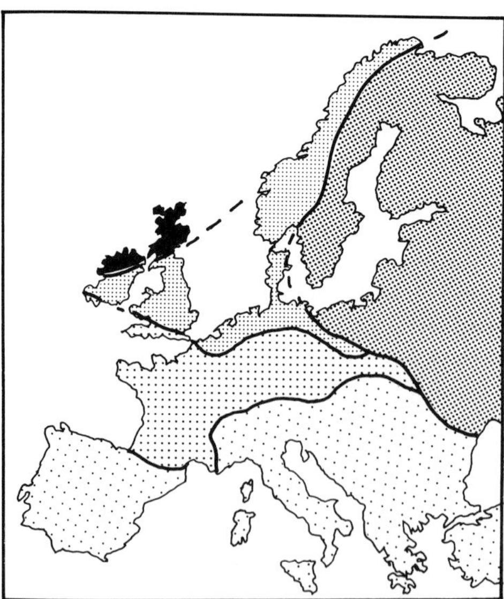

Figure 5.5. Tectonic framework of Europe, outlining the Laurentian portion of northern Britain (black), the craton of the Baltic Shield and Russian Platform (darker shading), the Caledonian mobile belt (medium shading), the Hercynian belt (lighter shading) and the Alpine belt (lightest shading), as used to select and classify paleopoles listed in Tables A2, A7, A8 and A11. The Alpine Belt may contain Hercynian Massifs, just as the Hercynian Belt may contain Precambrian elements. Figure from Van der Voo (1990a).

are treated separately (Chapter 7). The Alpine front runs through the Pyrenees between Spain and France and eastward along the Alps proper in southeastern France, Switzerland and northern Austria to continue into the Carpathian Arc of Czechoslovakia, southernmost Poland and onward along a sharp border with the Russian Platform to the Black Sea. This border is called the Scythian Line, which forms part of a major northwest–southeast lineament, the Tornquist–Tesseyre Lineament, delineating the SW border of the European craton (Figure 5.5). As already mentioned, part of Romania and Bulgaria (the Moesian Platform) has not been affected by Alpine deformation, but it has nevertheless been included in 'Alpine' Europe, because it may well have been displaced with respect to the Russian Platform along the Tornquist–Tesseyre Lineament as late as during Tertiary times. Results from Alpine Europe are not included as displacements continued for most of it well into Tertiary or even recent times; I will return to the terranes and blocks of Alpine Europe, when discussing the Western and Eastern Mediterranean areas in Chapter 7.

Greenland properly belongs in this section, but here we can be brief. Geologically, it is an ancient Precambrian craton which was part of Laurentia until the late Mesozoic, when the Labrador Sea rifted and opened. The eastern and northern margins of Greenland are disturbed by Silurian and Devonian to Early Carboniferous (Caledonian and Ellesmerian) orogenies. Table A3 lists the available paleopoles, which are unfortunately so scant that not much can be done to construct a separate APWP for Greenland. With exception of the poorly documented Triassic poles, the few results are in general agreement with those from North America after closure of the Labrador Sea. The more abundant Early Tertiary results will be compared to those from the other continents in Chapter 6, when I will discuss the opening stages of the Atlantic Ocean.

In this section, then, we have thus far seen the geological elements that make up Europe and North America, and I have explained the temporal limits used to exclude or include paleopoles from the mobile belts of these continents. The resulting APWPs (Figure 5.3) are the updated versions of 1989 that can be compared with the earliest ones from 1956 (Figure 5.1). It can be seen that, although much detail has been added, the overall shape and length is approximately the same. The new APWPs range in age only up to the Middle Jurassic, because after that time the early opening of the Atlantic will begin to play a role, as will be more fully explained in Chapter 6. However, for the Middle Ordovician through Middle Jurassic interval, the two APWPs of Figure 5.3 can be compared directly in terms of a closed Atlantic Ocean and, hence, the Pangea supercontinent. This comparison rather convincingly confirms the conclusions of Irving (1956) and Runcorn (1956) that continental drift between Europe and North America must have been a reality.

The oldest portion of the European APWP of Figure 5.3, which was constructed for the specific purpose of a comparison with the North American path, is based on paleopoles for pre-Middle Silurian times from the northern British Isles only. Thus, the comparison is for two continental elements both originally belonging to Laurentia that subsequently rifted apart. This is appropriate, because Baltica (i.e., Stable Europe) and Southern Britain drifted as parts of independent plates until they collided with Laurentia during the Silurian, and their early Paleozoic paleopoles cannot be used for a test of the reconstruction with the present-day Atlantic Ocean closed. There are few early Paleozoic paleopoles for Baltica (Table A2), although very recent work in progress by Hervé Perroud, Trond Torsvik and colleagues from the Universities of Rennes and Oxford, and the Norway Geological Survey, shows that Baltica underwent a large latitudinal and rotational motion with respect to Laurentia during the Ordovician (e.g., Torsvik & Trench, 1991a).

Ideally, cones of confidence (A_{95}) should be shown on the APWP comparison of Figure 5.3 as well, but this would rather clutter the figure. The two APWP paths have been used (Van der Voo, 1990a) to test different reconstruction parameters for the fit between Europe and North America, and I will return to this subject later in the chapter when discussing Pangea reconstructions. At that point, I will also illustrate the cones of confidence.

One final topic also properly belongs in this section, namely the proposal, made originally by Gordon, Cox & O'Hare (1984), that some segments of the APWP can be represented by small-circle trajectories on the globe. Gordon and colleagues correctly reasoned that the motion of the continent with respect to the pole, if relatively constant in direction and rate, would lead to an APWP segment that can be fitted by a small circle. As such, this concept is similar to that proposed by Morgan (1972, 1983) which states that, if hot-spot traces such as the Hawaii-Emperor Chain of seamounts (extinct volcanoes) reflect the motion of the Pacific plate with respect to a stationary 'plume' of upwelling mantle material fixed with respect to the whole earth, or at least the entire lower mantle (Figure 5.6), then these traces will be small circles. The APWP segments thus fitted by a small circle can be mathematically described by a pivot normal to the small-circle plane, called a Paleomagnetic Euler Pole (PEP), and a rotation rate (in degrees per unit of time). Gordon and colleagues

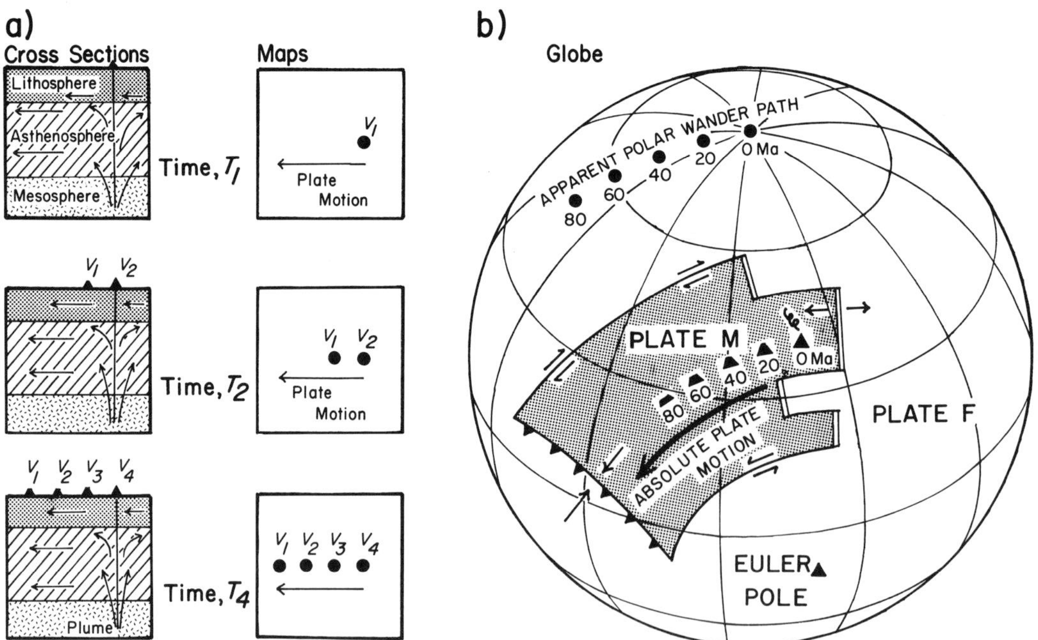

Figure 5.6. a) Diagrammatic representation of how a fixed hot-spot can produce a line of volcanoes increasing in age in a plate moving above the hot-spot. In the bottom frame, now-extinct volcano *V1* is the oldest and active volcano *V4* is the youngest. Figure redrawn from Wyllie (1976), and published with permission of the author and John Wiley and Sons, Publishers. b) Plate M, moving with respect to a hot-spot as well as the rotation axis, is characterized by a volcano chain and an apparent polar wander path, both increasing in age from 0 to 80 Ma and following a small-circle. The movement of Plate M with respect to the hot-spot and rotation axis frameworks can be described by the Euler Pole coordinates and a rotation angle. The (Paleomagnetic) Euler Pole vector (PEP axis) is normal to the small-circle planes. Figure redrawn after Butler (1991 page 252), and reprinted by permission of the author and Blackwell Scientific Publications Inc.

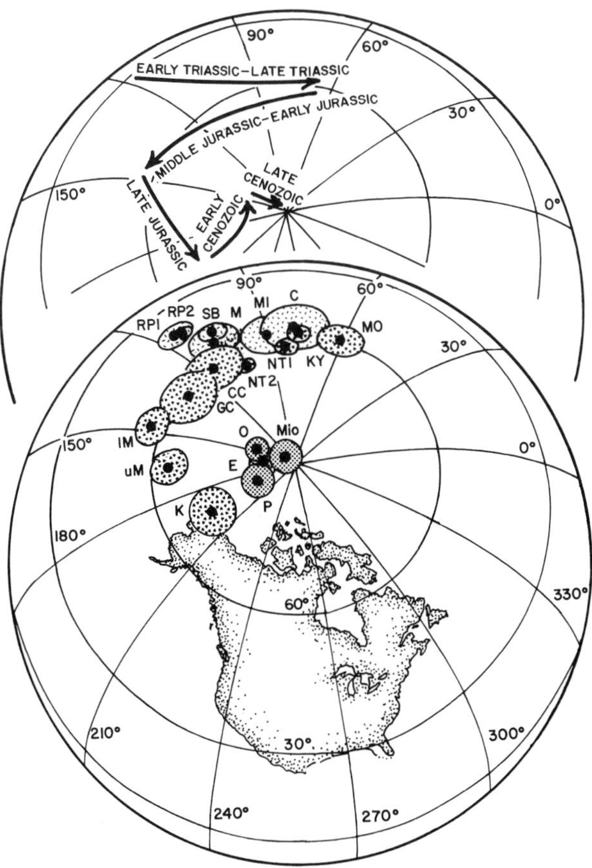

Figure 5.7. Apparent polar wander path (top) and individual paleopoles with their confidence ellipses (bottom) for North America that illustrate small-circle segments (called PEP tracks, where PEP is defined in Figure 5.6). Pole abbreviations are: RP = Red Peak, SB = State Bridge, M = Moenkopi, MI = Manicouagan, C = Chinle, MO = Moenave, KY = Kayenta, NT = Newark Dikes, CC = Coral Canyon, GC = Glance Conglomerate, lM = Lower Morrison, uM = Upper Morrison, K = Cretaceous, P = Paleocene, E = Eocene, O = Oligocene, Mio = Miocene. Figure redrawn after Butler (1991, page 254), and reprinted by permission of the author and Blackwell Scientific Publications Inc.

proposed that the 100 million year segment of the North American APWP between the Late Carboniferous and the Late Triassic to Early Jurassic represented such a 'PEP trace', followed by a younger Jurassic to Cretaceous PEP trajectory of about 80 million year duration. The implication of this method is, of course, that the movement of the continent with respect to the dipole axis is constant in direction for the duration of the trajectory. New results subsequently obtained for the Jurassic (e.g., May *et al.*, 1986) led to a slight modification whereby the younger trace was split up into two small circle segments (see Figure 5.7, from Butler, 1991; see also Kent & May, 1987).

The convincing part of the PEP proposal was that the two main trajectories make a sharp angle to each other, which results in a 'cusp' at about 200 Ma (Figure 5.7); the reality of the cusp was supported by Early Jurassic paleopoles for North America (e.g., see pole MO from Ekstrand & Butler, 1989, in Figure 5.7; see also Dooley

& Smith, 1982). In detail, however, there are some small, but – I believe – significant departures from the perfect small-circle trends. These departures are found in the Late Permian paleopoles (Peterson & Nairn, 1971; Molina-Garza, Geissman & Van der Voo, 1989) and perhaps also in the early Late Triassic (Carnian to early Norian) paleopoles for North America (e.g., Fang & Van der Voo, 1988; Symons, Bormann & Jans, 1989). Paleopoles from carbonate rocks remagnetized during the (Late?) Permian also show these departures (McCabe, Van der Voo & Ballard, 1984). When one examines some of the European Middle to Late Triassic paleopoles (Creer, 1959; Pohl & Soffel, 1971; Turner & Ixer, 1977; Storetvedt, Pedersen *et al.*, 1978a; see Table A2), a similar deviation can be seen, after correction for the opening of the Atlantic (Figure 5.8). These European Triassic paleopoles at one time led Roy (1972) to question whether Europe and North America changed their relative positions during the Triassic, something we now no longer need to assume because of the new and similar North American Triassic results.

My cautious reaction to the PEP hypothesis is that in principle it is correct, but that in practice the duration of intervals during which motion with respect to the pole is constant is rather shorter than the 80 million years or so implied by the original proposal of Gordon et al. (1984). Modern-day plate motions are not constant either for such long times, as revealed by sea-floor magnetic anomaly patterns. If the intervals of constant motion are short, say, some 20 million years, then the usefulness of the PEP technique is much diminished for the construction of APWP small circle segments, since the resolution of paleomagnetic data generally falls short of that which is needed to test paleopole trends of such short durations.

The Gondwana continents

The paleomagnetic data base for the Gondwana continents is not as extensive as that for Europe and North America, with a total of less than 350 paleopoles for the six elements combined (Tables A4 and A5 for Africa, South America, Antarctica, Australia, India and Madagascar). Thus there are, especially for the Paleozoic, many unsolved questions about its APWP.

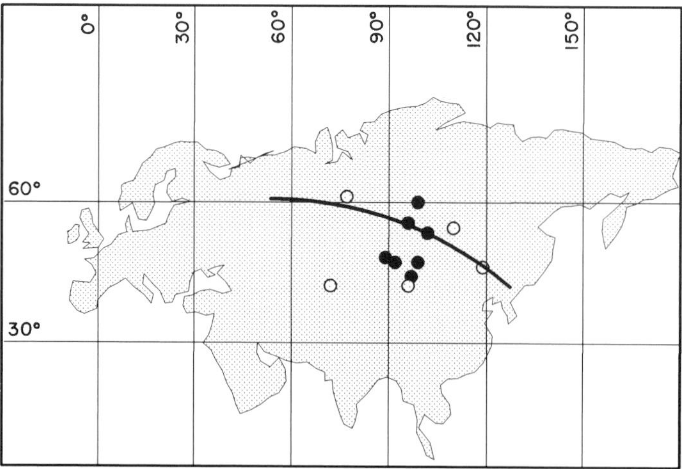

Figure 5.8. Paleopoles with Q ≥ 3 from Tables A1 and A2 for Europe (open circles) and North America (solid circles) for the Late Triassic in North American coordinates, together with the paleomagnetic euler pole-trajectory of Gordon, Cox & O'Hare (1984).

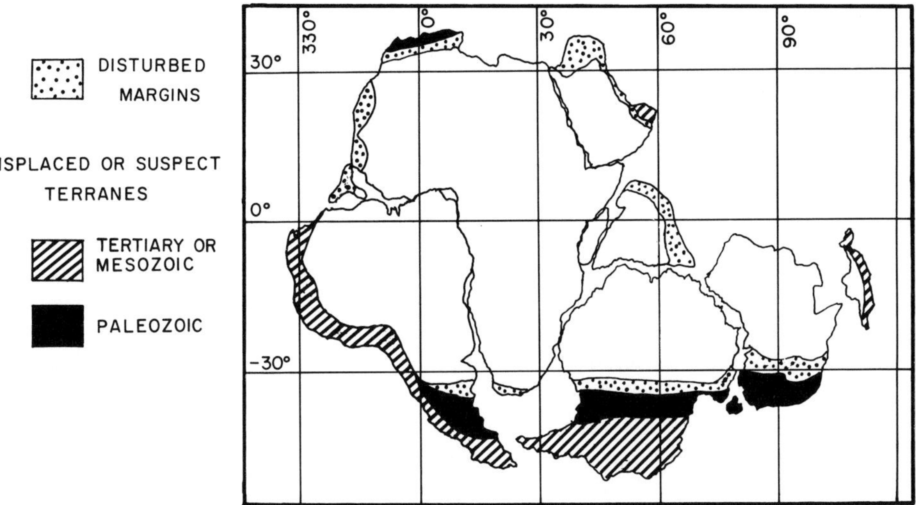

Figure 5.9. Tectonic framework of Gondwana in the fit of Lawver & Scotese (1988), with Antarctica left as one piece for recognition purposes. The figure outlines the cratonic parts (white), their disturbed margins (stipples), the Paleozoic mobile belts and possibly displaced or suspect terranes (black), and the Mesozoic–Tertiary mobile belts and possibly displaced or suspect terranes (oblique shading).

The Gondwana continents are shown in a recent continental reconstruction (Lawver & Scotese, 1988) in Figure 5.9; Antartica is shown as one piece for recognition purposes only. A major portion of this supercontinent is cratonic in nature. It includes most of Africa, with exception of the disturbed margins in southernmost and northwest Africa, most of South America, with exception of the displaced and suspect terranes of the Venezuelan–Colombian coastal belts, part of the Andes and Patagonia, all of Madagascar and India south of the Himalayan Front, all of Australia west of the eastern fold belts, and East Antarctica (Figure 5.9). Results from West Antarctica have been listed separately in Table A5 and are not necessarily representative of cratonic Gondwana. Although for several poles there is good agreement between East and West Antarctica (Watts, Watts & Bramall, 1984), Grunow, Kent & Dalziel (1991) have shown that West Antarctica consists of several displaced terranes, such as the Thurston Island Block, the Antarctic Peninsula, the Ellsworth–Whitmore Mountain Block and Marie Byrd Land. Two further issues need to be addressed before a supercontinent APWP can be constructed for cratonic Gondwana: (1) for which Phanerozoic time interval were the Gondwana continents together, and (2) what are the best-fitting reconstruction parameters.

The break-up of Gondwana is relatively well documented and thought to have been initiated in the Middle Jurassic. Thus, the paleopoles of the different continents that are younger than about 180 Ma can only be combined by taking the opening of the Atlantic and Indian oceans into account. This will be examined in Chapter 6. The earlier timing of the assembly of Gondwana, on the other hand, is not so easily determined. Much of cratonic Gondwana is last affected by the late Precambrian to early Paleozoic Pan-African orogeny, which forms mobile belts around and between the older Archean and earlier Proterozoic nuclei (e.g., Windley, 1977). Radiometric age dates for the Pan-African belts are known to range into the Early Ordovician, so a cautious approach would be to ignore the Cambrian period in the reconstruction of a common APWP. On the other hand, many scientists have argued that Gondwana was already assembled in the late Precambrian. In principle, a good Cambrian paleopole dataset could bring resolution, but this data set is rather poor.

Proposals for the rotation parameters that bring the Gondwana elements into proper juxtaposition are also rather varied, although with the generally limited resolution of the paleopoles, the differences in the APWP are small. In the period before the mid-1970s there were two principally different types of reconstruction: one in which Madagascar (with India, Australia and Antarctica attached as an East Gondwana Block) was fitted against the east African coastal bight off Mozambique, whereas the other placed Madagascar in the more northern coastal bight immediately south of Somalia. With better marine magnetic as well as paleomagnetic data from Madagascar (McElhinny *et al.*, 1976) this ambiguity was resolved in favor of the latter fit, modified from the earlier proposal (Smith & Hallam, 1970) by subsequent efforts (e.g., Lawver & Scotese, 1988; Scotese & McKerrow, 1991). The reconstruction parameters of East against West Gondwana, and between individual continents, are now a second-order matter, which at first glance does not appear to affect the combined Gondwana APWP greatly.

However, a further complexity occurs in Africa. When the South Atlantic Ocean opened between Africa and South America, a third arm of attempted rifting occurred in the vicinity of Nigeria, extending northwards as the Benue Trough. This arm subsequently failed, but a fair amount of extension occurred. Burke & Dewey (1974), Fairhead (1988), and others have argued that the African plate underwent extension followed by lesser compression in post-Lower Cretaceous times. Thus minor differential rotations should be undone between northwestern Africa, southern Africa and northeastern Africa for Early Cretaceous and older times. Rotation parameters have been compiled by Lottes & Rowley (1990); these have also been used in this book for the construction of the combined Gondwana APWP. For the Permian through Early Cretaceous mean poles of Africa, the average precision parameter (K) improves from 61 to 72 in making these internal adjustments to the African geometry; there is, therefore, some paleomagnetic support for them. From the Gondwana paleopoles of Tables A4 and A5 mean poles have been calculated for each continent (Table 5.2A), and these have also been rotated into the coordinates of northwest Africa (Table 5.2B), using the parameters of Lottes & Rowley (1990). Overall mean Gondwana poles resulting from the Scotese & McKerrow (1991), Smith & Hallam (1970), and the Lottes & Rowley reconstructions can be compared in Table 5.3.

Paleomagnetic results became available from the Gondwana continents very soon after the successes obtained in Europe and North America, but initially demagnetization techniques were not applied and tropical weathering apparently played a major role in obscuring the results. However, with the advent of limited demagnetization techniques, success was obtained in South America (Figure 5.10), first by Creer (1958, 1965, 1970) and later by him and his colleagues Embleton and Thompson (then) of Newcastle-upon-Tyne and Valencio and Vilas of Buenos Aires (e.g. Embleton, 1970; Thompson, 1972, 1973; Valencio, Rocha-Campos & Pacca, 1975a,b; Valencio *et al.*, 1977a, 1983; Vilas & Valencio, 1978a,b), as well as others of course.

In the meantime in the 1960s a large group of paleomagnetists had assembled in Africa in Harare (then Salisbury), Zimbabwe, to which at various times belonged (in alphabetical order): Jim Briden, Andrew Brock, Ted Evans, Ian Gough, Dai Jones, Mike McElhinny and Neil Opdyke, whereas earlier, Anton Hales, J. S. Van Zijl, and K. W. T. Graham had already obtained promising results from many rocks (e.g., Figure 5.11) in southern Africa (Graham & Hales, 1957, 1961; Van Zijl,

Table 5.2A. *Mean paleopoles for the Phanerozoic of Gondwana* ($Q \geq 3$)

Age Group	Africa				South America				India				Australia + foldbelt				E. Antarctica				Madagascar			
	Pole	N	K	A_{95}	Pole	N	K	A_{95}	Pole	N	K	A_{95}	Pole	N	K	A_{95}	Pole	N	K	A_{95}	Pole	N	K	A_{95}
Tl (37–66)	81, 185	4	77	11	79, 76	2			43, 285	10	24	10	61, 301	7	38	10								
Ku (67–97)	67, 245	12	60	6	85, 214	6	56	9	21, 295	7	47	9	56, 318	1							68, 231	6	103	7
Kl (98–144)	56, 263	6	105	7	84, 224	7	107	6	12, 299	2			41, 338	6	17	17								
Ju (145–176)	57, 251	7	22	13	85, 73	3	293	7	2, 310	2			48, 349	6	29	13	52, 30	6	89	7				
Jl (177–195)	72, 249	10	108	5									48, 357	3	562	5	49, 46	4	125	8				
Tru/Jl (196–215)	71, 214	2			71, 74	2			10, 310	1											74, 277	1		
Tru (216–232)	56, 253	2			79, 78	6	45	10					32, 350	1										
Trl (233–245)					63, 149	1			− 6, 305	3	321	7	31, 338	2							66, 292	1		
Pu (246–266)					80, 110	5	77	9	4, 283	1			36, 321	3	37	21								
Pl (267–281)	32, 246	5	51	11	62, 174	3	399	6					46, 302	1										
Cu (282–308)	40, 217	3	20	28	54, 165	9	42	8	−18, 291	1			62, 322	2										
Cl (309–365)	21, 227	4	20	21									84, 141	1										
Du (366–378)	5, 197	2											62, 203	6	19	16								
Dl (379–397)					12, 130	2							77, 80	3	5	60								
Su/Dl (398–414)																								
Sm-u (415–429)	43, 189	1											58, 173	3	4	68								
Om-Sl (430–467)	−35, 158	2											− 2, 215	2										
Ol (468–505)	−38, 189	3	15	33	− 4, 122	1							18, 199	3	22	27	9, 202	5	39	12				
Ꞓu (506–542)	−61, 162	5	3	50	−49, 194	3	15	33					18, 203	8	13	16								
Ꞓl (543–575)	−17, 90	2							35, 219	5	28	15	28, 178	4	12	27								

Mean paleopoles are given for each of the Gondwana continents, in their own coordinate system, calculated on the basis of the individual paleopole entries of Tables A4 and A5 with $Q \geq 3$. Age abbreviations as in Table 5.1. N is the number of paleopole entries and K and A_{95} are the statistical parameters associated with the mean (Fisher, 1953). Paleopoles are given in latitude (positive when North and negative when South) and East longitude.

Table 5.2B. *Rotated mean paleopoles for the Phanerozoic of Gondwana* ($Q \geq 3$)

Age Group	Africa				South America				India				Australia*				Antarctica				Madagascar	
	Pole	N	K	A_{95}	Pole	N	K	A_{95}	Pole	N	K	A_{95}	Pole	N	K	A_{95}	Pole	N	K	A_{95}	Pole	N
Jl (177–195)	72, 249	10	108	5									63, 259	3	562	5	74, 272	4	125	8		
Tru/Jl (196–215)	71, 214	2			71, 217	2			67, 251	1											68, 232	1
Tru (216–232)	56, 253	2			67, 236	6	45	10					52, 283	1								
Trl (233–245)					43, 222	1			50, 259	3	321	7	43, 276	2							65, 254	1
Pu (246–266)					61, 236	5	77	9	43, 224	1			34, 261	3	37	21						
Pl (267–281)	32, 246	5	51	11	36, 234	3	399	6					28, 242	1								
Cu (282–308)	40, 217	3	20	28	31, 225	9	42	8	33, 251	1			45, 230	2								
Cl (309–365)	21, 227	4	20	21									43, 182	1								
Du (366–378)	5, 197	2											17, 184	6	19	16						
Dl (379–397)					7, 177	2							53, 175	3	5	60						
Su/Dl (398–414)																						
Sm-u (415–429)	43, 189	1											19, 168	3	4	68						
Om-Sl (430–467)	−35, 158	2											−48, 189	2								
Ol (468–505)	−38, 189	3	15	33	− 2, 162	1							−26, 172	3	22	27	−30, 183	5	39	12		
Ɛu (506–542)	−61, 162	5	3	50	−74, 200	3	15	33					−26, 176	8	13	16						
Ɛl (543–575)	−17, 90	2							5, 162	5	28	15	−10, 157	4	12	27						

The mean paleopoles are all given in the coordinates of northwest Africa. The rotation parameters are from Lottes & Rowley (1990), and for pre-Middle Jurassic times are as follows: South America to northwest Africa, Euler pole at 53°N, 325°E, angle 51.01 counterclockwise; India to northwest Africa, Euler pole at 26.67°N, 37.29°E, angle 69.37 clockwise; Australia to northwest Africa, Euler pole at 28.13° S, 66.79°W, angle 52.06 counterclockwise; Antarctica to northwest Africa, Euler pole at 12.36°S, 33.81°W, angle 53.29 counterclockwise, and Madagascar to northwest Africa, Euler pole 14.9°S, 82.35°W, angle 15.7 counterclockwise.

*Australian means include Paleozoic results from the eastern foldbelt, listed separately in Table A5.

Table 5.3. *Overall mean poles for Gondwana in different reconstructions*

Age interval	Mean paleopole[1]					Mean paleopole[2]					Mean paleopole[3]				
	Lat.	Long.	N	K	A_{95}	Lat.	Long.	N	K	A_{95}	Lat.	Long.	N	K	A_{95}
Jl (177–195)	70,	260	3	138	11	65,	258	3	137	11	67,	250	3	162	10
Tru/Jl (196–215)	70,	230	4	163	7	67,	230	4	102	9	64,	256	4	23	20
Tru (216–232)	60,	261	3	32	22	58,	261	3	29	23	56,	260	3	75	14
Trl (233–245)	52,	253	4	20	21	50,	250	4	19	22	45,	264	4	13	26
Pu (246–266)	47,	242	3	17	31	44,	243	3	15	33	43,	256	3	48	18
Pl (267–281)	32,	241	3	156	10	30,	242	3	88	13	27,	244	3	176	9
Cu (282–308)	38,	231	4	37	15	34,	233	4	37	15	32,	239	4	14	25
Cl (309–365)	34,	207	2			31,	209	2			27,	210	2		
Du (366–378)	11,	191	2			9,	192	2			5,	196	2		
Dl (379–397)	30,	176	2			28,	179	2			25,	182	2		
Su/Dl (398–414)															
Sm-u (415–429)	31,	177	2			29,	179	2			25,	183	2		
Om-Sl (430–467)	−43,	172	2			−44,	170	2			−48,	175	2		
Ol (468–505)	−25,	176	4	19	22	−26,	175	4	19	22	−30,	180	4	12	27
Єu (506–542)	−55,	176	3	10	42	−55,	173	3	11	40	−58,	172	3	16	32
Єl (543–575)	− 9,	138	3	4	71	− 7,	140	3	4	73	− 1,	144	3	3	110

Age intervals, N, K and A_{95} are explained in Table 5.1. Mean paleopoles are calculated from the mean poles for each of the Gondwana continents (Table 5.2A) and are given for Gondwana in (northwest) African coordinates, after Euler rotations as detailed below.

[1]Rotation parameters of Lottes & Rowley (1990); Africa internally is treated as three plates as detailed in Table A4; for other continent rotation parameters see Table 5.2.

[2]Rotation parameters of Scotese & McKerrow (1991); Africa is internally treated as only one plate (unrotated poles in Table A4); all rotations are counterclockwise (ccw) and with respect to Africa: South America, Euler pole at 45.5°N, 327.8°E, angle 58.2; India, 28.1°S, 213.3°E, angle 66.5; Australia, 24.63°S, 297.36°E, angle 55.92; Antarctica, 9.68°S, 328.19°E, angle 58.54; Madagascar, 1.7°S, 272.2°E, angle 22.2.

[3]Rotation parameters of Smith & Hallam (1970); Africa internally is treated as only one plate; South America to Africa, Euler pole at 44°N, 329.4°E, angle 57 ccw; India to Africa, 25.5°S, 201.34°E, angle 44.97 ccw; Australia to Africa, 11.92°S, 294.59°E, angle 58.51 ccw; Antarctica to Africa, 1.3°N, 324°E, angle 58.4 ccw; Madagascar to Africa, 9°S, 313°E, angle 15 ccw.

Graham & Hales, 1962a,b). Subsequently, many different nationalities were represented among authors of African paleopoles, notably the French in northwest Africa (e.g., Bardon *et al.*, 1973; Conrad & Westphal, 1975; Daly & Pozzi, 1976, 1977a,b; Westphal *et al.*, 1979; Sichler *et al.*, 1980; Morel *et al.*, 1981).

The paleomagnetic data base of Australia had a rapid start in the late 1950s and continued strong, first with the work of Irving at the Australian National University in Canberra (Irving, 1963, 1966; Green & Irving, 1958; Irving, Stott & Ward, 1961; Irving & Parry, 1963) and also Robertson (1963, 1964; Robertson & Hastie, 1962), and later by the 'successor' groups in Canberra (e.g., McElhinny & Luck, 1970; Embleton, 1972; Luck, 1972, 1973; Embleton & Giddings, 1974; McElhinny, Embleton & Wellman, 1974a; Kirschvink, 1978; Klootwijk, 1980) and at CSIRO near Sydney (e.g., Schmidt, 1976; Schmidt, Currey & Ollier, 1976; Schmidt *et al.*, 1986; Schmidt, Embleton & Palmer, 1987; Embleton & McDonnell, 1981; Li, Schmidt & Embleton, 1988).

In the Indian subcontinent, Athavale, Verma, Bhalla and colleagues and also Dutch and Australian workers (Klootwijk, McElhinny, Wensink; see Table A.5) followed the footprints of the earlier workers such as Clegg, Deutsch and colleagues

Figure 5.10. Photograph of the Iguaçú Falls of the Paraná River where it descends over the Cretaceous Serra Geral volcanic flows at the border between Brazil and Argentina (photograph courtesy of Prof. Henry N. Pollack).

Figure 5.11. Photograph of the Jurassic Karroo Lavas at Victoria Falls where the Zambezi River forms the border between Zambia and Zimbabwe (photograph courtesy of Prof. Henry N. Pollack).

who did not yet have demagnetization at their disposal (e.g., Clegg, Deutsch & Griffiths, 1956; Clegg, Radhakrishnamurty & Sahasrabudhe, 1958). Madagascar and Antarctica were visited mostly in the 1970s for paleomagnetic sampling (Table A5).

What emerged, already early on in the 1960s, became one of the most convincing stories of paleomagnetism. The Phanerozoic paleopoles of the different Gondwana continents made sense only if the continents had been assembled as had been suggested by early 'drifters' such as Wegener and DuToit (see LeGrand, 1988), who had based their theories on paleoclimatic or faunal evidence as well as structural matching between continents (see also Hurley & Rand, 1969). And the rest, they say is 'history'! Or is it?

The combined Gondwana APWP, continent-mean poles and individual paleopoles are shown for latest Carboniferous to earliest Jurassic times in Figures 5.12 and 5.13, and for earlier Paleozoic times in Figure 5.14, including the Cambrian for which the paleopoles appear to follow a common looping pattern in defiance of the idea that Gondwana was finally assembled only by the earliest Ordovician. Tables 5.3 and 5.2 give mean paleopoles with their statistical parameters for Gondwana combined and for the individual continental elements, respectively.

Figure 5.13 shows individual latest Carboniferous to early Jurassic paleopoles (Q ≥ 3) in present-day coordinates as well as in the Gondwana reconstruction of Lotte & Rowley (1990). This figure illustrates most convincingly the large improvemen upon reconstruction in the clustering of paleopoles and strongly indicates the realit of the Gondwana Supercontinent.

However, two poorly determined segments of the combined Gondwana APWI play a role, still today, in major and insufficiently resolved issues. The first involve the Late Permian and Triassic segment, which is of importance in testing the many Pangea reconstructions. I will return to this matter later in this chapter. The secon concerns the APWP for Paleozoic times earlier than latest Carboniferous (Figur 5.14), which also illustrates that much remains to be learned about Gondwana's APWP and that 'history' (presumably) is not over yet.

In the Paleozoic there are two periods with very few paleopoles, the Silurian an Devonian. The Early Carboniferous is also poorly documented. At the same time the paleopoles that do exist, especially the apparently more reliable ones, show larg APWP swings, first suggested as a possibility by Morel & Irving (1978) and late repeated by Hargraves, Dawson & Van Houten (1987), Van der Voo (1988), Bachtadse & Briden (1990) and Schmidt *et al.* (1990). An APWP constructe entirely on the basis of paleoclimatic information has, moreover, indicated simila trends (Caputo & Crowell, 1985). On the other hand, Scotese & Barrett (1990) have followed only one component of the APWP swings discussed above in thei paleoclimatic/lithofacies analysis (independent of paleomagnetic data). Representative examples of these different paths are illustrated in Figure 5.15. Because of the enormous importance of the Paleozoic Gondwana APWP for models of Paleozoi plate movements which involve the demise of the Paleozoic oceans such as Iapetus

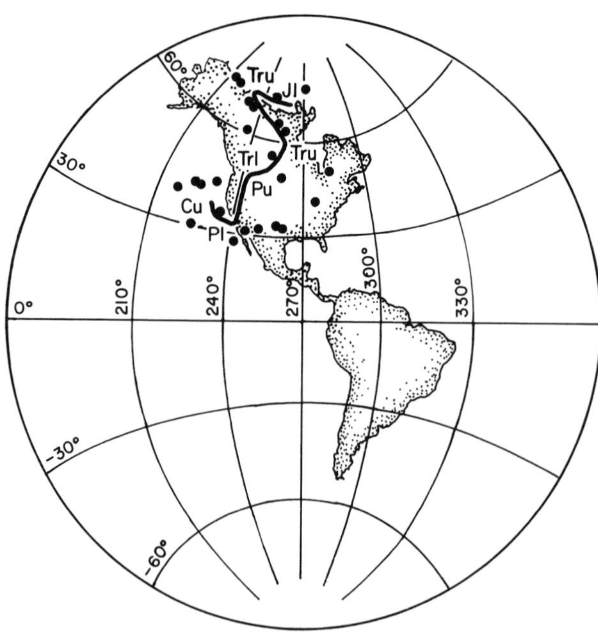

Figure 5.12. Mean paleopoles for each Gondwana continent and combined Gondwana apparent polar wander path for the Late Carboniferous through Early Jurassic interval (in northwest African coordinates; from Tables 5.2B and 5.3).

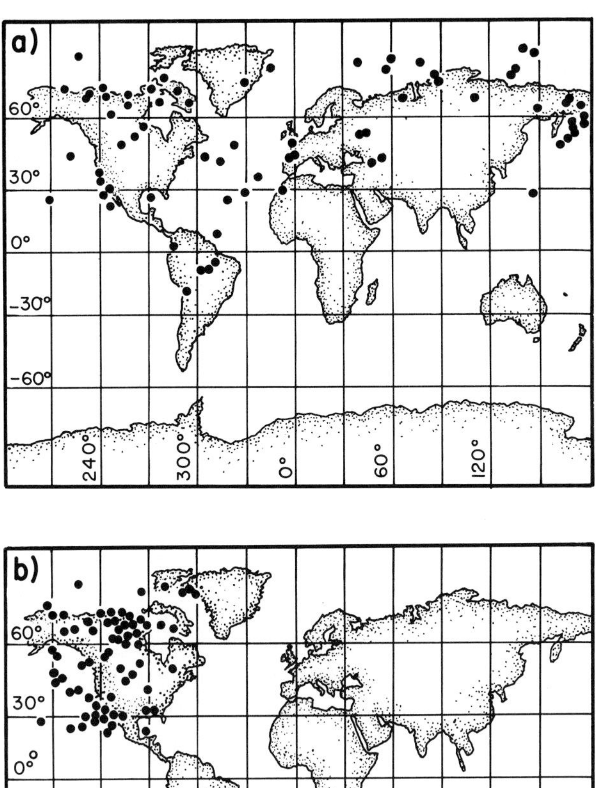

Figure 5.13. a) Individual paleopoles from each of the Gondwana continents in their own (present-day) coordinates for the Late Carboniferous through Early Jurassic interval. b) Individual paleopoles from each of the Gondwana continents rotated into West African coordinates for the Late Carboniferous through Early Jurassic interval, with the parameters of Lottes & Rowley (1990).

(Chapter 8) and the late Paleozoic assembly of Pangea, it is important to consider the Silurian–Devonian segments of the APWP of Gondwana in some detail.

Until 1987, there were no reliable Silurian paleopoles from cratonic Gondwana. A paleopole had been published for the Mereenie Sandstone of Australia, but the age of the rocks could range from Ordovician through Early Carboniferous. Paleopoles in this age range from South America were also poorly dated and paleomagnetically ill defined. However, for eastern Australia there were several results which increasingly supported a paleopole location to the west of southern Chile (e.g., Goleby, 1981), provided that this part of the mobile belt of Australia had not been displaced with respect to the craton since the Silurian to Early Devonian. Because support for this proviso was ambiguous, Morel & Irving (1978) offered two alternative APWPs, one of them passing in a loop through eastern Australian results and the other passing directly from the well determined and uncontested Ordovicia paleopoles in northern Africa to the Carboniferous paleopoles in southern Afric (paths X and Y in Figure 5.15a). The pre-Carboniferous results from eastern Aus tralia are listed separately at the end of Table A5 to highlight the uncertainty implie by its non-cratonic (possibly non-Gondwana) nature. However, in 1987 Hargrave and colleagues published a result from the Aïr Intrusive Complexes of Niger i cratonic western Africa, which yielded a paleopole similar to those from easter Australia upon reconstruction of the Gondwana continents. They also provide $^{40}Ar/^{39}Ar$ age dates that yielded a mean age of middle Silurian, whereas earlier ^{87}Rb ^{87}Sr ages had ranged from 480 to about 400 Ma. If the paleopole and its middl Silurian age are reliable, the conclusion is inescapable that the APWP of Gondwan moved rapidly from the Late Ordovician paleopole near Algeria to a middle Siluria paleopole near Chile. The paleomagnetically derived APWPs of recent vintag (Figure 5.15) all show this one way or another; moreover, the APWPs based o paleoclimate indicators and lithofacies (Caputo & Crowell, 1985; Scotese & Barrett 1990) also follow similar trends.

What is bothersome, however, about this Late Ordovician to middle Siluria trend, is that rocks of this age from South Africa (see Figure 5.16), the Pakhuis an Cedarberg Formations, do not reveal the trend (Bachtadse, Van der Voo & Hälbich 1987b). Moreover, taking the ages at face value, the trend implies an extraordinar rate of APWP and, hence, a very high velocity of Gondwana with respect to th pole. Estimates of this velocity range up to 40 cm/year, which is much higher tha the velocities inferred for the major continents in Mesozoic or Tertiary times. Thus there is some skepticism about either the paleopole location off southern Chile, o its age. Bachtadse & Briden (1990) have computed an APWP (Figure 5.15b) o

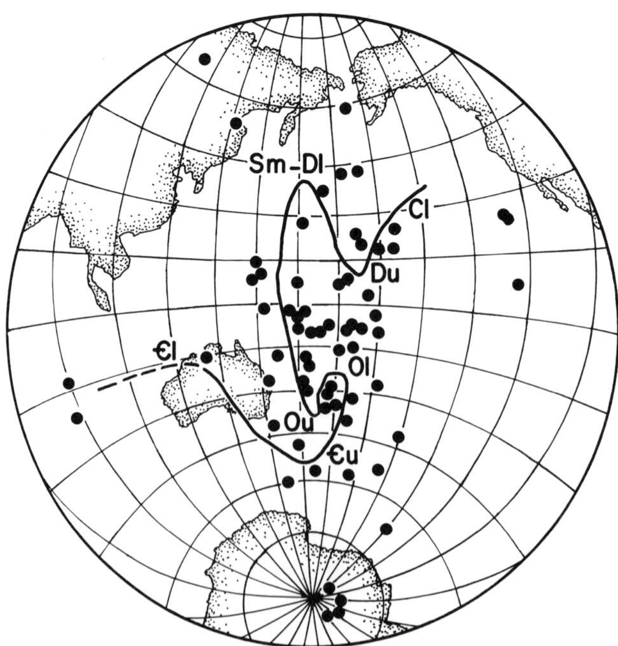

Figure 5.14. Cambrian through Early Carboniferous individual paleopoles (north poles) from eac of the Gondwana continents rotated into West African coordinates, combined with a best estima of the apparent polar wander path, with the parameters of Lottes & Rowley (1990).

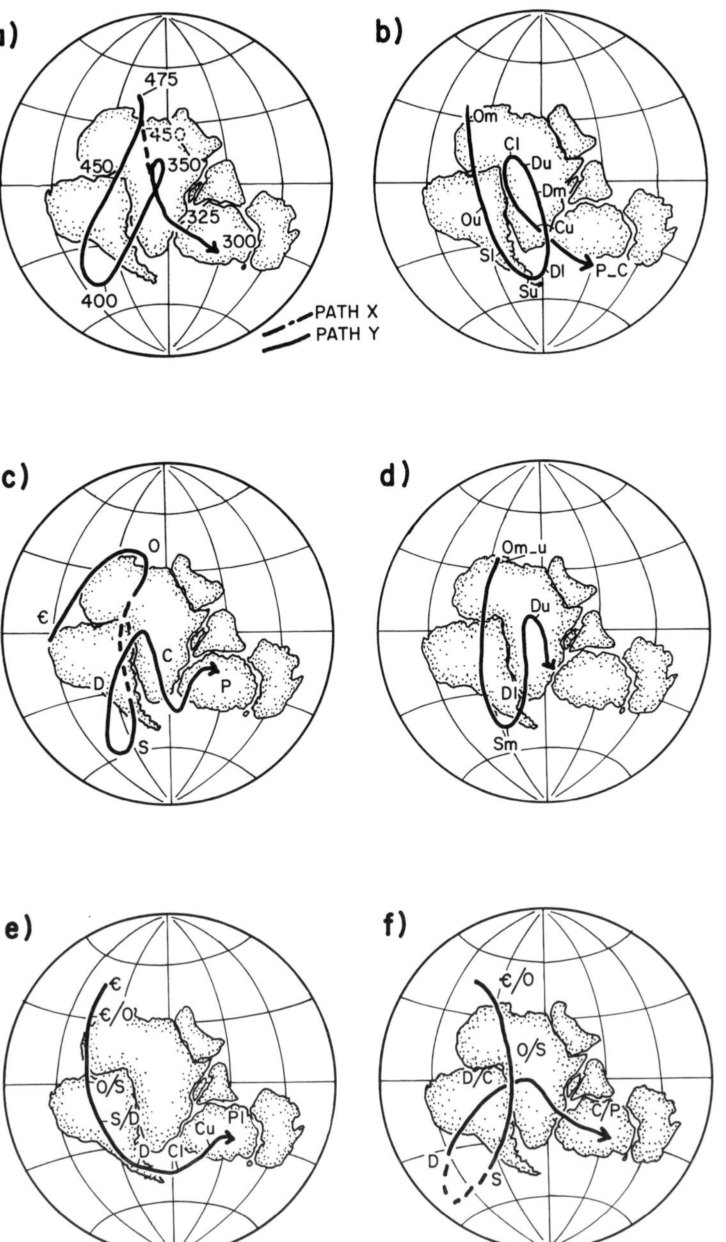

Figure 5.15. Various Paleozoic apparent polar wander paths (APWPs) for Gondwana (south poles; see text for discussion). Age abbreviations as in Table 5.1. a) Based on paleomagnetic poles available up to 1978, Morel & Irving (1978) proposed a path X and a more complicated path Y, with the latter including many paleopoles from eastern Australia's fold belts. b) A recent paleomagnetic APWP, with smoothing applied by a cubic-spline technique, from Bachtadse & Briden (1990). c) A paleomagnetic APWP with inclusion of recent Australian paleopole determinations, from Schmidt *et al.* (1990). d) Middle Ordovician through Late Devonian paleomagnetic APWP from Kent & Van der Voo (1990). e) APWP based on non-paleomagnetic information from Scotese & Barrett (1990), who used lithofacies indicators in an algorithm to find the best pole locations for a given period. f) APWP based on non-paleomagnetic information from Caputo & Crowell (1985), who used the occurrence or absence of glacial relicts to estimate the pole locations.

the basis of cubic splines; this method, as argued earlier, smoothes the time calibration and shape of the APWP. This can be seen clearly in the putative mean poles for the Late Ordovician or the middle Carboniferous of their APWP; the temporal 'smoothing' reduces the velocities of Gondwana to about 25 cm/year or less. Thus, a major consequence of the technique is that the APWP passes through poles with ages different from those assigned to that point by the smoothed time calibration. It appears to me that it is better to take the real ages of the rocks than to arbitrarily assign ages on the basis of some mathematical function.

The Devonian paleopole situation is equally complicated. Early paleomagnetic studies did not result in good paleopoles, with the possible exception of that of the Picos and Passagim Series of Brazil (Creer, 1970), until a result was published by Hailwood (1974) for the Msissi Norite in Morocco. The rocks were thought to be constrained in age by overlying Devonian sediments and the location of this paleopole was in central Africa. However, Salmon *et al.* (1986) revisited the outcrops, and found (1) that the rock type is not norite, (2) that the age of the rocks is probably Jurassic, and (3) that the paleomagnetic directions were very different from those of Hailwood. If Salmon and colleagues indeed sampled the same rocks as Hailwood, this is a major and rather startling revision. Be that as it may, in the meantime other paleopoles have become available (Hurley & Van der Voo, 1987; Bachtadse, Van der Voo *et al.*, 1987b; Li *et al.*, 1988; Aïfa, Feinberg & Pozzi, 1990) that support the central African paleopole location for the Late Devonian to earliest Carboniferous. Several of these results are representative of cratonic Gondwana (Figure 5.17), all are well dated, and together they appear to have yielded good paleomagnetic directions with antipodal reversals.

Figure 5.16. Photograph of the Ordovician–Silurian Table Mountain Group in South Africa, where the detrital sediments of the Graafwater, Pakhuis and Cedarberg formations have yielded valuable paleomagnetic results (Bachtadse, Van der Voo & Hälbich, 1987b). (Photograph courtesy of Prof. A. M. Kröner and Dr V. Bachtadse.)

Figure 5.17. Photograph of Late Devonian carbonate rocks in the Canning Basin in western Australia. The Frasnian–Famennian boundary is contained in this outcrop and is marked by many reversals (Hurley & Van der Voo, 1987). The rocks have yielded a valuable paleopole for cratonic Australia, and hence for Gondwana. (Photograph courtesy of Dr N. F. Hurley.)

Another Devonian paleopole (Kent, Dia & Sougy, 1984) falls near the Permian segment of the Gondwana APWP and may well be remagnetized. New Early Devonian paleopoles from eastern Australia (Schmidt *et al.*, 1986, 1987) are found in locations strung out between the paleopole from the Niger rocks and the Late Devonian paleopoles in central Africa, thus defining and supporting the more or less continuous trend (Figure 5.15c, Table A5). Once again, however, there is a lack of consensus among non-paleomagnetists about the Devonian loop of the APWP, mostly because it implies two important paleogeographic problems. As with the Silurian problem, the first concern is related to the rather high velocity with respect to the pole during the Middle Devonian to Early Carboniferous interval. The second concern relates to the high paleolatitudes implied for northernmost Africa (up to about 50° S), if the Late Devonian paleopole is located in central Africa. Morocco contains apparently warmer-water carbonates and other subtropical to temperate lithofacies indicators for the Late Devonian, which do not support such high paleolatitudes. Thus, Scotese & Barrett (1990) in their APWP based on such indicators, do not find the Devonian segment swinging through central Africa, but instead connect their Silurian position in central South America directly with the Late Carboniferous and Early Permian pole positions in or near Antarctica (Figure 5.15e). On the other hand, tillites and glaciofluvial sediments of earliest Carboniferous age are found in Western Africa in Niger (Lang *et al.*, 1991), suggesting intermediate to high paleolatitudes for the area at that time and a pole perhaps no farther away than central Africa.

There is no simple solution for this Siluro-Devonian ambiguity of the Gondwana APWP, until many more reliable paleopoles are obtained from cratonic Gondwana for these periods. Ignoring the available paleopoles leaves one either with a purely paleoclimatic APWP (e.g., Caputo & Crowell, 1985 versus Scotese & Barrett, 1990; see Figure 5.15) or without any data to connect the Ordovician and Late Carboniferous segments of the APWP. Accepting the available paleopoles for the Silurian and Devonian implies that the APWP shows two major swings, suggesting very high velocities with respect to the pole. I will return to the paleogeographic consequences of the Siluro-Devonian APWP for Gondwana in Chapter 8 when discussing the closure of Iapetus and other early to middle Paleozoic oceans.

Siberian realm

The geological situation of Asia is undoubtedly complicated. Siberia is shown in Figure 5.18 as a cratonic nucleus, but other more-or-less cratonic blocks exist in North and South China, which will be the subject of the next section in this chapter. In between these nuclei and to the north of the Indian craton, mobile belts and

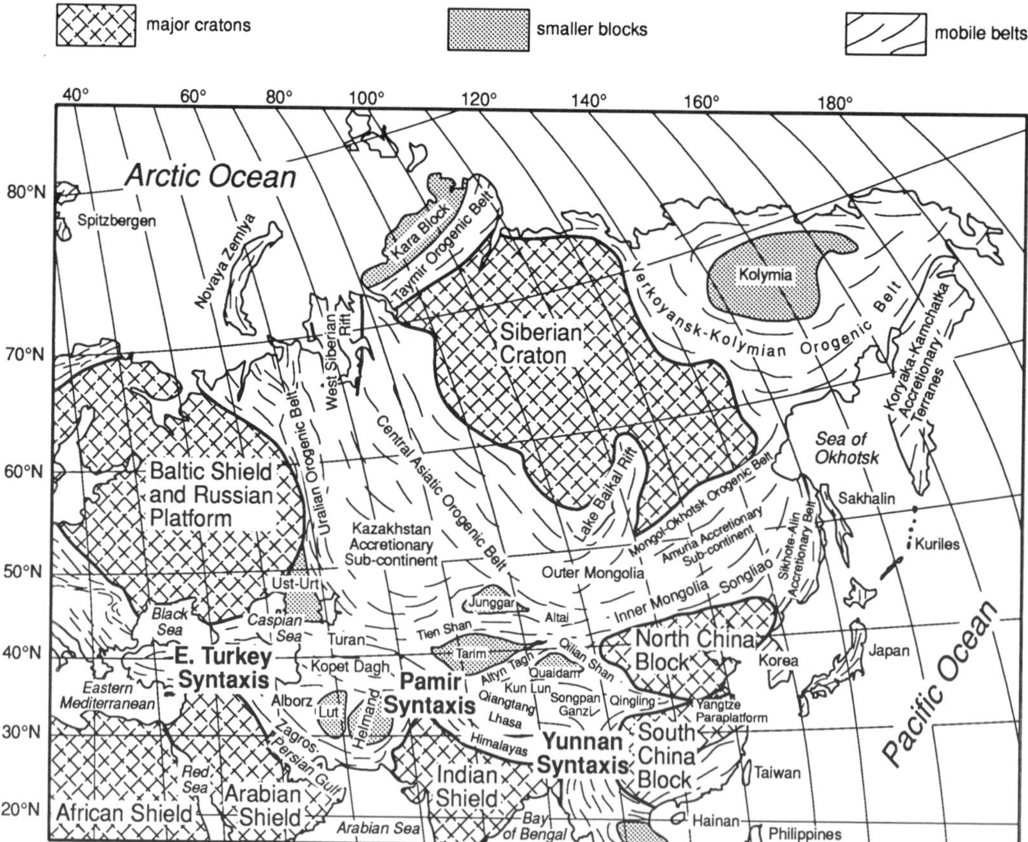

Figure 5.18. Location map of the Asian cratons and intervening mobile belts; outlines in northern Asia are after Zonenshain *et al.* (1991).

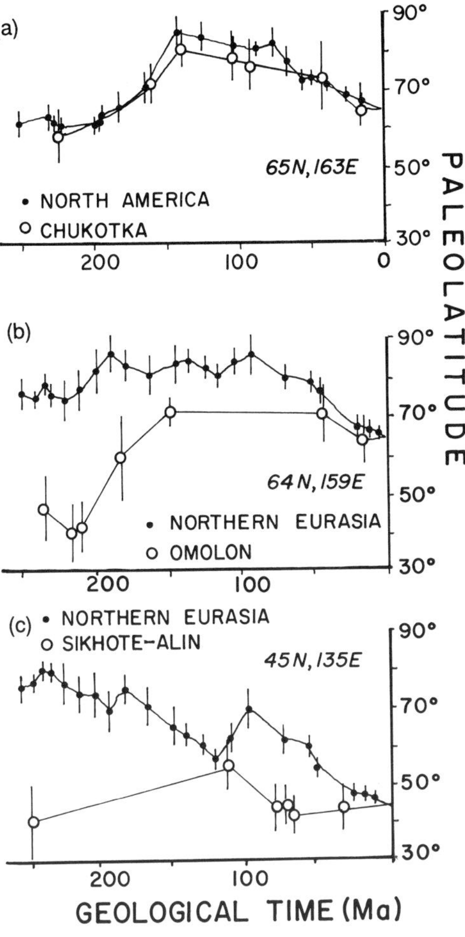

Figure 5.19. Observed (open circles) and predicted/extrapolated paleolatitudes for a) North Chukotka, b) Omolon Massif and c) Sikhote Alin. Paleolatitudes are extrapolated from North America in a) and from North Eurasia in b) and c). Figure redrawn after Khramov & Ustritsky (1990). Published with permission from Elsevier Science Publishers, BV.

displaced blocks or terranes (e.g., the Junggar Basin, Tarim, Lhasa and Qiangtang in Tibet, Mongolia) are found, which will be discussed in Chapter 7 dealing with the Tethyan domain. To the east of Siberia, several other blocks, accretionary complexes, and displaced terranes occur (e.g., Omolon and Chukotka to the east and north of the Verkoyansk-Kolymian Orogenic Belt, grouped together as 'Kolymia' in Figure 5.18; Koryaka-Kamchatka; Sikhote Alin), which will not be discussed in this book. Khramov & Ustritsky (1990) have examined some of the paleogeographic implications from paleomagnetic results of these areas. Examples of their comparative paleolatitude plots are shown in Figure 5.19. Some of the structural outlines of Figure 5.18, such as those of Kolymia, have been adopted from a recent map of northern Asia (Zonenshain *et al.*, 1991).

Late Cambrian and younger results for Siberia and the Kazakhstan Accretionary Subcontinent, which was originally not recognized as having been separate from Siberia for much of Paleozoic times (see Ziegler *et al.*, 1977), have thus far been

Table 5.4. *Mean paleopoles for Siberia*

Age or interval	Paleopole		N	K	A_{95}	Ref.
	Lat.	Long.				
Tl (45)	70,	162	4	39	11	1
Tl (59)	62,	146	5	46	9	1
Ku (88)	64,	151	4	108	7	1
Kl (121)	70,	167	5	234	4	1
Ju/Kl (142)	70,	150	6	42	9	1
Ju (151)	69,	131	5	49	9	1
Jm (173)	65,	132	6	29	11	1
Jl (195)*	63,	110	4	21	21	1
Tru/Jl (207)	56,	129	6	48	8	1
Tru (222)	55,	138	6	38	9	1
Trm (238)	52,	150	7	58	7	2
Trl (245)	52,	156	9	113	4	2
Pu (253)	42,	161	6	61	7	2
Pl (268)	38,	159	5	48	9	2
Cu/Pl (288)	34,	158	5	31	11	2
Devonian (360–408)	28,	151	7	65	8	3
Silurian (408–438)	−4,	121	8	—	19	4
M. Ordovician (458–478)	−22,	130	11	—	4	4
E. Ordovician (479–505)	−40,	132	12	—	7	4
L. Cambrian (505–523)	−36,	127	14	141	3	3
M. Cambrian (523–540)	−44,	157	4	61	12	3
E. Cambrian (540–575)	−35,	188	4	16	24	3

Age abbreviations as in Table 5.1; time in parentheses in Ma. N. K and A_{95} and mean paleopoles (given in (Eur-)Asian coordinates) as in Table 5.1. Mean paleopoles are calculated as detailed below.
References: 1 = Khramov (1988; pers. comm., 1990), with running mean averages in 20 Ma windows of all data from the USSR part of Eurasia; 2 = Khramov (1988; pers. comm., 1990) with running mean averages in 20 Ma windows of the data from the Siberian Platform only; 3 = from McElhinny (1973, table 20), conventional Fisher averages; 4 = Tarling (1983, table 9.3), conventional Fisher averages.
*Calculated without running mean average, for the time of the peak of the polar wander path cusp (see also the discussion of North American paleopoles).

obtained entirely by paleomagnetists from Russia. Since many of the original publications are not directly accessible in the literature, a compilation for the data tables of the appendices has not been possible. Instead, I have relied (Table 5.4) on summaries by others for the available paleopoles (e.g., McElhinny, 1973; Tarling, 1983; Khramov, 1988, and pers. comm., 1990).

The quality of the Siberian data set is an area of concern, since many paleopoles are apparently not fully based on directional analysis, and details of demagnetization and field tests are generally not reported. However, a partial way around this obstacle is to treat Eurasia (Stable Europe, Kazakhstan and Siberia) as a single continent for post-Permian times, and to let the European reference poles be valid for Siberia as well. This approach has been adopted later in this book, when I examine the results of the eastern Tethyan terranes with respect to Siberia. It derives its justification from our knowledge about the Urals, which have formed in a Paleozoic collision of

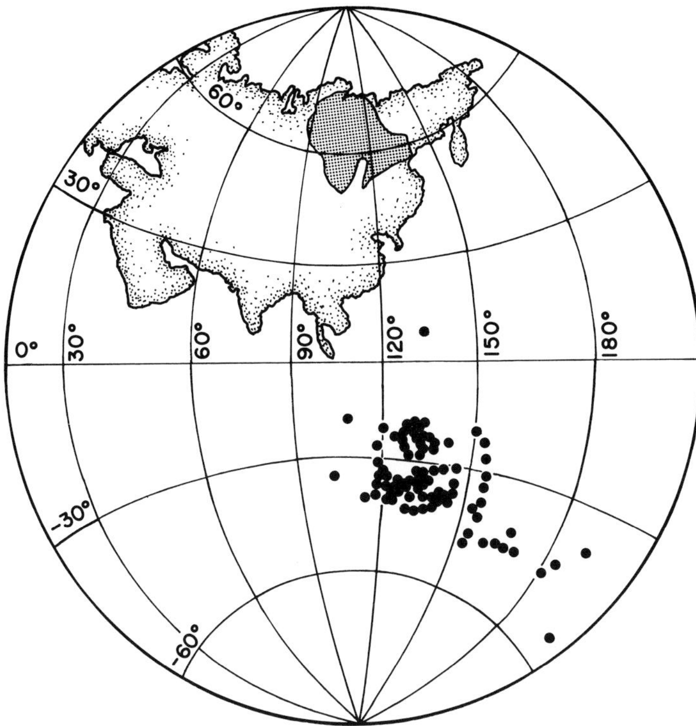

Figure 5.20. Cambro-Ordovician paleopoles for Siberia, compiled from the ORACLE data base (Lock & McElhinny, 1991).

Europe, Kazakhstan and Siberia no later than the Permian, and perhaps earlier (Hamilton, 1970; Zonenshain *et al.*, 1984; Ziegler, 1988).

This leaves the Paleozoic APWP segment for Siberia to be discussed, as it reflects independent movements of Siberia during the Paleozoic. Apparently reliable and well-clustered groupings of paleopoles are available for the Late Cambrian and Ordovician (Figure 5.20). These are geomagnetic north poles, falling in present-day coordinates well to the south of the Siberian craton (Table 5.4), indicating that paleogeographically in the early Paleozoic, Siberia was oriented 'upside down' in subtropical to tropical latitudes. The Silurian and Devonian paleopoles from Siberia suggest that it drifted northwards (i.e., towards the poles in Figure 5.20), but there is disagreement about the Silurian and Devonian paleolatitudes as derived from paleomagnetism and from paleoclimate and faunal considerations, as revealed by Silurian evaporites (Boucot, 1990), bryozoa (Tuckey, 1990), stromatoporoids (Nestor, 1990) and fishes (Young, 1990). The latter methods generally place Siberia not as far north, maintaining a more subtropical than temperate regime for this part of the world. Some Devonian paleopoles indicate that Siberia may be about to collide with Laurussia already at that time; compare, for instance, the mean Devonian paleopole for Siberia of McElhinny (1973) at 28° N, 151° E, or that of Tarling (1983) at 25° N, 155° E with those for the Devonian of Stable Europe (Table 5.1) at 24° N, 151° E and 27° N, 151° E. However, the Devonian paleopoles from Siberia may represent secondary magnetizations. Certainly, the Siberian and European continents were welded together by Triassic times (see the Siberian Triassic mean poles of Table 5.4 (52° N, 156° E; 52° N, 150° E; 55° N, 138° E) and the mean

Table 5.5. *Mean paleopoles for the North and South China blocks*

Time interval	South China Block					North China Block				
	Lat.	Long.	N	K	A_{95}	Lat.	Long.	N	K	A_{95}
Ku (67–97)	84,	213	2			88,	170	1		
Kl (98–144)	76,	201	2			77,	213	2		
Ju, uJm (145–176)						71,	224	7	102	6
Tru, uTrm (216–232)	45,	224	1							
Trl/m, Trl (233–245)	46,	215	8	26	11	42,	26	2		
Pu (246–266)	47,	232	12	30	8	46,	3	5	73	9
*N-polarity	(51,	241)	8	122	5					
*R-polarity only	(26,	216)	4	38	15					
Cu/Pl, Cu (282–308)	22,	225	1							
Dm, Dl (379–397)	− 9,	190	1							
Su, Sm (415–429)	5,	195	1							
Ol/m, Ol (468–505)	−39,	236	1			43,	333	1		
Єl (543–575)	37,	206	4	4	53					

Mean paleopoles are calculated from the entries (with Q ≥ 3), for each of the two blocks, as listed in Table A6. For abbreviations and other explanations see Table 5.1. *For the Emeishan Basalts (Late Permian) of the South China Block separate means have been calculated to highlight the differences between the results with predominantly normal (N) polarity and those with reversed (R) polarity only; for these mean basalt directions all results of Table A6 have been included, regardless of the Q-factor evaluation.

European Triassic poles of Table 5.1 (52° N, 150° E and 52° N, 133° E). Lottes & Rowley (1990) reached the conclusion that Siberia and Europe were together already by Early Permian time, on the basis of nearly identical mean paleopoles, despite evidence for significant crustal shortening in the Urals as late as Triassic time. They speculated that Permo-Triassic convergence between Siberia and Europe may have been offset by contemporaneous extension in the West Siberian Lowlands.

North and South China

These two Asian blocks (Figure 5.18) consist of Precambrian basement, but are perhaps not truly cratonic in the sense that the Phanerozoic cover rocks in most parts have been deformed (tilted and folded) by subsequent orogenies. Nevertheless, there seems to be little doubt about the autochthoneity of these cover rocks.

The North China Block has justifiably included Korea in previous publications, together forming what has been called Sino-Korea. There is indeed good geological evidence for their continued proximity in the past. However, I am treating Korea separately (Chapter 7), in part for paleomagnetic reasons; relative rotations of Korea with respect to North China are indicated for pre-Cretaceous times.

Paleomagnetic work in China before the 1980s concentrated mostly on late Mesozoic or younger rocks (e.g., McElhinny, 1973). However, in the last decade, a major increase has occurred in the rate of paleopole determinations. The available results are listed in Table A6 and mean paleopoles are given in Table 5.5. The North China paleopoles for pre-Jurassic times fall generally in Europe and northern Africa in present-day coordinates, whereas those for South China plot in the Pacific Ocean

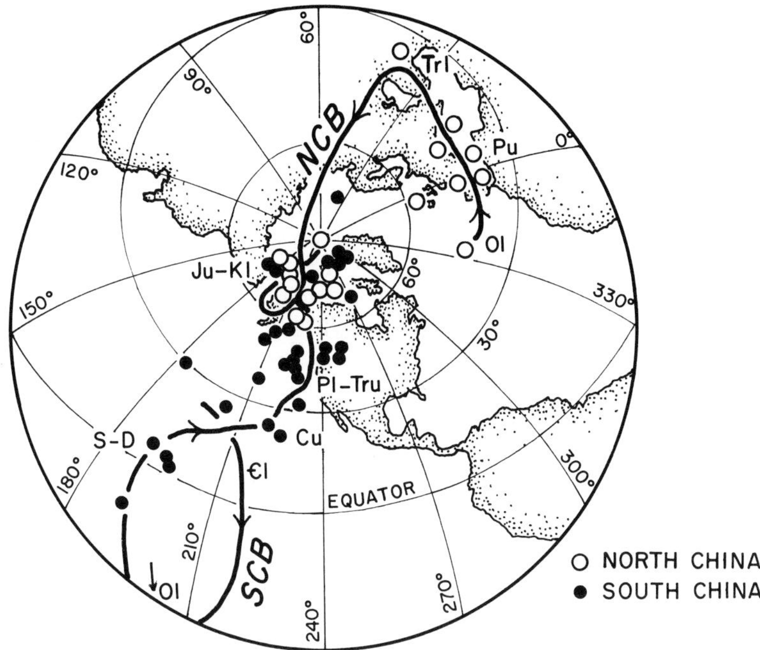

Figure 5.21. Individual paleopoles from South and North China as well as best estimates of their apparent polar wander paths. NCB = North China Block; SCB = South China Block.

(Figure 5.21). The Late Permian paleopoles of the two blocks appear to be well-determined, and have constituted some of the more convincing evidence for subsequent relative movements between the two blocks. These movements may have been as late as Late Triassic to Early Jurassic (Indosinian orogeny), although some geologists have maintained that the two blocks were already together in the Devono-Carboniferous (Li *et al.*, 1982; Wang, 1984). A solution to this discrepancy could be that after Devonian juxtaposition the two blocks behaved independently again.

A suggestion that the blocks were not far apart in the Permian, nevertheless, has led to models such as that of Figure 5.22 (from Zhao & Coe, 1987). In these models, the subsequent motions of the two blocks with respect to each other consist of a relative rotation of some 70° about a pivot at the adjoining eastern ends of the blocks (Figure 5.22). However, Permian paleolongitudes of the two blocks are, of course, indeterminate the two need not have been in contact and may have been separated longitudinally (e.g., Figure 2.9). Early and Late Triassic results (McElhinny *et al.*, 1981; Chan, Wang & Wu, 1984; Lin, Fuller & Zhang, 1985; Opdyke *et al.*, 1986; Heller *et al.*, 1988; Steiner *et al.*, 1989) confirm the relative rotation derived from the Permian results. Initially some of the latter had caused some uncertainties as to which were north poles and which south poles, when McElhinny *et al.* (1981) interpreted the South China data under the impression that the field would have been mostly reversed in the Permian Kiaman interval. However, we now know that the Chinese rocks are Upper to uppermost Permian, and that reversals began to occur in the Late Permian at about 260 Ma (e.g., Creer, Mitchell & Valencio, 1971; Molina-Garza *et al.*, 1989).

The Late Permian paleopoles for South China come mostly from the Emei Shan basalts, a series of flood basalts found over wide areas of Yunnan, Sichuan and

Guizhou. An anomaly in the results is noteworthy. Most of the Emei Shan basalts have yielded north–northeasterly directions interpreted to be of normal (N) polarity. However, reversed (R) polarity results have now been reported by several authors for the Emei Shan basalts (e.g., Huang *et al.*, 1986), as well as for slightly younger sediments (Heller *et al.*, 1988; Steiner *et al.*, 1989). What is anomalous in these directions is that systematically the R directions of Late Permian age appear to be more westerly to southwesterly and, hence, are not antipodal to the N directions, as can also be seen in the mean poles (Table 5.5). An explanation for this is lacking, but structural complexities and age differences of the rocks appear to be ruled out (Fang, Van der Voo & Liang, 1990a). This leaves a different age, not for the rocks, but for the magnetizations as a possible explanation, barring the *ad hoc* hypothesis of non-dipole field behavior. Whatever the cause of the non-antipodal R and N directions, the relative rotation shown in Figure 5.22 remains valid.

By Jurassic and certainly Cretaceous times, the paleopoles of the two blocks are in general agreement with each other and are beginning to resemble those from Siberia and Europe. This is in good agreement with geological estimates of the timing of consolidation of eastern Asia. Minor differences could be ascribed to imperfections in the reference paleopole determinations of Eurasia (Courtillot & Besse, 1986; Lee *et al.*, 1987); explanations in terms of large-scale, late Mesozoic relative movements (Lin & Fuller, 1986) are generally considered less likely. When the North American Late Jurassic and Cretaceous reference poles (Table 5.1) are rotated into Eurasian coordinates, they are similar to those from the China blocks and Korea. Thus, in general, it appears that this whole area of Asia coalesced by Jurassic to Early Cretaceous times, although subsequent transcurrent or transpressional movements are well demonstrated on structural and tectonic evidence (e.g., Tapponnier & Molnar, 1977).

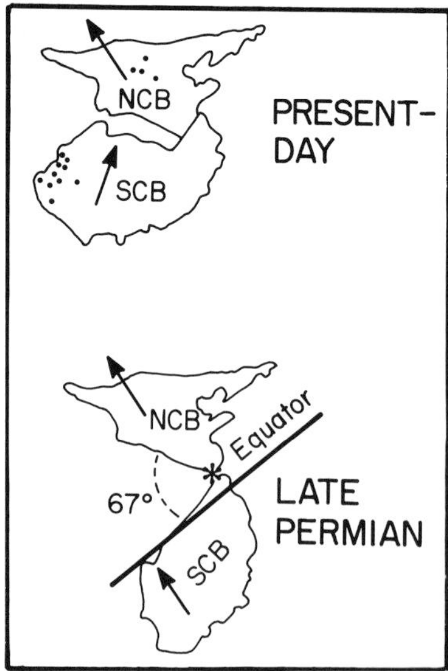

Figure 5.22. The North China and South China Blocks (NCB and SCB, respectively) today (top) and in their relative Late Permian positions, as proposed by Zhao & Coe (1987). Dots are Late Permian sampling localities and arrows represent magnetic Late Permian declinations. Published with permission from the authors and by permission from *Nature*. Copyright 1987, Macmillan Magazines Ltd.

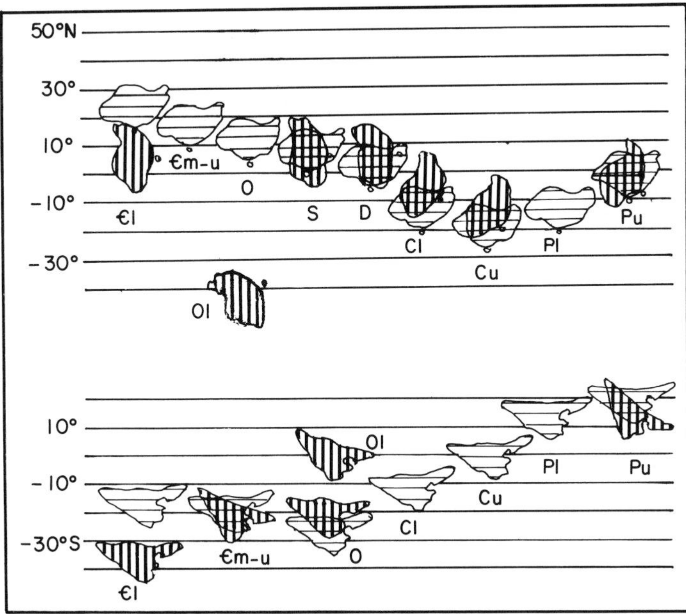

Figure 5.23. Paleolatitudinal positions of the North and South China Blocks deduced from lithofacies (horizontal shading) and paleomagnetic indicators (vertical shading), as compiled by Nie (1991), as a function of Paleozoic time. Published with permission from the author and Elsevier Science Publishers, BV.

For pre-Permian times, results for North China are virtually non-existent (three poles for the Cambrian and Ordovician), while for South China there are several paleopoles available that appear to support a long and sinuous APWP (Figure 5.21). Remagnetizations may have occurred in the South China sedimentary rocks, just as in the North American interior. Indeed some older Cambrian and Silurian sediments (Lin *et al.*, 1985; Opdyke, Huang *et al.*, 1987a) have yielded paleopoles very similar to the Devonian result (Fang, Van der Voo & Liang, 1989), whereas results from the Late Ordovician Daqing Formation resemble Permo-Triassic reference poles (Fang, Van der Voo & Liang, 1990b). Thus, the pre-Permian APWP for South China is still in its infancy, and many more paleopoles will have to be added before we can treat its Paleozoic APWP with the same confidence as for European and North American equivalents.

Most available Paleozoic paleopoles, however, indicate equatorial paleolatitudes for the two China blocks, and this is corroborated by a lithofacies study (Figure 5.23, from Nie, 1991). The only significant difference in paleolatitude between the paleomagnetic and lithofacies analyses is possibly for the Early Ordovician position of South China, for which the paleomagnetic latitude is intermediate (about 40°, likely to be in the southern hemisphere, but plotted at 40° N by Nie (1991)). Warm water faunas indicate mostly equatorial paleolatitudes throughout the Paleozoic as well. This strongly suggests that the China blocks cannot have been a permanent part of Gondwana for the entire Paleozoic, because there are no peri-Gondwana locations that remain equatorial from Cambrian through Permian. It is likely, furthermore, that the South China block was not contiguous with, or part of, Pangea during late Paleozoic to early Mesozoic times (Figure 2.9), but was instead in relative motion with respect to the supercontinent. The North China Block may well have been in contact with Siberia from Late Permian times onward (Figure

2.9), but it underwent a large Mesozoic relative rotation; as such, both China blocks could be considered as displacing terranes in the Tethys Ocean (Chapter 7).

For the earlier part of the Paleozoic, the paleolocations of the two China blocks are poorly constrained; most workers favor positions, separate from each other, in the vicinity of India or Australia (e.g., Burrett, Long & Stait, 1990) at the northern margin of Gondwana. When the Gondwana APWP reached central Africa and Antarctica in Devonian through Permian times (Figures 5.13–15), the paleolatitudes predicted for these positions became higher than those actually observed. It is therefore likely that during the Devonian the China blocks had begun to move away from the margin of Gondwana.

Pangea configurations

From Late Carboniferous through Middle Jurassic times, the Gondwana continents and the combined North American and European continents were assembled in a supercontinent called Pangea (from the greek words *pan* = all and *gea* = earth). Siberia and Kazakhstan joined this supercontinent in Permian times. Of the major continents, only North and South China were in relative rotational or latitudinal motion with respect to Permo-Triassic Pangea (see Figure 2.9).

Pangea is a classical concept dating back to Wegener (1915) or even earlier (e.g., Snider-Pelligrini, 1858). The evidence in those early days of continental drift hypotheses was based on coastline fitting, late Paleozoic glacial relicts now found widely dispersed in the Gondwana continents, and faunal similarities, but none of this evidence convinced the doubters of the reality of continental drift, not in the least because a mechanism was deemed inconceivable to let continents 'raft' across solid ocean floor. The subject was controversial and even ridiculed and remained contentious until the mid-1960s. For a detailed account of the scientific debates of the earlier part of this century, I refer to LeGrand (1988). One of the additional reasons that Pangea was not taken seriously early on was that without computers, maps of Pangea were flawed, as illustrated in Figure 5.24. But even today, there is no consensus about the exact configuration of the assembled Pangea continents.

Four different proposals have been made which deserve to be highlighted; these have subsequently been called Pangea A1, A2, B and C (Morel & Irving, 1981). Pangea A (1 or 2) configurations are closest to the original Wegener-type fits. Pangea A1 was introduced by Bullard *et al.* (1965), who minimized by computer the gaps and overlaps between the Atlantic continental margins at the depth contour of 500 fathoms (about 900 m). The reconstructions between Europe, Greenland and North America, on the one hand, and between South America and Africa, on the other hand, fit remarkably well; the few existing overlaps can be ascribed to continental margin growth since the opening of the Atlantic Ocean. The main uncertainty in the Pangea A1 fit is between the Gondwana continents and the combined northern continents, following the zones now represented by Central America, the Gulf of Mexico, the Caribbean, the Atlantic margin of the USA and their counterparts in northern South America, Africa, the Iberian Peninsula and the Mediterranean.

It is therefore not surprising that it is precisely this border zone which has been used as the principal zone of adjustment in the different Pangea reconstructions published subsequent to that of Bullard *et al.* (1965). Pangea A2, B and C were all proposed for similar reasons: the paleomagnetic poles simply did not agree well enough with Pangea A1, even when taking their sizeable cones of confidence into account. In essence, the paleopoles mandate that the continents are tighter together

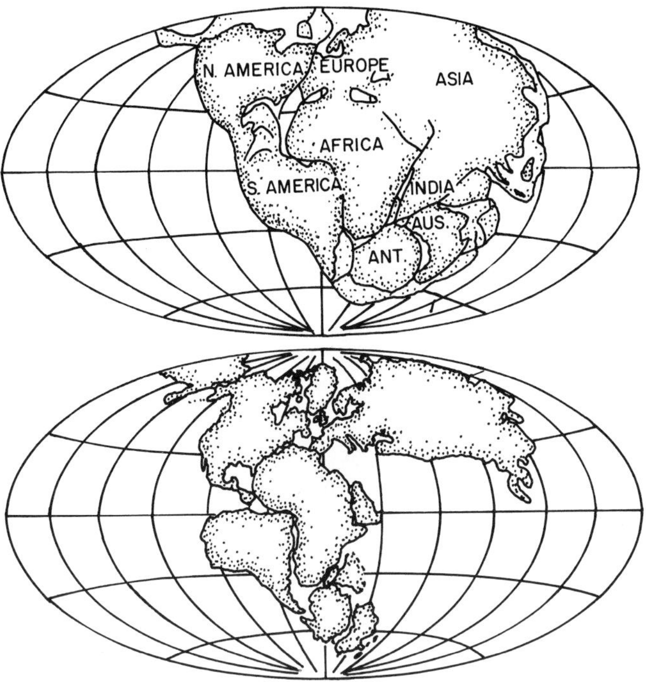

Figure 5.24. Pangea configuration of Wegener (1915; top) and one in which the actual shape of the continents is preserved (bottom). Figure adapted from A. Hallam ('Alfred Wegener and the hypothesis of Continental Drift', 1976; published originally in *Scientific American*, 1975; Copyright © February 1975 by Scientific American, Inc. All rights reserved; published with permission of Scientific American).

in a north–south sense than forseen by the Pangea A1 fit, unless one abandons the fundamental dipole assumption (e.g., Briden, Smith & Sallomy, 1971a). However, the room available to tighten the reconstruction is minimal, so some of the proposals used the typical paleomagnetic longitudinal indeterminancy (Chapter 2) to solve the problem. This explains then the difference between the Pangea A2, B and C reconstructions, involving placement of Gondwana at different longitudes with respect to the combined northern continents, so as to allow the latitudinal positions of the continents to be more compressed than with Pangea A1. Figures 5.24 and 5.25 illustrate some Pangea A1, A2, B and C reconstructions between West Gondwana and Laurussia.

Pangea A2 reconstructions involve the smallest departure from Pangea A1. I use the label 'A2' in this book for a type of reconstruction and not a specific one. In A2 reconstructions (first proposed by Van der Voo & French, 1974; Van der Voo, Mauk & French, 1976a), the Gulf of Mexico is tightly closed by letting the (cratonic) edge of South America snuggle up to the continental edge of the Gulf Coast of North America. In the A2 reconstructions, as well as in Pangea A1, parts of Mexico would overlap with South America, but considering the possibly displaced nature of the Mexican areas this may be permissible. Figure 5.26 shows some of the various (and minor) modifications of the A1 and A2 fits that have been published subsequently (from Klitgord & Schouten, 1986). The fit between Africa and North America involves the matching of continental edges that follow roughly a small circle on the globe; thus, a relative rotation about the Euler Pole (roughly located in the southern Sahara) pertaining to this small circle does not make much difference in the goodness of the fit. Pangea A2 reconstructions can be achieved by rotating Gondwana over about 20°in a clockwise sense with respect to its Pangea A1 position. The Tethys in the Pangea A2 reconstructions is thus wider than that of Pangea A1. I will examine below the improvement in the paleopoles.

Pangea B was proposed by Irving (1977) and Morel & Irving (1981). They considered that the available space in the Gulf of Mexico was too small to tighten their reconstruction sufficiently in order for their mean paleopoles to come together. Thus, they placed Gondwana at more easterly paleolongitudes with respect to the combined northern continents (Laurussia), because more space was available in the Tethyan domain to place Gondwana at the desired (more northerly) paleolatitudes.

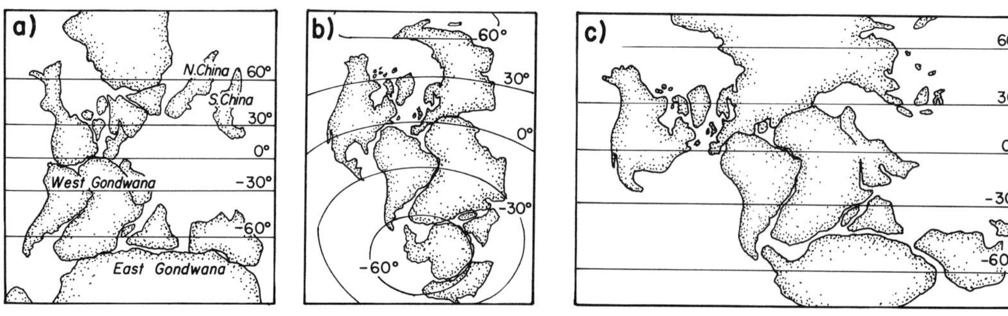

Figure 5.25. Three proposed Pangea reconstructions (with Permian paleolatitudes). a) One of the earlier Pangea A2 fits (Van der Voo & French, 1974; Van der Voo, Mauk & French, 1976a). b) Pangea B (Irving, 1977; Morel & Irving, 1981). c) Pangea C (Smith, Hurley & Briden, 1980). Redrawn after Hallam (1983). Reprinted by permission from *Nature*. Copyright 1983, Macmillan Magazines Ltd.

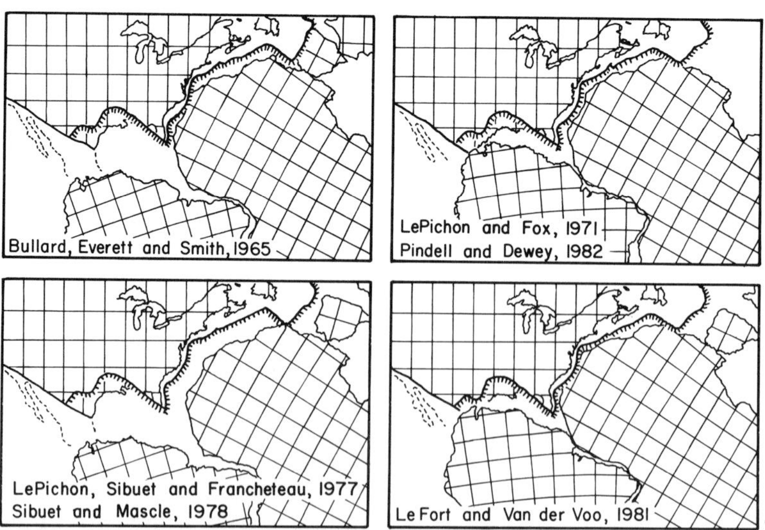

Figure 5.26. Atlantic closure reconstructions between West Gondwana and Laurentia, as minor modifications of the Pangea A1 (left) and Pangea A2 configurations (right), from original authors as indicated, with the South Atlantic closed according to Pindell & Dewey (1982) for the A2 reconstructions. The North American continental edge has been shown in all four frames. Figure redrawn after Klitgord & Schouten (1986).

Figure 5.27. Photograph of the main valley north of Trento in the Southern Alps of Italy, looking towards Bolzano. The Permian flows in the hillside on the right have yielded valuable paleomagnetic results (Van Hilten, 1960; Zijderveld, Hazeu *et al.*, 1970a).

The northwest coast of South America is placed against the Atlantic margin of North America in Pangea B, whereas Africa is now to the south of eastern Europe and southwestern Asia.

Pangea C, introduced by Smith, Hurley & Briden (1980), carries this one step further, such that South America is now to the south of Europe (see Figure 5.25). Obviously, the farther east Gondwana is placed with respect to Laurussia, the more room the Tethyan domain provides to adjust the paleolatitudes of Gondwana.

The more easterly longitudinal positioning of Gondwana in the Pangea B and C reconstructions had previously been proposed for somewhat different, albeit also paleomagnetic reasons. Early paleomagnetic work by Dutch students under the direction of Prof. R. W. Van Bemmelen in the Mediterranean area revealed Permian inclinations that indicated more northerly paleolatitudes for the northern part of the Italian Peninsula than were predicted from the available Permian poles of Stable Europe. The solution was to place Italy, together with other displaced Mediterranean terranes, in more easterly paleolongitudinal positions where the predicted paleolatitudes matched those observed. The preferred position (e.g., Van Hilten, 1964; De Boer, 1965; see also Irving, 1967) was south of Asia ranging from the Caucasus to the Himalayas. In order to let these Mediterranean terranes arrive at their present positions, a 'Tethys Twist' was proposed which in post-Permian times caused Gondwana (with the Mediterranean areas) to move in a major dextral shear pattern with respect to Laurussia over several thousands of kilometers.

The Tethys Twist was subsequently refuted (or, at the very least, made unnecessary) by better data: the Permian inclinations previously obtained for Permian flows in the Southern Alps of Italy (Figure 5.27) were found to be slightly on the high side (Zijderveld, Hazeu *et al.*, 1970a), but more importantly, the European reference poles for the Permian were found to be somewhat contaminated by recent magnetic overprints. The effects of improved demagnetization methods (Zijderveld, 1967a,b) and better age calibration of the Lower versus Upper Permian rocks involved (e.g., Zijderveld, 1967b, 1975; Merabet & Daly, 1986) have changed the European reference pole sufficiently to accommodate the Italian observations without requiring major lateral displacements.

However, half a decade later history repeated itself. The Permian (and Triassic) results from Gondwana seemed to require more northerly paleolatitudes, causing a significant overlap of Gondwana and Laurussia unless one of the two was moved laterally out of the way. It became therefore important to look, once again, at the accuracy of the paleopoles and their age calibrations (e.g., Van der Voo, Peinado & Scotese, 1984). I propose to take up this issue here in four steps: (1) an examination of the fits and paleopoles of Laurussia, (2) the same for the combined Gondwana continents, (3) a discussion of the critical Permo-Triassic segment of the geological time scale, and (4) a comparison of the mean Gondwana and Laurussia poles in terms of different Pangea fits. A critical discussion of the different reconstructions from a non-paleomagnetic viewpoint has been provided by Hallam (1983).

(1) Laurussia

At the beginning of this chapter, mean paleopoles and APWPs have been discussed for North America and Europe. In that section, it was mentioned that the entire Late Ordovician to Middle Jurassic segments of the two paths can be used to test fits with different reconstruction parameters between the two continents in order to obtain the best match of the APWPs (Van der Voo, 1990a). This can be seen in Table 5.6, where the difference between pairs of coeval paleopoles is expressed in degrees of latitude and longitude for seven different fits (Bullard *et al.*, 1965; Herron, Dewey & Pitman, 1974; Le Pichon, Sibuet & Francheteau, 1977; Sclater, Hellinger & Tapscott, 1977; Savostin *et al.*, 1986; Srivastava & Tapscott, 1986; Rowley & Lottes, 1988). All reconstructions have a negligible latitudinal misfit, but the longitudinal misfit ranges from 0.3° (for Bullard *et al.*, 1965) to some 20° (Rowley & Lottes, 1988). The larger values reflect a lesser rotation angle, or an Euler pole farther removed from the continental outlines (and at the same time closer to the paleopoles, causing less convergence), or both. The conclusion reached (Van der Voo, 1990a) was that the paleopoles are only brought into satisfactory coincidence by the tight fit of Bullard and colleagues. Figure 5.28 shows the paleopoles in the worst and the next best fits, whereas those of the best fit have been illustrated in Figure 5.3. In Figures 5.29 and 5.30 the same reconstructed mean poles are shown with their cones of 95% confidence

The less tight fits probably reflect the matching of features at or near the continent–ocean margins that do not accurately represent the true ancient edge of the continents. It is now well known that continental margins extend after break-up (e.g., Dunbar & Sawyer, 1989). This continental extension, or attenuation (e.g., Figure 5.31), means that the present-day boundary between continental and oceanic crust ideally should be corrected to restore the crustal thinning. However, the magnitude of the attenuation is typically not known and could range from a few tens to several hundreds of kilometers. Judging from the tests of the different Europe–North America reconstructions, it seems safe to argue that the 'correction' appears to be quite significant.

The excellent match of the paleopoles of Europe and North America in the

Table 5.6. *Mean paleopoles for Europe, rotated to North American coordinates in a test of different reconstructions*

Time interval	NAM mean pole	Rotated Poles Bullard			Rotated poles Herron			Rotated poles Le Pichon			Rotated poles Sclater			Rotated poles Srivastava			Rotated poles Savostin			Rotated poles Rowley		
		La Lo	Δ-La	Δ-Lo	La Lo	Δ-La	Δ-Lo	La Lo	Δ-La	Δ-Lo	La Lo	Δ-La	Δ-Lo	La Lo	Δ-La	Δ-Lo	La Lo	Δ-La	Δ-Lo	La Lo	Δ-La	Δ-Lo
177–195 (Jl)	68, 93	69, 89	−1	+4	72, 104	−4	−11	71, 106	−3	−13	72, 104	−4	−11	72, 119	−4	−26	71, 100	−3	−7	74, 122	−6	−29
196–215 (Tru/Jl)	61, 81																					
216–232 (Tru)	52, 96	51, 95	+1	+1	54, 108	−2	−12	52, 109	0	−13	53, 107	−1	−11	54, 118	−2	−22	53, 105	−1	−9	56, 118	−4	−22
233–245 (Trl)	52, 110	51, 112	+1	−2	53, 126	−1	−16	52, 126	0	−16	53, 125	−1	−15	52, 135	0	−25	53, 122	−1	−12	54, 136	−2	−26
246–266 (Pu)	52, 120	49, 122	+3	−2	51, 136	+1	−16	49, 136	+3	−16	50, 134	+2	−14	49, 145	+3	−25	50, 132	+2	−12	50, 145	+2	−25
267–281 (Pl)	45, 123	46, 126	0	−3	48, 140	−3	−17	46, 140	−1	−17	47, 138	−2	−15	46, 148	−1	−25	47, 136	−2	−13	47, 148	−2	−25
282–308 (Cu)	40, 128	40, 131	0	−3	42, 144	−2	−16	40, 144	0	−16	40, 142	0	−14	40, 152	0	−24	41, 141	−1	−13	40, 152	0	−24
309–365 (Cl)	29, 131	24, 121	+5	+10	26, 133	+3	−2	24, 133	+5	−2	25, 131	+4	0	25, 140	+4	−9	26, 130	+3	+1	26, 139	+3	−8
366–378 (Du)	30, 110	26, 113	+4	−3	28, 125	+2	−15	27, 126	+3	−16	28, 123	+2	−13	27, 132	+3	−22	28, 122	+2	−12	29, 131	+1	−21
379–397 (Dl)	24, 108	23, 113	+1	−5	25, 125	−1	−17	24, 125	0	−17	25, 123	−1	−15	24, 132	0	−24	25, 122	−1	−14	26, 131	−2	−23
398–414 (Su/Dl)	4, 97	2, 97	+2	−0	5, 108	−1	−11	3, 109	+1	−12	4, 106	0	−9	5, 114	−1	−17	4, 105	0	−8	7, 112	−3	−15
415–429 (Su)	18, 127	19, 123	−1	+4	21, 135	−3	−8	19, 135	−1	−8	20, 133	−2	−6	19, 141	−1	−14	20, 132	−2	−5	21, 140	−3	−13
430–467 (Om-Sl)	18, 146	12, 143	+6	+3	13, 154	+5	−8	12, 155	+6	−9	12, 153	+6	−7	11, 160	+7	−14	13, 151	+5	−5	11, 159	+7	−13
mean deviation		+1.7, −0.3			−0.4, −12.4			+1.1, −12.9			+0.3, −10.8			+0.7, −20.6			+0.1, −9.1			−0.7, −20.3		
rotation parameters		88.5, 27.7, −38.0			84.0, 166.7, −27.6			85.0, 129.9, −26.7			84.4, 146.6, −29.2			76.3, 144.7, −22.2			86.4, 158.5, −30.2			72.8, 154.7, −24.3		

Parameters for the rotation of Europe, with respect to North America (NAM) held fixed, are given in the bottom row, with the Euler pole as North Latitude (positive when North), Longitude, and rotation angle (clockwise when negative). La = latitude, Lo = East Longitude, Δ-La and Δ-Lo are the north–south and east–west differences in degrees, respectively, between the rotated European paleopoles and the coeval North American (NAM) mean poles of the first column. The mean deviations given at the bottom of the table are the arithmetic means of the 12 pairs of latitude (Δ-La) and longitude (Δ-Lo) differences.

reconstruction by Bullard and colleagues allows us to combine the dataset to construct a common APWP (Table 5.7). The combined mean paleopoles can be used as reference poles for displaced terrane analysis (e.g., Chapter 7) and for comparisons with other continents such as Gondwana, as will be done below.

(2) Gondwana

Before a common Gondwana APWP can be constructed, in order to test Pangea configurations for the interval of latest Carboniferous through earliest Jurassic time, a decision must be made which Gondwana reconstruction should be selected, and what the possible strengths and weaknesses are of the combined dataset.

Paleomagnetically, the different reconstructions tested (Table 5.3) are those of Smith & Hallam (1970), Lottes & Rowley (1990) and Scotese & McKerrow (1991), and do not appear to lead to significantly different statistical precision of the poles, even though the mean paleopole locations vary slightly. For the interval of interest (308–177 Ma), the average A_{95} of the mean poles is 17° to 18° for all three fits tested (Lottes & Rowley, Scotese & McKerrow, Smith & Hallam) with the average K ranging from 60 to 80 (Table 5.3). This lack of precision, that is, the large A_{95}s, undoubtedly results from the lower quality and greater scatter of the Gondwana results (Figure 5.13), and not necessarily from flaws in the reconstructions. The method used to calculate the mean may also play a role. Two options

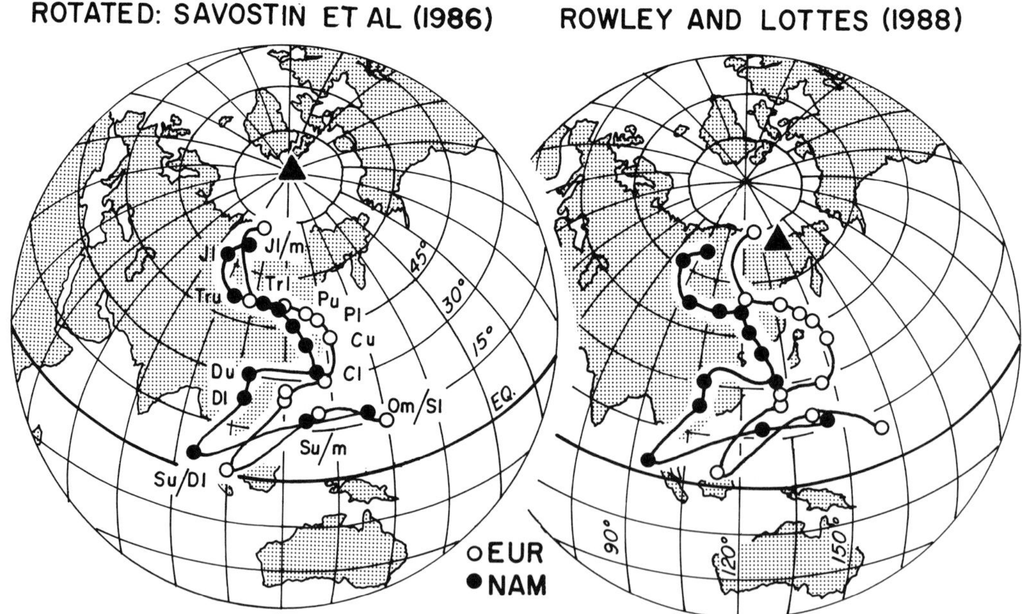

Figure 5.28. European and North American paleomagnetic apparent polar wander paths for the Middle Ordovician through Early Jurassic interval, with mean poles based on Q ≥ 3. In the plot on the left the Atlantic Ocean is closed with the parameters of Savostin *et al.* (1986), whereas on the right the parameters of Rowley & Lottes (1988) are used. The large triangles are the Euler poles of these rotations. For an explanation of age abbreviations, see Table 5.1; for pre-Middle Silurian times, only European paleopoles from northern Britain have been used. From Van der Voo (1990a).

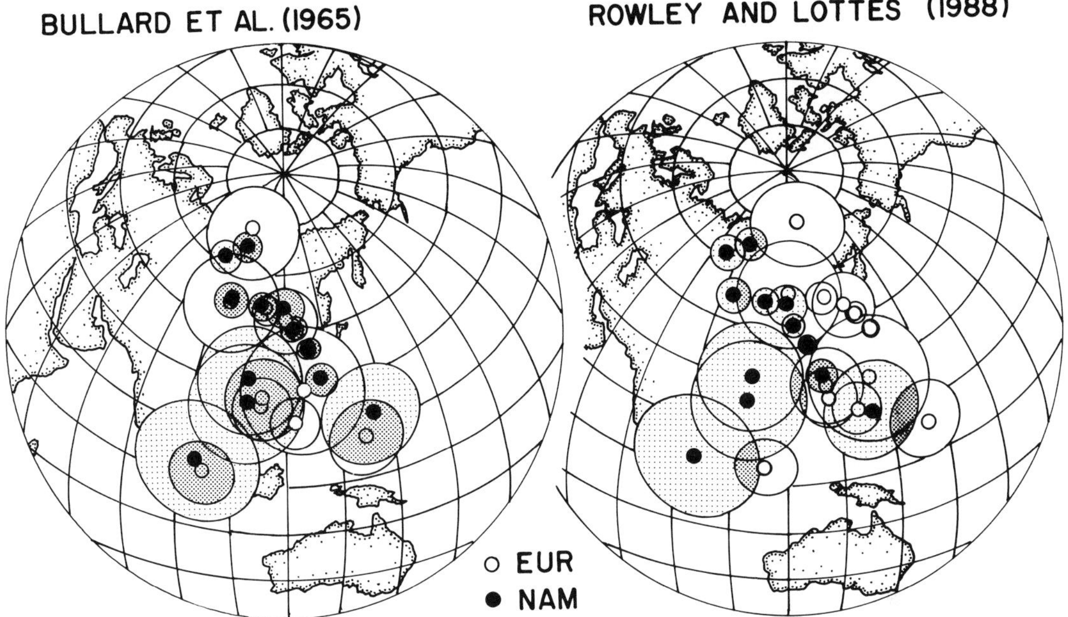

Figure 5.29. European and North American paleomagnetic apparent polar wander paths for the Middle Ordovician through Early Jurassic interval, as in Figures 5.3 and 5.28 but with the addition of their 95% cones of confidence. In the plot on the left the Atlantic Ocean is closed with the parameters of Bullard *et al.* (1965), whereas on the right the parameters of Rowley & Lottes (1988) are used. North American cones of confidence are lightly shaded and the European ones are left white; where coeval North American and European cones overlap, the area has been given a darker shading. The more the cones overlap (as is nearly completely the case for the Bullard fit), the better the agreement between reconstruction parameters and the mean paleopoles. From Van der Voo (1990a).

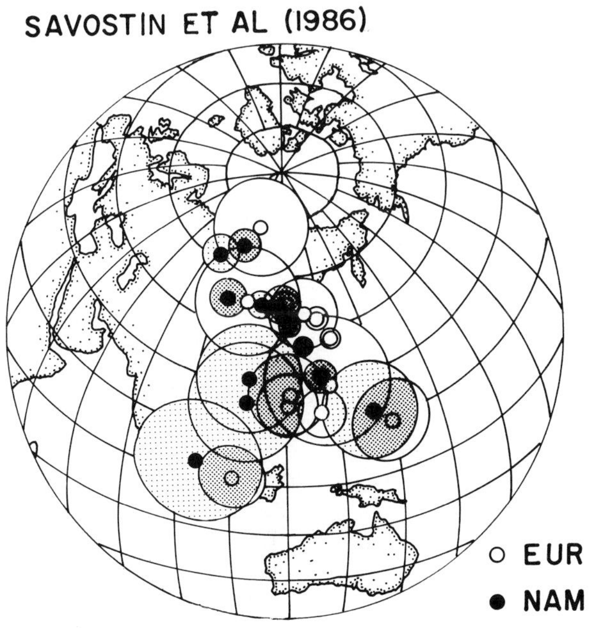

Figure 5.30. European and North American paleomagnetic apparent polar wander paths for the Middle Ordovician through Early Jurassic interval, as in Figure 5.29, but with the rotation parameters of Savostin *et al.* (1986), which is the next best fit following the Bullard fit in terms of the agreement between reconstruction parameters and paleopoles.

Table 5.7. *Combined mean paleopoles in North American and European coordinates*

Time interval	N. American coordinates		European coordinates		N	K	A_{95}
lJm, Jl (177–195)	68N,	91E	67N,	129E	14	156	3.2
Tru/Jl (196–215)	61N,	81E	60N,	119E	15	68	4.7
Tru, uTrm (216–232)	51N,	96E	50N,	134E	12	64	5.5
Trl/m, Trl (233–245)	52N,	111E	51N,	149E	24	134	2.6
Pu (246–266)	50N,	121E	49N,	159E	11	136	3.9
Pl (267–281)	46N,	124E	45N,	162E	35	161	1.9
Cu/Pl, Cu (282–308)	41N,	129E	40N,	167E	36	237	1.6
Cm, Cl, Du/Cl (309–365)	28N,	129E	27N,	167E	25	44	4.4
Du, Dm/Du (366–378)	28N,	111E	27N,	149E	14	26	7.9
Dm, Dl (379–397)	23N,	110E	22N,	148E	10	40	7.8
Su/Dl (398–414)	@3N,	97E	3N,	135E	@19	30	6.2
Su, Sm (415–429)	19N,	124E	18N,	162E	13	52	5.8
Ou/Sl, Ou, Om (430–467)	16N,	144E	15N,	182E*	18	19	8.1
Ol/m, Ol (468–505)	17N,	158E	16N,	196E*	4	27	18.1
Єu, Єm, Є (506–542)	9N,	158E	9N,	196E*	8	28	11.0
Єl (543–575)	5N,	170E	5N,	208E*	6	22	14.7

*Valid for Great Britain north of the Iapetus suture only.
Abbreviations as in Table 5.1.
@Also includes one result from Greenland.

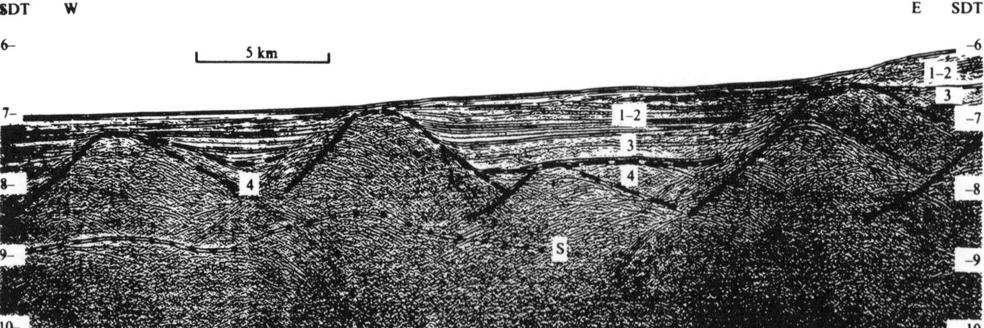

Figure 5.31. Seismic profile from immediately west of Galicia Bank (Spain), showing tilted basement blocks above normal faults. This section supports the idea of large-scale crustal extension occurring at times of continental break-up. Figure from Bond & Kominz (1988), after De Charpal *et al.* (1978), reprinted by permission from *Nature*. Copyright 1978, Macmillan Magazines Ltd.

Table 5.8. *Mean paleopoles for West Gondwana*

Time interval	N	K	A_{95}	Mean pole		
				NW AFR	NE AFR	SAM
Tl (37–66)	6	27	13	83, 181	83, 181	80, 91
Ku (67–97)	18	61	5	68, 246	68, 246	82, 214
Kl (98–144)	13	78	5	54, 260	50, 261	86, 243
Ju, uJm (145–176)	10	30	9	59, 250	53, 254	87, 140
Jl, lJm (177–195)	10	108	5	72, 249	66, 254	78, 44
Tru/Jl (196–215)	4	54	13	66, 227	61, 235	76, 87
Tru, uTrm (216–232)	8	45	8	64, 242	58, 247	82, 86
Trl/m, Trl (233–245)	1	—	—	43, 222	38, 227	63, 149
Pu (246–266)	5	77	9	61, 236	56, 242	80, 110
Pl (267–281)	8	60	7	34, 242	28, 245	63, 189
Cu/Pl, Cu (282–308)	12	33	8	33, 223	28, 227	56, 161
Cm, Cl, Du/Cl (309–365)	4	20	21	21, 227	16, 230	46, 174
Du, Dm/Du (366–378)	3	29	23	4, 200	1, 202	20, 153
Dm, Dl (379–397)	2	—	—	7, 177	6, 180	12, 130
Su/Dl (398–414)	0	—	—	—	—	—
Su, Sm (415–429)	1	—	—	43, 189	41, 196	47, 116
Ou/Sl, Ou, Om (430–467)	2	—	—	−35, 158	−33, 156	−34, 137
Ol/m, Ol (468–505)	4	9	33	−30, 181	−31, 180	−19, 152
Ꞓu, Ꞓm, Ꞓ (506–542)	8	5	28	−67, 173	−66, 161	−51, 177
Ꞓl (543–575)	2	—	—	−17, 90	−11, 92	−46, 54

The mean poles (given as latitude, longitude) were calculated by averaging the individual paleopoles of Africa and South America in the coordinate system of northwest Africa, as listed in Table A4; the means are presented above in the coordinate systems of northwest Africa (NW AFR), northeast Africa (NE AFR) and South America (SAM), after appropriate rotations with the reconstruction parameters of Lottes & Rowley (1990) and others as given in Table 5.2.
Age abbreviations as in Table 5.1. N is the number of individual paleopoles, K and A_{95} are statistical parameters associated with the mean (Fisher, 1953), and paleopole latitudes are positive (negative) North (South).

exist: one is to give unit weight to each individual paleopole, rotated into African coordinates, and the other is to give unit weight to averages for each continental element. Because the choice depends on what one intends to find out, namely whether the individual results show good coherence or whether the continental reconstruction is well supported by the mean poles for each block, I have used the latter method to calculate the values of Table 5.3.

There are further complications and possible permutations. If one compares the results from the West Gondwana continents (South America and Africa) with those from East Gondwana (Tables 5.2, 5.8 and 5.9), one can see that they are systematically offset from each other, with the East Gondwana mean poles generally more easterly in longitude (by 23 degrees on average for the Permian through Early Jurassic interval). Again, the quality of the data set may be the cause, but it is also possible that for a north–south intercontinental comparison the effects of non-dipole fields begin to play a role, or that here as well continental margins have attenuated so much that the reconstructions are not tight enough!

Table 5.9. *Mean paleopoles for East Gondwana*

Time interval	N	K	A_{95}	Mean pole
Jl, lJm (177–195)	7	89	6	70, 265
Tru/Jl (196–215)	2	—	—	67, 241
Tru, uTrm (216–232)	1	—	—	52, 283
Trl/m, Trl (233–245)	6	46	10	51, 265
Pu (246–266)	4	20	21	37, 253
Pl (267–281)	1	—	—	28, 242
Cu/Pl, Cu (282–308)	3	25	25	42, 239
Cm, Cl, Du/Cl (309–365)	1	—	—	43, 182
Du, Dm/Du (366–378)	6	19	16	17, 184
Dm, Dl (379–397)	3	5	60	53, 176
Su/Dl (398–414)	0			
Su, Sm (415–429)	3	4	68	19, 168
Ou/Sl, Ou, Om (430–467)	2	—	—	−48, 189
Ol/m, Ol (468–505)	8	31	10	−28, 179
€u, €m, € (506–542)	8	13	16	−27, 176
€l (543–575)	9	17	13	−1, 160

The mean poles (given as latitude, longitude) were calculated by averaging the individual paleopoles of India, Australia (including the eastern foldbelts), Antarctica and Madagascar rotated into the coordinate system of northwest Africa, with the parameters listed in Table 5.2 (Lottes & Rowley, 1990).
Age abbreviations as in Table 5.1. N is the number of individual paleopoles, K and A_{95} are statistical parameters associated with the mean (Fisher, 1953), and paleopole latitudes are positive (negative) North (South).

In general, though, the Gondwana mean poles have poor precision, with A_{95}s of some 15° on average. This will hamper, of course, precise comparisons with Laurussia poles.

The different half periods are unequally represented in the means of the different Gondwana continents. The best groups of results (with the lowest A_{95}s) are for the Early Jurassic (Table 5.3), and for the Early Permian. Late Permian and Triassic poles are sparse and do not cluster well upon reconstruction.

The Permo-Triassic poles from South America are numerous, but probably low in quality (Table A4). The difficulty with the paleopoles of South America is that its APWP segment (in South American coordinates) for this time is very close to the present-day pole (Table 5.2A). Recent remagnetizations and Permian or Triassic magnetizations therefore generally look alike, were it not that the former are generally of normal polarity and the latter may contain reversals. Even so, reversed-polarity remagnetizations of recent vintage have been observed in many rocks, and this effect, or alternatively the presence of overprints, may have seriously affected the South American results. For Africa, recent attempts to refine the Permo-Triassic poles have failed, because the results yielded only (Jurassic to Holocene) remagnetizations (Ballard, Van der Voo & Hälbich, 1986; Ghorabi, 1990), raising the suspicion that other poles in the data base are similarly contaminated. In summary, then, Gondwana's combined APWP has poor precision with especially the Late Permian and Triassic as 'weak' segments.

(3) *The Permo-Triassic time scale*

In the last two decades the geological time scale has undergone some revision. This is significant for intercontinental comparisons of paleopoles, if one considers that a paleopole from stratigraphically dated sedimentary rocks of one continent may need to be compared with a radiometrically dated paleopole for igneous rocks from another. Different time scales, therefore, also account for some of the past differences in the conclusions about Pangea reconstructions. Irving (1977), for instance, relied on the Van Eysinga (1975) time scale, whereas in more recent years the ages of Harland *et al.* (1982) or similar time scales have become prevalent. The ages for the Permo-Triassic boundary in these time scales differ by up to 15 Ma (Van der Voo *et al.*, 1984; Menning, 1989). When Creer *et al.* (1971) discovered normal polarities in the Quebrada del Pimiento rocks from Argentina with a radiometric age of 263 ±5 Ma, the time scale then in use indicated an Early Permian (early Artinskian) age, i.e., in the middle of the Kiaman reversed interval. With some of the more recent time scales, this age represents the Kungurian or even the uppermost Kazanian (i.e., middle Late Permian), in which normal polarities have been observed beginning at about 260–250 Ma, in reasonable agreement with the observations from Argentina.

(4) *Comparison of Gondwana and Laurussia poles*

Having selected the mean Laurussia poles (Table 5.7) and the best (?) mean Gondwana poles (Table 5.3), as well as the mean paleopoles for East and West Gondwana separately (Tables 5.8 and 5.9), we can test different Pangea reconstructions with the latest data. This turns out to be less than satisfactory.

None of the Pangea A1 type reconstructions tested bring the mean Gondwana and Laurussia poles into reasonable agreement. Pangea A1 (Bullard *et al.*, 1965) and its updated equivalents (e.g., Le Pichon *et al.*, 1977) leave the two datasets after reconstruction about 35° apart in longitude, or (on average) about 20° along a great circle (Table 5.10). Because these two reconstructions are similar, they do not show significant differences; Figure 5.32a displays a representative example using the parameters of Bullard *et al.* (1965). One can see in this figure why paleomagneticians thought they needed to resort to different reconstructions, such as Pangea A2, B or C, in order to close the systematic longitudinal gap of about 35° (which represents an average great-circle distance of some 20°).

Pangea A2 is the only other fit that can be truly tested, since Pangea B and C can always be made to fit the data by slight latitudinal adjustments of Gondwana during the course of the Permo-Triassic. Pangea A2 fares better in this test than A1 (Table 5.10), with the pole pairs of the two data sets apart (on average) by some 12° of arc (for the reconstruction of Lottes & Rowley, 1990; Figure 5.32b).

There are two ways to compare the locations of several pole pairs: one is to use the great-circle distance (p) between two poles of the same age, taking the arithmetic mean of all the p-values (as done in Table 5.10), whereas the other method is to use the east–west (Δ-Lo) and north–south (Δ-La) differences, taking the arithmetic mean of all Δ-Lo values as well as that of the Δ-La values (as done in Tables 5.6 and 5.11). At first glance, the first method seems preferable. However, the p-values are unidirectional and always positive; thus, with the typical paleomagnetic scatter of paleomagnetic data, they do not average out the fact that some deviations are to the north (or west) and others to the south (or east). By taking the east–west small-circle values and the north–south (great-circle) values separately, and

Table 5.10. *Laurussia and Gondwana mean poles compared in different reconstructions*

Age interval	Laurussia mean pole		Rotated[1] Gondwana pole			Rotated[2] Gondwana pole			Rotated[3] Gondwana pole			Rotated[4] Gondwana pole		
	Lat.	Long.	Lat.	Long.	p	Lat.	Long.	p	Lat.	Long.	p	Lat.	Long.	p
Jl (177–195)	68,	91	73,	100	5.8	72,	102	5.4	70,	66	9.1	73,	101	6.6
Tru/Jl (196–215)	61,	81	75,	112	17.7	74,	114	17.6	60,	73	4.0	63,	91	4.8
Tru (216–232)	51,	96	73,	141	29.1	73,	143	29.6	74,	95	23.0	74,	126	26.2
Trl (233–245)	52,	111	66,	168	31.0	67,	170	31.8	68,	118	16.3	64,	136	17.6
Pu (246–266)	50,	121	61,	159	23.7	62,	161	24.8	60,	121	9.9	57,	138	11.9
Pl (267–281)	46,	124	43,	158	24.2	43,	161	26.3	49,	140	11.1	45,	150	18.3
Cu (282–308)	41,	129	45,	149	15.1	45,	152	17.2	48,	125	7.5	44,	138	7.3
Average p					20.9			21.8			11.6			13.2

Mean poles are given in North American coordinates, after appropriate Euler rotations as detailed below. Lat. and Long, are latitude (positive North) and longitude in degrees. The distance p is measured in degrees of arc between the mean North American pole and the rotated mean Gondwana pole.

[1]Reconstruction parameters of Bullard, Everett & Smith (1965) of Africa with respect to North America (Euler pole 67.6°N, 346.0°E, angle 74.8 clockwise (cw), with the internal Gondwana reconstruction parameters of Smith & Hallam (see Table 5.3).

[2]Reconstruction parameters of Le Pichon, Sibuet & Francheteau (1977) of Africa with respect to North America (Euler pole 66.2°N, 347.6°E, angle 71.8 cw) together with the parameters of Smith & Hallam (1970) for Gondwana internally (see Table 5.3).

[3]Reconstruction parameters of Lottes & Rowley (1990), as given for the rotations internal to Africa and internal to Gondwana in Tables A4 and 5.2. Their Euler pole for Northwest Africa with respect to North America is 61.3°N, 343.18°E and angle 79.54 cw).

[4]Reconstruction parameters of Lefort & Van der Voo (1981) of Africa with respect to North America (Euler pole 65.95°N, 346.65°E, angle 76.74 cw) combined with the rotation parameters internal to Gondwana from Scotese & McKerrow (1991), as given in Table 5.3.

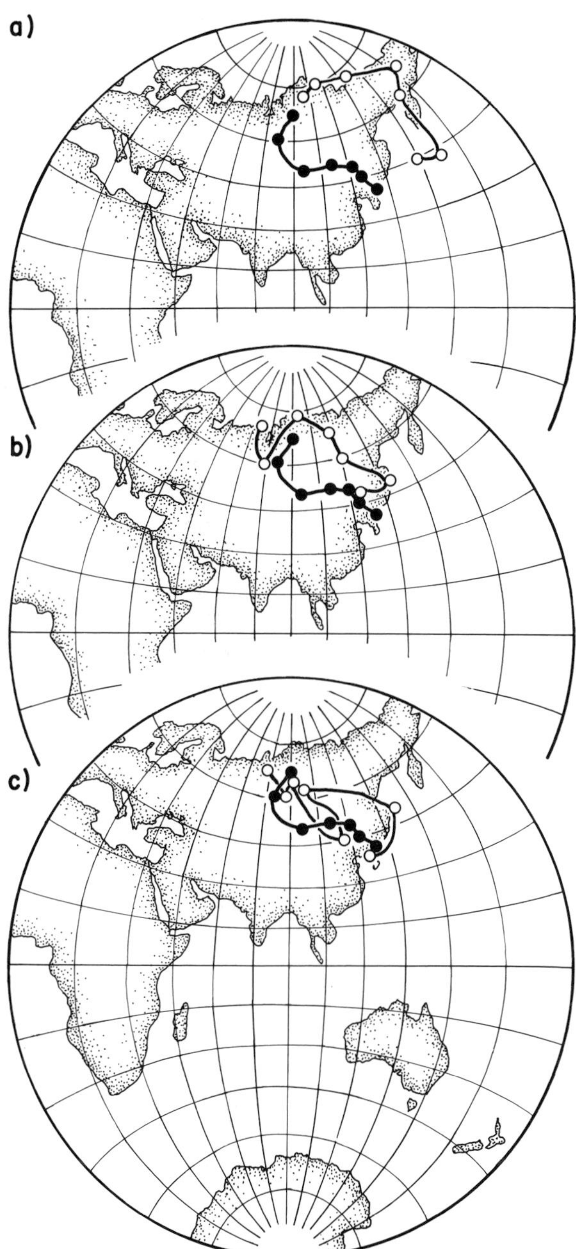

Figure 5.32. Laurussia (solid circles) and Gondwana mean poles (open circles), rotated with the parameters from different reconstructions. a) Bullard et al. (1965) Pangea A1-type fit (Table 5.10). b) Lottes & Rowley (1990) Pangea A2-type reconstruction (Table 5.10). c) Wissman & Roeser (1982) Pangea A2-type reconstruction with mean poles for West Gondwana only (Table 5.11).

given that these values can be positive or negative, one obtains a better averaging of the inevitable scatter in a data set as limited as that for Gondwana.

The discrepancies (resulting in large p or Δ-Lo and Δ-La values) can be made less by comparing Laurussia poles with poles for West Gondwana only (Table 5.11). Now the A2 type fits yield poles that are (on average) about 2 to 7° apart in longitude

Table 5.11. *Laurussia compared in different Pangea A2 reconstructions with West Gondwana only*

Age interval	Laurussia mean pole		Rotated[1] Gondwana pole				Rotated[2] Gondwana pole				Rotated[3] Gondwana pole			
	Lat.	Long.	Lat.	Long.	Δ-La	Δ-Lo	Lat.	Long.	Δ-La	Δ-Lo	Lat.	Long.	Δ-La	Δ-Lo
Jl (177–195)	68,	91	65,	66	+3,	+25	68,	72	0,	+19	70,	79	−2,	+12
Tru/Jl (196–215)	61,	81	58,	81	+3,	0	61,	86	0,	−5	62,	92	−1,	−11
Tru (216–232)	51,	96	64,	86	−13,	+10	66,	93	−15,	+3	67,	101	−16,	−5
Trl (233–245)	52,	111	46,	113	+6,	−2	46,	117	+6,	−6	46,	121	+6,	−10
Pu (246–266)	50,	121	62,	92	−12,	+29	63,	98	−13,	+23	64,	106	−14,	+15
Pl (267–281)	46,	124	52,	139	−6,	−15	50,	143	−4,	−19	48,	147	−2,	−23
Cu (282–308)	41,	129	40,	124	+1,	+5	39,	127	+2,	+2	38,	130	+3,	−1
Average					−2.6,	+6.0			−3.4,	+2.4			−3.7,	−3.3
Same, without Tru, Pu					+1.0,	+0.4			+0.6,	−1.3			+0.6,	−4.7

Mean poles are given in North American coordinates after appropriate Euler rotations of the West Gondwana mean poles of Table 5.8. For explanation of symbols see Table 5.10, except that, here, instead of p the difference in longitude and latitude is calculated for each pair (see text for reasons).

[1]Reconstruction parameters of Lottes & Rowley (1990), see Tables A4, 5.8 and 5.10.

[2]Reconstruction parameters of Wissman & Roeser (1982), with Euler pole 64.1°N, 344.26°E, angle 78.4 clockwise, combined with the mean West Gondwana poles (Lottes & Rowley reconstruction) of Table 5.8.

[3]Reconstruction parameters of Lefort & Van der Voo (1981), as given in Table 5.10, combined with the mean Gondwana poles (Lottes & Rowley reconstruction) of Table 5.8.

(Table 5.11), with an example shown in Figure 5.32c for the reconstruction of Wissman & Roeser (1982). More permutations are possible by, for instance, excluding the poorly determined Late Triassic Gondwana pole, and yes, the results are somewhat improved (Table 5.11), but I am sure that by now the readers are tired of this game.

My conclusions from this analysis are that (1) the Gondwana continents have, in general, poorly documented, low-precision paleopoles for the latest Carboniferous through earliest Jurassic, (2) Pangea A1 reconstructions do show systematic offsets between the reconstructed Laurussia and Gondwana means, ranging from 10 to more than 20° of arc, depending on the choice of comparison techniques and rotation parameters, (3) East and West Gondwana show similar offsets internal to the reconstructions of Scotese & McKerrow (1991) or Lottes & Rowley (1990), (4) the A2 fits of Lottes & Rowley (1990), Wissman & Roeser (1982) and Lefort & Van der Voo (1981) give acceptable agreements, provided that only West Gondwana poles are used, and (5) many of the discrepancies are within the large cones of 95% confidence. By definition, I should add, Pangea B and C reconstructions can be made to agree perfectly with the paleomagnetic data by letting Gondwana move episodically with respect to Laurussia during the interval of interest.

I am nevertheless not in favor of considering the Pangea B and C reconstructions seriously, and this for several reasons. All workers, including the protagonists of the Pangea B and C fits, agree that the Atlantic Ocean began opening from a Pangea A-type configuration. This implies that the Pangea B (or C) configuration must have evolved into Pangea A some time before the Early Jurassic, presumably during the Triassic. For the transition from Pangea B to A, this means a dextral strike slip (Tethys Twist) of some 3500 km with respect to Laurussia, requiring high velocities of up to 10 cm/year. Moreover, the strike–slip fault zone, running of necessity between Gondwana and Laurussia in a zone from the Mediterranean to the Gulf of Mexico, is by no means a simple small circle zone. This implies that the motion between the two sides must have involved transpressional and/or extensional parts. The geology of this zone reveals some extension for Late Triassic time, in the Pyrenees, in Morocco and in the Triassic rift grabens of eastern North America, but this does not seem to account for 3500km of displacement. Lastly, I do not believe that the paleomagnetic options (improvements of poor data quality, misfits generally within the A_{95}s, possibilities of non-dipole fields, etc.), or even the possibilities of restoring the continental margin attenuation, have been exhausted completely. A recent review (Smith & Livermore, 1991) also cites several of these issues, while concluding that a fit approximating Pangea A2 best matches the available geological and paleomagnetic constraints. For the near future, restudies of some of the rocks may yield insights into some of these possible causes. If only West Gondwana poles are used and the poorly determined Late Permian and Late Triassic Gondwana poles are excluded, then the A2 fits give very acceptable agreement.

Assembly of Japanese bicycle require great peace of mind

(Robert M. Pirsig, 1974, Zen and the Art of Motorcycle Maintenance.*)*

And so does the assembly of Pangea, I am inclined to add. These assemblies are not to be rush jobs, or several pieces from the assembly-kit will be left over at the end, and the ones that are attached to each other will look crooked. This is where quality again comes into consideration. And frankly, for the case of Pangea, some of the data (especially from the Gondwana continents for some geological periods) are pretty poor as well as scarce.

In the next chapter, the problems are not so severe, but even so there still are some considerable uncertainties and controversial issues. Generally, the more recent the geological period under discussion, the better the paleomagnetic resolution.

6 The opening of the Atlantic Ocean

After Early Jurassic time the Central Atlantic Ocean began opening between West Gondwana and Laurussia, whereas between Europe and North America the spreading began in the Cretaceous. The oldest seafloor in the Central Atlantic is estimated to be about 175 Ma (Klitgord & Schouten, 1986). However, it is not clear to what extent the two continental landmasses already had been moving apart before that time (through thinning of the continental crust without ocean floor formation). The Late Triassic to Early Jurassic dike swarms and sediment-filled grabens of eastern North America (e.g., Newark Basin) and in West Africa suggest that rifting began well before 175 Ma, but the amount of rift-related extension between the two continents is poorly known. From a paleomagnetic viewpoint, however, there is no serious discordance between the Early Jurassic Laurussia and West Gondwana paleopoles (Chapter 5, Tables 5.10 and 5.11), so only after 175 Ma do the paleopoles need to be compared with rotation parameters that take the ocean spreading into account.

The Atlantic Ocean is, of course, still widening today. Technological developments in the last decade have allowed precise measurements of this increase in width. Using geodetic techniques (e.g., Very Long Baseline Interferometry or VLBI) the distance between Onsala, Sweden and Westford, Massachusetts, USA is seen to be increasing by 13.4 ± 0.7 mm per year, that is, about 10 cm during the eight-year duration of the measurements (Figure 6.1, from B. O. Rönnäng, pers. comm., 1990). Similar measurements have documented higher plate velocities in the Pacific Ocean.

Using the 13.4 mm/year separation rate, one can make a quick calculation how long it has taken for the North Atantic Ocean to open between Europe and North America, assuming that this rate has remained constant (which, of course, has probably not been the case). The current distance between Onsala and Westford is 50.5°, whereas the distance between these localities before the opening of the Atlantic was 31.3°. This means that with constant spreading rates, the separation is to have begun 158 Ma; that is, Late Jurassic. Marine magnetic anomalies suggest a later beginning (Cretaceous), but as we will see below, earlier (Late Jurassic?) extension may well have occurred.

In this chapter, we will examine the available paleopoles from the Atlantic-bordering continents for Middle Jurassic and younger times and compare them with reconstructions based on marine magnetic anomalies and fracture zone orientations. For a valuable discussion of the well-known methodology used to calculate reconstruction poles of rotation (also called finite-difference poles of rotation), I refer to Klitgord & Schouten (1986).

In the last 20 years, these marine magnetic anomalies used to calculate rotation parameters for late Mesozoic and Cenozoic times have become very well documented. Thus, post-Early Jurassic reconstructions of the continents and parts of the oceans have become quite accurate, especially when compared to the lesser resolution provided by paleomagnetic poles. The first comprehensive study of the Atlantic Ocean spreading by Pitman & Talwani (1972) followed the pioneering studies of Le Pichon (1968), Morgan (1968), Heirtzler *et al.* (1968), Vogt, Anderson & Bracey (1971), and Le Pichon & Fox (1971). Subsequently, Sclater *et al.* (1977) and Olivet *et al.* (1984) have refined the scenario, followed by an excellent and up-to-date publication by Klitgord & Schouten (1986) which has been used generally for the calculations of the motions between West Gondwana and Laurussia described in this Chapter.

For the South Atlantic (between South America and Africa; see the legend of Table A4), Cande, LaBrecque & Haxby (1988) have documented plate kinematic parameters for the time since 80 Ma, while for times before 80 Ma interpolated rotations are based on Klitgord & Schouten (1986). For the North Atlantic (between Europe and North America) I have used the parameters of Rowley & Lottes (1988), while recognizing that some differences with other models (e.g., Srivastava & Tapscott, 1986; Srivastava *et al.*, 1990) remain unresolved.

Mean paleomagnetic poles for four intervals between 176 and 37 Ma have been compiled for North America (including Greenland), Europe, Africa and South America from the data in Tables A1–A4. Ideally, one should also add to this the mean paleopoles from East Gondwana continents, but since some uncertainties exist in the reconstruction parameters based on Mesozoic marine magnetic anomalies in the Indian Ocean, I have not done this. Moreover, our experience with comparisons between North American, European, East and West Gondwana paleopoles in Pangea reconstructions for pre-Jurassic times (Chapter 5) illustrates that the available data sets have to be very robust and well-characterized in order to achieve good precision. An examination of East Gondwana paleopoles (Tables A5 and 5.2) shows that the data are generally neither numerous (a total of 45 poles with $Q \geq 3$) nor well-clustered (arithmetic average of the A_{95}s is 11°) for each of the continents.

The purpose, then, of this chapter is to examine the available paleomagnetic data for each of the four selected periods, and to see whether they form coherent patterns

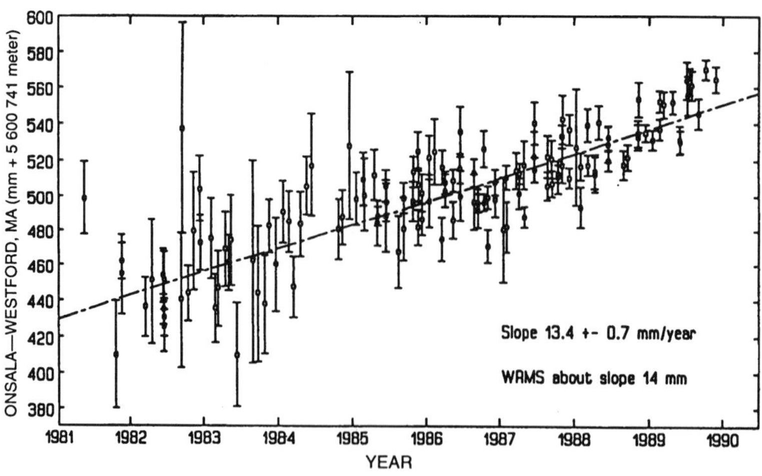

Figure 6.1. Change in distance (in mm) measured with very long baseline interferometry between Onsala (Sweden) and Westford (Massachusetts, USA) during the interval 1981–89. Figure courtesy of B. O. Rönnäng (pers. comm., 1990).

upon reconstruction of the Atlantic-bordering continents, and possibly to make choices between the different reconstructions proposed. I will proceed in this discussion from old to young.

The late Middle to Late Jurassic (176–145 Ma)

North America (including Greenland)

The clearest pattern of apparent polar wander in this interval is provided by the North American paleopoles. Because the mean paleopoles of the Early Jurassic (at 68° N, 93° E) and the Early Cretaceous (at 69° N, 194° E) are about 30° of arc apart (great-circle distance), it is clear that apparent polar wander during the intervening late Middle to Late Jurassic interval must have been significant. This can indeed be seen (Figure 6.2a) in the individual paleopoles for this interval for North America which range from 79° N, 90° E (169 Ma, Moat Volcanics; Van Fossen & Kent, 1990) to 64° N, 168° E (about 147 Ma, Upper Morrison Formation; Steiner & Helsley, 1975). Other poles for this interval form a scattered group between these extremes, roughly (but not exactly) from old to young towards the east.

There has been extensive debate about this track, which, for example, is of importance as a reference framework for determining terrane displacements in the western Cordilleras of North America. Until the 1980s, the Late Jurassic APWP segment for North America was very poorly defined. However, several paleopoles were published subsequently, owing for a large part to the efforts of the paleomagnetic team at the University of Arizona. These workers (e.g., May & Butler, 1986), following the PEP (Paleomagnetic Euler Pole) approach of Gordon *et al.* (1984), proposed that the Jurassic APWP segment consists of two tracks in the vicinity of the 60th parallel (Figure 5.7). They argued, with some apparently compelling arguments, that paleopoles located at higher present-day latitudes (e.g., the White Mountain Intrusions at 79° N, 167° E; Opdyke & Wensink, 1966; Van Alstine, 1979) were contaminated by recent overprints.

Very recently, however, Van Fossen & Kent (1990) have restudied some of the White Mountain rocks and argued that their higher-latitude pole (quoted above) is a reliable representation of the late Middle Jurassic geomagnetic field and that the Corral Canyon Paleopole (May *et al.*, 1986) is perhaps based on a Cretaceous, post-tilting, remagnetization; furthermore, the structural setting in the southeastern Arizona portion of the Basin and Range Province may be much more complicated than thought by May & colleagues. Recognized as such by Van Fossen & Kent (1990), and decidedly in favor of their alternative, is the fact that European and West Gondwana paleopoles, when rotated into North American coordinates, also fall at higher latitudes (Table 6.1, Figure 6.2). I will return to this below.

A few paleopoles calculated from the magnetic anomalies of western Central Atlantic Seamounts (see Table A1; Mayhew, 1986) are located generally within the streak of Middle to Late Jurassic poles. This is as expected from the estimated ages of the Seamount rocks, thought to be related to the same hot-spot track responsible for the White Mountain Intrusions. Paleopoles, previously thought to be Late Jurassic and obtained from carbonate and detrital sedimentary rocks in Wyoming, have been shown to be based on remagnetizations (McWhinnie, van der Pluijm & Van der Voo, 1990) and are marked by an asterisk in Table A1.

Europe

The late Middle and Late Jurassic paleopoles of Europe are also strung out over a large longitudinal range of some 80° along the 70th parallel (Table A2), comparable to the distribution of the North American poles. What is puzzling, however, is that there is no clear correlation from west to east with decreasing age. One of the older Middle Jurassic (Callovian) sequences, for instance, gives a pole at 74° N, 200° E (Ogg *et al.*, 1991), whereas other poles for Polish rocks of about the same age fall near 72° N, 140° E. At the young end of the interval, a similar situation is seen, when one compares the Late Jurassic poles from Spitsbergen near 62° N, 190° E with the Late Jurassic limestone poles from Germany and Switzerland near 70° N, 140° E. Several explanations may be invoked: (1) Spitsbergen was in a different position with respect to Stable Europe than today, (2) some of the rocks are remagnetized at younger times, (3) the Late Jurassic APWP for Europe shows no systematic progression from west to east, but is more complex, or (4) some of the age

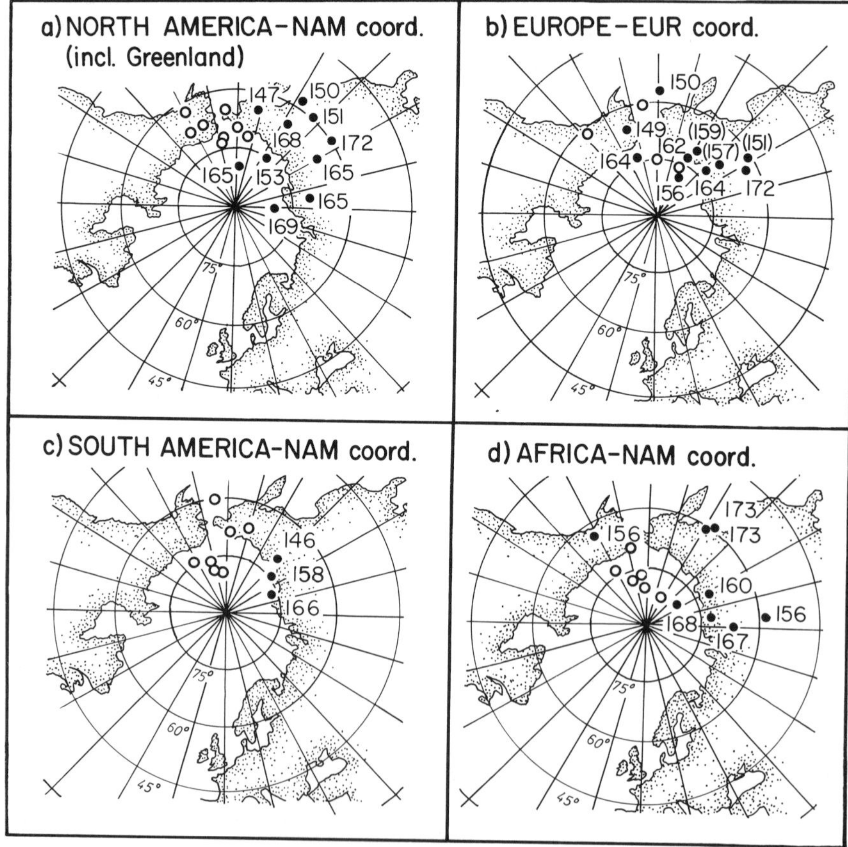

Figure 6.2. Late Middle and Late Jurassic paleopoles (solid circles) and Early Cretaceous paleopoles (open circles) for the Atlantic-bordering continents. Ages are given for the Jurassic paleopoles in Ma. a) North America. b) Europe, in European coordinates for reasons presented in the text; three limestone results without reversals or field tests have been placed in parentheses (and are given a '*' in the Q column of Table A2), because they are thought to be unrepresentative of the Late Jurassic geomagnetic field. c) South American results, rotated into North American coordinates. d) African results, rotated into North American coordinates.

Table 6.1. *Late Jurassic and younger mean paleopoles for the Atlantic-bordering continents in North American coordinates*

Time interval	North America			Europe			Africa			South America			Overall mean				
	Lat.	Long.	N	Lat.	Long.	N	Lat.	Long.	N	Lat.	Long.	N	Lat.	Long.	N	K	A_{95}
Tl (37–66)	77,	173	26	75,	181	20	80,	146	4	84,	114	2	80,	162	4	167	7.1
Ku (67–97)	68,	192	7	71,	158	5	72,	213	12	75,	214	6	73,	193	4	91	9.7
Kl (98–144)	69,	194	8	66,	185	4	76,	196	6	73,	186	7	71,	190	4	292	5.4
*Ju, uJm (145–176)	68,	137	10	71,	160	7	72,	148	7	75,	124	3	72,	143	4	219	6.2
**Ju, uJm (145–176)	68,	137	10	65,	151	3	72,	148	7	75,	124	3	70,	141	4	188	6.7

The mean poles for the four different continents have been taken from Tables 5.1 and 5.2; the North American means include the data from Greenland. Rotation parameters are given in Table 6.2. For explanation of age abbreviations, N, K and A_{95} see Table 5.1. For the Late Jurassic two options are given, one (*) in which the mean European pole is based on seven entries and one (**) in which the mean is based on only three entries, as explained in Table 5.1; each of these European means fits the overall grouping, provided different reconstruction parameters are used as detailed in Table 6.2.

dates are erroneous. Some of these possibilities will be further discussed below, after a description of the broader context of the dataset.

The bimodal grouping of the European Late Jurassic poles at the western and eastern ends of the APWP track (Figure 6.2b) between the average Early Jurassic and Early Cretaceous poles makes it very difficult to compare the European paleopoles with the North American ones. Not only are there several choices of mean paleopoles but also several options of closure reconstruction parameters (e.g., Bullard *et al.*, 1965, vs. Rowley & Lottes, 1988). This is why the European paleopoles of Figure 6.2b have not been rotated into North American coordinates. In Chapter 5, we have seen that the pre-Late Jurassic European and North American paleopoles best matched each other when using the reconstruction parameters of Bullard & colleagues, while the fit by Rowley & Lottes (1988) was not supported at all. However, these conclusions may be different for the Late Jurassic and Early Cretaceous, because some initial crustal rifting appears to have occurred inside Europe's western margin, as illustrated in Figure 6.3 (from Malod & Mauffret, 1990). The occurrence of significant extension beginning in the Late Jurassic between Stable Europe and Greenland has been documented in detail by Ziegler (1988), but there may have been also earlier extensions in Permo-Triassic time. Thus, continental extension itself is not in doubt, but its magnitude and timing remain very uncertain. The rifting, presumably without generating any true oceanic seafloor, would change the configuration of the two continents from a Bullard reconstruction to one much like that of Rowley & Lottes (Figure 6.3). I also note that the authors of this figure hypothesized paleo-north to south rifting between Spitsbergen and Norway, such that pre-rifting Spitsbergen results then would not be valid for Stable Europe. The data set from Spitsbergen included in Table A2 is not very large. With one (poorly dated and perhaps Late Carboniferous) exception, the Spitsbergen results integrate well with those from the rest of Europe. Other orientations of such rifting (e.g., between Great Britain and the continent and to the north of the Iberian Peninsula)

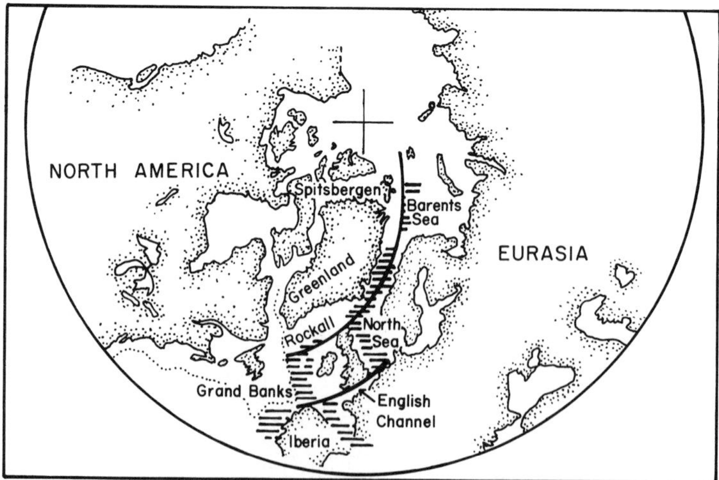

Figure 6.3. Proposed late Mesozoic rifting (horizontal ruling) between Europe and North America (from Malod & Mauffret, 1990), before the North Atlantic ocean floor began forming. The change in the continental reconstruction represents roughly the change from the configuration of Bullard *et al.* (1965) to that of Rowley & Lottes (1988). The North Sea and Bay of Biscay area extension is paleo-east to west directed, whereas that between Spitsbergen and Norway is paleo-north to south directed. Published with permission from Elsevier Science Publishers, BV.

Table 6.2. *Rotation parameters used to construct Table 6.1*

Time interval	Europe, Euler pole parameters	Africa and S. America Euler pole parameters
Tl (37–66)	65.6, 139.2, −10.5 (1)	74.5, −4.8, −15.3 (2)
Ku (67–97)	67.7, 149.5, −19.5 (1)	78.3, −18.3, −27.1 (2)
Kl (98–144)	72.0, 152.0, −23.0 (1)	66.3, −19.9, −54.2 (3)
*Ju, uJm (145–176)	72.8, 154.7, −24.3 (1)	67.1, −16.0, −64.7 (3)
**Ju, uJm (145–176)	88.5, 27.7, −38.0 (4)	67.1, −16.0, −64.7 (3)

Reconstruction parameters used to rotate the mean poles from Europe and Africa (from Tables 5.1 and 5.2) and from South America (in northwest African coordinates; Tables A4 and 5.2) into the North American coordinate frame (as listed in Table 6.1). The parameters are given as Lat., Long., and rotation angle – negative when clockwise – followed by reference to the footnotes below in parentheses.
(1) Rotation parameters from Rowley & Lottes (1988).
(2) Rotation parameters from Klitgord & Schouten (1986); for times after 80 Ma, South American paleopoles are rotated into African coordinates using the parameters of Cande *et al.* (1988).
(3) Rotation parameters from Klitgord & Schouten (1986).
(4) Rotation parameters from Bullard, Everett & Smith (1965).
For explanation of * and **, see Table 6.1.

would principally be paleo-east to west directed and would not be revealed clearly by the paleopoles.

In using paleopoles to test Late Jurassic reconstructions, we are thus faced with two choices of mean paleopoles (see Table 6.1 and footnote** in Table 5.1) because of the bimodal European distribution, and at least two sets of reconstruction parameters (see Table 6.2, Bullard *et al.*, 1965, Rowley & Lottes, 1988). It turns out that of the four possibilities, two are acceptable: (1) the Bullard *et al.* (1965) reconstruction with a more eastern Late Jurassic mean pole for Europe, that excludes several western paleopoles based on limestones, or (2) the Rowley & Lottes (1988) reconstruction combined with a more western overall Late Jurassic mean pole for Europe that also includes four of these limestone results (the ones which reveal reversals or positive foldtests). In terms of the problems with the European Late Jurassic poles mentioned earlier, we cannot a priori exclude any of the possibilities. The ambiguity is enhanced if one considers the uncertainties in the time scale for the latest Jurassic (e.g., Lowrie & Ogg, 1985/86). Below we will return to these matters after discussing the West Gondwana poles, which may provide better resolution.

South America

Figure 6.4a illustrates Jurassic and Cretaceous paleopoles from South America, including results from rocks in the Central Andes. The results with shaded cones of confidence are reference means for cratonic South America, which are very close to the present-day polar axis. On the basis of this figure, moreover, it has been argued that the Andean results show some minor rotations and cannot, therefore, be used for intercontinental comparisons (they have been marked by a * in Table A4). Within the Late Jurassic and Early Cretaceous intervals, individual cratonic paleopole entries are themselves distributed in a streaking distribution (Figure 6.4b) but over a much smaller distance than the trend of the rotated paleopoles from the Central Andes. There is therefore some support for the rotations suggested by Tanaka, Tsunakawa & Amano (1988). The paleopole distribution of Figure 6.4b was first noted by Schult & Guerreiro (1979).

In order to compare the South American cratonic paleopoles with those from North America and Europe, we have to rotate them (via Northwest African coordinates, as provided in Table A4) into North American coordinates. For these rotations, fortunately, there are few disagreements, assuming that the internal

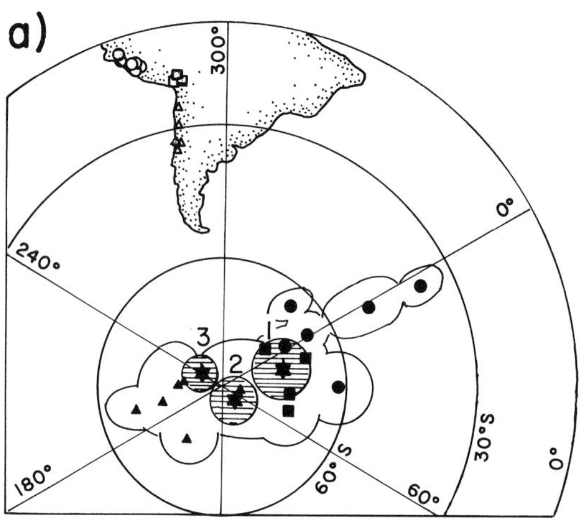

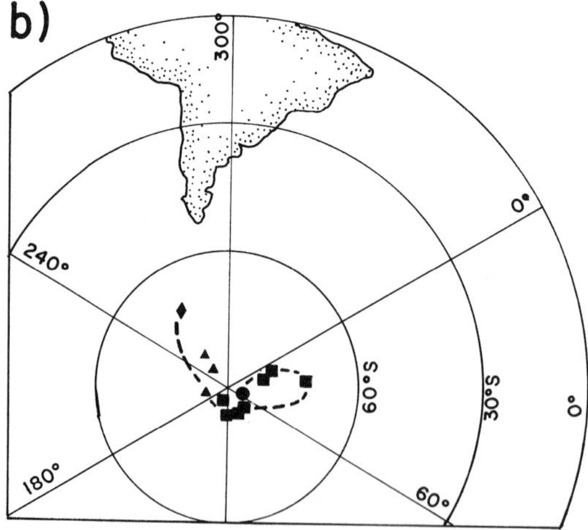

Figure 6.4. a) Mean Late Jurassic (1), Early (2) and Late Cretaceous (3) paleopoles for cratonic South America (with shaded cones of confidence; in South American coordinates) together with paleopoles from the Andes. The latter are thought to show relative rotations, with the circles representing the Peruvian Andes, squares the hinge zone, and triangles the Chilean Andes; corresponding site locations are marked by the same (open) symbols. Figure redrawn after Tanaka, Tsunakawa & Amano (1988). b) Individual cratonic South American paleopoles for the Late Jurassic (triangles) and Early Cretaceous (squares), as listed also in Table A4. Mean Late Cretaceous (circle) and Late Triassic to Early Jurassic (diamond) paleopoles and proposed Mesozoic apparent polar wander path are also shown (modified from Schult & Guerreiro (1979).

rotations within Africa, as discussed in Chapter 5 (e.g., Lottes & Rowley, 1990) are correct and well accounted for. I will return to the internal African rotations, when discussing the Cretaceous poles for Africa. In Figure 6.2c, the rotated individual South American paleopoles are shown in North American coordinates (calculated from Table A4 and the rotation parameters of Klitgord & Schouten (1986), as listed in Table 6.2). There are three late Middle to Late Jurassic paleopoles and these fall near the western end of the North American Late Jurassic APWP, whereas some earliest Cretaceous poles fall near the eastern end of the Late Jurassic track for North America, as would be expected from their younger ages. Thus, there are neither large surprises in the South American data set, nor any clues as to which of the two groups of European paleopoles has greater validity.

Africa

There are seven results available for the late Middle to Late Jurassic interval and these have been plotted, after rotation into North American coordinates, in Figure 6.2d. The individual paleopole locations streak in a by now familiar fashion, but again their ages are not systematically decreasing from west to east. Explanations for the distribution of the African paleopoles include similar arguments as used earlier: (1) some results may be based on remagnetizations and thus younger than the age of the rocks, (2) some rock units may be erroneously dated, (3) the APWP track may not be unidirectional, but could perhaps be backtracking along itself at least once, or (4) block rotations (e.g., in Morocco) could be held responsible.

It appears, then, that the paleopoles from all four continents generally agree, both in their overall mean locations (Table 6.1) and in the 'problems' revealed in more detail by the lack of systematic APWP progression in North American coordinates. This would be the end of the discussion, were it not that some interesting 'African' results are available which are not ordinarily included in compilations for this continent (Van Dongen, Van der Voo & Raven, 1967; Helsley & Nur, 1970; Gregor *et al.*, 1974; Ron, 1987). These results come from Israel and Lebanon (the Levant), situated on the tectonically disturbed western margin of the promontory of the Arabian Peninsula and, hence, of the Mesozoic African continent. When discussing the Eastern Mediterranean in Chapter 7, these results will be more fully discussed. Suffice here to note that these Late Jurassic and Early Cretaceous results ($Q \geq 3$) are well-documented in terms of (1) age of the stratified rocks, (2) normal and reversed polarities, (3) demagnetization analysis, (4) their paleohorizontal (bedding plane) and coherence with older basement of the Arabian (= African) plate.

The results from the Levant reveal local structural rotations due to their proximity to the Dead Sea transform fault, but these rotations do not affect the paleolatitudes, which range from -11 to $+6$ degrees for the Late Jurassic to Early Cretaceous, respectively (Table A8). Thus, these rocks were deposited slightly south of the equator in the Late Jurassic and near the equator in the Early Cretaceous. The constraints on the locations of the African equator provided by these Levant results are helpful in determining the best Late Jurassic paleopole location for Africa and, by inference, the other Atlantic-bordering continents.

Individual African late Middle to Late Jurassic paleopoles in North American coordinates show the full range of locations from west to east, but without a simple progressive temporal sequence (Figure 6.2d). The more westerly pole locations place the African equator to the south of the Levant, the more easterly locations have the African equator running to the north of the Levant (Figure 6.5). Thus, the Levant results suggest that the appropriate African reference paleopole (in North

American coordinates) is located at the eastern end of the APWP track. Because the age of the Late Jurassic Levant results is Kimmeridgian (152–156 Ma) and assuming that they are not based on younger (Cretaceous) remagnetizations (for which there is no evidence) we can now predict which paleopoles from Africa and the other continents are more representative of the Kimmeridgian paleofield.

The best way to illustrate the determination of the optimum location of the Kimmeridgian paleopole in North American coordinates is to plot the locus of possible paleopoles calculated on the basis of the observed average inclination in the Levant (Figure 6.6). This locus (heavy line) passes through the more easterly part of the population of paleopoles from the Atlantic-bordering continents. This population (from Figure 6.2) includes the European paleopoles rotated into North American coordinates with the parameters of Rowley & Lottes (1988); if instead the parameters from Bullard *et al.* (1965) are used, the conclusions do not change with respect to the optimum Kimmeridgian paleopole location, which falls near 66° N, 158° E in North American coordinates.

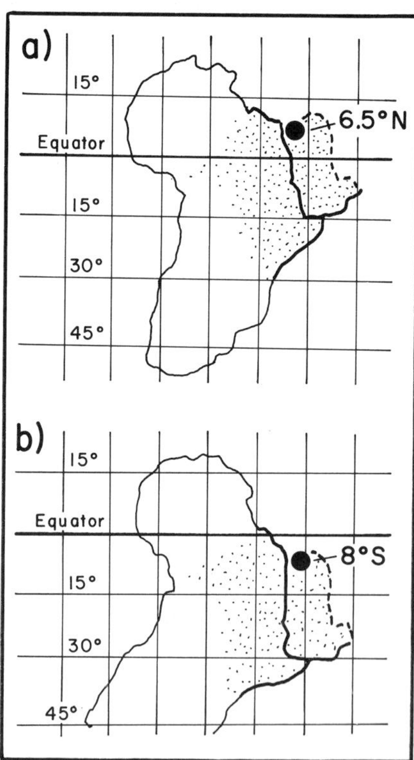

Figure 6.5. Kimmeridgian paleolatitude maps valid for the stippled area, that is, northeast Africa and the Arabian promontory (restored with respect to northeast Africa in order to close the Red Sea). a) Position calculated for a paleopole at 72° N, 102° E (North American coordinates; 62° N, 249° E in northeast African coordinates, or 68° N, 246° E in northwest African coordinates), which places the Levant area north of the equator. b) Position calculated for a paleopole at 66° N, 171° E (North American coordinates; 42° N, 258° E in northeast African coordinates, or 48° N, 255° E in northwest African coordinates), which places the Levant area south of the equator, in accordance with Kimmeridgian observations in Lebanon.

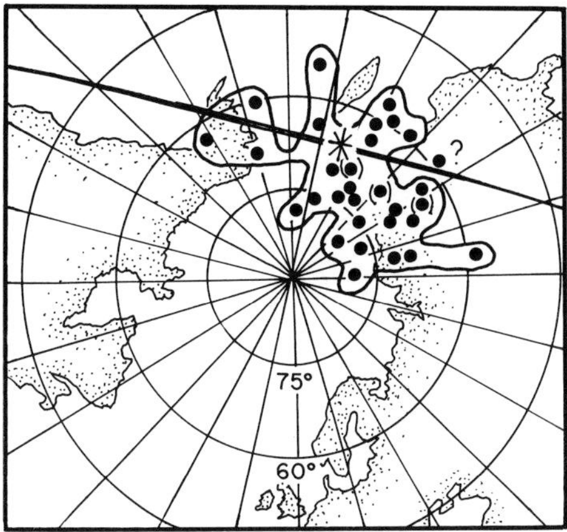

Figure 6.6. Locus of possible Kimmeridgian paleopole locations in North American coordinates, calculated from observed inclinations (−17°) in Lebanon, compared with Late Jurassic paleopoles from Figure 6.2. The European paleopoles have been rotated into North American coordinates with the parameters of Rowley & Lottes (1988). The question mark indicates the Corral Canyon paleopole (May *et al.*, 1986) which has been disputed by Van Fossen & Kent (1990). The asterisk marks the optimum location of the Kimmeridgian paleopole in North American coordinates, at 66°, 158° E.

Mean late Middle to Late Jurassic paleopoles

It is clear from the foregoing discussion that there is considerable scatter in the available data base, even after rotation into North American coordinates, but that the aggregate of data (including those from the Levant) provides a pattern. It is very likely that between the mean paleopoles for the Early Jurassic (at about 70° N, 90° E, see Chapter 5) and the Early Cretaceous (at 71° N, 190° E, see Table 6.1) the APWP path for North America passes through a late Middle Jurassic paleopole near 70° N, 120° E at about 170 Ma and a Kimmeridgian mean paleopole near 65° N, 160° E at about 154 Ma. These putative mean paleopoles are listed in Table 6.3 in the coordinates of each of the Atlantic-bordering continents.

There is no reason to assume that the Late Jurassic APWP backtracked onto itself. The scatter in the groupings of Late Jurassic paleopoles from each of the continents, as illustrated in Figure 6.2 is likely caused by inaccurate assignments of the ages of magnetization or imprecise paleopole determinations. The worst cases are for Europe and Africa (Figure 6.2b,d). The Beni Mellal poles from Morocco are perhaps younger than their age assignment (thought to be about 173 Ma). The limestone paleopoles from Europe (two sets of results, one without any fieldtests or reversals, marked by a * in Table A2 and shown in parentheses in Figures 6.2 and 6.6; the other set with fieldtests or reversals and marked by Q* in Table A2) match the overall APWP rather poorly, as they plot in locations that appear older than the well-dated rocks, i.e., near the Bathonian–Callovian end of the path. Alternatively, these poles are not too far removed from some Early Tertiary European paleopoles; it has been argued that they may represent pre-folding, but secondary, Tertiary remagnetizations (Halvorsen, 1989; see also Gehring & Heller, 1989). If the Atlantic Ocean is closed with the parameters of Bullard *et al.* (1965), then these paleopoles fall near the Late Triassic and Early Jurassic mean poles for North America in a position that is certainly 'too old'. Thus, one could conclude that all the European limestone paleopoles (including the results with Q* in Table A2) must be excluded

Table 6.3. *Mean reference paleopoles for the Atlantic-bordering continents*

Time interval	N	K	A_{95}	North America		Europe*		Europe**		Africa		South America	
				Lat.	Long.	Lat.	Long.	Lat.	Long.	Lat.	Long.	Lat.	Long.
Tl (37–66)	4	167	7.1	80,	162	81,	146			80,	200	82,	115
Ku (67–97)	4	91	9.7	73,	193	78,	190			69,	230	81,	173
Kl (98–144)	4	292	5.4	71,	190	76,	194			52,	258	85,	219
Ju, uJm (145–176)	4	219	6.2	72,	143	71,	145	73,	183	58,	249	86,	150
Kimmeridgian	Best estimate			65,	160	66,	168	66,	200	49,	249	78,	192
Bathonian	Best estimate			70,	120	66,	128	71,	159	63,	237	81,	97

The reference mean poles for the four different continents have been taken from the overall means of Table 6.1 and rotated into the different coordinates by the inverse rotations of Table 6.2. For explanation of age abbreviations, N, K and A_{95} see Table 5.1. For the Jurassic of Europe, two options are given, one (*) in which the rotation parameters of Rowley & Lottes (1988) are used, the other (**) in which the parameters of Bullard, Everett & Smith (1965) are used.

from consideration, so that only the three paleopoles remain that agree well with the North American results in the Bullard fit. The scenario of Figure 6.3 then begins to play a role only in the Cretaceous (Halvorsen, 1989; Srivastava *et al.*, 1990), if at all.

On the other hand, one could also conclude that the Europe–North America closure parameters of Rowley & Lottes (1988) are better already for the Late Jurassic, and by implication also the scenario of Figure 6.3. This implies that all pre-Cretaceous poles from Spitsbergen need to be rotated into North American coordinates with the parameters of Bullard *et al.*, whereas with respect to Stable Europe those for the Late Jurassic and older will need a differential rotation. For the moment, this ambiguity in the Late Jurassic limestone poles for Europe and in the North Atlantic paleogeography cannot be solved with the available data. Returning briefly to the uncertainties in the time scale for the Jurassic, noted earlier, it is worth emphasizing that alternatives to the DNAG (Palmer, 1983) time scale used in this book, such as the age assignments of Lowrie & Ogg (1985/86), do not resolve the European paleopole disagreements. Lowrie & Ogg, for instance, assign ages of 143 and 146 Ma to the upper and lower boundaries of the Kimmeridgian; these alternative ages would make the discrepancies between the Spitsbergen poles (about 150 Ma) and the Late Jurassic (e.g., Oxfordian–Kimmeridgian) limestone poles even larger. In the summary tables of Chapters 5–7, I have included both the options given above for the Late Jurassic reference pole for Europe.

Lastly, the higher-latitude Jurassic APWP track for North America, as advocated by Irving & Irving (1982) and Van Fossen & Kent (1990) is supported by the paleopoles from the other continents when rotated in North American coordinates. The track along the 60th parallel (e.g., Figure 5.7) suggested by May & Butler (1986) is not supported.

The Early Cretaceous (144–98 Ma)

Figure 6.2 includes the available Early Cretaceous paleopoles from the Atlantic-bordering continents as open symbols. The individual groupings show reasonably good clustering and the four continental means compare well, such that the overall mean pole in North American coordinates (71° N, 190° E) is accompanied by the highest statistical parameters for post-Early Jurassic time ($K = 292$ and $A_{95} = 5.4°$, see Table 6.1). It is noteworthy that the Early Cretaceous paleopoles from the Levant (Chapter 7) appear to agree well with this mean.

The main ambiguity in the Early Cretaceous analysis is, therefore, not paleomagnetic but tectonic, and relates to the plate boundaries within Africa. Already in Chapter 5 the case was made that the three different parts of Africa (northeast, northwest and southern Africa) should be fitted together for pre-Cretaceous times with small but significant rotations, as proposed by Pindell & Dewey (1982) and Lottes & Rowley (1990). The reason is that Africa, as it is constituted today, does not fit precisely against South America (Chapter 5; Figure 6.7). Thus, it has been proposed that the Benue Trough in Nigeria and its possible northward and eastward branching continuations formed briefly a set of plate boundaries within Africa. There is perhaps little disagreement about this, but there is considerable uncertainty about the timing and magnitude of successive movements necessary to explain the geological observations. These movements are of two contrasting types: (1) an earlier extensional phase, in which the Benue Trough widened and (2) a later (and lesser) compressional phase in which the previously formed rift valleys deformed and partially closed. The constraints are: (1) that the movements had ceased in Africa by the Santonian (about

84 Ma; Burke & Dewey, 1974) and that Africa's outline then looked like it does today, and (2) that Africa, before rifting started sometime in the Early Cretaceous (at about 130 Ma), fitted snugly against South America (Figure 6.7). Thus, we know the net total effect of the two phases of deformation, but not the magnitude or timing of each. This makes a comparison of Early and early Late Cretaceous paleopoles from West Gondwana continents and Laurussia somewhat uncertain; Early Cretaceous African paleopoles have been partially rotated, but Late Cretaceous paleopoles have not been rotated (see Table A4). Despite the uncertainties in this, the paleopoles for the Early Cretaceous interval agree well with each other.

In the earlier section on Late Jurassic results and their interpretation, it was noted that continental extension (and pre-ocean-spreading separation) between Europe and North America could have begun in the Late Jurassic, but that the paleomagnetic evidence was ambiguous. However, for the Early Cretaceous this is no longer the case; from paleomagnetic as well as marine geophysical evidence we can conclude that the scenario of Figure 6.3 applies to Early Cretaceous (if not earlier) times. The Early Cretaceous mean poles for Europe and North America are brought into better agreement by the rotation parameters of Rowley & Lottes (1988) than by the parameters of Bullard *et al.* (1965). Figure 6.8 shows reconstructions of North America and the Iberian Peninsula relative to a fixed Eurasian plate from Srivastava & colleagues (1990). These authors infer continental separation between Europe and North America to begin after Chron M0 (i.e., about 118 Ma, during the Aptian–Albian). No ocean floor is thought to have been generated in this separation phase, although the absence of marine magnetic anomalies during the Cretaceous Quiet Zone (118–84 Ma) hampers detailed analysis of plate kinematics in this interval.

Between Iberia and North America separation is proposed to have started earlier (after Chron M25 or Late Jurassic); as we will see in Chapter 7, the Iberian Jurassic and Cretaceous paleomagnetic data are insufficient to test this idea. What is clear from the model of Srivastava *et al.* (1990), however, is that the plate boundary or boundaries between Africa and Europe were located at various times to the north and/or to the south of the Iberian Peninsula, illustrating the microplate nature of Iberia in the Cretaceous and part of the Tertiary. I will return to this issue in Chapter

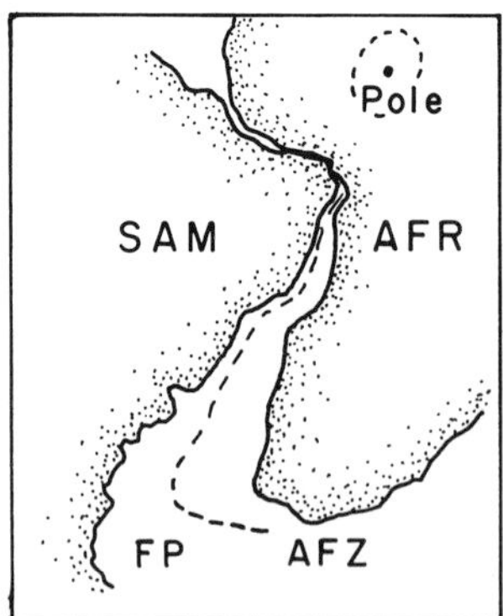

Figure 6.7. Mismatch between Africa (AFR) and South America (SAM) (solid lines, as presently constituted), which must be corrected by postulating plate boundaries within Africa, as discussed in the text of Chapters 5 and 6, so that Africa's outline for pre-Cretaceous times is represented by the dashed line. Figure redrawn after Pindell & Dewey (1982). The Euler pole of rotation has subsequently been modified by Lottes & Rowley (1990). FP = Falkland Plateau; AFZ = Aguilhas Fracture Zone.

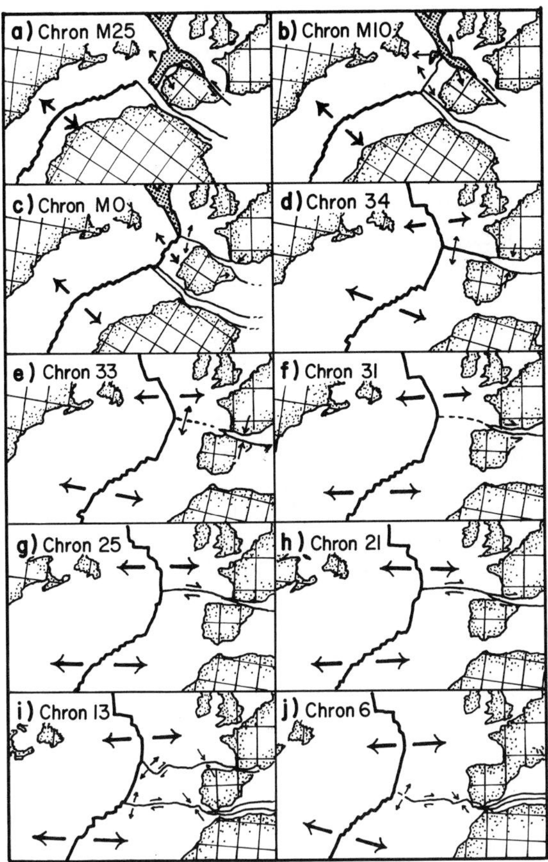

Figure 6.8. Reconstructions (from Srivastava *et al.*, 1990) of North America and the Iberian Peninsula relative to a fixed Eurasian plate. Motions along boundaries between plates from previous time are shown by small arrows, whereas large arrows show the direction of relative plate motion. Shaded areas are the regions of overlap between plates implying later (continental) extension in those regions. The times of the reconstructions are a) Kimmeridgian, about 157 Ma, b) latest Valanginian, 132 Ma, c) Aptian, 118 Ma, d) Campanian, 80 Ma, e) Early Maastrichtian, 74 Ma, f) middle Maastrichtian, 69 Ma, g) Late Paleocene, 59 Ma, h) Lutetian, about 50 Ma, i) earliest Oligocene, about 36 Ma, and j) Burdigalian, 20 Ma. Note that extension (and the beginning of separation) between the European and North American continents is inferred by Srivastava *et al.* (1990) to begin after the Early Aptian. Figure published with permission from the authors and Elsevier Science Publishers BV.

7, when discussing the Western Mediterranean and its complex plate boundaries which form the earlier divergence and later convergence zones between Africa and Europe. At this point, however, it may be useful to briefly mention the plate kinematic methods used to determine the relative motions between Africa and Europe.

When two plates converge, the resulting plate boundaries (e.g., subduction with or without some transform zones) do not allow the kinematic parameters to be determined directly. This is true for the interactions between the Pacific and Americas plates or the Indian–Asian plate boundaries, as well as for the Mediterranean. However, the nature of the passive Atlantic margins of Europe, Africa and North America, and the divergent plate boundaries along the Mid-Atlantic Ridge, allow the motions between Africa and Europe to be determined indirectly through vector subtraction of the Europe–North America and Africa–North America motions. This means that for the Mediterranean area, even for post-Jurassic times when Europe, North America and Africa each were separate plates, the relative motion regime between the major continents to the north and south is relatively well known.

For the Early and early Late Cretaceous, as can be seen in Figure 6.8 by comparing frames c and d, Africa rotated counterclockwise with respect to Europe, and the separation between the two (and between Africa and Iberia) in the western, but not the eastern, part of the Mediterranean increased slightly. As Africa rotated, so did Iberia although not by the same amount according to Srivastava *et al.* (1990). This subject will be discussed further in Chapter 7.

The Late Cretaceous (97–67 Ma)

The Late Cretaceous paleopole data set is the least well grouped of all four intervals discussed in this Chapter (Table 6.1), but even so the agreement is good compared to the Triassic or Paleozoic mean paleopoles discussed in Chapter 5, with $K = 91$ and $A_{95} = 9.7$ degrees based on four entries from each of the Atlantic-bordering continents. As before, rotations used to reconstruct the continents and their mean poles are listed in Table 6.2.

In the Late Cretaceous the North Atlantic spreading was well underway between Europe and North America (e.g., Figure 6.8d). Greenland remained close to Europe at first, with spreading, therefore, predominantly in the Labrador Sea. After the Early Oligocene North Atlantic spreading occurred primarily between Greenland and Europe (e.g., Rowley & Lottes, 1988; Olivet *et al.*, 1984), when the spreading in the Labrador Sea ceased.

In the Central and South Atlantic, spreading continued progressively as documented by Klitgord & Schouten (1986) and Cande *et al.* (1988) for the region between Iberia, Africa and North America (Figure 6.8). Africa and Iberia continued their rotation with respect to Eurasia with the greater part of this movement occurring in the earlier part of the Late Cretaceous.

The Early Tertiary (66–37 Ma)

The mean paleopoles for the four Atlantic-bordering continents are well grouped for this interval (Figure 6.9). Rocks formed during the extensive Paleocene volcanism in the British Province and in Greenland have yielded many individual paleopoles, while from western North America a fair number of paleopoles have been determined as well. Thus, the Early Tertiary data come primarily from the northern continents.

On the basis of younger, Neogene–Quaternary paleopoles, many analyses have attempted to assess the relative magnitudes of the dipole and non-dipole fields of the long-term geomagnetic field. This analysis can also be extended to earlier Tertiary or even Mesozoic time, as attempted by Coupland & Van der Voo (1980), Lee (1983) and Livermore, Vine & Smith (1984).

However, a quick check of the Early Tertiary mean poles of Figure 6.9 shows that the non-dipole pattern recognized for younger times is not a priori obvious. This pattern, first recognized by Wilson (1970a, 1971), would show up as far-sided paleopoles as viewed from the continent where the samples were collected, with respect to the 'global' mean for the time interval in question. This far-sidedness can

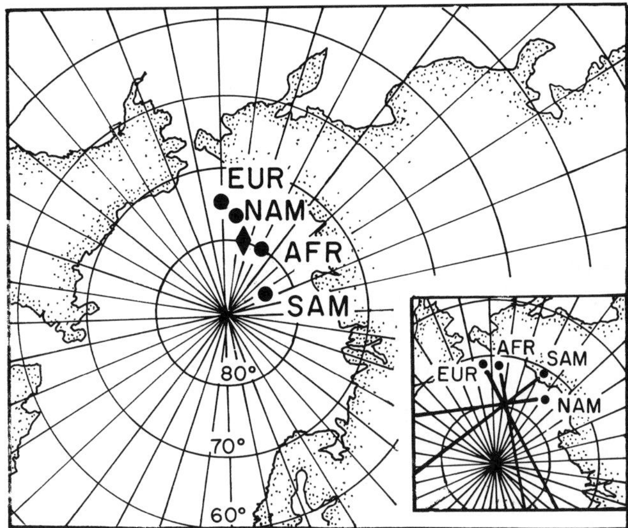

Figure 6.9. Mean Early Tertiary paleopoles for the four Atlantic-bordering continents in North American coordinates and their mean (diamond). The inset shows a hypothetical distribution of such mean poles in the case of a significant quadrupole field contribution to the total field. The observed situation does not resemble the hypothetical model. If there were any long-term quadrupole components in the total Early Tertiary field, then they are likely to have been insignificant for paleomagnetic purposes. EUR = Europe; NAM = North America; AFR = Africa; SAM = South America.

be explained by a significant quadrupole field, superimposed on the predominant dipole field; in terms of local geomagnetic field directions, the inclinations would appear to be systematically less positive/more negative for the whole world (this is for a positive quadrupole field; the inclinations would be less negative/more positive for a negative field, which is not observed). Recent analyses of paleomagnetic data for Pliocene–Quaternary times (e.g., Schneider & Kent, 1990a) have confirmed the presence of a quadrupole field, but with a magnitude (2.6 to 4.6% of the total field) that is not very large. While it is possible that the non-dipole field in pre-Pliocene times occasionally may have been larger, determinations of the quadrupole or octupole fields for earlier times are subject to very large uncertainties. The small magnitude of Pliocene–Quaternary non-dipole fields is reassuring for tectonic purposes (errors generally less than 4° according to Schneider & Kent, 1990a, p. 82).

The inset in Figure 6.9 illustrates the hypothetical effect on the paleopole distribution of such a quadrupole field superimposed on the dominant dipole field. A comparison between this theoretical distribution and the mean paleopoles actually observed shows that there is no indication of such a long-term non-dipole field influencing the data set. Inspection of the older mean poles (Table 6.1) shows that there is also no reason to infer the presence of a quadrupole field for the Early and Late Cretaceous results discussed earlier.

Summary

The Late Jurassic and younger paleopoles for the Atlantic-bordering continents illustrate that the paleomagnetic data generally agree well with the continental recon-

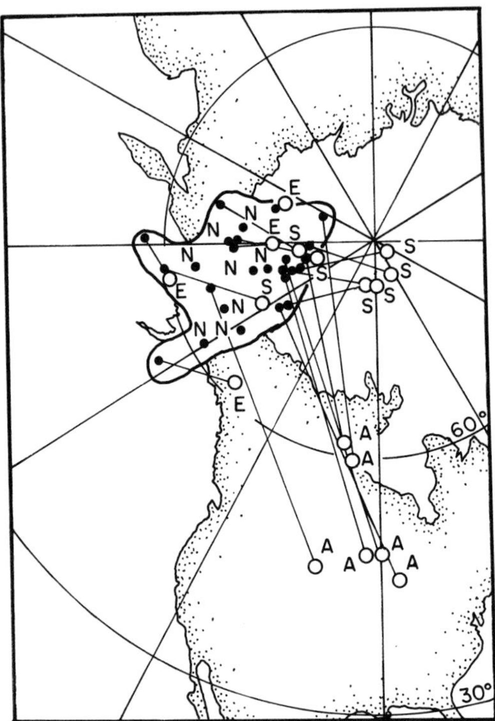

Figure 6.10. Early Cretaceous individual paleopoles in present-day coordinates (open circles) and after rotation into North American coordinates (solid circles), using the rotation parameters explained in the text and tables, as an example of a well-determined paleomagnetic dataset that confirms the continental reconstructions based on marine geophysical data. N = North American, E = European, S = South American, and A = African paleopoles.

structions inferred from marine geophysical data, and that the paleopoles show an impressive improvement in their clustering upon such reconstructions. As an example, individual Early Cretaceous paleopoles are shown in Figure 6.10 before and after rotation into North American coordinates. Together with the robust European and North American Phanerozoic data sets compared in Chapter 5 and some of the earlier Mesozoic paleopoles for the Gondwana continents (Figure 5.11), this forms one of the more convincing examples of paleomagnetism applied to global tectonics. In contrast, some of the Permian and Triassic uncertainties in the Pangea reconstructions form examples of data sets that are still poorly defined and offer challenges to future paleomagneticians.

You are never dedicated to something you have complete confidence in. No one is fanatically shouting that the sun is going to rise tomorrow. They *know* it is going to rise tomorrow. When people are fanatically dedicated to . . . dogmas or goals, it's always because these dogmas or goals are in doubt.

(Robert M. Pirsig (1984), Zen and the Art of Motorcycle Maintenance.*)*

This also seems true in paleomagnetism. When colleagues shout arguments to each other, it is almost always because they have – deep inside – some doubts. Controversies about Jurassic reference poles, displacements of Tethyan elements or polarity choices for a given result, all hinge on hidden doubts, uncertainties, concerns about quality (the other person's data, not one's own, of course!).

Time, however, has a way of letting these issues sort themselves out. A controversial issue attracts new researchers who tackle the old rocks with new vigor. I know. With a colleague, I once proposed that the Great Glen Fault in Scotland had a displacement of more than 2000 km in the Carboniferous. The data then available supported the idea, I shouted to my learned opponents. Subsequent work (more than 20 different publications) proved me wrong.

7 The Tethys blocks

We have seen in Chapter 5 that an eastward widening oceanic gap existed between Eurasia and the Gondwana continents in a Pangea configuration. This ocean is called the Tethys. In post-Triassic time, when the Gondwana continents dispersed and the Atlantic Ocean began opening, the Tethys Ocean gradually began closing. As the Gondwana continents collided with Eurasia in Tertiary times (Africa–Arabia against southern Europe, Turkey and Iran; India against Tibet; and Australia against Indonesia), the resulting orogenic belts formed some of the most impressive mountain ranges of the world, such as the Alps and the Himalayas.

The overall orogenic belt, marking the closure of the Tethys, is one of extreme complexity. Rather than a simple belt of continent–continent collision, it includes ancient microplates and displaced terranes, upthrusted oceanic remnants (ophiolites), successor ocean basins, and plutonic/volcanic complexes related to multiple subduction zones. On the northern side, this broad zone contains mountain belts ranging from the Pyrenees in the west through the Alps, Carpathians, the Caucasus, and on to the Tien Shan and Far Eastern mountain ranges in Asia (Figure 7.1). These belts form the deformed southern margin of the Eurasian continent. On the southern side, one finds the Betic Cordillera in Spain, the Riff, Tell and Atlas Mountains in northwest Africa, the thrust belt of Sicily and the mainland Italian Apennines, Greece, the Taurides in Turkey, the Zagros Belt in Iran, the Oman mountains, the Salt Range in Pakistan, the Himalayas, Indo-China's mountain ranges in Burma, Thailand and Malaya, the western Indonesian (the Sunda Block) and eastern Indonesian tectonic zones, and New Guinea. In part, these belts form the deformed northern margin of the Gondwana continents. In between, there are yet more mountain ranges in Yugoslavia, Hungary, Bulgaria, around North and South China, Korea and the Philippines (Figure 7.1). Some of these define the intervening blocks of the Tethys Ocean (e.g., Iberia, Corsica–Sardinia, the Moesian Block in northern Bulgaria, North and South China). To unravel the history of plate motions of the many Tethyan blocks and terranes during the past 300 million years in such a wide and very long geo-suture, is a daunting task.

Paleomagnetists have made their sample collections in this belt in great numbers, so there is an enormous amount of paleomagnetic information available, which will form the subject of this chapter. But even before the 1960s, when the mobility of the continents and the plate tectonic paradigm gradually became established, and when paleomagnetism in mobile belts became an increasingly popular research field, there were several publications that speculated on the mobility of the major continents around the Tethys and on the ancient locations of terranes now found in the mobile belts. As early as 1924, Argand discussed his ideas about the mobility of such crustal elements as the Iberian Peninsula (Spain and Portugal) and several other Mediterranean and Asian blocks in *La tectonique de l'Asie*. Paleontologists had long recognized that floras in northern Asia (called 'Angaran') differed from those in China ('Cathaysian') and from typical Gondwana assemblages, including such famous Permian fossil plants as *Glossopteris* (Figure 7.2, from Şengör *et al.*, 1988). They proposed, therefore, that these different floral assemblages were separated by ancient oceanic domains, collectively called the Tethys.

Carey (1958) was the first to present a detailed paleogeographical scenario based on his analysis of the basins and surrounding mountain ranges in the Mediterranean (Figure 7.3). He proposed that triangular deep basins (either oceanic or foundered and thinned continental crust) were formed by extensional rotation of one continental side with respect to the other about a nearby pivot, and called these deep basins 'sphenochasms'. Examples include: the Bay of Biscay, and the Ligurian, Balearic, and Tyrrhenian Seas in the western Mediterranean (see Figure 7.3). 'Rhombochasms' were similar basins with parallel sides, formed as gaps by lateral displacement of one side with respect to the other (pivot far away). Lastly, he proposed that now curved orogenic belts (e.g., the Alps) were 'oroclines' (*oros* = mountain, and *clino* = to tilt or turn), which were originally straight(er).

In retrospect, Carey's proposals are quite remarkable if one considers that at the time continental drift was ridiculed in many parts of the world, and that sea-floor spreading, and plate tectonics in general, had yet to be invented, let alone be documented.

In the last 20 years, especially, enormous progress has been made in furthering our understanding of this complex belt, and certain elements of Carey's hypotheses are now part of conventional wisdom, although other parts have been replaced

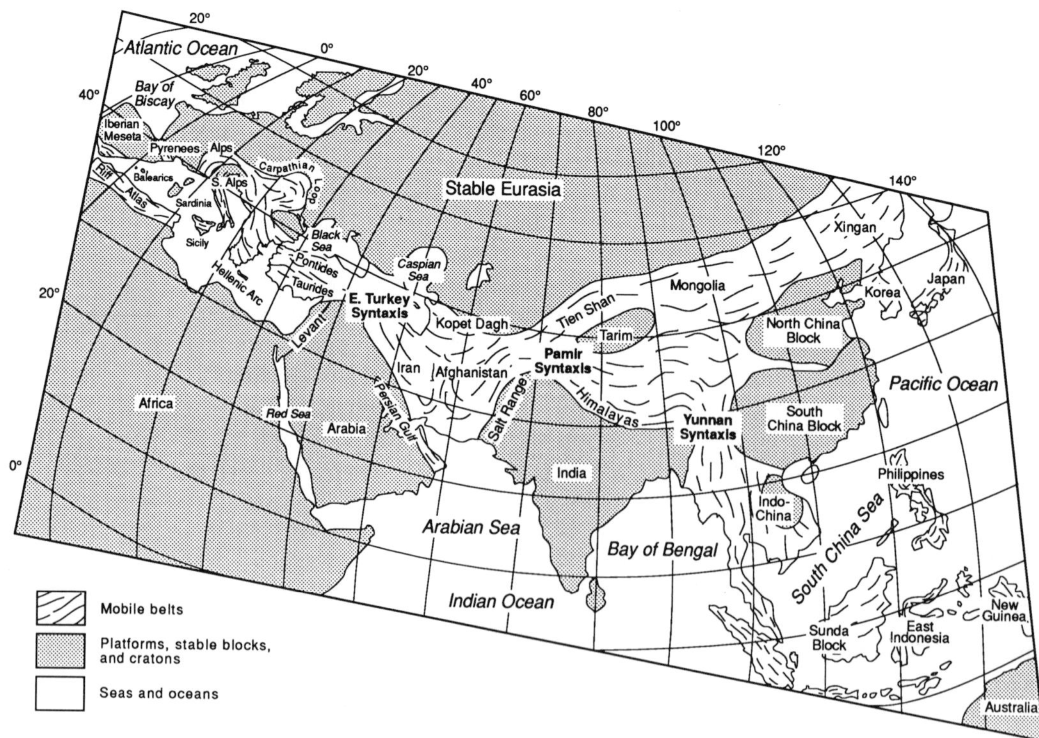

Figure 7.1. Location map of the Tethyan domain, from Spain to Indonesia, showing the blocks, mobile belts and syntaxes discussed in Chapter 7.

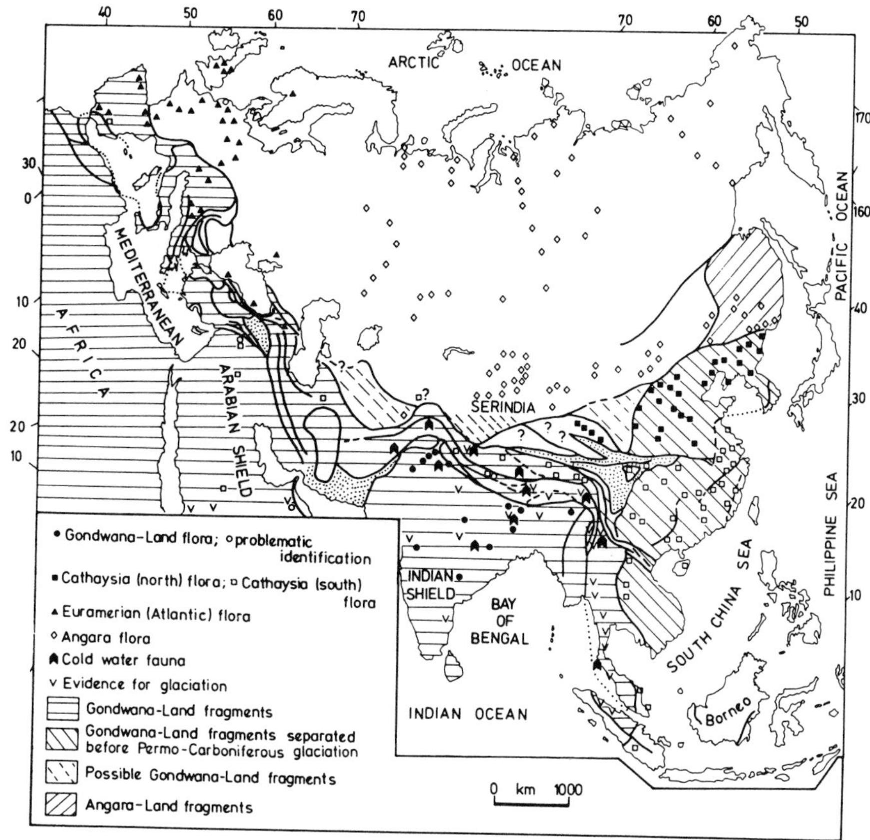

Figure 7.2. Distribution of Late Carboniferous to Early Permian floras, cold water faunas, and traces of glaciation in and around the Tethysides, from Şengör *et al.* (1988). Published with permission from the authors and the Geological Society Publishing House.

by plate tectonic aspects. Yet another major early idea, developed mostly from Mediterranean geology, deserves mention (although it too has been greatly modified by plate tectonic scenarios), namely the idea of 'geosynclines'. Geosyncline theory, after its introduction in the nineteenth century by Hall, was developed especially by French geologists (see Aubouin, 1965). The theory interprets the initial sedimentation patterns and facies, the fossil ridges and the basins that were recognized by them in the now highly deformed settings of the belt. In a typical transect through a mountain belt from one side to the other, the geosynclinal model holds that one encounters: (1) a continental platform and shelf with shallow-water sedimentation (the 'miogeosyncline', today called the miogeocline), (2) a transition zone to the basin with deeper-water sediments (the continental rise), (3) a deep basin with pelagic sediments overlying mafic extrusive and intrusive rocks ('eugeosyncline'), (4) a submerged ridge, sediment-starved, and consisting of mafic rocks, and – if the belt is symmetrical – (5) again a deep basin and (6) the continental edge of the other side with its rise and shelf. With our knowledge of more modern plate tectonic ideas, we can recognize in this the deep basins of the abyssal ocean floor, and the mid-ocean ridges.

Ophiolites, as originally proposed by Steinmann (1905), have long been known.

In geosynclinal theory, they were considered to be now-dismembered elements of the eugeosynclinal basins and ridges, consisting (if complete, as in the 'Steinmann trinity') of (1) pelagic sediments and radiolarian cherts, (2) pillow lavas (spilites) overlying mafic sheeted dikes and gabbro, and (3) ultramafic rocks, now often serpentinized. This is compatible with the plate tectonic explanation of ophiolitic rocks, although the displacements are now generally thought to have been much larger than foreseen by the geosyncline modelers. Figure 7.4 illustrates the Alpine belt of ophiolites in Switzerland that marks the remnants of a part of Tethys (from Hsü, 1989). Any student of geology who grasps the essentials of this map and its legend will also have achieved an understanding of the plate tectonic regime of the Alps.

Subsequent to this development of geosynclinal theory and the recognition of the significance of ophiolites, and with the advent of the plate tectonic paradigm, major and now-classical analyses of the geological evolution of the Tethyan domain were made by McKenzie (1972, 1978), Dewey *et al.* (1973), Laubscher & Bernouilli (1977), and Şengör (1979, 1987) among others. Many other articles, books and symposium volumes devoted to Tethyan geology fill entire shelves of any geology

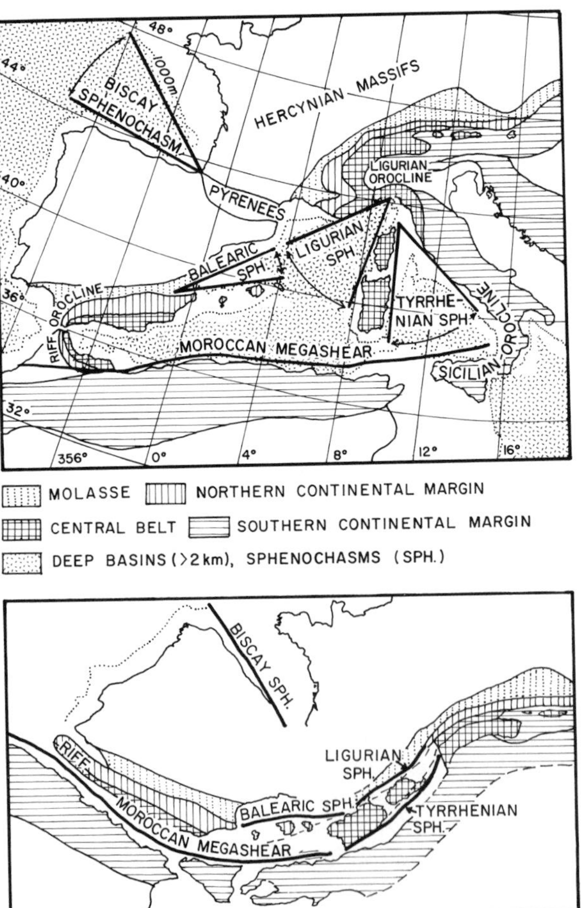

Figure 7.3. Tectonic development of the western Mediterranean between early Mesozoic times (bottom situation) and today, according to the model of Carey (1958). Figure modified and redrawn after Carey (1958) and reproduced with permission from the author.

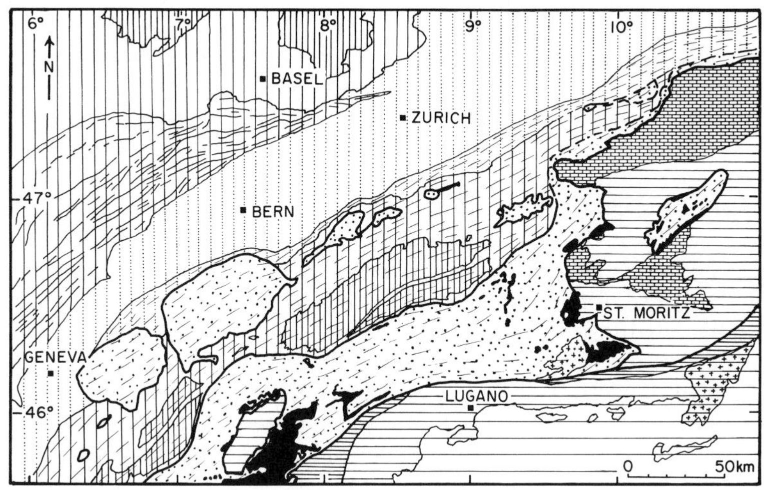

Figure 7.4. Tectonic map of the Swiss Alps, showing as main divisions the areas previously located on the southern shores of the Alpine Tethys, those on the northern side, and remnants of the ocean itself. Figure modified and redrawn after Hsü (1989). Published with permission from the author and the Geological Society Publishing House.

library in the world. The purpose of this chapter, however, is not to give a complete treatment of the geology of Alpine orogenic belts; that would clearly be impossible. Rather, it is to discuss the available paleomagnetic data and the use that can be made of them, as well as an evaluation of their limitations. Where relevant and appropriate, information from other geological disciplines will certainly be included, but it will not be the main focus.

The methodology selected for this chapter is similar to that used earlier: published paleopoles are compiled in the Appendix (Tables A7–A10), and the text will contain summary tables and figures based on these compilations. In contrast to the treatment of the major cratons, however, construction of APWPs for mobile belts is impractical, because of the rotations about a vertical axis (including oroclinal bending) that may occur there. Instead, I have selected to plot paleolatitudes, as calculated from the observed inclinations, and declinations, as a function of geological time. In the same diagrams, extrapolated paleolatitudes and expected declinations will be plotted for the area under discussion, as calculated from the mean reference paleopoles of neighboring continents. This way, relative north–south displacements can be recognized in the paleolatitude differences, if any, and relative rotations can be deduced from the declinations. Earlier, Westphal *et al.* (1986) used a similar approach, although their data base was not published and many data points in their figures were based on unidentified averages. In all such figures in this book, the declination and paleolatitude values are calculated from the paleopoles (from either the reference continent or the block itself) for the same single point, e.g., near Madrid at 40° N, 4° W, for Iberia, so that they can be directly compared. If the blocks are small, this hardly introduces any 'errors' due to extrapolation. If the blocks are sizeable, extrapolating from a location that underwent a large rotation may introduce a small error, as illustrated in Figure 7.5, but this will be generally only a few degrees. There is, moreover, no way in which this can be circumvented, unless one presupposes a certain configuration or assumes the amount of rotation. Because such knowledge is generally based on the very same paleomagnetic data that are being analyzed, it could lead to circular reasoning to make such corrections.

The declinations and paleolatitudes predicted from the reference paleopoles of a neighboring continent are, of course, only as 'good' as the reference poles themselves. For example, the Late Permian and Late Triassic paleopoles from Gondwana

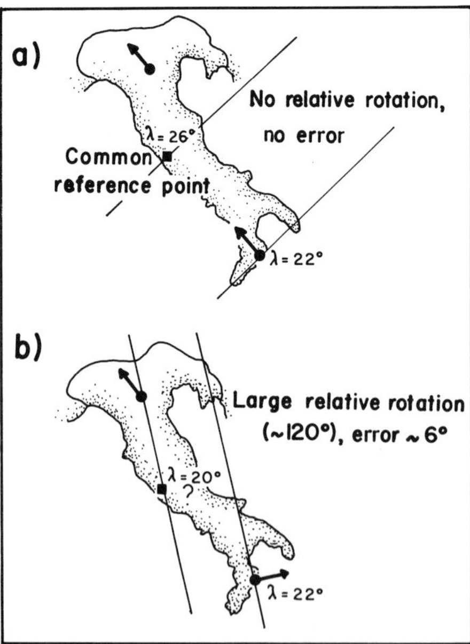

Figure 7.5. Illustration of the possibility that a small error may be introduced by block or thrust sheet rotations, when a paleolatitude is calculated from a paleopole for a common reference site. In the case shown, the reference site is Rome, and the (hypothetical) paleopoles are obtained from rocks in Calabria and the (unrotated) Southern Alps. a) Situation in which the observed Calabrian paleolatitude of 22° agrees with the one expected for Rome because the (hypothetical) declination in Calabria shows no rotation with respect to the rest of Italy. b) Situation in which the paleolatitude calculated for Rome shows a difference of about 6° with the one expected there (i.e., as given in a), because the declination in Calabria shows a large rotation with respect to the rest of Italy.

are, in my opinion, rather suspect and so, therefore, will be the predicted declinations and paleolatitudes based on these reference paleopoles. But there is no way in which this problem can be avoided unless one bases predictions on 'ideal' reference paleopoles that are *not* based on real data. The use of a global APWP derived from all continents, combined in reconstructions based on non-paleomagnetic information, is in this case not preferable, because globally averaged APWPs (1) still are contaminated by the erroneous contributions from a given single continent, (2) depend on the correctness of the reconstructions, which in some cases may be controversial, and (3) make the analysis less purely paleomagnetic. For these reasons, I am using only the reference paleopoles for single continental units (e.g., Stable Europe, West Gondwana, India and Australia as part of East Gondwana, Siberia, North or South China).

In this chapter, we will travel from the apex of the Tethyan domain in the Western Mediterranean towards the east, through the complicated Middle Eastern and Himalayan/Tibetan mountain chains, to end up in the Far East, when we will discuss the mountain belts surrounding the China Blocks, and in Indo-China, Korea, the Philippines and Indonesia.

Western Mediterranean

A more detailed map than that of the overview presented in Figure 7.1 is provided in Figure 7.6. The partly oceanic basins of the western Mediterranean are surrounded by mountain belts, almost all of Alpine (Tertiary) age, and these belts define several intervening continental blocks, such as Iberia (Spain and Portugal), the islands of Corsica and Sardinia, Sicily and the Italian Peninsula. All are separated by these belts from the cratonic margins of Stable (also called in this context 'extra-Alpine') Europe and Africa.

Paleomagnetism made much progress early on in this area, during the 1960s, because of a fortuitous circumstance: Dutch geology students from the University of Utrecht generally did their graduate research in Mediterranean areas (not only because of the nice climate, but also because of the interesting pre-Neogene structural and stratigraphical problems, an aspect rather lacking in the geology of the Netherlands!). Under the supervision of Profs Rutten and Van Bemmelen, these students were generally expected to include paleomagnetic sampling and analysis of suitable lithologies in their thesis work. Thus, a wide-ranging set of results, generally on Permian and Triassic rocks, came into existence rather coincidentally, without advance knowledge about their interesting implications (Van der Lingen, 1960; Van Hilten, 1960, 1962; Schwarz, 1963; De Boer, 1963, 1965; Guicherit, 1964; Van Dongen, 1967). Later augmented by work of the Dutch specifically directed at paleomagnetism (e.g., by De Jong, Klootwijk, VandenBerg, Van der Voo, Wonders & Zijderveld), as well as by many others (e.g., Alvarez, Channell, Edel, Freeman, Gregor, Heiniger, Heller, Horner, Lowrie, Manzoni, Napoleone, Parès, Perroud, Schott, Schult, Soffel, Storetvedt, Tarling, and Westphal; for references, see Table A7), a large data set has been accumulating that documents rotations, with respect to Stable Europe, of all these western Mediterranean blocks.

The first, nevertheless, to claim evidence for such rotations were Clegg *et al.* (1957), who studied the Triassic rocks of northern Spain near Villaviciosa and who obtained a present-day field direction (northerly with moderate to steep inclination). They interpreted the direction as Triassic, however, and compared it with the north-easterly Triassic directions obtained in Great Britain, with the result that they claimed a counter clockwise rotation of the Iberian Peninsula in agreement with the proposal of Carey (1958). We now know this idea of a rotation to be essentially correct, even though the interpretation of Clegg and his colleagues was based on erroneously interpreted data, because they mistook a recent remagnetization for an ancient, Triassic, direction. The rotations of, and paleomagnetic data from, the blocks will be discussed below individually.

Iberia

The first paleomagnetic results, after the paper by Clegg and his colleagues, were obtained by Van der Lingen (1960) and Schwarz (1963) in the Spanish Pyrenees of the province of Huesca, where their mapping areas included Permo-Triassic red beds and volcanics. The declinations were southeasterly (of reversed polarity) in contrast to the typical expected directions as extrapolated from Stable Europe, which would be south–southwesterly. Thus, their directions supported Carey's model of the rotation of the Iberian Peninsula, although it could not be precluded (especially not in hindsight) that the rotations were only local in nature, considering that Alpine thrusting of the Pyrenean areas over regionally extensive detachment surfaces has been pervasive. It became therefore important to collect rocks from the stable interior part of Iberia.

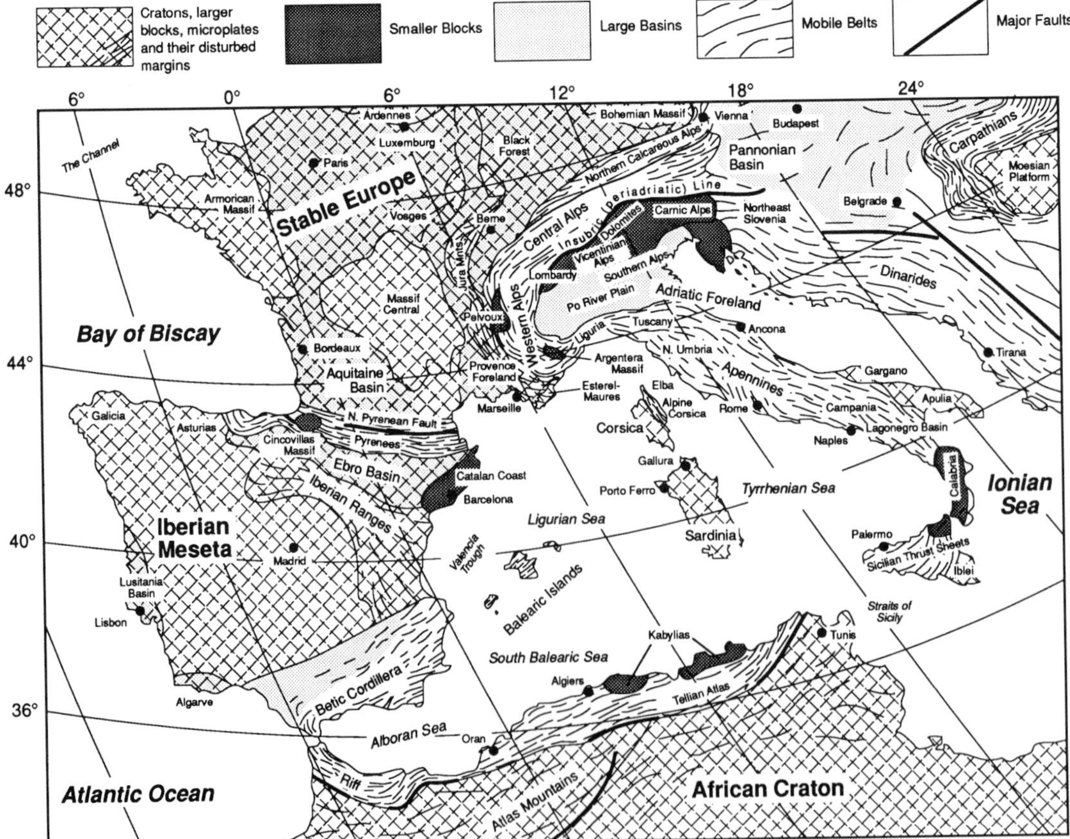

Figure 7.6. Tectonic map of the western Mediterranean crustal and tectonic elements discussed in the text.

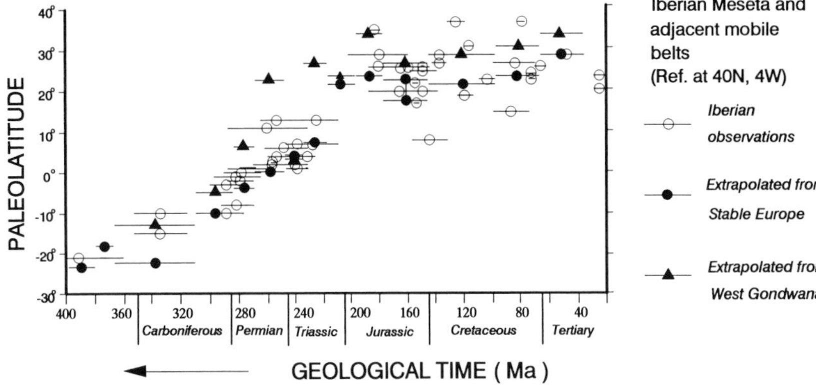

Figure 7.7. Plot of observed and predicted paleolatitudes as a function of geological time for results from Iberia, calculated for a common reference site at Madrid (40° N, 4° W). The predicted values on the basis of African and Stable European paleopoles (solid symbols) are given in Table 7.1 and the paleomagnetic observations (with $Q \geq 3$) from Iberian rocks, including those in the Pyrenees and Betic Cordillera, are listed in Table A7.

The geology of Spain and Portugal includes a large massif, called the Iberian Meseta, which was consolidated in the Hercynian Orogeny during the Late Carboniferous (Figure 7.6). To the northeast of this massif, the Iberian Ranges (also called Celtiberian or Hesperic), the Ebro Basin, and the Catalan Coastal Ranges are found, followed by the major Alpine mountain belt of the Pyrenees. To the south of the Meseta, the major belt of the Betic Cordillera is found, whereas in Portugal, the coastal Lusitanian Basin forms its western margin (Figure 7.6) as an on-shore Triassic–Jurassic rift basin related to the opening of the Atlantic between Iberia and Newfoundland (Sopeña *et al.*, 1988; Coward & Dietrich, 1989). The Pyrenees and Betic Cordillera contain many allochthonous thrust sheets, but what about the Iberian Ranges? Very recent interpretations of this belt includes detachment surfaces at depth (Guimera & Alvaro, 1990), so the area of the Iberian Ranges and regions to the northeast of it may not have retained coherence with the Iberian Meseta. I will return to this below.

The available paleomagnetic data (Table A7) are subdivided into three parts: Meseta and adjacent stable (?) and foreland areas, the Pyrenees and the Betic/Balearic Chain. The criterion of structural control for these results is generally only met by results from the Meseta and its western margin, with the other areas possibly rotated as already mentioned. Figures 7.7 and 7.8 give paleolatitudes and declinations derived from all appropriate results with the quality factor, Q, equal to or greater than three, as well as the declinations and paleolatitudes expected near Madrid (at 40° N, 4° W) if Iberia had remained rigidly attached to either Stable Europe or West Gondwana (i.e., Africa). These predicted values are extrapolated from the mean paleopoles of the major cratons, and are listed in Table 7.1. We know of course that Iberia did not remain rigidly attached to Stable Europe (nor to Africa), so it is the deviations between predictions and observations that we are interested in: these deviations are a measure (taking the paleomagnetic uncertainties into account) of the relative displacements or rotations of Iberia with respect to Stable Europe or Africa. It also must be noted that the predictions from Stable Europe and Africa would not be identical for the location of Madrid, even in the case of 'perfect' reference paleopoles, because these values are based on the (false) premise

Table 7.1. *Declinations and paleolatitudes for Iberia, at 40° N, 4° W, from European and West Gondwana reference poles*

Age interval	Europe		W. Gondwana	
	D	Lat.	D	Lat.
Tl (37–66)	360	28	359	33
Ku (67–97)	7	23	336	30
Kl (98–144)	354	21	318	28
Ju, uJm (145–176)	354	17	327	26
**Same	5	22		
lJm, Jl (177–195)	18	23	339	33
Tru/Jl (196–215)	27	21	340	23
Tru, uTrm (216–232)	26	7	334	26
Trl/m, Trl (233–245)	17	4	328	3
Pu (246–266)	11	0	333	22
Pl (267–281)	10	−4	310	6
Cu/Pl, Cu (282–308)	7	−10	322	−5
Cm, Cl, Du/Cl (309–365)	9	−22	312	−13
Du, Dm/Du (366–378)	25	−18	328	−41
Dm, Dl (379–397)	28	−23	359	−43

Age abbreviations as in Table 5.1. D is declination. The reference poles are taken from Tables 5.1 (post-Early Jurassic) and 5.7 (pre-Late Jurassic) for Stable Europe and from Table 5.8 for West Gondwana (West African coordinates). Declination and paleolatitude values are calculated for a common location in Iberia (40° N, 4° W), that is, by extrapolation.
**For the Late Jurassic there are two options for the European reference pole (see Table 5.1). This entry is based on the mean pole at 72° N, 162° E.

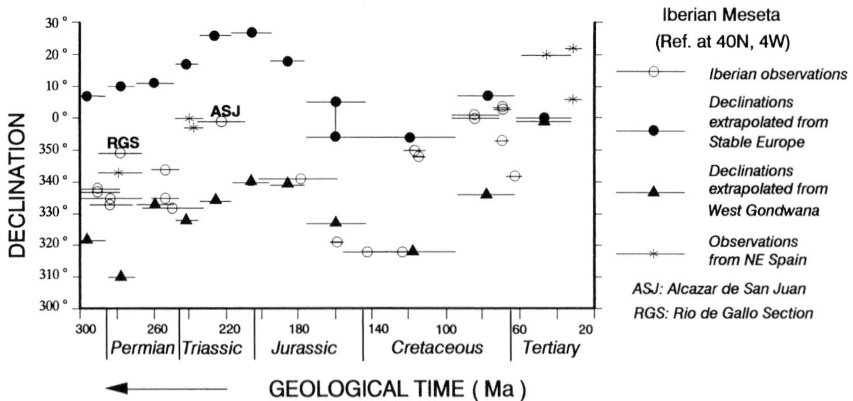

Figure 7.8. Plot of observed and predicted declinations as a function of geological time for results from Iberia, calculated for a common reference site at Madrid (40° N, 4° W). The predicted values on the basis of African and Stable European paleopoles (solid symbols) are given in Table 7.1 and the paleomagnetic observations (with $Q \geq 3$) from Iberian rocks (excluding those in the Pyrenees and Betic Cordillera, where rotations are likely) are listed in Table A7. Results from Catalunya (northeastern Spain) are marked by an asterisk. ASJ and RGS are results discussed in the text.

that Iberia remained rigidly attached to Africa *and* Stable Europe and, by implication, that Africa has not moved with respect to Europe since the Permian. Iberia was located in a more westerly position (in Pangea reconstructions) than today with respect to Europe and obviously Africa was also significantly displaced with respect to Iberia and Europe (e.g., Figures 5.26 and 6.8).

It can be seen in Figure 7.7 that the paleolatitudes (both observed and extrapolated) are in very good agreement. This is to be expected: in Pangea configurations, Iberia is solidly wedged at the western apex of the Tethys between Africa and Europe and, hence, paleolatitudes of the three should show agreement, because there has been no room for significant north–south (latitudinal) relative displacement. The paleolatitudes are plotted for the Iberian Meseta, as well as the Pyrenees and Betic-Balearic chains. They show a gradual shift from equatorial to about 30° N during the Permian to Paleogene interval. African (West Gondwana) predictions for Late Permian (266–246 Ma) and Late Triassic time (232–216 Ma) are on the high side, however, compared to those observed in Iberia, as well as those predicted from Stable Europe. This illustrates once more that these two mean poles from West Gondwana are probably flawed, as already discussed in Chapter 5, when I evaluated Pangea reconstructions.

Pre-Permian paleolatitudes are equatorial to subtropical in the southern hemisphere, and were probably near-south-polar in the Ordovician (not plotted in Figure 7.7). In Chapter 8, when discussing the Paleozoic history of the Iapetus Ocean, I will return to the pre-Permian results for Iberia and other central/southern European areas.

The declinations, as shown in Figure 7.8, are plotted only for the Iberian Meseta and adjacent areas, exclusive of those from the Pyrenees and Betic-Balearic belt. They are less well clustered than the inclinations but also follow a regular progression. First, I note that pre-Tertiary West Gondwana and European reference directions are rotated with respect to each other by about 40° or more, as is to be expected from Pangea configurations. In the Early Tertiary, this declination difference disappears. Second, most of the Iberian declinations follow the West Gondwana pattern, but with considerable scatter and a tendency to deviate somewhat less from those of Stable Europe than the reference directions from West Gondwana, indicating a relative rotation of only some 30°–35° with respect to Europe. Third, declinations from the Catalan Coastal Ranges (* in Figure 7.8) are closest to those of Stable Europe, with the difference being only about 20°-25° for the Permian and Triassic. In the Early Tertiary, the Catalan declinations deviate clockwise from those of Europe as well as West Gondwana, suggesting a separate Late Paleogene to Early Neogene clockwise rotation for this area (Parès, Banda & Santanach, 1988a) with respect to Europe as well as perhaps Iberia.

In detail, however, there are some complexities, which at this time cannot be completely resolved because of the scarcity of data. The oldest Catalan result (for Permian dikes) reveals a declination of 343°, whereas the typical Iberian Meseta declinations for the Permian are 332°–338°. However, another Permian result from the Iberian Ranges (Turner *et al.*, 1989; Rio de Gallo Section, marked RGS in Figure 7.8) reveals a declination of 349°. We can only speculate about this apparent discrepancy; either the Rio de Gallo Section rotated (together with the Catalan area) with respect to Iberia, or the differences in declination are fortuitous and there have been no rotations inside the Iberian Peninsula. For Triassic results, a similar situation is found for the Catalunya red beds and the Alcázar de San Juan red beds from the Meseta (marked ASJ in Figure 7.8; the rocks are now dated as Late Triassic (Sopeña *et al.*, 1983) which is more precise than could be stated in the original paleomagnetic study). The declination of the latter result is 359°, whereas those from the Catalan Coastal Ranges are very similar, namely 357° and 360°. However, there is a small but significant age difference, precisely at the time that the European reference declinations are increasing from 15° to nearly 30° (Figure 7.8). This increase is related to the cusp at about 200 Ma seen so clearly in the North American APWP at 60° N, 60° E (Gordon *et al.*, 1984; Ekstrand & Butler, 1989, see Chapter 5; Figure 5.7). Calculating the expected declination in Iberia for the sharp endpoint of the cusp yields a predicted European reference value of about 34 degrees. The newly determined age of the Alcázar de San Juan red beds (Late Triassic) may well correspond to the age of this cusp, in which case its declination (359°) would correspond to a rotation of precisely 35 degrees. In that case, there is no longer any disagreement between the declinations from Catalunya and the Meseta, provided that the age difference is taken into account and a correction is made for a small clockwise rotation of the Catalan Coastal Ranges with respect to the Meseta, as argued above.

Early Cretaceous declinations show a rather large spread, suggesting either relative rotation between the sites (in Portugal, the Cantabrian north coast of Spain, and the Iberian Ranges), or flaws in the dataset, or an extremely rapid (and somewhat early!) rotation of Iberia. Without further studies, this cannot be resolved; there are, for instance, several suggestions of remagnetization in the collection from Portugal which includes two of the four results plotted in Figure 7.8 (Galdeano *et al.*, 1989). What is clear, however, is that in the Late Cretaceous the Iberian declinations begin to resemble those from Stable Europe, indicating that by that time the rotation of Iberia and the related opening of the Bay of Biscay were nearing completion. The timing of this rotation agrees with the ages (120–80 Ma) of marine magnetic anomalies of the Bay of Biscay (C. A. Williams, 1975; Srivastava *et al.*, 1990). When this was first suggested as the age of the rotation of Iberia, however, it posed a dilemma (Van der Voo, 1969). Carey's original model suggested that the opening of the Bay of Biscay and the peak of the Pyrenean Orogeny in the Eocene were of the same age and were due to the rotation of Iberia about a pivot in the western Pyrenees. However, if the rotation is Late Cretaceous, this idea is not supported. To resolve this, Le Pichon & Sibuet (1971) suggested a two-phase model in which (1) Iberia rotated during the Cretaceous about a pivot near Paris, such that the motion between Iberia and Stable Europe was then of pure strike – slip nature along a small circle passing through the Aquitaine Basin in southwest France and through the eastern half of the Pyrenees, followed by (2) a later north–south convergence in the Eocene between Africa and Stable Europe, which pushed Iberia northward causing the Pyrenean compression. Through the use of Late Permian paleomagnetic data from the western Pyrenees (Figure 7.9) this model was later modified; Van der Voo & Boessenkool (1973) found 'European' declinations in and near the Cincovillas Massif, which indicated that the area belonged to the European side of the suture between Iberia and Stable Europe. Thus, this suture could not run through the Aquitaine Basin, but stepped from the major North Pyrenean Fault (Figure 7.9) in en-echelon fashion to the south-side of the Pyrenees. They postulated that this east–west oriented suture was the (deformed) result of a transtensional gap caused by the rotation of Iberia about a pole location such as that of LePichon & Sibuet. The thick Late Cretaceous–Paleogene sedimentary series (e.g., Puigdefabregas & Souquet, 1986), for example 5 km of Eocene flysch in Huesca (Ten Haaf, Van der Voo & Wensink, 1971), would in this model form the basin-fill of the transtensional grabens, in partial accordance with a model proposed earlier by Mattauer & Seguret (1971). The Eocene or later deformation in the Pyrenees is therefore unrelated to the main rotation phase of Iberia.

The pivot about which Iberia rotated actually remains unknown. While Carey's

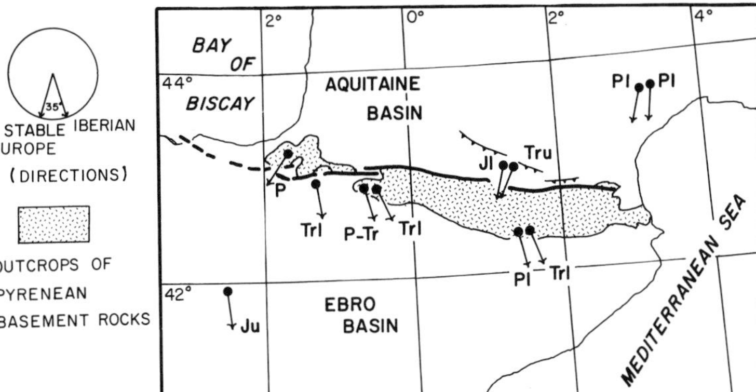

Figure 7.9. Schematic tectonic map of the Pyrenees and adjacent areas, showing Permian, Triassic and Jurassic reversed-polarity paleomagnetic declinations (arrows). Directions representative of Stable Europe are south–southwesterly, those from the rotated Iberian block are south–southeasterly. A major Europe–Iberia fossil plate boundary can be inferred (heavy line) on the basis of these results. Figure redrawn after Van der Voo & Boessenkool (1973).

preferred location in the Pyrenees is unlikely, any Euler pole farther removed from Iberia towards northern or central Europe (by about 5° to 10°) would satisfy all constraints as long as the total post-Triassic rotation amount is of the order of 35 degrees. Perroud (1982) has attempted to determine the best location for this pivot, but found that the available paleomagnetic data were too sparse to make this attempt precise enough. He also speculated that the movements of Iberia with respect to Stable Europe may have progressed in three separate phases: an early rotation during the Jurassic to Early Cretaceous, a later rotation during the Late Cretaceous, and finally, an Early Tertiary northward convergence. If the rotation proceeded in different phases and with different pivots each time, it will be very difficult to calculate their optimum locations from the available data. However, from the declination plot of Figure 7.8 there is no unambiguous indication of more than one phase of rotation. Moreover, the kinematics of Srivastava *et al.* (1990) as displayed in Figure 6.8 do not show more than one significant phase of rotation, which took place in the interval of 120–80 Ma.

The Pyrenean deformation continues westward through eastern Cantabria and onward off-shore along the Asturian–Galician Coast, further indicating that the Pyrenean compression could not be related to the opening of the Bay of Biscay. An interesting but unresolved question is to what extent the Pyrenean convergence between the north coast of Spain and the Bay of Biscay led to subduction or underthrusting of Biscay ocean floor. A further possibility, albeit entirely speculative, is that during the opening of the Bay of Biscay, Iberia, actually rotated (counter clockwise) by more than about 35°, with the subsequent convergence being accompanied by a lesser, but now clockwise, rotation which partially undid the earlier counterclockwise one. Puzzling results obtained by Storetvedt *et al.* (1987) on Portuguese Late Cretaceous and Early Tertiary rocks, with northeasterly deflected declinations (Table A7) could perhaps be explained in this way. However, the kinematic (non-paleomagnetic) scenario for the Iberian movements (Srivastava *et al.*, 1990) does not allow the Pyrenean convergence to involve a clockwise component of rotation (Figure 6.8).

Storetvedt (1990) and Storetvedt *et al.* (1990) have amplified on this theme by arguing for rapid and large clockwise as well as counterclockwise rotations of all of

Iberia during the 68–90 Ma interval. However, they have – at least to my taste – taken rather extreme liberties with the available data; the questionable northwesterly and shallow directions in the Burgau Intrusion (Table A7), which are obtained from only one hand sample(!), and similar directions interpreted as overprints in some samples of the Lisbon Volcanics, are given much weight, whereas all Permian and Triassic directions (also northwesterly and shallow, i.e., as normal polarity) are labeled as Late Cretaceous remagnetizations. In their treatment, it is completely overlooked that several Permian and Triassic results are based on positive fold tests, with the folding of at least one result (the Buçaco Red Beds; Van der Voo, 1969) constrained in age to the Permian (Saalian folding) by an angular unconformity with overlying Triassic strata (Figure 4.20).

At this point, it could be questioned whether Iberia was actually a microplate at all, despite the appeal of that hypothesis. By this I mean that its motions may have been alternatingly coupled with those of Europe or of Africa, such that at no time was there an entirely independent movement of Iberia with respect to either one (Schouten, Srivastava & Klitgord, 1984). From the outset, however, it must be clear that we should regard Iberia as neither a permanent part of Africa nor of Europe. With respect to coupling with Africa, there simply is not enough room to the east of Newfoundland to rotate Iberia into a Pangea fit with the same parameters as those of Africa. So, although the Iberian declinations and paleolatitudes resemble those of Africa, there must have been a relative paleo-east to west displacement (see, for instance, Figure 6.8). With respect to permanent coupling with Europe, the declinations clearly mandate a significant relative rotation. On the other hand, the hypothesis that Iberia first moved with Africa, then with Europe, perhaps again with Africa and again with Europe until today, remains a possibility that needs further research. Recent kinematic analyses on the basis of seafloor magnetic anomalies, sedimentary basin evolution, and structural geology, have generally favored the idea that Iberia at one time or another was an independent microplate (e.g., Malod, 1989; Srivastava *et al.*, 1990).

The rotations internal to the Pyrenees (see declination column in Table A7) have generally not been very large, in contrast to the Betic Cordillera where large rotations, both clockwise and counterclockwise, have been observed. The Balearic continuation of the Betic Chain is marked by north–northeasterly declinations, which have been interpreted as showing a clockwise rotation of the islands of about 15–20° with respect to the Catalan Coast (Parès, Banda & Santanach, 1988a; Freeman *et al.*, 1989). This is approximately the amount needed to open by crustal extension the Valencia Trough between the Balearic islands and the coast (the Balearic sphenochasm of Figure 7.3). The rotation and this extension are thought to have occurred in Miocene time, at the same time or just after the opening of the Ligurian Sea between France–Iberia and Corsica–Sardinia. However, an alternative explanation exists for the paleomagnetic declination deviations in terms of rotations caused by thin-skinned crustal detachments. The paleolatitudes (and, hence, the inclinations) of all these rotated areas agree very well with those from Iberia and the European and African cratons; thus, significant north–south displacements are not indicated by the paleomagnetic data.

Corsica–Sardinia

The paleomagnetic results are listed in Table A7 and the paleolatitudes and declinations are shown in Figures 7.10 and 7.11, treated in identical fashion as those from Iberia. The results are very abundant for the Tertiary of Sardinia and several

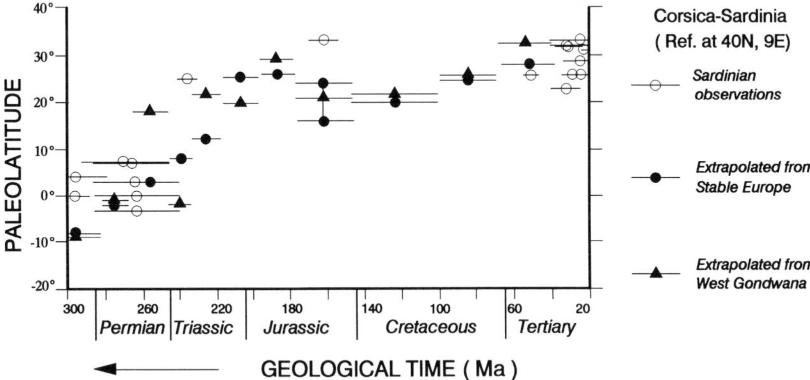

Figure 7.10. Plot of observed and predicted paleolatitudes as a function of geological time for Corsica and Sardinia, calculated for a common reference site at 40° N, 9° E. Predicted values have been calculated from the African and Stable European paleopoles (solid symbols), whereas the paleomagnetic observations (with $Q \geq 3$) from Corsican and Sardinian rocks are taken from Table A7.

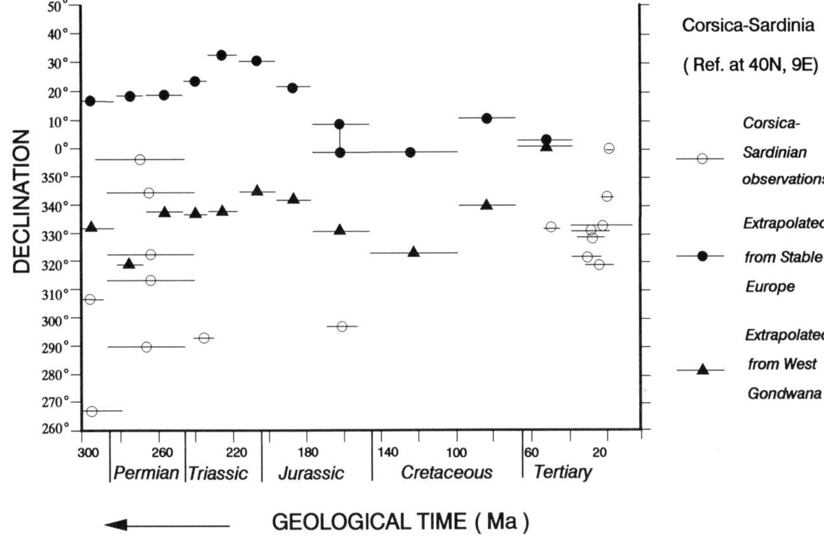

Figure 7.11. Plot of observed and predicted declinations as a function of geological time for results from Corsica and Sardinia, calculated for a common reference site at 40° N, 9° E. For other explanations see Figures 7.7 and 7.10.

Permian directions, of varying reliability, augment this set. In addition, there is one Triassic and one Jurassic result from Sardinia obtained by Horner & Lowrie (1981).

The paleolatitudes (Figure 7.10) reveal good agreement, given the typical resolution of paleomagnetic data, although the Mesozoic results suggest paleolatitudes that are a bit too far north with respect to the reference values of either Stable Europe or West Gondwana. The declinations, on the other hand, show wide scatter, although all deviate counterclockwise from the declinations predicted from Stable Europe. It is not completely clear what causes the scatter, but at the very least relative rotations internal to the islands should be examined. Sardinia and the major part of Corsica (except for its northeastern Alpine overthrust belt, see Figure 7.6) consist of basement, consolidated during the late Paleozoic Hercynian Orogeny. Although normal faults associated with Tertiary rifting and extensional volcanism are pervasive, these can hardly have caused the large rotations that would be indicated by the Permian declinations which vary by up to 95°. However, it is possible that the Hercynian orogenic deformation continued into the Permian, as also suggested for other parts of Europe and for eastern North America (e.g., Sacks & Secor, 1990).

Until very recently, most paleomagneticians considered the data sets of Sardinia and Corsica separately, because the results appeared to indicate different amounts of rotation (e.g., Westphal, Orsini & Vellutini, 1976), even though the Corsican data were not of very high reliability (see also Table A7). On the other hand, Arthaud & Matte (1977) have presented convincing geological evidence that northern Sardinia and southern Corsica shared the same orientation of several features which argued against relative rotations. This has finally been corroborated paleomagnetically in a study of Permian dikes (Vigliotti, Alvarez & McWilliams, 1990). The declinations on both islands are statistically identical (about 314° for normal polarity), and hence the two islands must have rotated the same amount. This does not resolve the discrepancies with the Permian declinations of 267°–307° from the Barbagia, Nurra and Porto Ferro areas (Zijderveld, De Jong & Van der Voo, 1970b; Edel, Montigny & Thuizat, 1981a), nor with the Permian declinations of 357° from Osani (Corsica), as obtained by Nairn & Westphal (1968). It cannot be precluded either that these directions are possibly based on paleomagnetic results contaminated by later overprints. The one deviating Permian declination obtained from Gallura by Storetvedt & Markhus (1978) is not supported by other results from the same location (Table A7).

If we were to take the combined dike declination as the best Permian direction (Vigliotti *et al.*, 1990), a relative ccw rotation of about 60° is indicated with respect to Stable Europe, whereas with respect to West Gondwana the rotation is small and of the order of 5°–20° ccw. However, considering the Triassic and Jurassic directions (Horner & Lowrie, 1981) it appears that the rotations were larger ccw, and the same is true for the Early Tertiary directions. In conclusion, the Permian reference direction should perhaps be an average of all the values between 265° and 325°, for an approximate average of about 305°. In that case, a relative ccw rotation of about 70° is indicated with respect to Stable Europe, whereas with respect to West Gondwana the rotation is about 30° ccw.

The rotation of 70° of Corsica–Sardinia with respect to Europe may have occurred in two phases, one involving a rotation of 35° ccw during the Cretaceous, presumably together with the rotation of Iberia, the other during the Miocene and amounting also to about 35° ccw. The latter is very precisely dated (Bellon, Coulon & Edel, 1977; Montigny, Edel & Thuizat, 1981) as between 21 and 15 Ma, during the Early to earliest Middle Miocene, and involved the opening of the Ligurian Sea (Figures 7.3 and 7.6) between France–Iberia and the islands. Since this rotation is synchronous with or postdates the main compressional phases of the Betic, Pyrenean and Alpine orogenies, constraints are provided on the paleolocation of Corsica and Sardinia, which are affected only by Alpine deformation in the NE corner of Corsica. Thus the islands must have been located to the north of the Betic–Balearic belt, to the south of the Pyrenean belt (if any existed there), and at the northwestern edge of Alpine deformation. The best Early Tertiary fit is, therefore, offshore from western Provence and against the northern Catalan coasts of France and Spain. Permian volcanics on Corsica are very similar to the ones in Estérel and the Maures Massif along the southeast French coast, suggesting ancient proximity to some workers (e.g., Westphal *et al.*, 1976), but this may be fortuitous in view of the earlier (Cretaceous) rotation.

The earlier rotation phase, if true, suggests that the main Cretaceous plate boundary between Iberia–Corsica–Sardinia and Europe continued eastward from the North Pyrenean fault towards the coast of Provence. The Miocene rotation, on the other hand, demonstrates that the islands moved briefly as an independent microplate, because they clearly were displaced with respect to all neighboring blocks and continents. At the end of this section, after a discussion of the Italian results, I will return to these matters.

Finally, it has been suggested (Alvarez *et al.*, 1974) that the Calabrian displaced terrane of southern Italy and the Kabylie allochthonous massifs of northern Africa (Figure 7.6) were located immediately adjacent to Sardinia and the Balearic islands before they were transported in Tertiary times across the Tyrrhenian and Algerian Seas to the southeast. This aspect will also be discussed further below.

Italy and Adjacent Areas

In Italy, a number of areas along the Adriatic coast of the peninsula, in the Southern Alps, and in southeast Sicily are thought to be autochthonous (see Figure 7.6), i.e., not underlain by major detachment zones, in contrast to the areas in central zones of the Alps and in the Apennines and northwest Sicily. In Table A7 of the Appendix, I have called these areas 'the Southern Alps and Autochthonous Italy and Sicily'. By that I do not a priori mean to imply that they formed necessarily one rigid block; the different autochthonous areas may well have moved with respect to each other, and one issue that will be examined in this subsection is whether paleomagnetic data support such movements (we shall see that there are hardly any).

In addition, the areas in the Alps near the French/Italian border, in southern Switzerland, Austria and Italy north of the major Insubric Fault (also called, variously, Peri-Adriatic, Tonale, Canavese, Giudicaria, Pustertal or Gailtal Line, etc.), and in the Sicilian and Apennine fold and thrust belts can be analyzed with autochthonous Italy in terms of their paleolatitudes, but not their declinations, because rotations are very likely to have occurred in all these mobile belts.

The paleolatitudes and declinations for Italian paleopoles are listed in Table A7 of the Appendix and are plotted in Figures 7.12 and 7.13, together with the reference values extrapolated from Stable Europe and West Gondwana (see Table 7.2). The paleolatitudes form a dense network that shows the normal paleomagnetic scatter (about 10° bandwidth) and generally good agreement between reference poles and observations (Figure 7.12). Paleolatitudes changed rapidly from equatorial in the Permian to about 30° N in the Early Jurassic and varied between 15° and 30° N since then until the Neogene. There is one anomaly, that of a Triassic result from the Verrucano from the Tuscany Apennines (at 40° N) that suggests that this paleopole may not be very reliable. Referring back to the Tethys Twist model (Van Hilten, 1964; De Boer, 1965; Irving, 1967) discussed in Chapter 5 when evaluating Pangea configurations, there are no longer strong suggestions that the Italian Permo-Triassic paleolatitudes indicate an (impossible) overlap between Italy and Stable Europe. In the Jurassic and, especially, the Late Cretaceous the observed Italian paleolatitudes are a bit farther south than those extrapolated from Europe, possibly suggesting some separation between the two before Alpine north–south convergence. The abundant Late Cretaceous and Early Tertiary data from all of Italy indicate also, on average (but significantly), a more southerly paleolatitude than would be expected from extrapolation of the West Gondwana reference paleopoles.

This undoubtedly indicates that these reference values are imprecise by about 10°.

The Italian declinations (Figure 7.13) also show a reasonably well-grouped and regular progression from about 320° in the Early Permian to about 345° in the Triassic to Early Jurassic, then back again to about 320° in the Early Cretaceous. During the Late Cretaceous and Early Tertiary, the data indicate a rotation to the present-day situation with northerly declinations. The declinations clearly track the West Gondwana (i.e., African) reference curve. With respect to Stable Europe, a rotation of about 45° to 50° ccw is indicated. It should be kept in mind that these reference poles are often determined with a precision no better than 5° to 15° in declination (Chapter 5, see also Heller, Lowrie & Hirt, 1989).

At first glance, there are no indications that any of the 'autochthonous' areas used for the declination plot (Iblei in Sicily, the Gargano Peninsula in Apulia, the

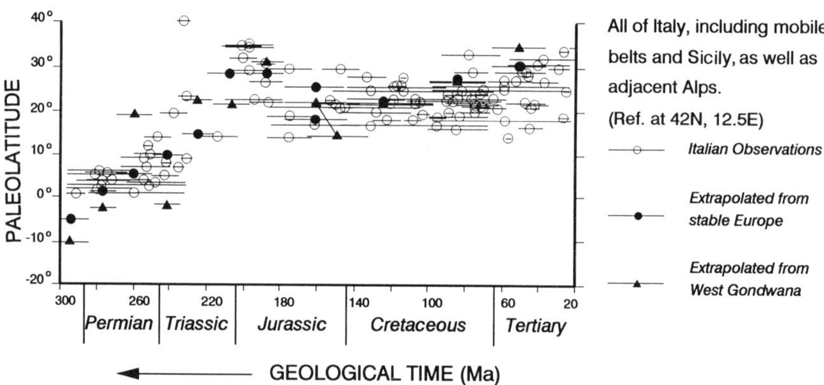

Figure 7.12. Plot of observed and predicted paleolatitudes as a function of geological time for results from Italy, calculated for a common reference site near Rome (42° N, 12.5° E). The predicted values on the basis of African and Stable European paleopoles (solid symbols) are given in Table 7.2 and the paleomagnetic observations (with $Q \geq 3$) from Italian rocks, including those in thrust sheets and rotated blocks, are listed in Table A7. For the Late Jurassic, different reference paleolatitude options (as explained in Chapter 6) are connected by a line.

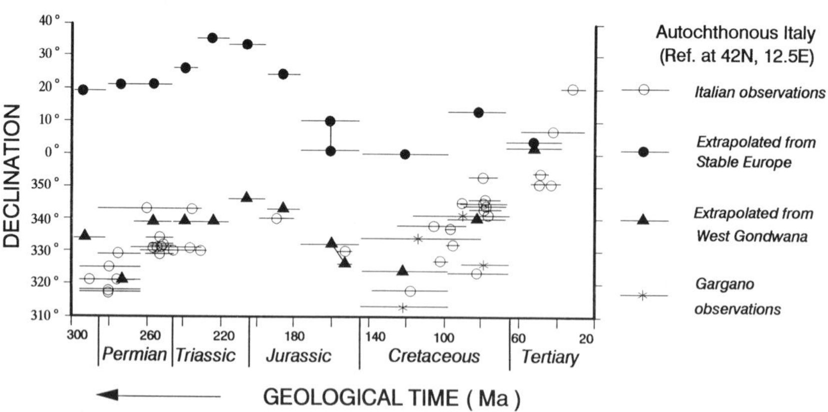

Figure 7.13. Plot of observed and predicted declinations as a function of geological time for results from the areas of Italy that can be considered autochthonous (i.e., coherent with basement), calculated for a common reference site at 42° N, 12.5° E. For other explanations see Figures 7.8 and 7.12.

Table 7.2. *Declinations and paleolatitudes for Italy, at 42° N, 12.5° E from European and West Gondwana reference poles*

Age interval	Europe		W. Gondwana	
	D	Lat.	D	Lat.
Tl (37–66)	4	30	2	35
Ku (67–97)	13	27	340	27
Kl (98–144)	0	22	324	22
Ju, uJm (145–176)	1	18	332	22
**Same	10	26		
*Kimmeridgian best Estimate			326	14
lJm, Jl (177–195)	24	29	343	31
Tru/Jl (196–215)	33	29	346	21
Tru, uTrm (216–232)	35	15	329	23
Trl/m, Trl (233–245)	26	10	339	−1
Pu (246–266)	21	6	339	19
Pl (267–281)	21	1	321	−2
Cu/Pl, Cu (282–308)	19	−5	334	−10

*For the Kimmeridgian, a 'best estimate' pole is used from Table 6.3, rotated into northwest African coordinates. For other explanations see Table 7.1, including the note on **. West Gondwana reference poles, used for the calculations, are those in northwest African coordinates (Table 5.8).

Southern Alps, including the Vicentinian and Lombardy Alps, and the Istrian area in Yugoslavia, to the east of Trieste) show systematic differences. An earlier Cretaceous study in Istria (Marton & Veljovich, 1983), which gave slightly more westerly declinations, has now been updated with new and fully concordant results (Marton, Milicevic & Veljovic, 1990a). This changes the conclusions reached about Istria by Lowrie (1986) and Westphal *et al.* (1986). To the best of our 'paleomagnetic knowledge', i.e., within typical paleomagnetic resolution, all these areas were coherent. However, I will return again to the special case of Gargano with respect to the other areas.

A similar coherence, on the other hand, cannot be postulated for the mobile areas, where detachments from underlying and deeply buried basement are generally well-established features. As an example, I have plotted the declinations from the Umbrian Apennines to examine further this complicated and, in the past, somewhat controversial issue. The Umbrian area has been very extensively studied, beginning in the mid-1970s with the work of Channell & Tarling (1975), Lowrie & Alvarez (1975, 1977), VandenBerg & Wonders (1976) and Roggenthen & Napoleone (1977). That so many different groups, from the UK, Switzerland, the USA, the Netherlands and Italy all converged at about the same time on the charming cypress-dotted Italian countryside may seem surprising, but as Walter Alvarez once remarked (with a nod to Voltaire), paleomagnetic work on the rather complete sections of Umbria 'was clearly an idea whose time had come'; moreover, the advent of cryogenic magnetometers made possible the study of these weakly magnetic carbonate rocks.

The goals of the first studies included mostly the determination of paleomagnetic directions for tectonic and structural purposes, but the work by Lowrie and Alvarez rapidly began to focus on magnetostratigraphy, including that of the now famous rocks of the Gubbio section (which, as a consequence of all this paleomagnetic attention have begun to look like gruyère cheese!). The Mesozoic and Early Tertiary pelagic limestones are ideally suited for this: as demonstrated by many different field tests, the magnetizations are primary, they contain many reversals, the rocks are generally very rich in microfossils so that paleontological collaborators like Wonders and Premoli-Silva could date the samples precisely, the Vecchia Romagna has a delicate taste, and Gubbio . . . , well, it is just a magnificent Mediterranean town. The correlation between the Umbrian magnetostratigraphy and that established from marine magnetic anomalies is a spectacular confirmation of both data sets (e.g., Figure 7.14). Unforeseen but important further implications for science were identified when the Cretaceous–Tertiary boundary was established and docu-

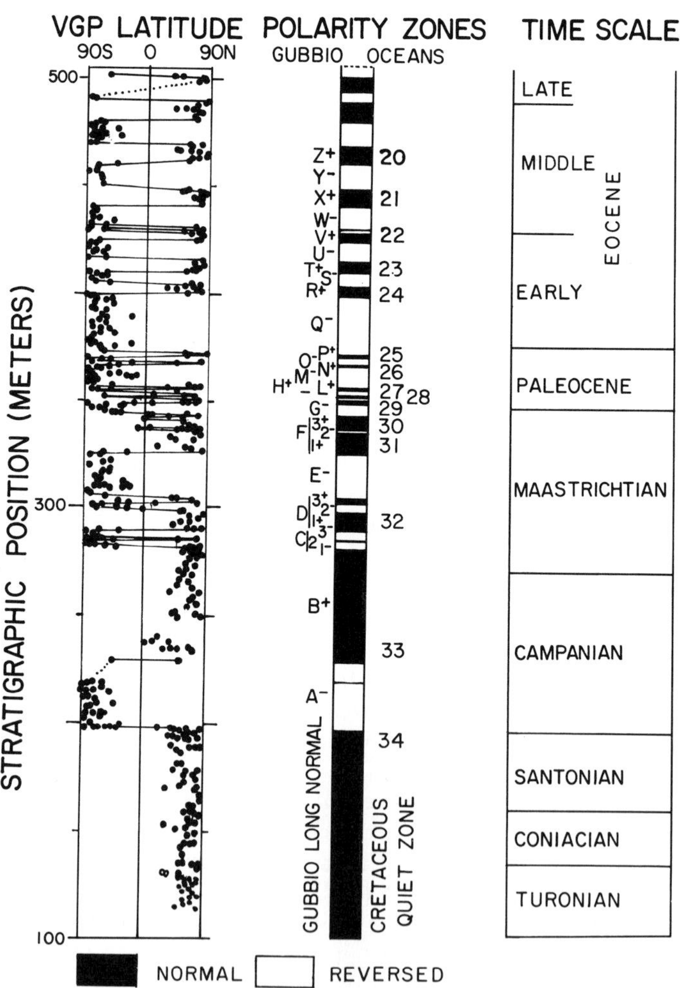

Figure 7.14. Correlation of the reversal sequence observed in the Late Cretaceous to Early Tertiary rocks from Gubbio, Italy, with the reversal sequence inferred independently from marine magnetic anomalies (middle column), plotted alongside the stratigraphical information for the rocks. Virtual geomagnetic pole latitude is calculated from the virtual geomagnetic poles for each sample and reflects the geometry of these individual poles with respect to the overall formation-mean paleopole. Figure redrawn after Lowrie & Heller (1982).

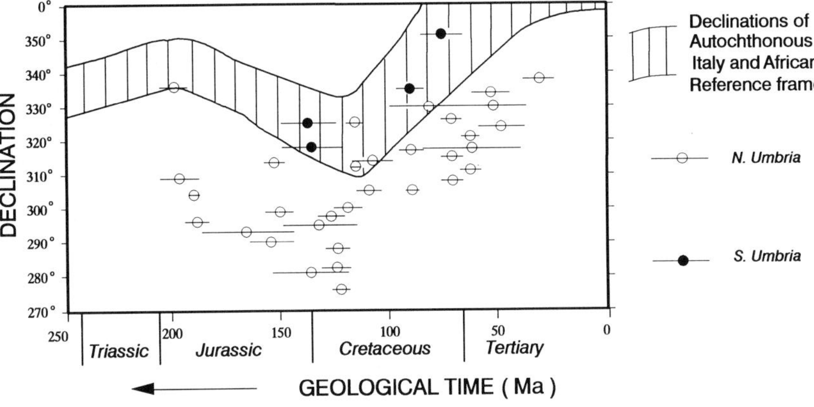

Figure 7.15. Observed declinations for Umbrian areas as a function of geological time plotted with respect to the envelope (shaded area) of declination values, both predicted and observed for autochthonous Italy as shown in Figure 7.13, illustrating the relative rotations of North Umbria.

mented magnetostratigraphically in the sequence. This led to the discovery by Alvarez *et al.* (1980b) of the iridium anomaly at the boundary, which in turn triggered an ongoing and contentious debate concerning the cause of the latest Cretaceous extinction (impact versus flood basalt volcanism, that is, extraterrestrial versus internal origin).

In the last decade, however, the emphasis of Umbrian paleomagnetic work has partially swung back to structural applications (e.g., Channell *et al.*, 1978, 1984; VandenBerg, Klootwijk & Wonders, 1978; Hirt & Lowrie, 1988; Jackson, 1990), with the main goal an examination of thrust sheet rotations relative to each other and to autochthonous Italy (Lowrie, 1986). In essence, the controversial nature of the publications on this issue resulted from an unresolvable question: namely to what extent do the thrust sheets show rotations with respect to the underlying and deeply buried basement from which they are separated by sometimes multiple detachment zones. Not surprisingly, therefore, many arguments have been based on the paleomagnetic directions, often not too well determined, from the autochthonous foreland area, for example, in the Gargano Peninsula.

The Gargano rocks, moreover, are very weakly magnetized and show rather large scatter in their directions, and they are not as easily dated by paleontological means as the pelagic limestone of Umbria. There are four results (Table A7) now in the literature for Gargano (Channell & Tarling, 1975; Channell, 1977; VandenBerg, 1983) with declinations 326°, 341°, 334°, and 313° for the Late Cretaceous, the late Early to Late Cretaceous, the Early to early Late Cretaceous, and the Early Cretaceous, respectively. The fourth result in this list has been recalculated from the original data of VandenBerg (1983), whose mean declination was reported erroneously as 301°. Inspection of Figure 7.13 (symbols marked by *) shows that the first three of these roughly match the other Italian as well as the reference directions from Africa, whereas the fourth is somewhat apart from the rest.

In Figure 7.15, the declinations from autochthonous Italy are represented by a swathe, forming the envelope of the (Italian and West Gondwana) values of Figure 7.13, with the Umbrian declinations plotted as individual entries. The North and South Umbrian values have been given different symbols, moreover, to distinguish between those from the northern NW–NNW striking part of the arc-like outcrop area (see Figure 7.16) and those from the NNE striking southern part. The difference in structural trends has naturally given rise to speculations about differential rotations between the two. This additional complexity will also be examined below.

With exception of only a few data points, the North Umbrian declination values deviate systematically from those of autochthonous Italy (Figure 7.15). Thus, this area is rotated, but we can state for certain only that this is with respect to Iblei, the Southern Alps, Istria, etc. The South Umbrian values, averaged by half periods or less, do not appear to be rotated with respect to autochthonous Italy. Thus, North and South Umbria, according to this simplified analysis, have undergone relative rotations. The latter conclusion, however, was not fully adopted by one of the latest studies on these rocks (Hirt & Lowrie, 1988), in part because the authors looked at structural trends for each site and found, given the typical paleomagnetic scatter, an inconclusive correlation.

On the other hand, it is understandable that with the previously anomalous declination of 301° from Gargano (now known to be erroneous), some argued that

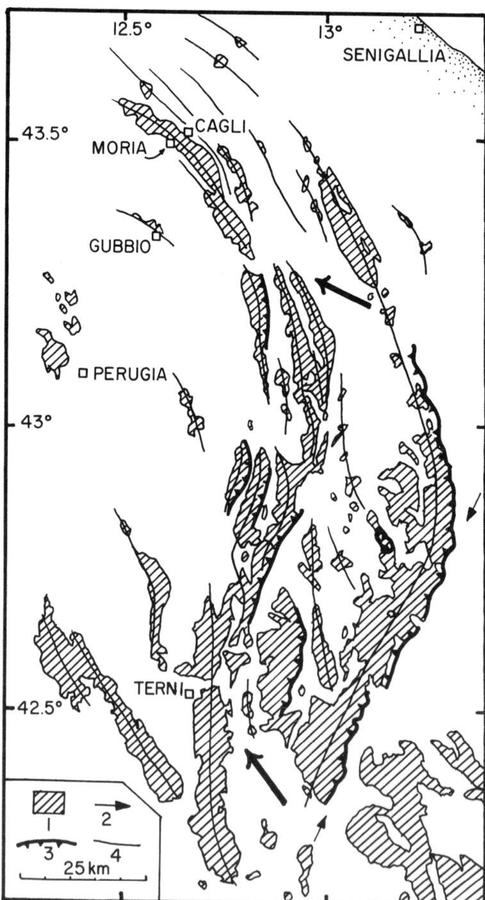

Figure 7.16. Map of the Umbrian Apennines, illustrating the arc-like outline of the area. The two heavy arrows represent the average declinations for northern and southern Umbria for latest Jurassic to earliest Cretaceous times (see Figure 7.15). Shaded areas (feature 1) represent the outcrops of rocks with Upper Triassic to Lower Cretaceous ages. Feature 2 is the Anzio–Ancona line, feature 3 shows major reverse or thrust faults, and feature 4 represents anticlinal fold axes. Figure redrawn after Channell *et al.* (1978). Published with permission from Elsevier Science Publishers, BV.

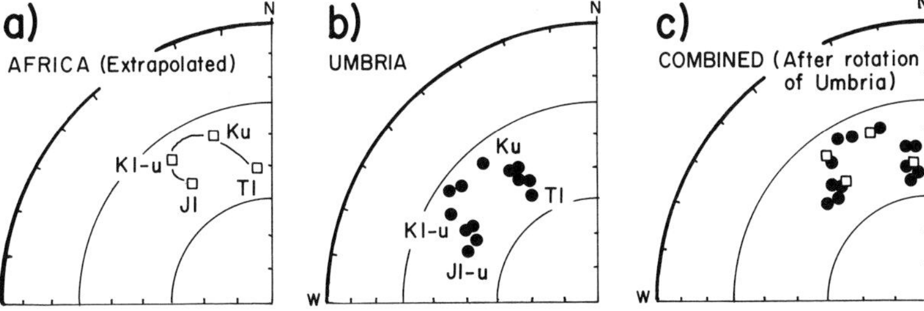

Figure 7.17. A comparison of a) paleomagnetic directions expected in Umbria, if it had remained fixed with respect to Africa, calculated from the mean poles listed in Van der Voo & French (1974); b) directions actually observed in Umbria; and c) the directions of (b) after a 25° clockwise rotation and those of (a) superposed, illustrating that a rotation of Umbria is needed to explain the observations. Figure redrawn after VandenBerg, Klootwijk & Wonders (1978).

the relative rotation of North Umbria was not so clear, and that the overall data set contained more complexities than assumed thus far. The uncertainties in the Late Jurassic to Early Cretaceous paleopoles for Gondwana (Chapter 6) indicate that our knowledge about the reference poles may still be incomplete. Nonetheless, I conclude that the preponderance of Italian data of Figures 7.13 and 7.15 indicates that autochthonous Italy seems fairly coherent, and that, when averaged, the North and South Umbrian data sets show relative rotations. However, the last word has not been written on the subject and many more holes will be drilled in the pelagic carbonates of Italy before we can all agree on these issues.

There is a consensus that the Umbrian rocks are detached from basement and have been transported over large distances. The controversy centers on the choice whether the thrusts have also rotated with respect to their own basement (Kligfield & Channell, 1979) or only (as all agree; see Figure 7.17) with respect to Africa, the Southern Alps and other more remote areas (VandenBerg, 1983; Lowrie, 1986).

This issue is of greater importance than it may seem at first. If all of autochthonous Italy remained together as one block (called Adria), and given the good coherence with the African predicted pattern (Figure 7.13), the proposal by Channell, D'Argenio & Horvath (1979) that Adria was a (fixed) promontory of the African continent is viable. In that case, the kinematics of the continent–continent collision in the Alps are directly related to the relative movements between Europe and Africa (see also Laubscher, 1988; Coward & Dietrich, 1989). On the other hand, if the Apulian (and Umbrian) basement rotated with respect to Africa, Iblei and the Southern Alps, then Italy consists of many small blocks that did not retain coherence (VandenBerg, 1983; Lowrie, 1986), such that there was no Adriatic promontory, or at least not a permanent one. Both sides have agreed that the Mesozoic movements of Italy were in concert with those of Africa, so the disagreement is only over the issue whether the promontory was a permanent part of Africa or a part subsequently disrupted. The possibility also exists that Adria underwent a small, paleo-east to west displacement during the Mesozoic with respect to Africa along a fault located in the straits between Sicily and Tunisia (e.g., Dercourt *et al.*, 1986; see also the 'M-plate' model of Hsü, 1989) but this cannot be tested with paleomagnetic data.

In the thrust belts of allochthonous Sicily, tectonic transport and its accompanying rotations are very clear. Results obtained by Schult (1976), Channell, Catalano & D'Argenio (1980) and Channell *et al.* (1990a) indicate increasingly larger clockwise rotations the higher up one goes in the stack of thrusts and nappes. In contrast, the southern Apennines (Manzoni, 1975; Channell & Tarling, 1975; Catalano *et al.*, 1976; see also Van der Voo & Channell, 1980) show increasingly larger ccw rotations, in symmetry with an axis through the intervening Calabrian Massif. The results have been derived from deposits on alternating platforms and basins that form a pre-deformation paleogeographic setting of a rifted, horst-and-graben type continental margin (Figure 7.18a). The timing of this extension has been related to the early opening of the Central Atlantic and the commensurate rifting that began to take place in the Late Triassic between Europe and Africa (i.e., between Spain and Italy) in the western Tethys. The elevation contrast in this platform and basin pattern is thought to have caused, during deformation, the different nappes that rotated by successively greater amounts during transport (Figure 7.18b).

Lastly, rotations are also clear in Calabria and in the Alps north of the Insubric Line. Calabria consists of Paleozoic–Mesozoic, in part metamorphic, terranes (e.g., the Sila Massif and nappes), which have yielded southwesterly declinations that indicate large ccw rotations during tectonic transport. This transport is thought to have been over a large distance, possibly arriving all the way from a location adjacent to Sardinia (Alvarez *et al.*, 1974; Dewey *et al.*, 1989), passing over the sedimentary nappes illustrated in Figure 7.18.

In the Alps, typical Italian/African directions are found in many places, but clockwise rotations are also indicated by the declinations in several places (e.g., the Northern Calcareous Alps in Austria, parts of the Carnic Alps, the Gailtal area near the Insubric Line ('Drauzug'), and NE Slovenia in Yugoslavia). The more reliable results ($Q \geq 3$) from the French Alps indicate Italian type directions for the Pelvoux

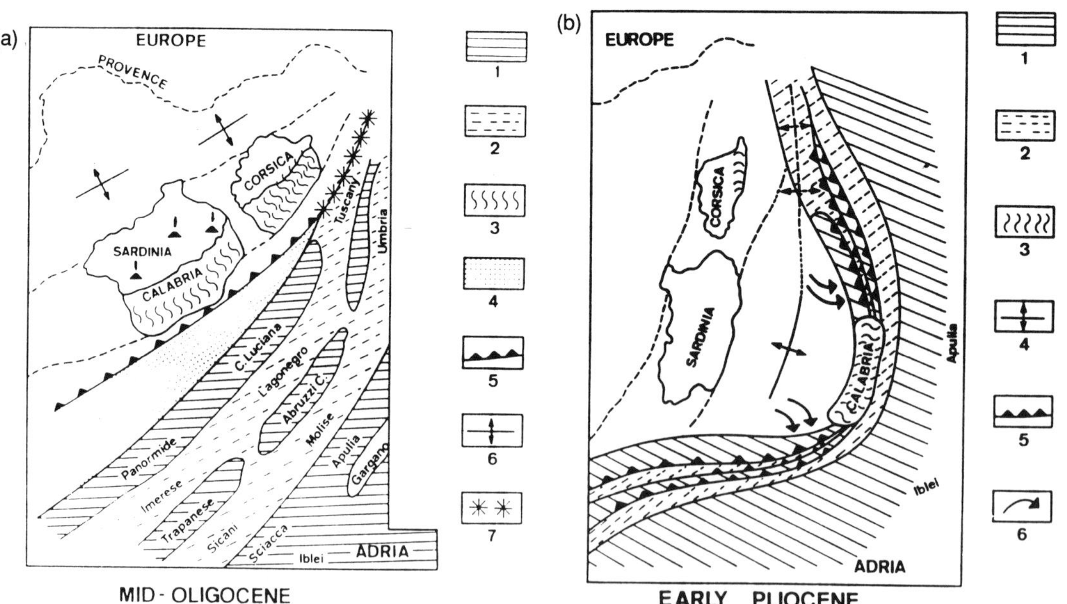

Figure 7.18. Schematic reconstructions of the west (Tyrrhenian) side of the Adriatic promontory for a) Mid-Oligocene and b) Early Pliocene times. Explanation of symbols: 1 = carbonate platforms, partially undergoing deformation in b); 2 = marginal basins, partially undergoing deformation in b); 3 = regions affected by earlier Late Cretaceous deformation; 4 = Mesozoic Tethyan crust in a) and regions of Pliocene (back-arc?) extension in b); 5 = subduction zones in a) and crustal-level thrusts in b); 6 = Oligocene–Miocene back-arc extension in a) and sense of paleomagnetically determined rotations of thrust sheets in b); and 7 = zone of continent-continent collision. Figure from Van der Voo & Channell (1980).

area and the Guil valley in the Briançonnais Zone, and European type directions for the Argentéra Massif (Bogdanoff & Schott, 1977). The few available paleomagnetic declinations do not provide support for Carey's orocline model for the Western Alpine arc (Ligurian orocline in Figure 7.3), because the rotations do not appear to correlate with structural trends (but see also Vialon, Rochette & Menard, 1989). For an excellent review of the paleomagnetic results from the Alps *sensu stricto*, see Heller *et al.* (1989); their data base includes several results from the 'grey' literature, which have not been included in Table A7.

Western Mediterranean overview

Clearly this area is extremely complex, although in comparison to the eastern Mediterranean and the less well studied Middle Eastern areas it may seem relatively simple. Paleomagnetically, the results, at least in large part, seem to make sense and indicate ccw rotations with respect to Stable Europe for all the blocks discussed. While many of the rotations occurred in the Cretaceous to Early Tertiary, at the same time as the rotation of Africa, additional movements took place independently of Africa when the western Mediterranean basins of the Ligurian and Balearic seas opened in the Miocene. These basins, then, are not Tethyan remnants! They formed during the northwesterly subduction of older ocean floor that was responsible for the Oligocene–Miocene arc-type volcanism of Sardinia, by a southeasterly movement of the subduction hinge and may be thought of as back-arc basins. The Tyrrhenian Sea, which is underlain by thinned continental and some oceanic crust (Rehault, Boillot & Mauffret, 1985; Lavecchia & Stoppa, 1990), likely contains small continental fragments from the original Sardinia–Calabria basement and may not be a sphenochasm (Carey, 1958), whereas the Ligurian and probably also the Balearic basins are formed by relative rotations between the two sides and thus are better examples of sphenochasms.

For pre-Tertiary times, it is tempting to relate the movements of Adria, the islands of Corsica and Sardinia, and Iberia directly to those of Africa, but as we have already seen some of these blocks (e.g., Iberia and the islands) cannot have been permanent parts of Africa, whereas Adria possibly was.

The opening phases of the Atlantic (North and Central) that I discussed in Chapter 6 indirectly prescribe the relative movements of Europe and Africa. In convergent zones, such as the Mediterranean, the paleopositions of the plates cannot be constructed directly, as described in Chapter 6. However, by completing the finite-difference circuit across spreading centers (the passive margins between Europe–North America and North America–Africa), relative motions can be obtained for Africa with respect to Europe. A drawback of this technique is that errors and uncertainties are cumulative. In several papers, Smith (1971), Dewey & colleagues (e.g., 1973, 1989), as well as many others in the intervening years, have used the parameters of these relative movements to aid in the kinematic analysis of the plate tectonics of the Mediterranean (e.g., Figures 7.19 and 7.20). In such syntheses, ophiolites, volcanic arc and rift-related rocks, metamorphic pressures and temperatures, pelagic and miogeocline sedimentary facies, structural (including seismic) and rock-age information, ages of deformation and paleomagnetic directions are all integrated. Assumptions used in such models are generally based on the idea that the relative motions of Africa and Europe prescribe similar relative motion regimes (divergent, convergent, strike–slip) for the different locations within the Mediterranean. The sense of these relative movements at any point can be directly calculated from the instantaneous Euler poles describing the overall Africa–Europe movements, and proposed plate boundaries must be in accord with these Euler poles. A comparison of Figures 7.19 and 7.20 illustrates how in the last two decades the analysis of ocean-floor magnetic anomalies, and the more refined kinematics from it, have changed the implications for areas such as the Mediterranean. The large dextral shear regime of Late Cretaceous to Paleocene times (e.g., Dewey *et al.*, 1973) surprisingly has disappeared entirely in the newer analyses of Livermore & Smith (1985) and Dewey *et al.* (1989)!

Examples of some of the resulting paleogeographies are shown in Figure 7.21 (from Dewey *et al.*, 1989) for Eocene–Oligocene and Oligocene–Miocene boundary times. In this figure, one can recognize the Paleogene configuration of Iberia, Adria, Sardinia and Corsica and the subsequent Early Miocene rotation of Corsica and Sardinia as in agreement with the paleomagnetic data discussed earlier, although in this model the rotation occurs slightly earlier than suggested by the paleomagnetic

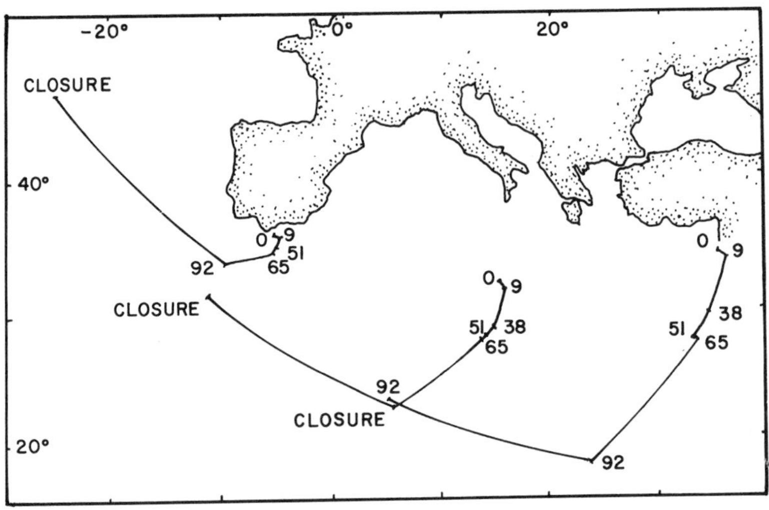

Figure 7.19. Smoothed flow lines depicting the path of Africa's motion with respect to a fixed Stable Europe, as calculated from the opening history of the Central and North Atlantic Ocean. Figure redrawn after Dewey *et al.* (1989). Published with permission from the authors and the Geological Society Publishing House.

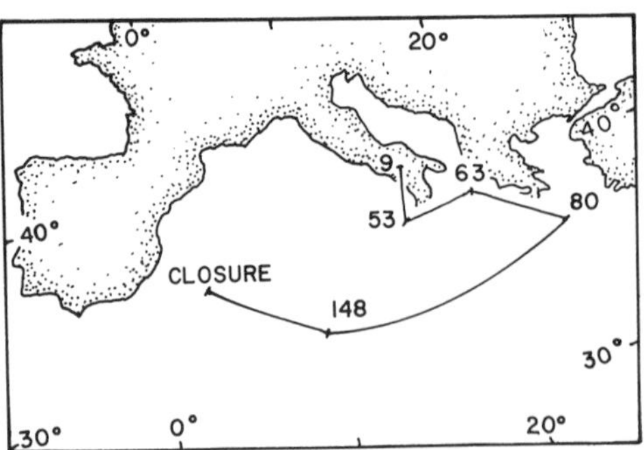

Figure 7.20. A smoothed flow line depicting the path of Africa's motion with respect to a fixed Stable Europe, as calculated from a previous model for the opening history of the Central and North Atlantic Ocean; a comparison with Figure 7.19 shows how this model has changed. Figure redrawn after Dewey *et al.* (1973). Published with permission from the authors.

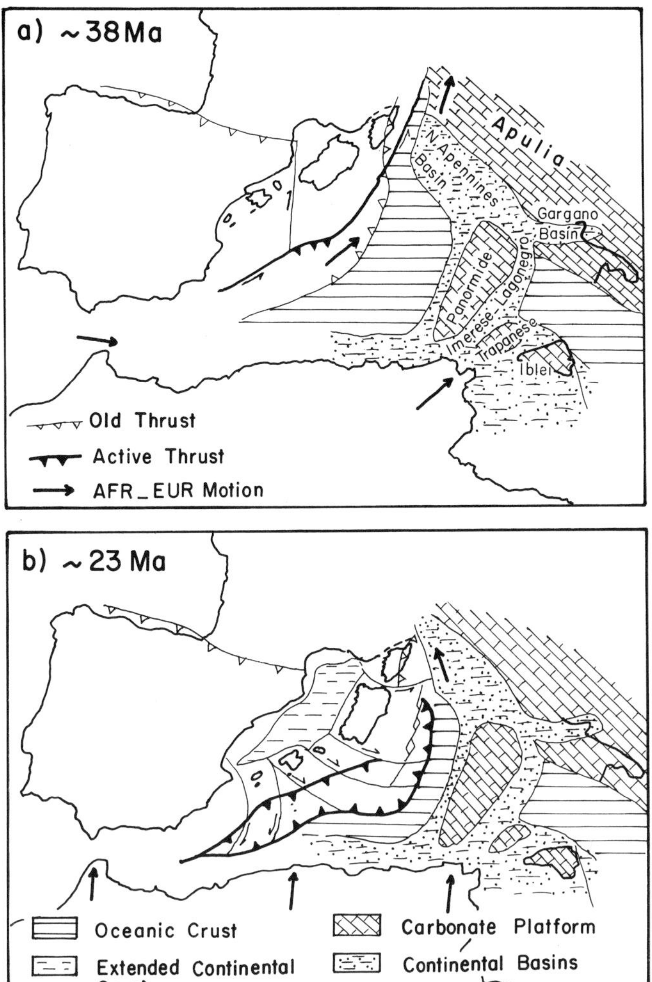

Figure 7.21. Schematic reconstructions of the western Mediterranean area for a) latest Eocene and b) earliest Miocene times. Figure redrawn after Dewey *et al.* (1989). Published with permission from the authors and the Geological Society Publishing House.

results. Figure 7.22 presents Early and Late Miocene equivalents. These illustrations focus on the basins and the developments in the Adriatic promontory, and do not display the Betic–Balearic deformation that must have been going on at the same time. It is also thought (Burrus, 1984; Rehault *et al.*, 1985) that the extent of ocean floor is greater than that shown in Figure 7.22. Be that as it may, the two figures illustrate very well the tectonic evolution of the Balearic, Ligurian and Tyrrhenian basins during the Neogene. A (transform) fault is shown between the Balearics and Sardinia (Figure 7.21) in order to solve the 'problem' that the earlier Pyrenean deformation is not, or hardly ever, seen in and between Sardinia, Corsica and the south coast of France.

Earlier, the Late Cretaceous and Paleogene saw the development of the Pyrenean and earlier phases of the Alpine and Betic mountain building (e.g., De Ruig, 1990), and the counterclockwise rotations of Iberia–Corsica–Sardinia and Adria–Africa.

In contrast to the partly compressional tectonics of the Late Cretaceous and Paleogene, the Late Triassic to Early Cretaceous interval before that was one of extension and some strike–slip (transtension) between Africa and Europe, and presumably therefore also in the areas surrounding Iberia–Corsica–Sardinia and Adria. The Late Triassic to Early Jurassic extension is seen in the Atlantic bordering margins, in Iberia and the Pyrenees (Germanic Keuper facies), and especially, in the French, high-elevation Swiss, the Southern and Austrian Alps (Alpine carbonate facies developing into deeper-water sedimentation in the Jurassic). Some of these areas (Algarve in Portugal, Pyrenees, and Southern Alps) display Late Triassic volcanism thought to be related to this extension. Figure 7.23 illustrates this early extension of the Western Mediterranean and Figure 7.24 shows a model for the Early Cretaceous (Dewey *et al.*, 1973) when the European and African sides are maximally apart. I note that Dewey & colleagues propose many more small microplates for the western Mediterranean area than are necessary to explain the paleomagnetic data: Morocco and Adria, for instance, are broken up into several parts in Figures 7.23 or 7.24, while the paleomagnetic data show no relative movements between these parts.

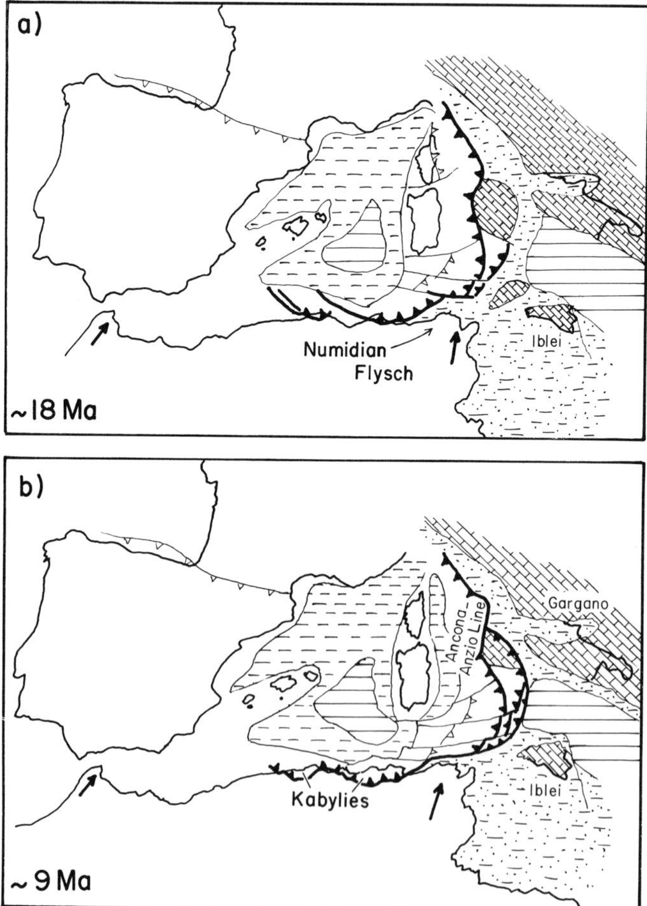

Figure 7.22. Schematic reconstructions of the western Mediterranean area for a) Early Miocene (Burdigalian) and b) Late Miocene (Tortonian) times (see also Figure 7.21 for explanation). Figure redrawn after Dewey *et al.* (1989). Published with permission from the authors and the Geological Society Publishing House.

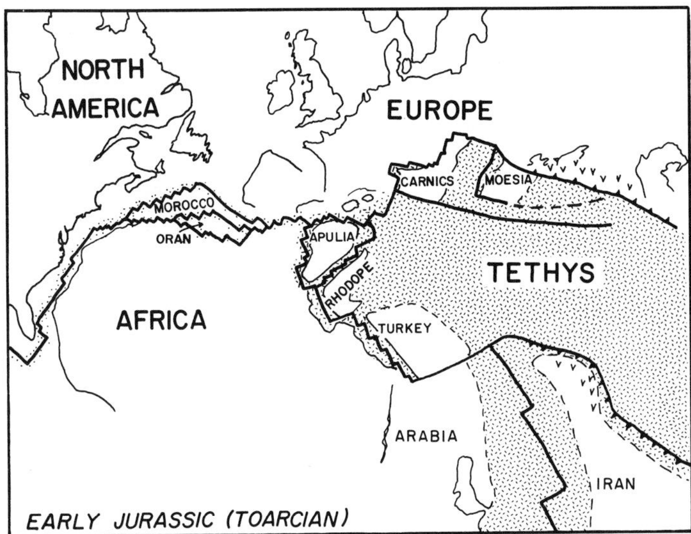

Figure 7.23. Reconstruction of the break-up of Pangea and the Mediterranean area with incipient plate boundaries as inferred for Early Jurassic time. Oceanic areas of the Tethys thought to have been in existence at about 190 Ma have been shaded. Figure redrawn and modified after Dewey *et al.* (1973). Published with permission from the authors.

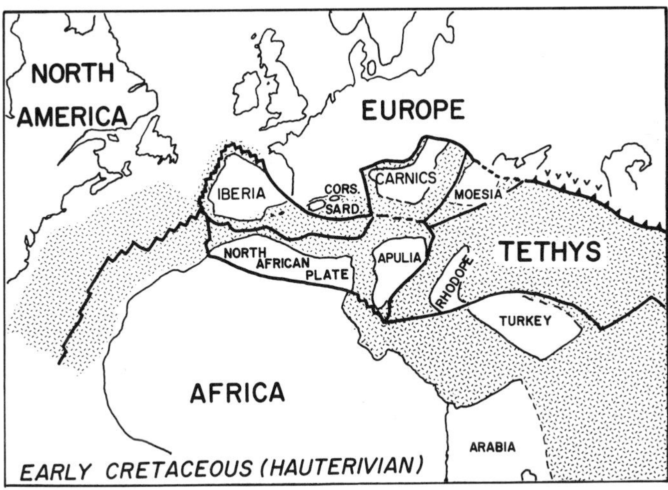

Figure 7.24. Reconstruction of the major continents and the Mediterranean blocks as inferred for Early Cretaceous time. Oceanic areas of the Atlantic and Tethys thought to have been in existence at about 125 Ma have been shaded. Figure redrawn after Dewey *et al.* (1973). Published with permission from the authors.

As with the present West Mediterranean basins, the older ocean floor created in this Mesozoic extension is not, strictly speaking, (Paleozoic) Tethys Ocean either. As can be seen in Figure 7.23, the (Paleo-)Tethys existed only to the east of Adria (= Apulia) in pre-Jurassic times. In some publications, the ocean floor created during the Jurassic–Cretaceous interval has been called Neo-Tethys. In the Eastern Mediterranean, Middle Eastern and Asian areas we will see more of this Neo-Tethys.

Eastern Mediterranean

Cursorily examining a geological map of the eastern Mediterranean, one is struck by the abundant occurrences of ophiolites (Figure 7.25) in the mobile zones of Yugoslavia, Greece, Cyprus, and Turkey and in the northern Levant, which constitutes a promontory of Africa, that is, the Arabian Peninsula, comprising the countries of Israel, Lebanon, and western Jordan and Syria. In the preceding chapters we have seen that a large oceanic gap existed between southeastern Europe and northeastern Africa in early Mesozoic Pangea A-type configurations (the Paleo-Tethys). At first glance, therefore, the present-day situation of the Eastern Mediterranean can be explained by the closure of this Paleo-Tethys in post-Triassic times such that the ophiolites are the obducted remnants of this ocean and the broad mobile belts are the deformed continental margins produced by the continent–continent collision of Africa and Europe–Southwest Asia. Nothing, however, could be further from the 'truth' as revealed by the ages and locations of the ophiolites and the complicated geological situations in the mobile belts.

The overwhelming complexity of these belts has allowed a plethora of contradicting paleogeographic models to be proposed, with many arguments focusing on structural and tectono-stratigraphical details. Only rarely can these arguments be settled by paleomagnetic analysis, which for these purposes has to be based on reliable paleopoles. We will see in this section that thus far high reliability has generally not been achieved and that much work remains to be done in the Eastern Mediterranean.

As in other parts of this book, I will focus primarily on those problems and analyses where paleomagnetic results have been or can be of importance, while ignoring many other equally interesting controversies where that is not possible because of the limited resolution of paleomagnetic data. Again, it is not within the scope of this work to give a complete geological picture of the areas under dis-

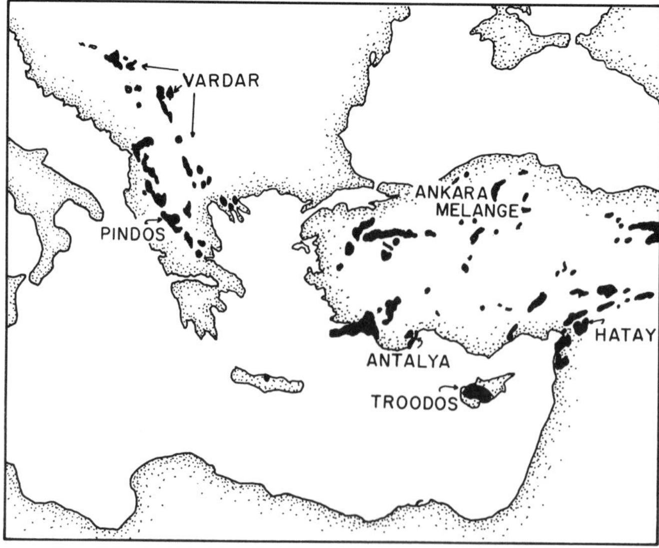

Figure 7.25. Ophiolite occurrences in the eastern Mediterranean. Figure redrawn after Robertson & Dixon (1984). Published with permission from the authors and the Geological Society Publishing House.

cussion; instead attention is generally more focused on broad-scale paleogeographic topics.

In the present section, I will start by outlining some of these topics ('problems'), then proceed by describing the available results, and finish by attempting to draw some conclusions with suggestions for future research.

The Paleo-Tethys problem

The Pangea A-type configurations favored in Chapter 5 would prescribe an Early Jurassic (Liassic) or older age for the oceanic gap between Africa and Eurasia. However, ocean-floor relicts with these older ages have not been sufficiently documented according to some scientists (see several articles in Dixon & Robertson, 1984). Moreover, the sedimentary rocks of these ages in the eastern Mediterranean are generally not even of deeper-water facies. Thus, it is not surprising that some authors have questioned the existence of a Paleo-Tethys oceanic gap (e.g., Stöcklin, 1984; see also Robertson & Dixon, 1984). *Ad hoc* and alternative suggestions, though not always satisfactory solutions to the aggregate of related problems, have been (1) to assume that the disappearance of Paleo-Tethys left few traces, or (2) to adopt a different Pangea reconstruction (e.g., Pangea B of Chapter 5) in which this gap does not occur, or (3) to attempt to find ancient continental areas that could fill it. However, there is a shortage of continental material, given that the gap is thought to have been at least 2500 km wide between the Caucasus and the northern part of the Arabian Promontory: only continental elements such as the China blocks are large enough to provide reasonable options, and the distinct faunal provincialism of the latter blocks makes this an unsatisfactory solution. Pangea B or C configurations create their own problems in areas to the west of Italy and are generally not favored either (see Chapter 5).

The Paleo-Tethys problem does not occur in discussions about the western Mediterranean, because this area in a Pangea A-type fit was completely continental in Permo-Triassic times, as can be seen in Figure 7.26 (from Şengör, Yilmaz & Sungurlu, 1984), with Adria (called Apulia in the figure), Corsica–Sardinia and Iberia snugly fitted against each other and with Africa and Europe. This western Mediterranean assembly has already been discussed earlier in this chapter.

Can it be that Paleo-Tethys ocean floor was subducted without leaving many clues? Şengör & co-workers have argued that, in fact, there are sufficient clues. Theirs is an imaginative model in which the post-Liassic ages of the ophiolites are explained by the formation of a Neo-Tethys which simultaneously caused the disappearance of Paleo-Tethys. The temporal consequences of this model are still being disputed (e.g., Robertson & Dixon, 1984), but it has strong appeal and can be paleomagnetically tested. In this model, shown in Figure 7.27, a strip of continental landmasses called the Cimmerian Continent detached itself from northern Gondwana in Permo-Triassic times, drifted northward over and across the Paleo-Tethyan domain like the single motion of a windshield wiper and collided at the Eurasian side. Cimmeria left the newly created Triassic–Liassic or younger Neo-Tethys Ocean in its wake, as well as several other continental elements that would follow the northward motion of Cimmeria at a later time (e.g., Anatolia, the Tauride Platform and the Sakarya (SC) microcontinent, all now part of Turkey; see Figure 7.27).

In contrast to the above model, but with some similarities nonetheless, Robertson & Dixon (1984) have argued that the Paleo-Tethys Ocean disappeared gradually but later, during Jurassic and Cretaceous times (Figures 7.28–7.30), by northward subduction under the Eurasian (Caucasus–Pontide) margin as well as internally within the Tethys. Northward drifting elements of Turkey are shown in Figures 7.29 and 7.30 as the Sakarya microcontinent and the Kirşehir, Menderes and Tauride blocks, in analogy with the model of Şengör *et al.* (1984). The model of Robertson & Dixon has the advantage that it links the timing of the creation of Neo-Tethys directly to the post-Liassic ages of the ophiolites (Thuizat *et al.*, 1981; Spray *et al.*, 1984; Yilmaz, 1984), although older, pre-Jurassic, pelagic sediments may suggest an earlier beginning of sea-floor spreading in some cases (e.g., Yilmaz, 1984; Whitechurch, Juteau & Montigny, 1984). The Robertson & Dixon model also has the appeal of making the initiation of sea-floor spreading in the Central Atlantic and the Eastern (and Western) Mediterranean simultaneous (i.e., about 175 Ma). The differences between the models of Şengör *et al.* (1984) and Robertson & Dixon (1984) lie primarily in the temporal paleolatitude evolution of the Cimmerian continental elements (e.g., Pontides, Iran) and these differences can be tested with good paleomagnetic data.

Problems related to the paleopositions of the Balkan elements, Greece and Turkey as a function of time

The problem of the uncertain paleolocations of several continental elements now incorporated in the Balkan countries, and in Greece, Cyprus and Turkey, is related to the problem of the creation of Neo-Tethys ocean floor. Again, paleomagnetic data, if of sufficient reliability, can allow choices to be made between initial (pre-Middle Jurassic) paleopositions at the northern versus those at the southern margin

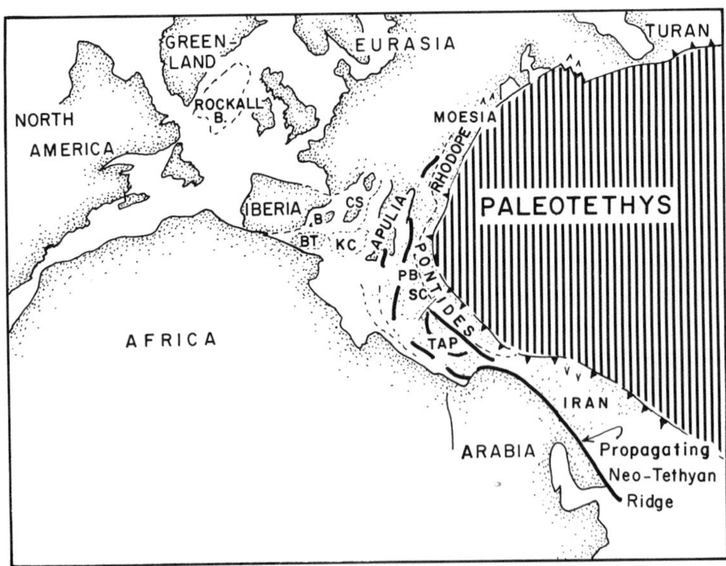

Figure 7.26. Triassic configuration of Pangea and Paleo-Tethys, including the continental elements now reconstituted in the eastern Mediterranean land areas. CS = Corsica–Sardinia, B = Balearics, BT = Betics, KC = Kabylia-Calabria, PB = Pindos Basin, SC = Sakarya Continent, TAP = Tauride–Anatolian Platform. Note the hypothesized initiation of the Neo-Tethyan spreading ridge between Arabia and Iran plus the Pontide Fragment. Figure redrawn after Şengör, Yilmaz & Sungurlu (1984). Published with permission from the authors and the Geological Society Publishing House.

Figure 7.27. Liassic reconstruction of Pangea and the Tethys, divided into Neo-Tethys and a Paleo-Tethys remnant to the north of the Cimmerian Continent which includes the Pontides of Turkey and Iran. Other continental elements now incorporated into Turkey (Sakarya = SC, Menderes and Kirşehir Massifs = Anatolia, Taurides) are following the northward drifting Cimmerian Continent. CI = Central Iran, HB = Helmand Block (Afghanistan). Figure redrawn after Şengör *et al.* (1984). Published with permission from the authors and the Geological Society Publishing House.

of the Tethys, and could potentially follow their motions with respect to the neighboring cratons as a function of later Mesozoic and Tertiary times.

Problems related to neotectonic movements in the eastern Mediterranean

The convergence between Africa and Eurasia initiated in the late Mesozoic is continuing today, as can be deduced from the very active seismicity and the subduction of eastern Mediterranean ocean floor under the Sicily–Calabria–Apennine and Hellenic arcs (e.g., McKenzie, 1972). Overriding the two trenches, both associated with strongly curved subduction zones, the arcs appear to bulge gradually outwards as a function of time, leading to rotations of the lateral arc-flanks (e.g., the clockwise rotations in Sicily and the counterclockwise rotations in the Southern Appenines discussed in the first part of this Chapter; Figure 7.18). Paleomagnetic data are of value in determining the magnitude and sense of such rotations, as we shall see below for the Hellenic Arc running through mainland Greece, Crete and western Turkey.

Paleomagnetic results from the northern Balkan Countries

This area constitutes a transition zone between the western and eastern Mediterranean, the former being characterized by paleomagnetic studies of crustal block rotations, whereas the latter derives its interest from north–south displacements, as well as rotations, of course. A recent review of Marton & Mauritsch (1990) summarizes many of the available data for the Balkan area inside the Carpathian loop that runs from northern Austria through southeastern Czechoslovakia and southernmost Poland into Romania. Because several of the results from this part of the world (as used by these authors) are unpublished or not fully accessible, I have not compiled the available data in appendix tables, but Figures 7.31 and 7.32 illustrate some of the conclusions that can be drawn from the data base.

The area in and around Hungary is separated by the Mid-Hungarian Mobile Belt into two parts (Figure 7.31). Paleomagnetic results from the northwestern area (including the Transdanubian Mountains, the Buekk and Matra Mountains and the western Carpathians) reveal northwesterly declinations (e.g., Figure 7.31 for Cretaceous and early Cenozoic time) in agreement with those from the adjacent

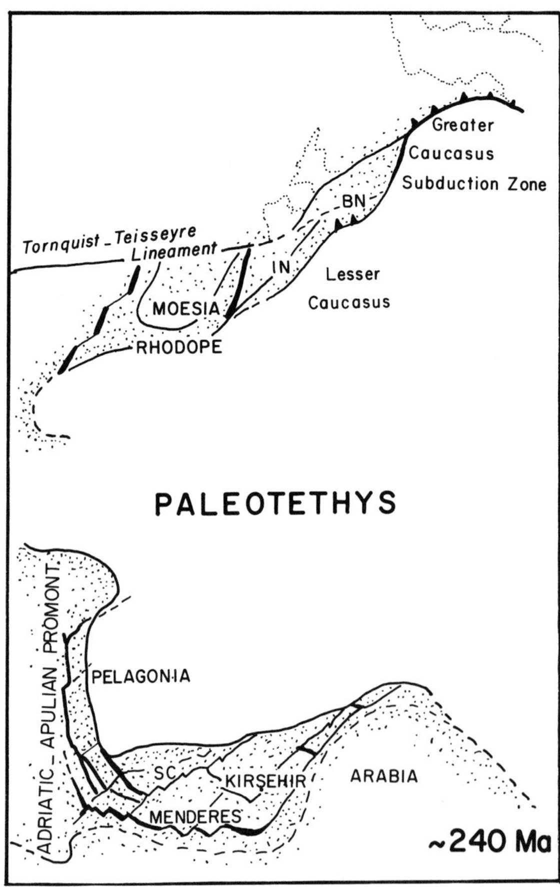

Figure 7.28. Triassic configuration of Paleo-Tethys according to Robertson & Dixon. The Pontides (IN, BN) are located in this model already on the northern (Eurasian) margin of the Tethys; SC = Sakarya Continent. Figure redrawn and modified after Robertson & Dixon (1984). Published with permission from the authors and the Geological Society Publishing House.

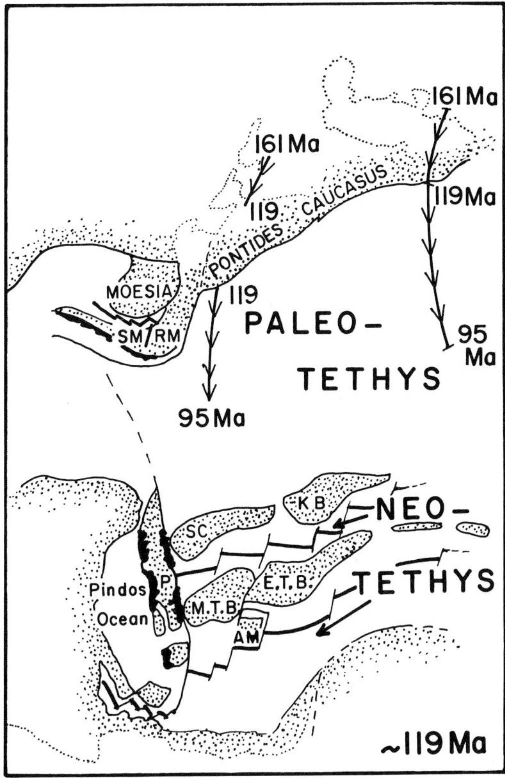

Figure 7.29. Early Cretaceous configuration of Paleo- and Neo-Tethys according to Robertson & Dixon. Arrowed lines indicate preceding and subsequent relative motions; P = Pelagonian Block, SC = Sakarya, KB = Kirşehir Block, MTB = Menderes – western Taurides, ETB = eastern Taurides, AM = Antalya Massif, SM/RM = Serbo-Macedonia/Rhodope Massif; ophiolite obduction is shown in black. Figure redrawn and modified after Robertson & Dixon (1984). Published with permission from the authors and the Geological Society Publishing House.

areas in the Alps, and in Italy. The paleolatitudes of the Transdanubian Mountains are also in excellent agreement with those from Italy (Figure 7.32), thus suggesting that the northwestern part of Hungary moved together with Adria. This northeastern Transdanubian 'bulge' of Adria is included in the paleogeographic outline of the Promontory in Figures 7.28–7.30.

The pre-Cretaceous paleogeographic location of the Transdanubian area with respect to Stable Europe was to the south of its present-day position, giving rise to the question as to what occupied the region within the modern-day Carpathian loop in the late Paleozoic and early Mesozoic, but paleomagnetism has not contributed to this matter. Robertson & Dixon (1984) appear to shift the Moesian Platform (northern Bulgaria) into the void (compare Figures 7.28 and 7.29), whereas Şengör et al. (1984) do not provide details (e.g., Figure 7.26).

To the south of the Mid-Hungarian Mobile Belt few paleomagnetic results are available for the Balkan countries, so the next area to be discussed is Greece to southern Bulgaria.

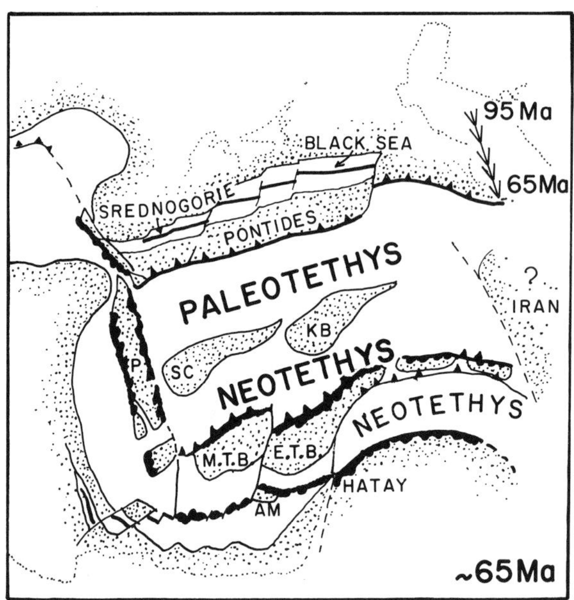

Figure 7.30. Latest Cretaceous configuration of Paleo- and Neo-Tethys according to Robertson & Dixon. Arrowed lines indicate preceding relative motions; P = Pelagonian Block, SC = Sakarya, KB = Kirşehir Block, MTB = Menderes – western Taurides, ETB = eastern Taurides, AM = Antalya Massif; ophiolite emplacement is shown in black. Figure redrawn after Robertson & Dixon (1984). Published with permission from the authors and the Geological Society Publishing House.

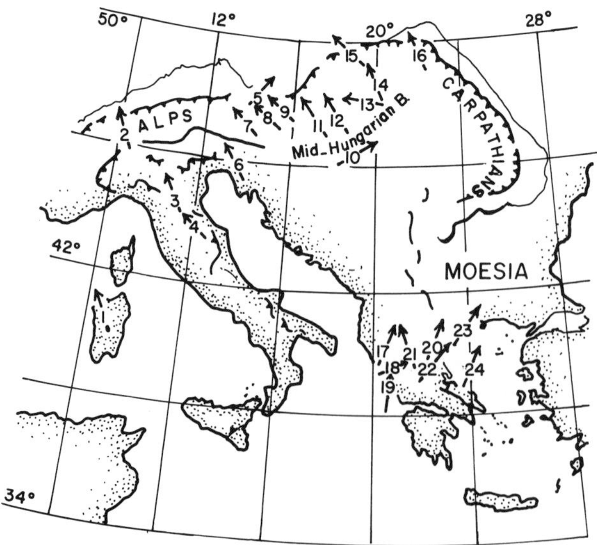

Figure 7.31. Paleomagnetic declinations in the Balkan countries and adjacent areas for Cenozoic and Late Cretaceous times. Figure redrawn and modified after Marton & Mauritsch (1990). Numbers refer to their data base index (not used in this book). Published with permission from Elsevier Science Publishers, BV.

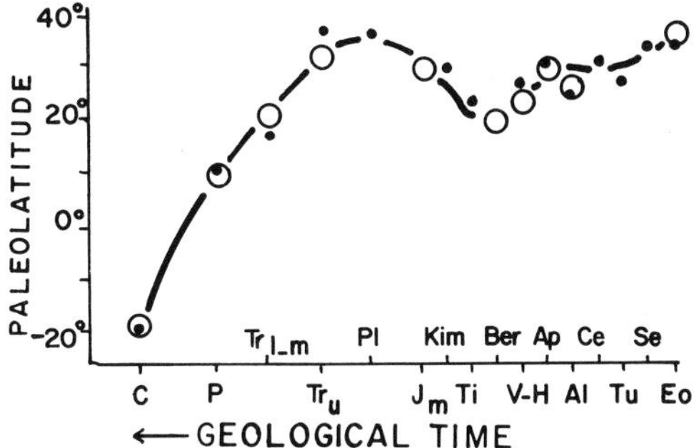

Figure 7.32. Paleolatitudes versus time for the Transdanubian Mountains in Hungary (open circles) and the Adriatic Promontory (solid circles). Age abbreviations are: C = Carboniferous, P = Permian, Tr = Triassic, Pl = Pliensbachian, J = Jurassic, Kim = Kimmeridgian, Ti = Tithonian, Ber = Berriasian, V–H = Valanginian–Hauterivian, Ap = Aptian, Al = Albian, Ce = Cenomanian, Tu = Turonian, Se = Senonian, Eo = Eocene, l = lower, m = middle and u = upper. Figure redrawn and modified after Marton (1984). Published with permission from the Geological Society Publishing House.

Greece and southern Bulgaria

Results for this area have been compiled in Table A8, following the usual format, including a column which indicates the declination and paleolatitude derived from a result for a common location at 40° N, 22° E. The paleolatitudes for results with Q ≥ 3 have been plotted in Figure 7.33 together with expected paleolatitudes extrapolated to this location from the mean paleopoles of Stable Europe and Africa.

In Greece several subparallel zones occur that were named initially by French geologists who in pre-plate tectonic days developed their geosynclinal models from this 'type locality' (e.g., Aubouin, 1965). From west to east, these comprise in abbreviated form: (1) the pre-Apulian and Ionian Zones which form the eastern margin of the Adriatic Promontory, (2) the Pindos Zone with its Jurassic ophiolites (see also Figure 7.25) and the Gavrovo–Tripolitsa Zone, and (3) the sub-Pelagonian and Pelagonian Zones (Figure 7.34), followed (4) by the Vardar Zone with its ophiolites (shown in Figure 7.25 but not in Figure 7.34) and by the Serbo–Macedonian, Rhodope and Thrace zones in northern and northeastern Greece and southern Bulgaria. These zonal indications have been included, where appropriate, in Table A8. The Rhodope Zone (or Block) is an ancient microcontinent, today separated by the Srednogorie mobile belt (presumably initiated as a Cretaceous rift) from the Moesian Platform of northern Bulgaria (e.g., Figures 7.28–7.30).

For Greece only four pre-Tertiary results (with Q ≥ 3) are available, whereas for southern Bulgaria (symbols marked by *) two older Late Cretaceous and two recently published Jurassic paleolatitude determinations from Surmont *et al.* (1991) and Peybernes *et al.* (1989) are included (Figure 7.33). Another recent paleomagnetic determination on Jurassic rocks in southern Bulgaria (Kruczyk *et al.*, 1990) has been interpreted by the authors as a complete remagnetization, so this is not included here. The pre-Tertiary paleolatitudes follow, with considerable scatter, the predictions from Europe and Africa and illustrate that there is not sufficient resolution to decide for or against north–south displacements of the areas with respect to the major continents.

The Tertiary and Quaternary paleolatitudes appear to illustrate values that are too far south with respect to Europe as well as Africa. While a somewhat more southerly Early Tertiary paleolocation with respect to Africa is possible, this is unlikely to have been as much as the 10° indicated by Figure 7.33. Either the reference value from Africa is subject to considerable uncertainty (there are only four Early Tertiary results with Q ≥ 3 from Africa, as discussed in Chapter 6) or the results from Greece display inclinations that are too shallow, for example, because of unrecognized structural tilt or inclination error.

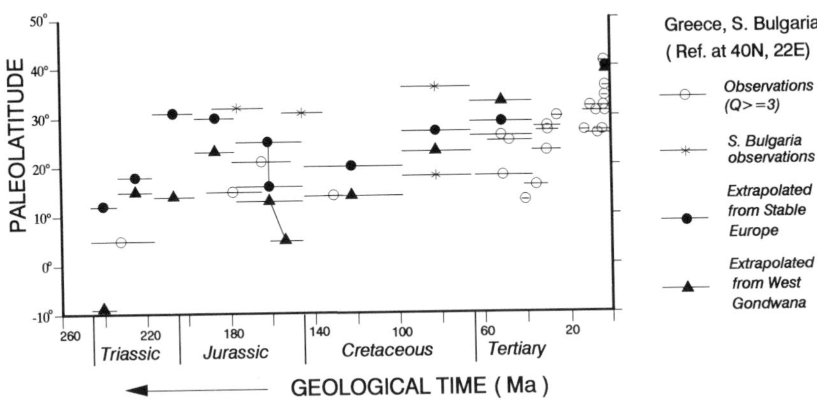

Figure 7.33. Paleolatitudes versus time for Greece (open circles) and southern Bulgaria (*), together with the values expected for the common reference site at 40° N, 22° E as extrapolated from the Stable European (dots) and African paleopoles (solid triangles) of Tables 5.1, 5.7 and 5.8. For the Late Jurassic, different reference paleolatitude options (as explained in Chapter 6) are connected by a line.

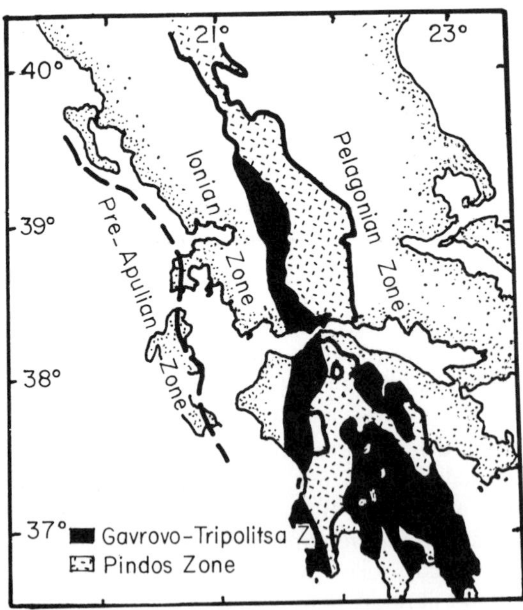

Figure 7.34. Locations of the Pre-Apulian, Ionian, Pindos, Gavrovo–Tripolitsa, and Pelagonian Zones in Greece. Figure redrawn after Pe-Piper & Piper (1984). Published with permission from the Geological Society Publishing House.

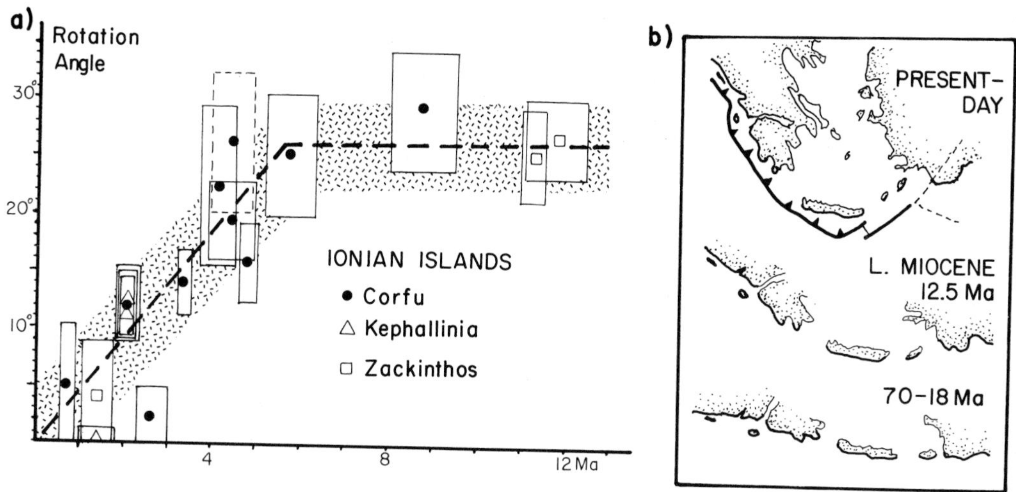

Figure 7.35. a) Declinations as a function of Neogene time for the Ionian Zone in Greece, illustrating the (ongoing) clockwise rotation of the area. Figure redrawn after Laj *et al.* (1982). b) Reconstruction of the Aegean arc for the present-day (top), the Late Miocene and the Late Cretaceous to Early Miocene (bottom), illustrating the oroclinal bending of the arc in two, pre-Late Miocene and post-Late Miocene, phases. Figure redrawn after Kissel & Laj (1988). Published with permission from the authors and Elsevier Science Publishers, BV.

The declinations, on the other hand, provide an interesting pattern, as revealed by the extensive research program of the paleomagnetic team at Gif-sur-Yvette in France (Laj *et al.*, 1982; Kissel, Laj & Muller, 1985; Kissel, Laj & Mazaud, 1986a; Kissel *et al.*, 1986b, c; Kissel & Laj, 1988). Figure 7.35a shows the clockwise rotation that can be inferred from the declinations obtained in the Ionian islands (Ionian zone of western Greece), which occurred during the last 5 Ma and presumably is continuing today. Note that the value listed in Table A8 for these results is an average and does not reflect the temporal changes in declination. Older (Miocene) formations in the Ionian Zone record an even greater amount of subsequent rotation. In contrast, the Pliocene and Late Miocene declination results from Crete, at the center of the Hellenic Arc, do not reveal any rotations (Table A8, Laj *et al.*, 1982). In northeastern Greece, the interior of the arc does display rotations, but not to the systematic extent of the exterior zone of the arc. In a logical extension of this research, Kissel & Poisson (1986, 1987) collected sedimentary rocks farther east near Antalya in southern Turkey, and found convincing indications for a symmetrical completion of the pattern in counterclockwise rotations there; only the central (Crete) portion of the arc did not rotate (Figure 7.35b). However, in more northerly zones of the East Aegean and in Turkey the pattern of declination deviations is unsystematic. Although some northwesterly declinations were observed (Kissel *et al.*, 1986c; Kissel *et al.*, 1987), the pattern is far from clear with an array of presumably local clockwise and counterclockwise rotations (see Table A8, included in part II for Turkey). Some of these results are still unpublished and were, in fact, labeled 'unpublishable' by the author at an international meeting (Kissel, 1989). Thus, the exterior zones of the neotectonically interesting Hellenic Arc reveal an oroclinal evolution as a southward 'bulge' with increasing curvature (Figure 7.35b), but the data from the interior zones are much less clear.

Cyprus and Turkey

The structural elements of this complicated area of the eastern Mediterranean are shown in Figure 7.36. Along the Black Sea Coast the Pontide Zone occurs, comprising from west to east: the Istanbul Nappe, the Kure Nappe and the Bayburt Nappe. The present-day North Anatolian Fault zone dissects parts of the Pontides. To the south of the Istanbul Nappe in the Pontides, the Sakarya microcontinent is located. Farther south and to the east in western and central Turkey, the Menderes and Kirşehir Massifs constitute older continental nuclei, bordered to the north by ophiolites and mélanges (Figure 7.25). Southern and eastern Turkey contain the Tauride Mobile Belt, with many allochthonous sequences including ophiolites (e.g., Figure 7.25) overlying the presumably autochthonous Calcareous Axis in southern Turkey (for locations see Figure 7.36).

The paleomagnetic results are compiled in Table A8 and include a column listing the declination and paleolatitude calculated from the results for a common location at 38° N, 32° E. In Table 7.3, the reference declinations and paleolatitudes expected for this location are given, calculated on the assumption that the area did not move with respect to Stable Europe or Africa. A large part of the Turkish provinces is allochthonous, so that the declinations are not representative for much more than the immediate sampling areas and have not been plotted.

The paleolatitudes, on the other hand, could be very interesting in view of the large-scale problems noted earlier; they are shown in Figure 7.37 for those results with $Q \geq 3$. A quick inspection of Table A8 reveals that many results exist with $Q \leq 2$, and as in earlier chapters of this book, these are deemed to be less reliable. In Figure 7.37 different symbols are used for Cyprus (*), for the more numerous results from the Pontides (squares) and for the results from the rest of Turkey and the East Aegean area (open circles).

The observed Neogene paleolatitudes are too shallow by about 5° to 10°, as they were in Greece. For many of these results, a combination of some northward drift and a significant inclination error in the sedimentary rocks may have played a role. For the Late Cretaceous and the Paleogene, a somewhat scattered grouping occurs which coincides roughly with the reference paleolatitudes extrapolated from Stable Europe and Africa (from Table 7.3). Because Africa and Europe were converging

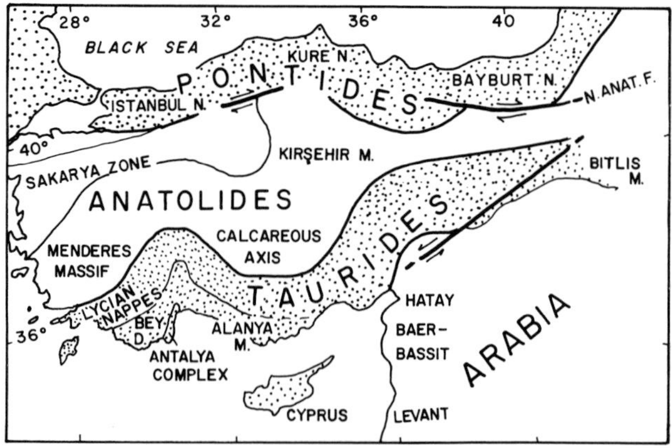

Figure 7.36. Geological zones and divisions of Turkey and adjacent areas. The Pontide and Tauride zones are stippled for contrast. Figure redrawn after Dixon & Robertson (1984). Published with permission from the authors and the Geological Society Publishing House.

Table 7.3. *Declinations and paleolatitudes for Turkey, at 38° N, 32° E from European and West Gondwana reference poles*

Age interval	Europe		W. Gondwana	
	D	Lat.	D	Lat.
Tl (37–66)	8	28	4	32
Ku (67–97)	17	27	347	19
Kl (98–144)	7	19	331	8
Ju, uJm (145–176)	9	15	336	8
**Same	15	25		
*Kimmeridgian Best Estimate			332	−1
lJm, Jl (177–195)	27	32	343	19
Tru/Jl (196–215)	37	34	349	10
Tru, uTrm (216–232)	43	22	342	10
Trl/m, Trl (233–245)	35	15	348	−13
Pu (246–266)	32	9	344	7
Pl (267–281)	33	4	330	−17
Cu/Pl, Cu (282–308)	33	−2	346	−22
Cm, Cl, Du/Cl (309–365)	40	−12	339	−33
Du, Dm/Du (366–378)	53	−2	16	−50
Dm, Dl (379–397)	57	−5	41	−37
Su/Dl (398–414)	80	−8		
Su, Sm (415–429)	50	−17	12	−10
Ou/Sl, Ou, Om (430–467)			101	−45
Ol/m, Ol (468–505)			85	−63

*For the Kimmeridgian, a 'best estimate' pole is used from Table 6.3 rotated into northeast African coordinates. For other explanations see Table 7.1, including the note on **. West Gondwana reference poles, used for the calculations, are those in northeast African coordinates (Table 5.8).

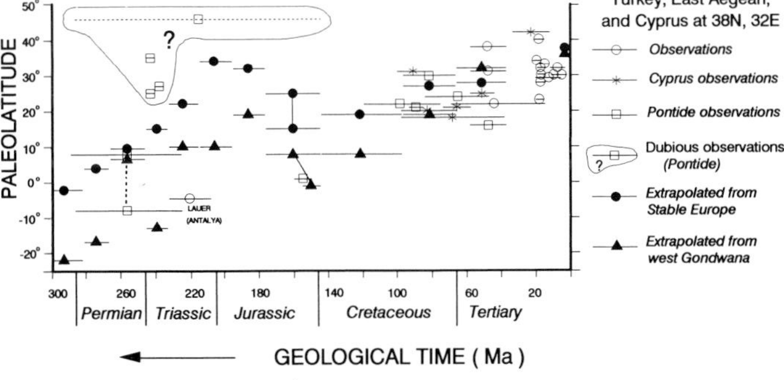

Figure 7.37. Paleolatitudes versus time for Turkey, the Greek islands in the Eastern Aegean and Cyprus (*), together with the values expected for the common reference site at 38° N, 32° E as extrapolated from the Stable European (solid circles) and African paleopoles (solid triangles) according to Table 7.3. For the Late Jurassic, different reference paleolatitude options (as explained in Chapter 6) are connected by a line. Open squares represent results from the Pontides (dotted in the envelope where dubious). A Permo-Triassic result from the Pontides presents an ambiguity in terms of normal versus reversed polarity; both paleolatitude options are shown, connected by a dashed line. The result labeled 'Lauer (Antalya)' is discussed in the text.

in the later Cretaceous, the younger reference paleolatitudes do not show significant differences, whereas for early Late Cretaceous and older times they are about 10° to 15° apart. With paleomagnetic observations from Turkey of very high precision and reliability, this difference could be detected, that is, one could potentially distinguish between a provenance of the Turkish elements from the southern Tethys border and one from the northern Eurasian border. Such a distinction has been foremost in the minds of the paleomagnetists working in the area (e.g., Lauer, 1981a, b; Saribudak, 1989; Evans & Hall, 1990; Evans *et al.*, 1991).

The pattern obtained from the various blocks, however, is far from unambiguous for numerous reasons. As already noted, the Late Cretaceous results (from the Pontides as well as from Cyprus) show some scatter and range between 20° and 31° without any systematic regional pattern. This paleolatitude range coincides with that of the reference values from Europe and Africa, making it impossible to conclude to a northern versus southern paleolocation. Only one result exists from the Pontides for the Jurassic and at first glance this seems to favor a southern paleolocation (Lauer, 1981a). However, this conclusion is tenuous: the Late Jurassic result is an average of three less reliable individual entries (compiled by Lauer, 1981a).

For the same Pontide region, a Permo-Triassic result from the Amasra region may be of either normal or reversed polarity, resulting in a paleolatitude ambiguity of 8° North or South. Three Early Triassic results from the Istanbul Pontides (Saribudak *et al.*, 1989) and one poorly dated result from the Kure Pontides (Guner, 1982) present a special problem: their observed paleolatitudes are far higher than those expected for the Eurasian Margin for the Early Triassic, which would be paleogeographically implausible. These results are outlined by an envelope in Figure 7.37; either they are representative of a younger age (remagnetizations), or they are of dubious value. Suspecting the latter, I have marked them with a * in the Q-column of Table A8. The three Triassic results from the Istanbul area also have very discordant declinations (Table A8).

Lastly, a set of pre-Permian results from the western Pontides, while in general indicating equatorial paleolatitudes, is of very low reliability (Evans *et al.*, 1991). Only three (Late Devonian) results have $Q \geq 3$, but these have internally discordant paleolatitudes of 10°, 14° and 40° South. Inspection of Table 7.3 for the Late Devonian reveals that this range corresponds precisely to the values expected for the equatorial Eurasian margin and the temperate Arabian margin at about 35° to 50° South, respectively.

Many of Lauer's results (1981b) have not been published and thus have not been included in Table A8. In a separate paper (Lauer, 1981a), a Late Triassic result from the Taurides is mentioned without details (e.g., no declination or paleopole values), other than that it reveals a paleolatitude of 5° South. This result, obviously not included in Table A8, is labeled 'Lauer (Antalya)' in Figure 7.37 and would indicate a southerly, that is, near-African, provenance for the Antalya Taurides.

In summary, no clear provenance indications exist for any of the Turkish areas. For the areas to the south of the Pontides, too few published data are available, whereas the more numerous Pontide data set is fraught with problems, such as low reliability (Late Jurassic), a lack of resolution (Late Cretaceous), polarity ambiguity (Permo-Triassic) and lack of coherence (Triassic and Devonian). Of these, the Late Jurassic result is perhaps the best: it would indicate a southerly paleolocation for the Pontides, presumably in the vicinity of the Levant or Egyptian coasts of Africa and Arabia. This, however, conflicts strongly with the paleolocation suggested by the earlier-mentioned geological models for the provenance of the Turkish areas. While Robertson & Dixon (1984) propose that the Pontides were at the Eurasian margin throughout Mesozoic time (Figures 7.28–7.30), the model of Şengör *et al.*

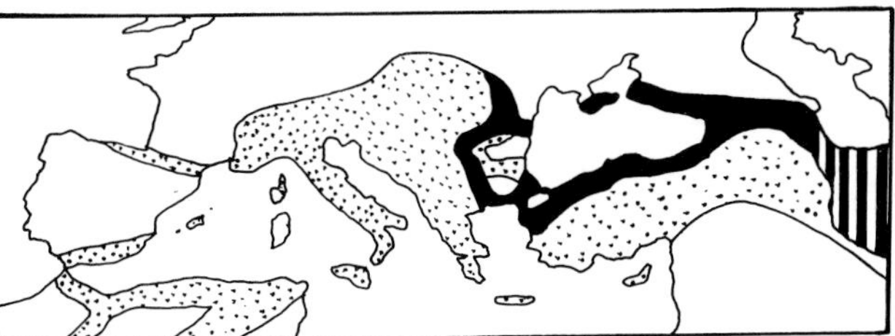

Figure 7.38. Outline of the Cimmerian Continent (solid black in the eastern Mediterranean and striped in Iran), together with the rest of the Alpine belts. Figure redrawn after Şengör *et al.* (1984). Published with permission from the authors and the Geological Society Publishing House.

(1984) includes the Pontides in the Cimmerian Continent just to the west of Iran on the southern Paleo-Tethys margin during the early Mesozoic; in Figure 7.38 the Mediterranean elements included in the Cimmerian Continent are outlined in black. However, for post-Liassic time both models place the Pontides near the Crimea Peninsula on the northern Black Sea coast of the Ukraine (Figures 7.27 and 7.29), which is the type area of Cimmerian deformation. Despite the tentative paleomagnetic indications for a southerly paleolocation during the Late Jurassic, more work is needed on pre-Tertiary rocks from the Pontides before it can be concluded that these models must be modified.

On the basis of his Jurassic and Late Triassic results, Lauer (1981a) concluded that the Turkish elements were, in fact, better placed even farther south than the Levant or Egyptian coasts, arguing instead for a paleolocation to the (south-)east of the Arabian Peninsula. On the basis of improved African reference poles (as used for Figure 7.37 and Table 7.3) this hypothesis is not supported. The Kimmeridgian best-estimate pole for Africa, as deduced from the discussion in Chapter 6, eliminates the need for a relative displacement of such large magnitude between Turkey and Africa, whereas the Early and Late Triassic reference poles for Africa (as discussed in Chapter 5) are fairly imprecise with their average, at any rate, agreeing fairly well with Lauer's Late Triassic paleolatitude value for the Antalya Taurides (see Figure 7.37).

For the Sakarya Continent and the other areas just to the south of the Pontides, many more high-quality data are needed as well before any paleolocation models can be substantiated. The discussions about a northern versus southern Tethyan paleolocation of the Sakarya Continent (e.g., Saribudak, 1991) are therefore somewhat premature.

Lastly, the interesting declination results from the Troodos Ophiolite in Cyprus deserve to be mentioned. It has long been known that the Troodos Massif has been rotated counterclockwise by about 90 degrees since its formation in the Late Cretaceous. Clube, Creer & Robertson (1985) have documented that this rotation of the entire complex probably took place during the latest Cretaceous and Paleocene. The present-day north–south orientation of the dikes in the Troodos sheeted dike complex was therefore paleo-east to west at the time of sea floor intrusion, which fits the Late Cretaceous spreading trends inferred for the Neo-Tethys in this area (Whitechurch *et al.*, 1984). Bonhommet, Roperch & Calza (1988) and Morris, Creer & Robertson (1990) have further refined the paleomagnetic picture by documenting local rotations near a presumably fossil transform fault (paleo-north to south) within the complex, allowing insights into ridge-transform crustal accretion processes that are very inaccessible in modern-day situations.

Table 7.4. *Declinations and paleolatitudes for the Levant at 33°N, 36°E from European and West Gondwana reference poles*

Age interval	Europe		W. Gondwana	
	D	Lat.	D	Lat.
Tl (37–66)	8	23	5	27
Ku (67–97)	17	23	349	13
Kl (98–144)	8	14	333	2
Ju, uJm (145–176)	10	11	338	2
**Same	16	21		
*Kimmeridgian Best Estimate			334	−7
lJm, Jl (177–195)	27	29	345	13
Tru/Jl (196–215)	36	32	351	5
Tru, uTrm (216–232)	43	20	344	5
Trl/m, Trl (233–245)	36	12	351	−18

*For the Kimmeridgian, a 'best estimate' pole is used from Table 6.3 rotated into northeast African coordinates. For other explanations see Table 7.1, including the note on **. Reference poles from West Gondwana (Table 5.8) are those in northeast African coordinates.

Paleomagnetic results from the Levant

Several Neogene, Cretaceous and Jurassic results are available from Israel and Lebanon with Q ≥ 3 (Table A8). The expected declinations and paleolatitudes for a site at 33° N, 36° E in the area have been calculated from the European and African reference poles and are listed in Table 7.4. These can be compared with the observed declinations and paleolatitudes in Figures 7.39 and 7.40

The observed declinations show sizeable rotations with respect to the expected African pattern (Figure 7.39). The sampling locations, to the west of the Dead Sea Transform Fault, have clearly been affected by local rotations, thought to be due to a local high-angle fault pattern and en-echelon movements of the blockfaulted segments (see Quennell, 1984; Ron, Nur & Eyal, 1990; Chaimov *et al.*, 1990). The observed paleolatitudes, on the other hand, should not be affected by such rotations, since inclinations are invariant, provided that paleohorizontal is well determined. This is the case for the Jurassic and Cretaceous stratified volcanic rocks, intercalated in well-dated gently dipping sedimentary sequences in the Lebanon and Anti-Lebanon ranges (Figure 7.41). These sequences, moreover, retain coherence with older basement (Garfunkel & Derin, 1984) and are thus autochthonous, even if locally rotated. The aggregate data set (Table A8) shows normal and reversed polarities and their paleomagnetic directions appear to be well determined.

The observed paleolatitudes are plotted in Figure 7.40. The Early Cretaceous and Late Jurassic results range from 11° South to 6° North; their relationship to the African predicted pattern was already mentioned in Chapter 6. A result dated as Late Cretaceous from Israel (Helsley & Nur, 1970) has a paleolatitude that is too low, but it is perhaps possible that these Carmel Region Basalts are Early Cretaceous, given the general ages of the abundant volcanism in the area. In previous compilations, the Late Jurassic to Early Cretaceous Levant results as well as those from Turkey discussed above were generally unexplained or questioned (e.g., Westphal *et al.*, 1986). With a better understanding of the internal plate boundaries

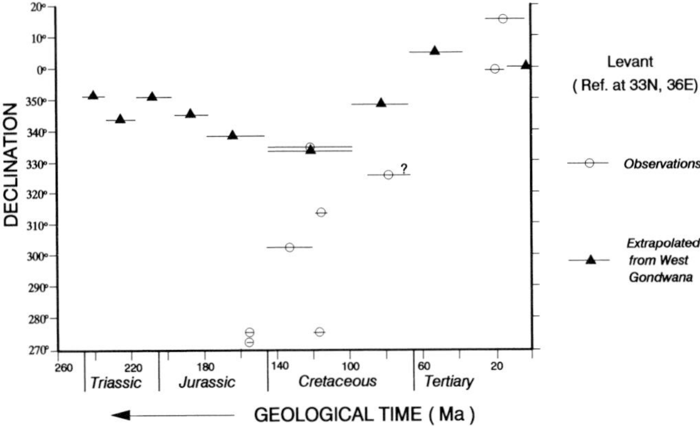

Figure 7.39. Declinations versus time for the Levant, together with the values expected for the common reference site at 33° N, 36° E as extrapolated from the African paleopoles (solid triangles) according to Table 7.4.

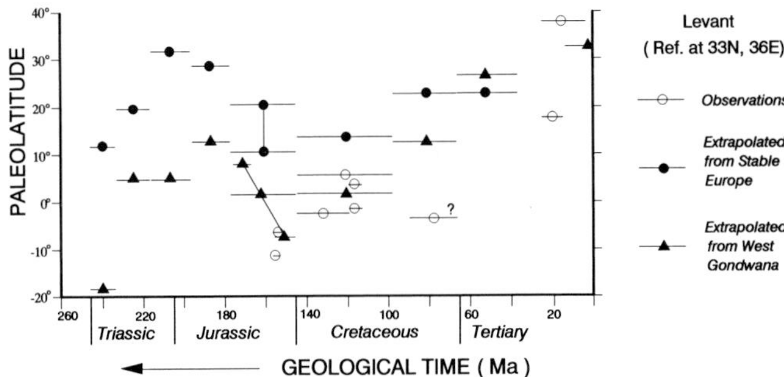

Figure 7.40. Paleolatitudes versus time for the Levant, together with the values expected for the common reference site at 33° N, 36° E as extrapolated from the Stable European (solid circles) and African paleopoles (solid triangles) according to Table 7.4. For the Late Jurassic, different reference paleolatitude options (as explained in Chapter 6) are connected by solid lines.

within Africa, an adjustment to close the Red Sea and several new paleopole determinations (Chapter 6), these results are now compatible with the overall Atlantic–Mediterranean framework.

Discussion

From the preceding descriptions of the available results it can be seen that several of the main paleogeographic hypotheses (e.g., Şengör *et al.*, 1984; Robertson & Dixon, 1984) remain unsubstantiated by the scarce paleomagnetic database. Where a somewhat larger data set exists, such as for the Pontides, the poor quality of many of the results hampers successful evaluation, although preliminarily it appears that the paleomagnetic paleolatitudes indicate a southern Tethyan origin for the Pontides in conflict with all geological models. Clearly much more work needs to be done, especially on pre-Late Cretaceous rocks, with sufficient samples and statistical precision. It would also be helpful to observe reversals in the studied sequences and to obtain positive field tests.

The age of the Neo-Tethyan spreading is hypothesized to be either largely pre-Liassic or Late Jurassic to Cretaceous (Şengör *et al.*, 1984 versus Robertson & Dixon, 1984). These models were conceived to explain the disappearance of the Paleo-Tethys and the ophiolite formation in Greece and Turkey, which remains one of the principal unresolved questions for the eastern Mediterranean. In more detail, there certainly have not yet been any paleomagnetic contributions either to the problem of whether the ophiolites each represent their own spreading basin or whether they were transported from a common root zone over large distances (e.g., Ricou, Marcoux & Whitechurch, 1984; Whitechurch *et al.*, 1984; see also Robertson & Dixon, 1984, pp. 25–31).

Iran, central Afghanistan and Oman

The Iran and Afghanistan blocks and mobile belts are separated from the Arabian Platform to the south by a remarkable zone of ophiolites (the *Croissant ophiolitique péri-arabe* of Ricou, 1971). The area is confined in the west and east by the Turkish and Pamir Syntaxes (Figure 7.1), which are the narrow constrictions between the Eurasian foreland, on the one hand, and the Arabian and Indian subcontinental indentors, on the other. The ophiolites (Figure 7.42) are well described by Ricou (1971) who documents their allochthonous nature, their emplacement age as latest Cretaceous, and their association with radiolarian cherts, carbonate rocks and colored mélanges with ages between Triassic and Late Cretaceous. They are overthrust in the Zagros Zone upon the Arabian Platform as remnants of a Mesozoic Neo-Tethyan Ocean and continue for 3000 km from the Mediterranean southeastward

Figure 7.41. Photograph of the nearly horizontal Jurassic and Cretaceous strata in the Lebanon, indicating that structural control of the paleomagnetic sampling sites (Van Dongen, Van der Voo & Raven, 1967) is adequate. Prof. Teddy Raven is standing on the right in the foreground, and Pieter Van Dongen is just to the right of the Land Rover.

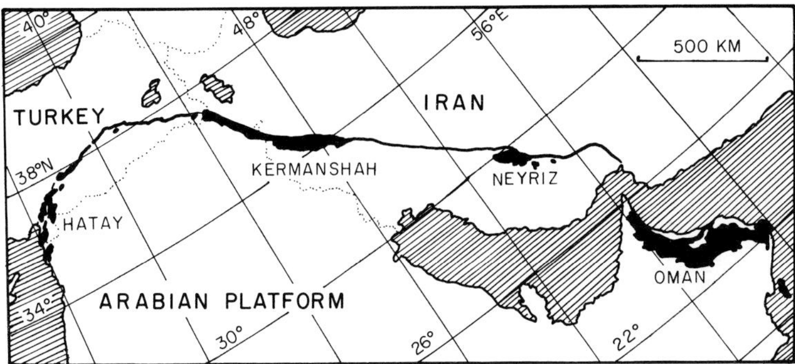

Figure 7.42. The peri-Arabic ophiolite belt between the Mediterranean and Oman. Figure redrawn after Ricou (1971).

into Oman (Figure 7.42), where studies by Reinhardt (1969) and his colleagues of Shell Oil Company have documented their well-exposed allochthonous structural and stratigraphic setting.

Most workers argue for a paleogeographic setting in common between the different areas of Iran and central Afghanistan (e.g., Stöcklin, 1977; Davoudzadeh, Soffel & Schmidt, 1981; Wensink, 1981; Şengör *et al.*, 1984), although the presence of major fault zones, ophiolites, and mélanges between the more stable blocks of Alborz–Great Kavir, Lut and Sistan-Helmand (Figure 7.43) and paleomagnetic documentation of relative rotations between them demonstrate that the area is not one single structural entity. Boulin (1991) has recently published a comprehensive summary of the tectonic evolution of this area.

Paleomagnetic results for the area are listed in Table A8 (Part IV) and their paleolatitudes have been plotted in Figure 7.44, together with the reference values extrapolated for a common location at 35° N, 60° E from the Eurasian (i.e., Stable European) and African continents, as provided in Table 7.5. There are very few results published for Afghanistan, so the following discussion centers mostly on Iranian data.

The paleolatitudes reveal an interesting pattern, which is in remarkable contrast with the ambiguity of the data from Turkey discussed above. First, I note that because we are progressing eastward in the (Paleo-)Tethys, we see greater separation during Mesozoic time between the Eurasian and Arabian–African margins at the longitudes of Iran; this separation is of the order of 30° until it begins to diminish in Early Cretaceous (about 20°) and, especially, latest Cretaceous time when it reduces to about 5°. There is, therefore, potentially better resolution to determine north–south motions in this area than there was in the Eastern Mediterranean. Second, I recall the hypothesis of the Cimmerian Continent, discussed earlier and illustrated in Figure 7.27, in which Şengör *et al.* (1984) included the Iranian and Afghanistan (= Helmand) areas. It turns out that the observed paleolatitudes vindicate, in a very convincing way, the model of the Cimmerian continental motion of Şengör & colleagues, at least insofar as Iran is concerned.

Figure 7.45 is a simplified version of Figure 7.44 in which the principal relationships between Eurasia, Africa and Iran can be seen at a glance. During Silurian to earliest Devonian time, Eurasian as well as African paleopoles predict equatorial paleolatitudes for Iran, which accords with the observation of Wensink (1983) on the Kerman Old Red Sandstone. However, in the Late Paleozoic and Mesozoic,

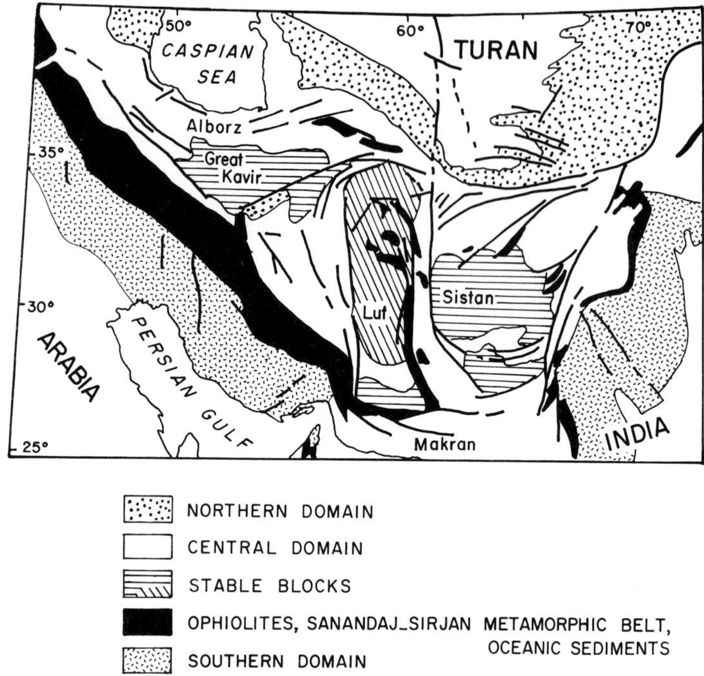

Figure 7.43. Structural sketch map of Iran, Afghanistan and adjacent areas, illustrating the paleogeographic (but not structural) entity existing between the Eurasian (Turan) and Arabian-Zagros borders in this area. Figure redrawn and modified after illustrations from Stöcklin (1977) and Davoudzadeh, Soffel & Schmidt (1981; *N. Jb. Geol. Paläontol. Mh.*, Heft 3, pp. 180–92). Reproduced by permission of the authors and E. Schweizerbart'sche Verlagsbuchhandlung.

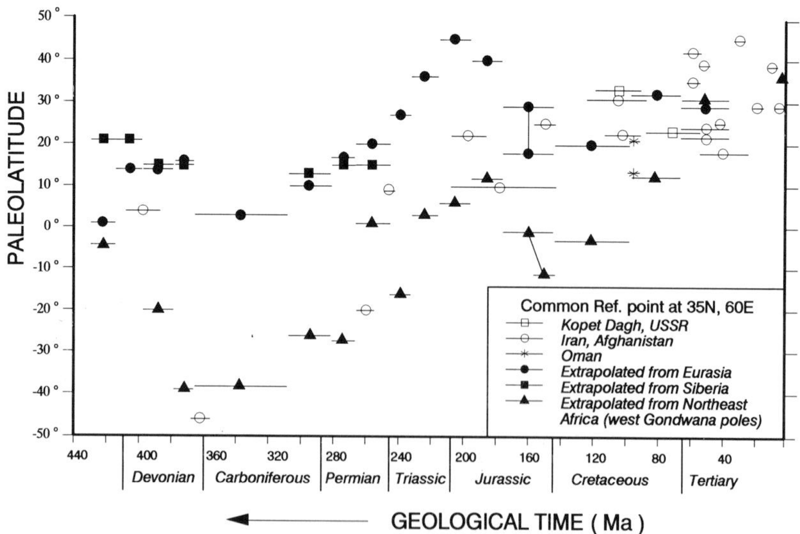

Figure 7.44. Paleolatitudes versus time for Iran, Afghanistan, Kopet Dagh (open squares) and the Oman Ophiolite (*) together with the values expected for the common reference site at 35° N, 60° E as extrapolated from the Stable European (dots), Siberian (solid squares) and African paleopoles (solid triangles) according to Table 7.5. For the Late Jurassic, different reference paleolatitude options (as explained in Chapter 6) are connected by a line.

Table 7.5. *Declinations and paleolatitudes for Iran, Afghanistan, and Oman at 35°N, 60°E from European, Siberian, and West Gondwana reference poles*

Age interval	Europe		Siberia		W. Gondwana	
	D	Lat.	D	Lat.	D	Lat.
Tl (37–66)	12	29			7	31
Ku (67–97)	21	32			358	13
Kl (98–144)	16	20			347	−3
Ju, uJm (145–176)	19	18			352	−1
**Same	20	29				
*Kimmeridgian Best Estimate					351	−11
lJm, Jl (177–195)	28	40			354	12
Tru/Jl (196–215)	37	45			2	6
Tru, uTrm (216–232)	50	36			356	3
Trl/m, Trl (233–245)	45	27			11	−16
Pu (246–266)	44	20	49	15	359	1
Pl (267–281)	46	17	54	15	355	−27
Cu/Pl, Cu (282–308)	49	10	57	13	13	−26
Cm, Cl, Du/Cl (309–365)	59	3			12	−38
Du, Dm/Du (366–378)	68	16	66	15	53	−39
Dm, Dl (379–397)	73	14	66	15	67	−20
Su/Dl (398–414)	96	14	111	21		
Su, Sm (415–429)	69	1	111	21	32	−4

*For the Kimmeridgian, a 'best estimate' pole is used from Table 6.3 rotated into northeast African coordinates (location at 43°N, 252°E). For Paleozoic time, Siberian reference poles are taken from Table 5.4. For other explanations see Table 7.1, including the note on **. Reference poles from West Gondwana (Table 5.8) are those in northeast African coordinates.

the reference paleopoles from Eurasia and Africa would predict very different paleolatitudes for Iran. The paleomagnetic observations on Iranian rocks allow us to make a clear choice about their pre-Jurassic affinity with Gondwana. This successful application of paleomagnetic studies is due largely to the well-designed research program of Wensink & colleagues, which – surprisingly – has not achieved the widespread recognition it deserves. The good to high-quality observations of Wensink & colleagues for the Late Devonian and the Middle Permian reveal that Iran tracks the African predictions rather precisely (Figures 7.44 and 7.45). Then, in the Triassic and Jurassic, the Iranian data indicate latitudes more northerly than those predicted by Africa; they begin to resemble Eurasian ones (Wensink, 1979, 1981, 1982; Soffel & Förster, 1981). By Late Jurassic and Cretaceous time the Iranian values are in good agreement with those from Eurasia (Davoudzadeh *et al.*, 1981) and in clear discordance with the African predictions (Figure 7.45).

In detail there are still some nagging problems, however, because a Middle to Late Jurassic result from the Bidou Beds (Wensink, 1982) presents an 'African' paleolatitude of 13° S. However, in the absence of a fold test, and given the uncorrected, *in situ* direction that is close to directions expected for much younger times, it is quite possible that this result is based on a remagnetization. It has, moreover, a Q value of only 1 in Table A8.

Thus far, the discussion has centered on the paleolatitudes, but the declinations are also of interest: for the Lut Block, as part of a larger Central to East Iran

Microplate, a sizeable (135°) counterclockwise rotation has been documented by Soffel and colleagues of the University of Munich (e.g., Davoudzadeh *et al.*, 1981). This rotation is with respect to the Turan area of Eurasia (see Figure 7.43) and occurred in Jurassic through Paleogene time.

In the compilation of Table A8, two results have been included from the Kopet Dagh, located just to the north of the Iran–Turkmenistan border in an area that could well have drifted together with Iran (see Wensink, 1981). The ages of these observations with paleolatitudes of 23° to 33° N (squares in Figure 7.44), however, are Late Cretaceous to Early Tertiary so that the results are too young to test this idea.

Two late Cretaceous results have become available from the Semail Ophiolite in Oman (Luyendijk *et al.*, 1982), which give slightly lower paleolatitudes than those observed in Iran as well as those predicted for today's location with respect to Eurasia. The values, on the other hand, are slightly higher than those predicted from Northeast Africa, suggesting that at the time of their (late) Neo-Tethyan formation these ophiolites originated about midway between Iran and Arabia.

In summary of this section, then, it appears that Iran is one of the better examples of paleomagnetic success in its confirmation of a major geological model, namely that of the Cimmerian Continent (Şengör, 1979; Şengör *et al.*, 1984). The Cimmerian deformation of Late Triassic to Early Jurassic age, which gave its name to the continental phalanx that swept across the Paleo-Tethys (Figure 7.27), is reportedly very clear at the southern Eurasian border in the Caucasus, and east of the Caspian Sea in northern Iran and Turkmenia (Turan in Figures 7.27 and 7.43). Previously, such early Mesozoic orogenic deformation was often called 'late Hercynian', or also 'early Alpine', which gave rise to confusion about its causes (e.g., Şengör, 1987). This Cimmerian deformation and plutonism (Berberian & Berberian, 1981) can now be linked to a Jurassic continent–continent collision between Eurasia and Iran, and to a disappearance of the Paleo-Tethys in this area, contemporaneous with the formation of a Triassic–Cretaceous Neo-Tethys Ocean, itself now preserved only as the ophiolitic remnants illustrated in Figure 7.42.

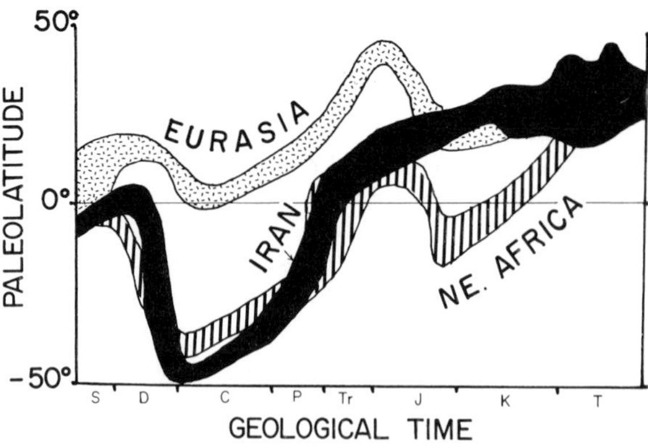

Figure 7.45. Simplified paleolatitude versus time plot for Iran, together with the values predicted for the area if it had remained coherent with Eurasia (stippled pattern) or Africa (striped pattern) derived from Figure 7.44 by marking the envelope of the individual data points.

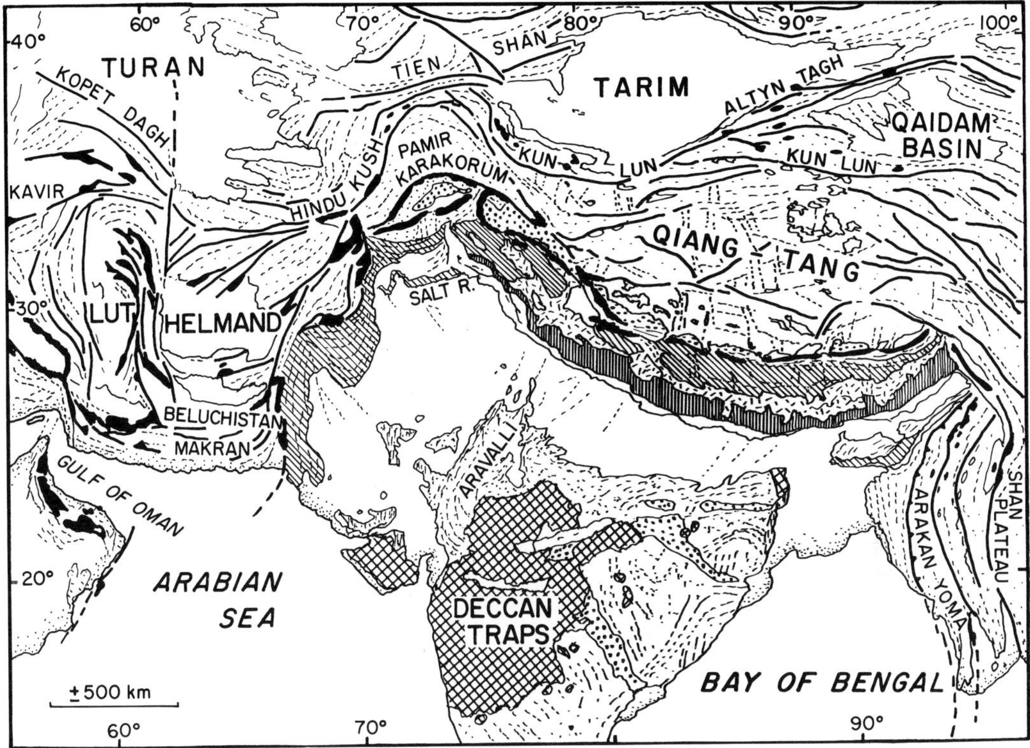

Figure 7.46. General tectonic map of Central Asia. Ophiolites and mélanges are indicated in black. Figure redrawn after Gansser (1981).

The central Asian blocks and mobile belts

Like the previous section, the present one also concerns an area between two major syntaxes of the Alpine–Himalayan mountain belt: this area is located between the Indian subcontinent to the south and the Kazakhstan, Junggar, and Siberian continental blocks of 'Stable Eurasia' to the north, and between the Pamir Syntaxis in the west and the Yunnan Syntaxis in the east (Figures 7.1 and 7.2), with the North and South China blocks forming its eastern boundaries. From south to north this area includes the Himalayas, Tibet (Lhasa and Qiangtang, the 'roof of the world'), and Tarim (called Serindia by Şengör in Figure 7.2).

In the Pamir Syntaxis (also called the Pamir Knot) the Indian subcontinent and Kazakhstan are separated over a relatively short distance by the mountain ranges of the Kashmir and Kohistan Himalayas, the Hindu Kush and Karakorum, the Pamirs and western Tian Shan (Figure 7.46), which radiate from the 'Knot' in all directions like a bunch of long straws tightly held in a hand. In contrast, the eastern or Yunnan Syntaxis resembles the heavy curtains of an opera house draping down over the eastern end of the Indian subcontinent (Figures 7.2 and 7.46) and separating latter from Tibet to the north and the South China and Indo-China blocks to the east.

The Tibetan Plateau in the middle of this area is probably the most remarkable feature of the Earth as viewed from space. Its general elevation of 5 km or more over an immense area (Figure 7.47) makes it unique, intriguing geoscientists for more than a century. The underlying crust is double the thickness normally found elsewhere, suggesting that either this crust underwent enormous lateral contraction or that it has been doubled by large-scale continental underthrusting. As we will see, paleomagnetists have attempted to constrain hypotheses about this enigma.

In detail, the high Tibetan Plateau to the north of the even higher Himalayan

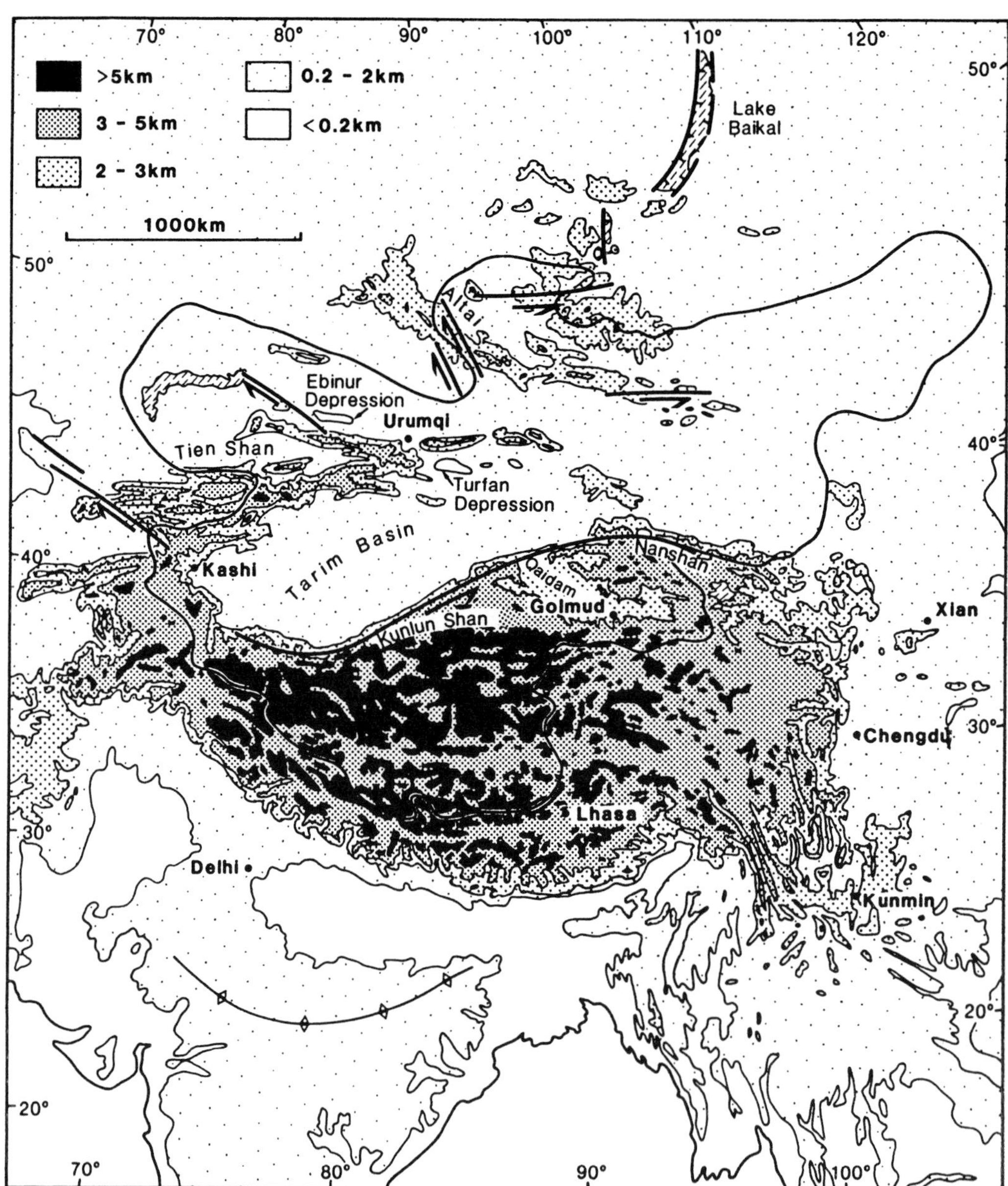

Figure 7.47. Simplified topographic map of the India–Eurasian collision zone. The solid outline indicates an internal drainage basin in and around Tarim. In central India, in the area of the Deccan Traps, a line with anticline symbols indicates the (positive) foreland flexural bulge of the Himalayan load, whereas the 'moat' occurs along the plains west–northwest and east–southeast of Delhi. Lake Baikal in the north is a Tertiary rift. Figure from Dewey *et al.* (1988), reproduced with permission from the authors and the Royal Society of London.

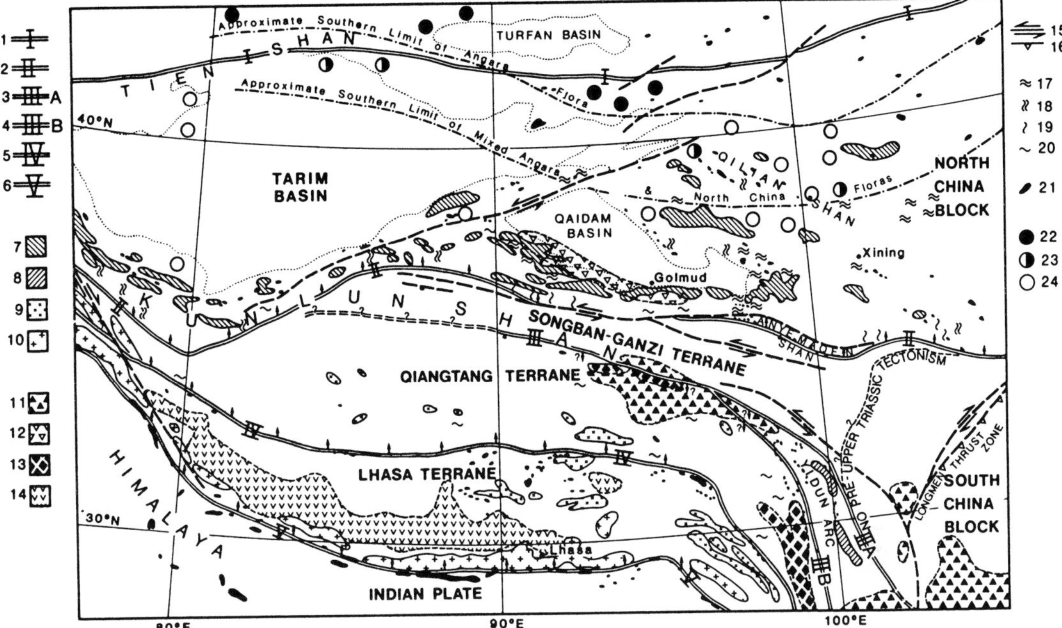

Figure 7.48. Tectonic map of Tibet, indicating the Indian Plate and the Lhasa and Qiangtang terranes and the Songpan-Ganzi accretionary belt. Paleomagnetic sampling sites in the Kunlun Mountains are between Golmud and the Songpan-Ganzi Terrane, to the south of the Qaidam Basin and to the southwest of the North China Block. Sutures are indicated by roman numerals: I = Tian Shan, II = Kunlun – Qingling, IIIA = Jinsha-Litang, IIIB = southeast Jinsha, IV = Banggong Co (Co = Lake), V = Indus-Zangpo. Other symbols are: 7 = Upper Paleozoic plutons, 8 = Triassic plutons, 9 = Cretaceous plutons, 10 = Late Cretaceous-Tertiary plutons, 11 = Permian rift-related rocks, including the Emei Shan Basalts of the western South China Block, 12 = Permian subduction-related rocks, 13 = Triassic subduction-related rocks, 14 = Tertiary subduction-related rocks, 15 = strike-slip faults, 16 = thrusts, 17 = Early Paleozoic tectonism, 18 = Carboniferous tectonism, 19 = Permian tectonism, 20 = Late Triassic-Early Jurassic tectonism, 21 = ophiolites, 22 = Angaran flora, 23 = mixed Angaran and North Cathaysian flora, 24 = North Cathaysian flora. Small arrows on sutures indicate inferred direction of subduction. Figure from Dewey *et al.* (1988), reproduced with permission from the authors and the Royal Society of London.

peaks of the Indian Plate is thought to consist of several more or less parallel zones (e.g., Figure 7.48 from Dewey *et al.*, 1988), which include from south to north: the Lhasa Terrane, the Qiangtang Terrane, the Songpan-Ganzi accretionary belt, the Qaidam (Tsaidam) Basin separated from the adjacent North China Block by the Qilian Shan (also Nan Shan; *Shan* = mountain in Chinese) and offset from Tarim by the Altyn Tagh Fault. These zones are separated by tectonic lineaments that are often marked by ophiolites (Pearce & Deng, 1988) and which have been interpreted as sutures by Dewey *et al.* (1988). A mountain belt separating Qiangtang in the south from Tarim and Qaidam in the north is called the Kunlun Shan (Figures 7.46 and 7.48). In some publications (e.g., Lin & Watts, 1988) this belt has erroneously been given separate terrane status (the 'Kunlun Terrane'); this area, together with Songpan-Ganzi (Figure 7.48), is better considered as part of a subduction-accretion complex (D. Rowley, pers. comm., 1992).

The paleogeography and paleolatitudes of the different terranes have long been a subject of study and speculation. Equivalents of the characteristic Gondwana (coldwater) fauna and flora of the Indian subcontinent and its Late Carboniferous to Early Permian glacial relicts are also found in the Lhasa and West Qiangtang terranes (Figures 7.2, 7.49) but apparently not in the Songpan–Ganzi accretionary belt. It has, therefore, been speculated that these more southern Tibetan terranes derived from the northern margin of Gondwana after the Early Permian, whereas the more northern and eastern terranes and blocks (Tarim, South and North China Blocks) either departed from Gondwana before the Late Carboniferous or were never part of it. The ophiolites between the Himalayas and Tibet are taken in most models to be remnants of a Neo-Tethys ocean which opened when the Tibetan terranes separated from Gondwana, whereas the ophiolites within Tibet are thought to be remnants of the Paleo-Tethys. Şengör & coworkers (1984, 1988; see also Şengör, 1987) include all Tibetan terranes in the Cimmerian Continent (see Figure 7.27). Because, clearly, these models can be tested by paleomagnetic data I will examine the evidence below in a similar fashion as used in preceding sections of this chapter. I will proceed with descriptions of the results for the different areas going from south to north.

The Himalayan suture area

The available paleopoles for the Himalayas, to the south of the Lhasa Block and its boundary of the Indus–Yarlung Zangpo (= *Tsangpo*, meaning river) suture, are compiled in Table A9. The paleolatitudes from the Himalayas can be compared with those predicted from the reference paleopoles of India (Table 7.6) in Figure 7.50: clearly, the match between the Himalayan observations and the Indian reference paleolatitudes is excellent. Many of these results have been obtained by the extensive efforts of Chris Klootwijk of Australia. I have deviated from my usual practice not to include paleopoles based on secondary magnetizations in the case of the Himalayas (Table A9). Although not all of them are precisely dated, they reveal (Figure 7.50) the remarkably fast northward drift of the Indian subcontinent in the

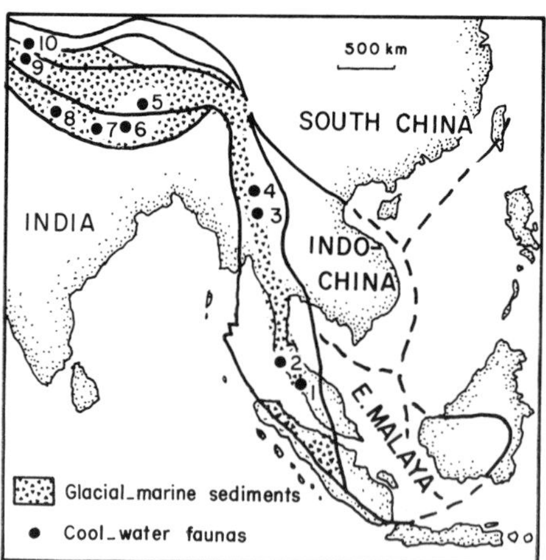

Figure 7.49. Distribution of Late Carboniferous to Early Permian glacial-marine sediments and Early Permian cold water faunas. Note the extension of these high-latitude indicators from India into the Himalayas, southern Tibet (Lhasa Block and Qiangtang Terrane) and into the outlined Sibumasu Block, extending from western Yunnan (China) through Burma, western Thailand, western Malaya and western Sumatra. Figure redrawn after Metcalfe (1988). Published with permission from the Geological Society Publishing House.

Table 7.6. *Reference paleopoles for East Asian terranes*

Age interval	India		Australia		Siberia		N. China		S. China	
	Lat.	Long.	Lat.	Long.	Lat.	Long.	Lat.	Long.	Lat.	Long.
Tl (37–66)	43,	285	61,	301	78,	177				
Ku (67–97)	21,	295	56,	318	72,	154	80,	170	84,	213
Kl (98–144)	12,	299	41,	338	70,	193	77,	213	76,	201
Ju, uJm (145–176)	2,	310	48,	349	66,	191	71,	224		
**Same					72,	162				
lJm, Jl (177–195)	11,	316	46,	8	67,	129				
Tru/Jl (196–215)	13,	307	55,	4	60,	119				
Tru, uTrm (216–232)	−9,	320	32,	350	50,	134			45,	224
Trl/m, Trl (233–245)	−7,	309	42,	342	51,	149	42,	26	46,	215
Pu (246–266)	−16,	295	43,	320	42,	161	46,	3	47,	232
Pl (267–281)	−17,	281	46,	302	38,	159				
Cu/Pl, Cu (282–308)	−5,	289	55,	320	34,	158			22,	225
Cm, Cl, Du/Cl (309–365)	32,	270	84,	137	31,	154				
Du, Dm/Du (366–378)	22,	243	62,	203	28,	151				
Dm, Dl (379–397)	37,	281	77,	80	28,	151			−9,	190
Su/Dl (398–414)					−4,	121				
Su, Sm (415–429)	37,	238	58,	173	−4,	121			5,	195
Ou/Sl, Ou, Om (430–467)	−14,	187	−2,	215	−22,	130				
Ol/m, Ol (468–505)	3,	200	17,	206	−40,	132	43,	333	−39,	236
C–u, C–m, C– (506–542)	6,	199	18,	203	−40,	142				
C–l (543–575)	34,	212	37,	177	−35,	188			37,	206

Age abbreviations as in Table 5.1. The reference poles are given in the coordinates of each continent and are taken from the following sources: India and Australia – Table 5.2A for Late Jurassic and younger, Table 5.9 rotated into Indian or Australian coordinates for Early Jurassic and older; Siberia – Table 5.1 for Late Jurassic and younger, Table 5.7 for the Triassic and Early Jurassic intervals, Table 5.4 for the Paleozoic; North China and South China Blocks – Table 5.5.

**For the Late Jurassic there are two options for the Siberian reference pole (see Table 5.1).

Early Tertiary (65–50 Ma) with velocities of 10–15 cm per year, slowing down substantially after the Middle Eocene when the subcontinent of India had fully collided and began to indent the central Asian areas (Molnar & Tapponnier, 1978; Tapponier *et al.*, 1986).

The other remarkable feature of Figure 7.50 is the Late Carboniferous to Late Permian interval characterized by high (>50°) southerly paleolatitudes. Results from the Blaini (glacial) beds in the Himalayas show paleolatitudes of about 55° to 60°, in good agreement with the coldwater facies and fauna and the glacial relicts in this part of the world (Klootwijk, Jain & Khorana, 1982).

One Late Devonian result is included from the eastern Hindu Kush (* in Figure 7.50), which is not, strictly speaking, part of the Himalayas; nevertheless, its paleolatitude agrees well with results from India.

In Figure 7.51 the declinations of India and the allochthonous Himalayas can be compared. Klootwijk (1981) has demonstrated that, with only a few exceptions, the Himalayan results are rotated clockwise by about 15° with respect to India. The exceptions are labeled in Figure 7.51 and include results from locally rotated rocks in the Indus Suture Zone, the eastern Hindu Kush and in Kashmir. Figure 7.52 also illustrates this rotation in a comparison between the reference paleopoles from India and Australia (rotated into Indian coordinates) connected by the solid line and those obtained from the Thakkola region in the Nepal Himalayas connected by the dashed line. Bingham & Klootwijk (1980) have argued that a large rotated portion of the Himalayas, when restored to a pre-collision position, reflects the extent of 'Greater India'. According to their model, the lower crustal basement of the Himalayan nappes around Thakkola and elsewhere has become incorporated (by underthrusting?) in the thickened Tibetan Plateau. In a later paper, Klootwijk, Conaghan & Powell (1985) have extended their analysis to include Tibet and have argued for large-scale oroclinal bending. I will return to the questions of crustal thickening and oroclinal bending after discussing the results from Tibet and more northerly terranes.

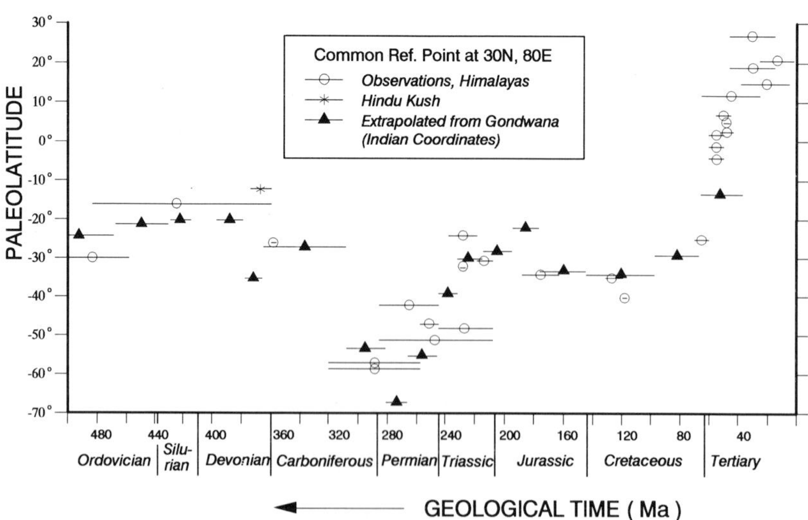

Figure 7.50. Paleolatitudes versus time for the allochthonous Himalayas and eastern Hindu Kush (*), together with the values expected for the common reference site at 30° N, 80° E as extrapolated from the mean East Gondwana (India) paleopoles of Table 7.6.

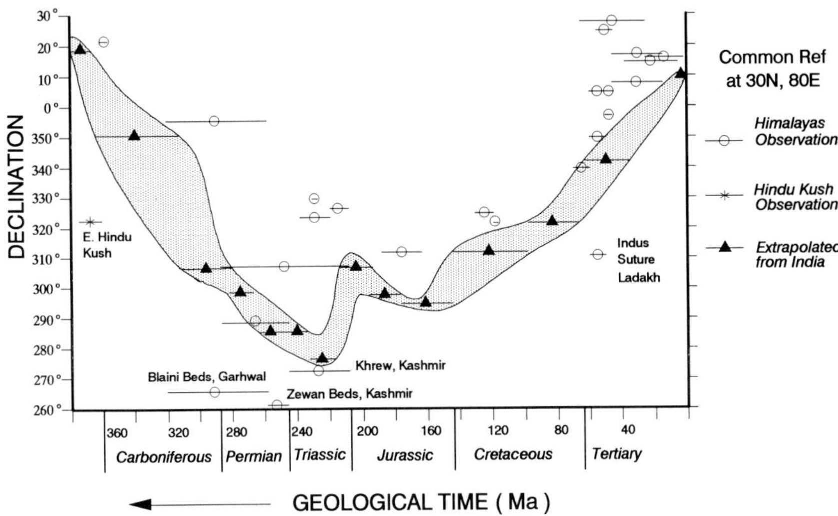

Figure 7.51. Declinations versus time for the allochthonous Himalayas and eastern Hindu Kush (*), together with the values expected for the common reference site at 30° N, 80° E as extrapolated from the mean East Gondwana (India) paleopoles of Table 7.6. Many observed declinations deviate in a clockwise sense; the ones deviating counterclockwise have been labeled.

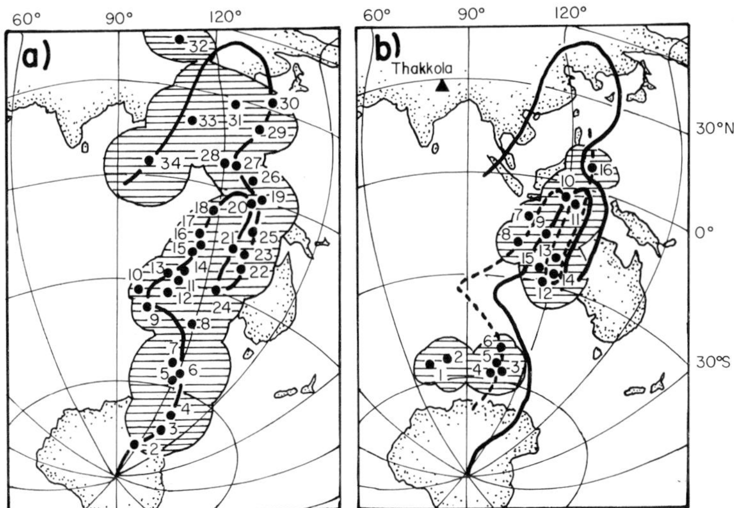

Figure 7.52. Paleopoles from a) Stable India, including results from the Indo-Australian plate transferred to Indian coordinates and b) the Thakkola region in the allochthonous Himalayas of Nepal. The solid line in b) repeats the apparent polar wander path of the plot a), whereas the dashed line connects the Nepal paleopoles. Numbers refer to individual poles listed in the original publication. Figure redrawn after Klootwijk (1981).

Tibetan terranes

In order to compare the available paleomagnetic results from Tibet (Table A9) with results from the surrounding continental blocks, the reference paleopoles from Siberia, North China, South China and India (Table 7.6) have been used to compute

Table 7.7. *Declinations and paleolatitudes for Tibet at 35° N, 90° E from the reference paleopoles of Table 7.6*

Age interval	India	Siberia	N. China	S. China
	D Lat.	D Lat.	D Lat.	D Lat.
Tl (37–66)	349, −11	15, 35		
Ku (67–97)	333, −29	22, 41	12, 36	6, 32
Kl (98–144)	324, −36	22, 28	12, 27	15, 29
Ju, uJm (145–176)	306, −37	27, 27	15, 21	
Same**		22, 39		
lJm, Jl (177–195)	308, −27	23, 51		
Tru/Jl (196–215)	317, −31	28, 59		
Tru, uTrm (216–232)	287, −38	51, 55		31, 0
Trl/m, Trl (233–245)	299, −45	50, 45	298, 41	35, 5
Pu (246–266)	304, −61	60, 36	309, 26	25, −1
Pl (267–281)	329, −69	65, 36		
Cu/Pl, Cu (282–308)	325, −55	70, 35		44, −19
Cm, Cl, Du/Cl (309–365)	360, −23	75, 37		

Age abbreviations as in Table 5.1. The declinations (D) and paleolatitudes (Lat.) are calculated for the common site at 35° N, 90° E from the reference poles given in Table 7.6.

**For the Late Jurassic there are two options for the Siberian reference pole (see Table 5.1).

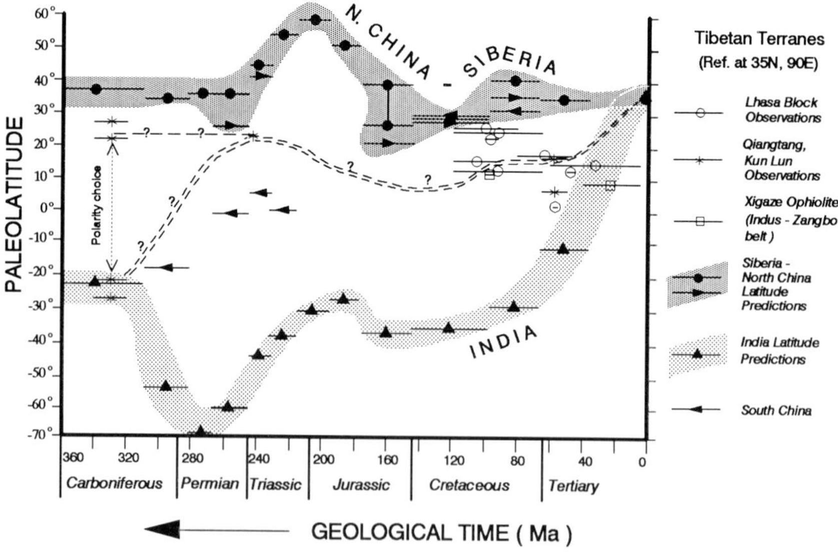

Figure 7.53. Paleolatitudes versus time for the Tibetan terranes (Indus-Yarlung Zangpo Suture ophiolite = open squares; Lhasa = open circles; Qiangtang and elements of Kunlun = *), together with the values expected for the common reference site at 35° N, 90° E as extrapolated from the Siberian (solid circles), Indian (solid upright triangles), North China (right-pointing triangles) and South China paleopoles (left-pointing triangles) according to Tables 7.6 and 7.7. For the Late Jurassic, different reference paleolatitude options for Siberia (as explained in Chapter 6) are connected by a line. The Early Carboniferous Kunlun observations have a polarity ambiguity, so that the results can indicate northern or southern hemisphere provenance; the double-dashed line is the preferred paleolatitude–time trajectory of the Kunlun areas studied.

the declinations and paleolatitudes predicted for a common site at 35° N, 90° E (Table 7.7). The paleolatitude comparison is shown in Figure 7.53.

The northerly paleolatitude band, ranging from 20° to 60° N, includes predicted values for Tibet if it had remained in its present-day position with respect to Siberia and North China. In contrast, the southern limit is provided by the Indian predictions, which range from 20° South in the Early Carboniferous, through the high southerly values of the Permo-Carboniferous glacial interval, to a Mesozoic 'plateau' of about 30° to 40° S, followed by the equatorial Early Tertiary values associated with the rapid northward drift of India after the latest Cretaceous. The paleopoles from South China would predict – if it had remained together with Tibet – equatorial paleolatitudes for Tibet during the late Paleozoic and values similar to those of North China and Siberia in the later half of the Mesozoic.

The ages of the observations from Tibet cluster mostly in the Cretaceous to Early Tertiary interval, where the paleolatitudes are between those from India on the one hand, and those from the Siberian and China blocks, on the other. These low paleolatitudes with respect to Eurasia are not likely to be due to inclination errors, given that the magnetizations of the widely studied Takena Red Beds are carried by hematite, as a CRM. Moreover, there are several results derived from volcanic rocks (Table A9), which confirm the paleolatitudes obtained from the sedimentary rocks. Clearly one can conclude that northward displacement of Tibet of some 10° to 15° occurred with respect to northern and eastern Asia after the Eocene. An even greater Late Tertiary displacement of about 20° is seen in the paleolatitudes of India itself and of the Himalayas (Figure 7.50). This will be of importance for the discussion below of the mechanisms for Tibetan crustal thickening and Neogene tectonics.

For pre-Cretaceous time, only three results have $Q \geq 3$ and they all come from what Lin & Watts (1988) have called the Kunlun Terrane, but this area should probably be called the Songpan-Ganzi accretionary belt or, by its geographical name, the Kunlun Shan. They are of no value for testing the Cimmerian Continent model of Şengör. The result of greatest interest is probably that of the Early Triassic Kunlun Dikes (Lin & Watts, 1988), which reveals a paleolatitude of about 20° N (Figure 7.53). Two Early Carboniferous results from the Dagangou Formation could be of normal or reversed polarity, although the declination values suggest a reversed polarity and, hence, a southerly paleolatitude (Lin & Watts, 1988; see also Dewey *et al.*, 1988); the latter choice would imply that the area formed part of the northern margin of Gondwana until the Middle Carboniferous, whereupon it may have separated and drifted northward. From Early Triassic through Early Tertiary times, the area resided south of the Eurasian margin in a location about 15° to the south of its present latitude. Obviously, paleolongitudes are not constrained, so the area may well have been to the west or east of its present longitude with respect to Eurasia.

Dewey *et al.* (1988), in a synthesis of the 1985 Anglo-Chinese Tibetan Geotraverse, have constructed a similar diagram as that of Figure 7.53, but one in which the paleolatitudes are not reduced to the same reference point; they propose, albeit in this case not on the basis of paleomagnetic data (Figure 7.54), that the Qiangtang Terrane drifted across the Paleo-Tethys in the Permo-Triassic, followed by a similar journey of the Lhasa Terrane during the Triassic and Jurassic. A paleontological analysis of the Tibetan terranes (Smith & Xu, 1988) appears to support somewhat older (Permian) ages for these crossings. In terms of models for the evolution of Tethys, the Lhasa and Qiangtang terranes, therefore, may well have belonged to the Cimmerian Continent of Şengör, whereas the Kunlun Terrane of Lin & Watts (1988) did not, given its pre-Permian crossing (if any took place at all).

The model of Dewey & colleagues includes the Kunlun Shan, Tarim, and North

China in the same tectonic element. We will see shortly that there is some paleolatitudinal evidence from Tarim for this association, although an earlier (Devonian) time of crossing might be indicated by the data from Tarim; moreover, the declination differences between Tarim and North China, to be discussed below, clearly indicate relative rotations. Alternatively, one could conclude from Figure 7.53 that the transfer of the Kunlun Shan and Songpan-Ganzi areas from Gondwanan to Eurasian shores took place roughly at the same time as the late Paleozoic northward movement of the South China Block. This possible association merits further paleomagnetic investigation if other suitable Paleozoic-Early Mesozoic rocks can be found in the Kunlun Shan and Songpan-Ganzi accretionary belt.

The suture between the Himalayas (i.e., the India Plate) and the Lhasa Block is marked by ophiolites, one of which (Xigaze) has been studied paleomagnetically by Pozzi *et al.* (1984). Early Cretaceous (Albian) and secondary Neogene paleolatitudes are around 10° N, indicating that the ophiolite formed near the northern margin of the Neo-Tethys, slightly to the south of the Lhasa Block (Figure 7.53).

Tarim and Junggar

Figures 7.46 to 7.48 illustrate the location of the Tarim Basin, which is separated on its northern side from the Junggar region by the Tian (= Tien) Shan, along 'suture I' in Figure 7.48. The Junggar Basin in northwestern-most China is wedged between Siberia to the northeast, Kazakhstan to the northwest and Tarim to the

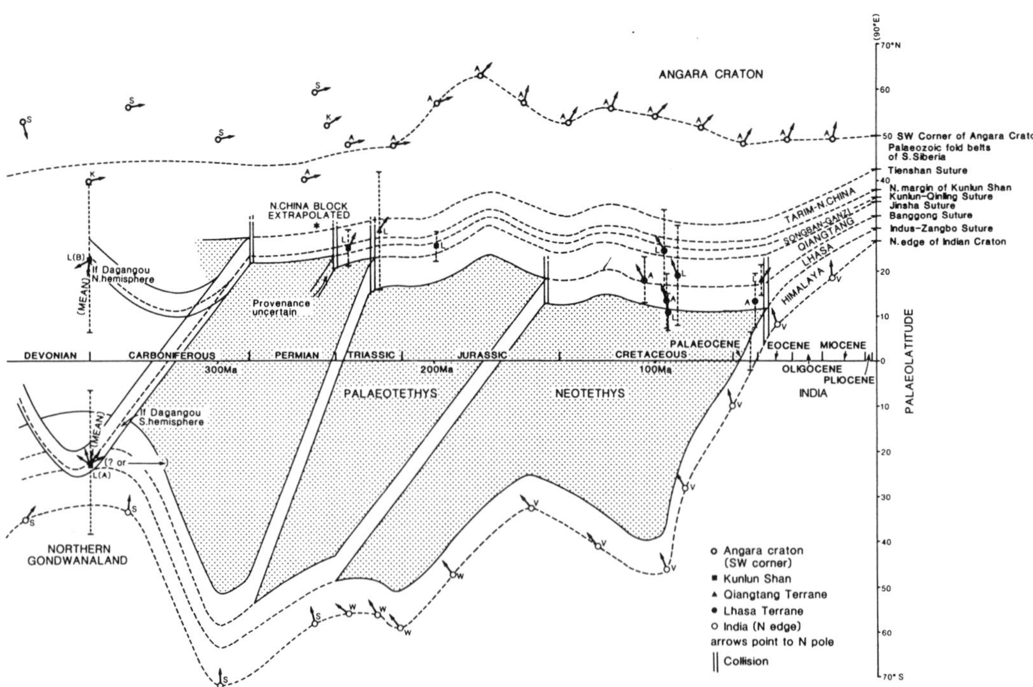

Figure 7.54. Paleolatitudes versus time for the Tibetan terranes, together with the paleolatitude trajectories of points located at 90° E on the present-day margins of India and Siberia, indicating the transfer from Gondwana to Eurasian shores of the various terranes during the course of the late Paleozoic and Mesozoic. Figure from Dewey *et al.* (1988), reproduced with permission from the authors and the Royal Society of London.

Table 7.8. *Declinations and paleolatitudes for Tarim at 42° N, 80° E from the reference paleopoles of Table 7.6*

Age interval	India		Siberia		N. China		S. China	
	D	Lat.	D	Lat.	D	Lat.	D	Lat.
Tl (37–66)	342,	−2	16,	39				
Ku (67–97)	326,	−19	25,	44	13,	41	6,	38
Kl (98–144)	317,	−25	22,	32	11,	32	15,	34
Ju, uJm (145–176)	301,	−27	26,	30	12,	26		
Same**			24,	42				
lJm, Jl (177–195)	302,	−16	30,	54				
Tru/Jl (196–215)	311,	−20	39,	60				
Tru, uTrm (216–232)	284,	−28	59,	52			25,	3
Trl/m, Trl (233–245)	295,	−34	54,	43	289,	51	30,	7
Pu (246–266)	300,	−50	60,	32	302,	37	19,	2
Pl (267–281)	318,	−59	65,	32				
Cu/Pl, Cu (282–308)	327,	−48	70,	30			34,	−18
Cm, Cl, Du/Cl (309–365)	351,	−15	75,	31				
Du, Dm/Du (366–378)	17,	−24	79,	32				
Dm, Dl (379–397)	343,	−9	79,	32			83,	−21
Su/Dl (398–414)			130,	31				

Age abbreviations as in Table 5.1. The declinations (D) and paleolatitudes (Lat.) are calculated for the common site at 42° N, 80° E from the reference poles given in Table 7.6.
**For the Late Jurassic there are two options for the Siberian reference pole (see Table 5.1).

south and southwest. Carroll *et al.* (1990) have speculated that Junggar is a basin on top of trapped Paleozoic ocean-floor (or island arc) crust. It resembles the island-arc and ophiolite-studded geological setting of southern Kazakhstan and is likely its eastern continuation.

The paleolatitudes and declinations of the available results (Table A9) can be compared with the reference values from the neighboring continents (Table 7.8) in Figures 7.55 and 7.56. The Permo-Triassic is very well represented and fortunately a few good results are also available for the Cretaceous and the Devonian, owing in large part to the systematic travail of Yianping Li of Stanford University and colleagues. The Late Carboniferous and younger paleolatitudes from both Tarim and Junggar track those predicted by Siberia and North China, although many are slightly more southerly by about 5° to 10°, as if – even this far north in Asia – the India-Eurasia convergence has still considerable influence.

The two Devonian paleolatitudes, on the other hand, deviate from the Siberian predictions. I recall, however, that the Devonian and Silurian paleolatitudes from Siberia have been questioned by paleontologists (see Chapter 5) as being too far north. If their arguments are valid, it may well be that the Siberian reference values should be rather in the near-equatorial range observed in Tarim. On the other hand, and taken at face value, one could see in these results the justification for models that have Tarim move across the Paleozoic Tethys in advance of the Qiangtang and Lhasa (= Cimmerian) and Indian blocks.

The paleolatitudes predicted from Siberia and the North China Block for the

Tarim and Tibet areas (Figures 7.53 and 7.55) are very similar and might create the mistaken impression that these two continents remained joined together. This, however, is definitely not the case, as revealed by an inspection of the declination results (Figure 7.56) and the mean paleopoles (Table 7.6). It is simply a coincidence that the predicted paleolatitudes coincide at the western end of the North China Block and at the southern end of Siberia, as if to indicate that the pivot about which North China moved to close the Mongolian Ocean between it and Siberia is located in this area.

The plot of Figure 7.56 also shows that for late Paleozoic time, the declinations of Tarim and Junggar are different from those of Siberia and those of North China, indicating post-Middle Triassic rotations of the various areas of Tarim and Junggar with respect to both major continents (Li, 1990). It has been suggested that Siberia and Tarim-Junggar were approximately in the relative positions they have today only in later Tertiary times (Li, 1990), but there are few paleomagnetic data to constrain this.

Li (1990) has attempted to find the optimum Euler pole to bring the Tarim and Siberia paleopoles into agreement, and concluded that a clockwise rotation of 26° about a pivot at 60° N, 14° E best matched the data. This rotation would imply a startling post-Cretaceous movement of about 2000 km of Tarim to the northeast, with respect to Siberia, which Li relates to the India–Asia impingement tectonics. If this were so, it would amplify the above conclusion (Li, 1990) for the lack of pre-Tertiary coherence between Siberia and Tarim, but Li's interpretation seems to be rather unlikely.

Mongolia

There are four late Paleozoic results with $Q \geq 3$ from Inner and Outer Mongolia (Pruner, 1987; Zhao *et al.*, 1990). These are listed separately in Table A9, whereas

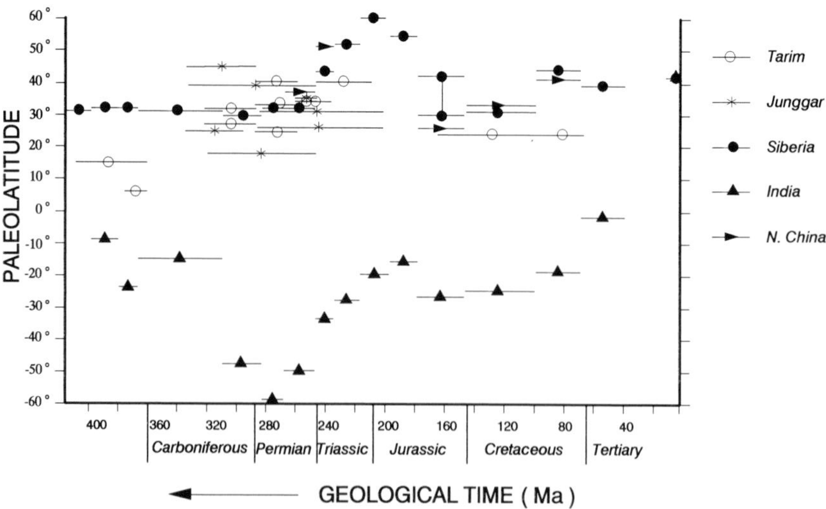

Figure 7.55. Paleolatitudes versus time for Tarim (open circles) and Junggar (*), together with the values expected for the common reference site at 42° N, 80° E as extrapolated from the Siberian (solid circles), Indian (solid upright triangles), and North China paleopoles (right-pointing triangles) according to Tables 7.6 and 7.8. For the Late Jurassic, different reference paleolatitude options for Siberia (as explained in Chapter 6) are connected by a line.

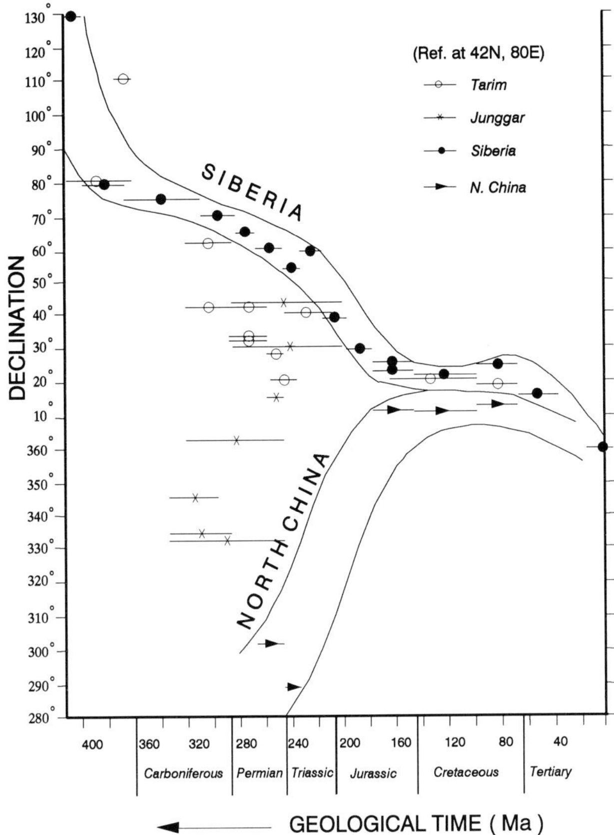

Figure 7.56. Declinations versus time for Tarim (open circles) and Junggar (*), together with the values expected for the common reference site at 42° N, 80° E as extrapolated from the Siberian (solid circles) and North China paleopoles (right-pointing triangles) according to Tables 7.6 and 7.8. For the Late Jurassic, different reference declination options for Siberia (as explained in Chapter 6) are given.

the Jurassic and younger results have been included in the list for North China (Table A6). This temporal separation may be artificial, because it is thought that the southeast Mongolian and North China areas were adjacent since at least the Late Permian. This is indeed supported by late Paleozoic results, because the observed northwest declinations and subtropical northerly paleolatitudes listed in Table A9 for the Late Permian of Mongolia match those predicted from the North China results. In contrast, Mongolia and Siberia were far apart because late Paleozoic Siberian paleolatitudes for the Mongolian area would be near 55° N. Thus, the North China–Siberia suture runs to the northwest of the Mongolian areas studied.

Discussion

Paleomagnetic investigations in Central Asia have become much more numerous in the last decade, and substantial progress has been made in the documentation of pre-Tertiary paleogeographic relationships among the major blocks. The successive northward movements of various terranes, as foreseen in the Cimmerian Continent model of Şengör, have been documented in a few cases and are generally accepted nowadays. Nevertheless, the timing of some terrane journeys remains to be resolved in several cases with more paleopoles for Early Mesozoic and older times.

The Neogene tectonics of the area is quite spectacular, as documented by the studies of Molnar, Tapponnier and numerous others. They have argued that strike–slip movements form a major component in the adjustments of the Asian crust to the continued northward convergence between India and Siberia (e.g., Tapponnier, Peltzer & Armijo, 1986). The indentation model (Molnar & Tapponnier, 1978) hypothesizes that two major and stable cratons like Siberia and India can continue to converge after initial collision because the 'softer' (i.e., younger) intervening crust escapes sideways out of the vise.

There are thus three principal models for the tectonics of Central Asia: (1) the indentation model just discussed, (2) the crustal thickening model for Tibet in which the crust laterally contracted by basement reactivation and internal strain, and (3) the underplating model in which Indian continental crust has been underthrusted (subducted?) below Tibetan crust. The high average elevation of Tibet must, of course, also be explained by that combination of models that is deemed most valid.

We have seen earlier in this section that northward convergence is not limited to the India–Tibet boundary alone; instead, it appears that the post-50 Ma northward displacement of India is about 20°, and that of Tibet about 10° to 15°, with respect to Siberia. Moreover, it can be argued – as always – that the paleolatitudinal data for the different blocks show considerable scatter and that resolution is lacking to make precise calculations. Be that as it may, the India–Siberia convergence appears to be taken up gradually over a wide Central Asian distance. This argument, if true, favors the indentation model over the other two for at least 50% of the total post-Eocene shortening, i.e., for about 1200 km of the total of more than 2400 km convergence after complete collisional contact was made between India and Tibet in the Eocene. The same point, in different form, has been made by Klootwijk (1981), who illustrated that in the Pamir Syntaxis as well, northward convergence is distributed over the largest possible distance between India and Eurasia. He based his arguments on the paleomagnetic data from the northern Pamir and western Tian Shan obtained by Bazhenov, Burtman & Gurariy (1978), which show, on average, about 10° northward movement with respect to Eurasia. The other 1200 km of post-Eocene shortening would be 'available' for the thickening/doubling of Tibetan crust. Because the pre-collisional crustal thickness of Tibetan crust is very poorly known, the adequacy of this amount remains rather unconstrained.

But this is only part of the story. An additional and interesting complication has been raised by Rage (1988), who noted that a species of frog occurs in India in beds with ages close to the Cretaceous–Tertiary boundary. This species was, until at least that time, not known from Africa, but it did occur in Eurasia. How did the freshwater animal possibly arrive in the island (?) of India? Could it be that even greater shortening took place than the post-Eocene 2400 km?

Klootwijk *et al.* (1985) have addressed this issue by proposing that initial contact was made between India and Tibet already in the Paleocene at about 60 Ma (if we add about 5–10 million years more, we can perhaps solve the frog problem). Subsequent convergence between India and Siberia proceeded, they claim, in three stages (Figure 7.57). Stage 1, from the Late Paleocene through the Early Eocene involved indentation of India into Tibet near the western Pamir Syntaxis, and subduction of the last of the Neo-Tethyan ocean floor between them in the east. Stage 2 lasted from the Middle Eocene through the Early Miocene and saw lateral extrusion (the escape tectonics of Tapponnier *et al.*, 1982) of southeast Asian elements. Stage 3, continuing today from the Early Miocene onward, includes the

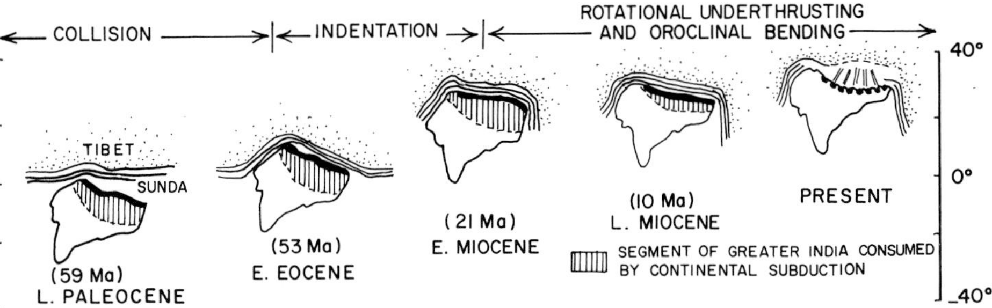

Figure 7.57. Three stages (collision, indentation, rotation) in the evolution of the India–Tibet–Eurasia collision in a paleolatitude framework. The segment of Greater India that was underthrusted in Tibet is marked by vertical shading, whereas the post-Late Miocene rotations of the allochthonous Himalayas and southern Tibet are indicated in the far-right diagram. Figure redrawn after Klootwijk, Conaghan & Powell (1985). Published with permission from Elsevier Science Publishers, BV.

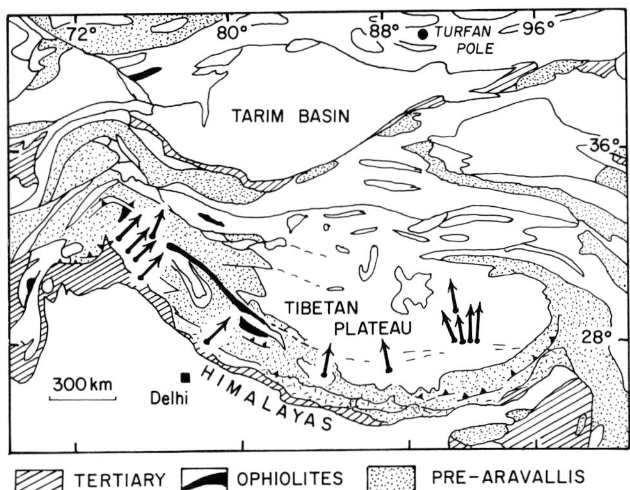

Figure 7.58. Map of the Himalaya–Tibet area with paleomagnetic declinations indicating systematic relative rotations. The Turfan Pole is proposed as a pivot point for the rotations, interpreted as showing oroclinal bending (but see text for discussion). Figure redrawn after Lowrie & Hirt (1986), as modified after Klootwijk *et al.* (1985). Published with permission from Elsevier Science Publishers, BV.

intracontinental underthrusting of Tibet mentioned above and oroclinal bending of the Himalayan–Tibetan area (Figure 7.58) since the Late Miocene. Thus, in this model, the high Tibetan elevation is a feature of only the Neogene. Geological observations, however, suggest that high Tibetan elevations have evolved progressively from the Eocene onwards (D. Rowley, pers. comm., 1992), thus casting doubt on the model of Klootwijk & colleagues.

A comment on the oroclinal bending (Figure 7.58) is in order. There is little doubt that the western and Nepal Himalayas rotated with respect to the Indian craton, as can be seen also in the data of Figure 7.51. And the Tibetan post-Cretaceous declinations are generally northerly (Table A9), deviating in a clockwise sense from those in India. Thus, the picture of Figure 7.58 is correct, but to call this 'oroclinal bending' is perhaps not warranted, because the entire Tibetan area may well have undergone a systematic, uniform-sense rotation.

A corollary of the indentation model is that the previously intervening blocks caught in the India-Siberia vise are driven sideways leading to rotation of such blocks in the ensuing 'escape tectonics'. Given that the indentation is asymmetric, these blocks have moved primarily towards the east and would, today, be found in South China, Indo-China and Indonesia, which then would have undergone a clockwise rotation (Tapponnier *et al.*, 1982). We will in the next section examine this hypothesis.

The Far Eastern blocks and mobile belts

This large area, ranging from eastern Tibet to the Philippines and from Korea to New Guinea, consists of many different blocks, the details of which remain yet to be sorted out. Paleomagnetism, it must be said, has not yet reached its fullest potential in this area, despite several valiant and sustained research efforts by François Chamalaun, Vincent Courtillot, Mike Fuller, Neville Haile, Bob McCabe, Y. Otofuji, S. Sasajima and Hans Wensink, to name a few. The tropical weathering, structural complexities, and inaccessability of the area undoubtedly play a role, but it must also be noted that several studies have failed to include full demagnetization analyses and have often been based on too few samples. And, as always, the problems of remagnetization have reared their ugly head, as is evident from the first block to be considered below.

Indo-China

Separated from the South China Block by the Song Ma–Song Da suture zone and from the Indian subcontinent by the two elongated terranes of West Burma and Sibumasu beyond the Uttaradit-Nan suture (Figure 7.59), the Indo-China Block contains the Khorat Plateau in eastern Thailand, the subject of many paleomagnetic investigations. However, surprisingly, all the strata give more or less the same paleomagnetic pole positions (Figure 7.60) despite the fact that their ages range from Paleozoic to Cretaceous. The suggestion that all these beds were remagnetized at the same time became gradually more and more obvious and has been confirmed by the study of Chen Yan & Courtillot (1989). Thus, all pre-Tertiary results, regardless of their apparent reliability, are useless for paleogeographic reconstructions; this is indicated by the * in the Q column of Table A10 – part I.

The Neogene results from Indo-China offer no surprises, but I note that the older results show northeasterly declinations (as do all the remagnetized results; Table A10), indicating a clockwise rotation of late Neogene vintage. Such a rotation accords well with the indentation/extrusion model of Tapponnier *et al.* (1982), or, for that matter, with many other models of the India–Asia collision. This rotation, with respect to the rotation axis and Siberia, appears to have occurred also with respect to the South China Block, although good Neogene reference poles from the latter are scarce. If true, this would mean that (sinistral?) displacements have occurred in the last few tens of million years along the Indo-China to South China boundary, i.e., along the Red River Fault and its continuation towards the South China Sea.

Given the complete lack of unremagnetized pre-Neogene paleopoles for Indo-China, paleomagnetism has yet to contribute to our knowledge about its Mesozoic

Figure 7.59. Map of the southeast Asian terranes and sutures. The connection between the Lhasa and Sibumasu blocks, the latter extending from western Yunnan (China) through Burma, western Thailand, western Malaya and western Sumatra, is poorly documented. Dashed lines represent sutures inferred for offshore, covered, or poorly studied areas. Figure redrawn after Metcalfe (1988). Published with permission from the Geological Society Publishing House.

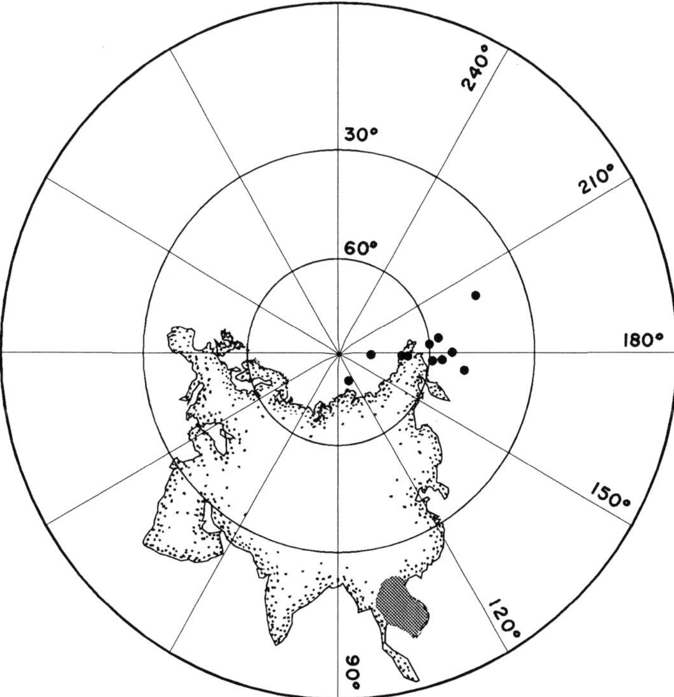

Figure 7.60. Paleopoles for all pre-Neogene results (including those with low reliability) from the Khorat Plateau in Thailand (Indo-China Block). The similarities between these poles, ranging in age from Paleozoic to Cretaceous, argue for remagnetization in the Tertiary (e.g., Chen Yan & Courtillot, 1989).

or Paleozoic history. The late Paleozoic Cathaysian flora of Indo-China (Figure 7.2) suggests that it was not (then) part of East Gondwana; rather it may have been close to the South China Block (Metcalfe, 1988) although the intervening Song Ma–Song Da sutures are early Mesozoic in age (e.g., Mitchell, 1981; Klimetz, 1987).

Sibumasu

This terrane derives its name from the regions of Siam (Thailand), Burma, Malaysia and Sumatra that are included in its customary outline (e.g., Figures 7.49 and 7.59). Mitchell (1981) has called this the Western Southeast Asia Block, whereas the northern end of the terrane in Yunnan Province in southwest China and Burma is often called the Baoshan or Shan-Thai Block. Figure 7.49 reveals that this terrane is characterized by a typical colder water Gondwana fauna during the Early Permian (e.g., Ridd, 1971; Stauffer, 1983), whereas in the Late Permian the fauna became Tethyan, i.e., warmer water, in aspect (Metcalfe, 1988).

Many maps (e.g., Figure 7.59; Mitchell, 1981) show a continuity between the Sibumasu terrane and the Lhasa terrane of Tibet (discussed earlier), but connections between the two in the Yunnan Syntaxis are unlikely because the terranes are thought to have sutured at different times.

The paleomagnetic results for the Sibumasu terrane and the adjacent areas of east Malaya and east Sumatra are compiled in Table A10, part II. There are several older results of great interest, because they reveal the higher paleolatitudes that one would expect in the vicinity of the Gondwana continents of India and northwest Australia during the Late Paleozoic. Table 7.9 lists the declinations and paleolatitudes that one would expect for Sibumasu if it had not moved with respect to South China and Australia, respectively.

Paleolatitude observations that deviate significantly from these 'predictors' indicate either northward or southward subsequent displacements with respect to the nearby major continents. The results (with $Q \geq 3$) plotted in Figure 7.61 show that such displacements of Sibumasu have indeed occurred with respect to the South China Block and Australia. For Late Paleozoic times, the observations reveal more southerly paleolatitudes than those predicted, indicating subsequent northward displacements with respect to South China and even more so with respect to Australia (Fang, Van der Voo & Liang, 1989; Huang & Opdyke, 1991). In the Early Triassic, observed and predicted paleolatitudes are in agreement, suggesting a situation not unlike today (although paleo-east to west displacements cannot be detected by paleomagnetism), whereas during the Cretaceous the data appear to indicate a position of Sibumasu with respect to South China as well as Australia that is more northerly than is seen today. This is in conflict with the unorthodox model of Late Jurassic rifting of Sibumasu from Australia presented by Audley-Charles (1988). Some of these Cretaceous results are not from Sibumasu *sensu stricto*, but from adjacent east Malaya. However, by that time (and possibly already since the late Paleozoic according to Harbury *et al.*, 1990) the suture line dividing east and west Malaya, called the Bentong–Raub line or the Medial Malaya Zone, is thought to have become inactive (e.g., Klimetz, 1987; Metcalfe, 1988).

The declinations from the Sibumasu results have not been plotted, in part because the data from older rocks are not systematic. This indicates that rotations are likely to have happened internal to the terrane (e.g., Huang & Opdyke, 1991), but it may also suggest that the Sibumasu block as a whole underwent rapid rotations during northward motion. In fact, a Devonian result from the western Yunnan portion of

Table 7.9. *Declinations and paleolatitudes for the Sibumasu Block at 18° N, 95° E from the reference paleopoles of Table 7.6*

Age interval	S. China		Australia	
	D	Lat.	D	Lat.
Tl (37–66)			348,	−8
Ku (67–97)	6,	15	337,	−8
Kl (98–144)	14,	14	317,	−7
Ju, uJm (145–176)			320,	3
lJm, Jl (177–195)			314,	15
Tru/Jl (196–215)			324,	14
Tru, uTrm (216–232)	34,	−12	305,	−3
Trl/m, Trl (233–245)	37,	−6	317,	−4
Pu (246–266)	29,	−14	327,	−16
Pl (267–281)			340,	−21
Cu/Pl, Cu (282–308)	53,	−27	336,	−8
Cm, Cl, Du/Cl (309–365)			4,	22
Du, Dm/Du (366–378)			27,	8
Dm, Dl (379–397)	97,	−7	356,	30

Age abbreviations as in Table 5.1. The declinations (D) and paleolatitudes (Lat.) are calculated for the common site at 18° N, 95° E from the reference poles given in Table 7.6.

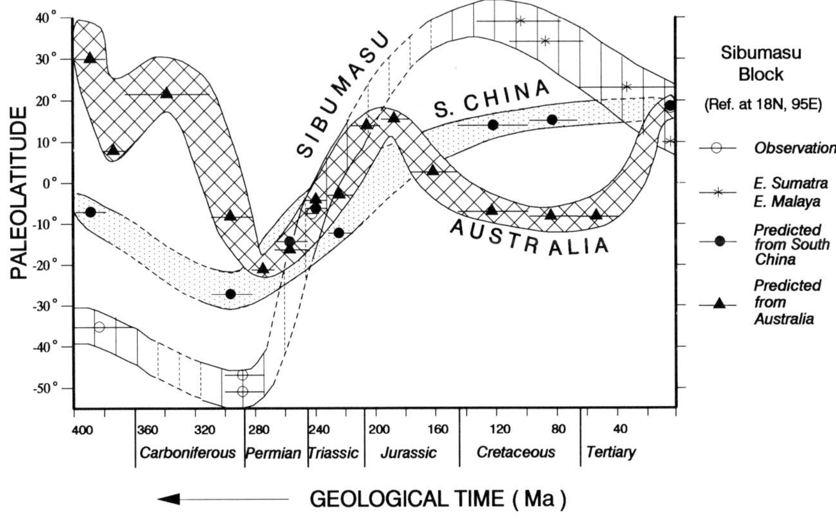

Figure 7.61. Paleolatitudes versus time for the Sibumasu Block, together with the values expected for the common reference site at 18° N, 95° E as extrapolated from the South China paleopoles (solid circles) and the Australian results (triangles) according to Tables 7.6 and 7.9. Symbols marked by '*' represent results from eastern Malaya and Sumatra; however, these areas are thought to have moved together with Sibumasu after Early Cretaceous or earlier time.

Sibumasu with paleolatitude of 42° S suggests that the present-day southern (Malayan) end of the elongated terrane was pointing towards the Devonian north pole, as illustrated in the paleogeographic map of Figure 7.62 (from Fang *et al.*, 1989). In this map the present-day western margin of Sibumasu (STM) is located adjacent to northwest Australia, the Lhasa Block (LS) and northern India in a Gondwana configuration. There are many other non-paleomagnetic arguments for precisely such an orientation in this location (e.g., Mitchell, 1981; Audley-Charles, 1988; Stauffer, 1983; Scotese, 1986; Metcalfe, 1988; Burrett *et al.*, 1990; see also Figure 2.9). Cocks & Fortey (1988) also use this location but orient the terrane differently.

The Cretaceous location of Sibumasu with respect to South China, more northerly by about 15° to 20° than that of today (see Figure 7.61), is somewhat surprising and needs further corroboration. If true, the subsequent southward displacement must be considered a very significant component of the extrusion model. However, we will see below that results from southwestern Borneo, thought by many workers to be part of the same block as Sibumasu at least since the Cretaceous (e.g., Figure 7.59), do not indicate a more northerly position with respect to South China. Either Borneo and Sibumasu underwent differential movements in the Tertiary (e.g., Nishimura & Suparka, 1990), or the Cretaceous data from Malaya need reassessment. Cretaceous declinations from Sibumasu, as well as those from western Borneo, indicate counterclockwise rotations since that time with respect to South China and Siberia for both blocks. These rotations are not in the right sense for the model of Tapponnier & co-workers (1982), but it may well be that this far away from the India–Tibet–Eurasia collision the effects are not systematic; Schmidtke, Fuller & Haston (1990), for instance, have argued for an irregular Asian margin that included the southeast Asian blocks, before India collided.

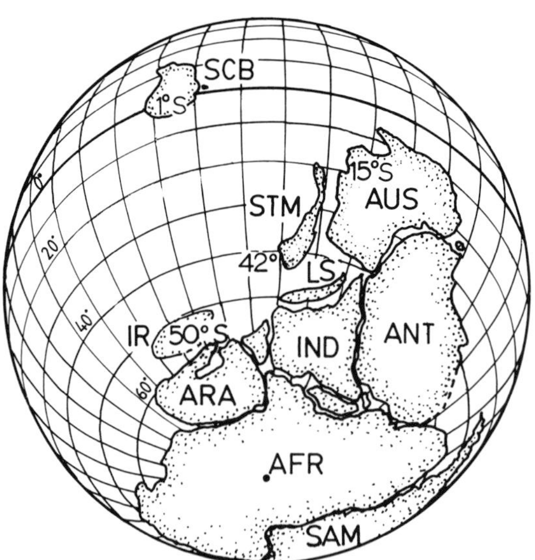

Figure 7.62. Devonian paleogeographic map showing Gondwana and some of the Asian displaced blocks (IR = Iran; LS = Lhasa, STM = Shan-Thai-Malay or Sibumasu; SCB = South China). The Sibumasu Terrane is oriented with its present-day northern end pointing towards the south pole, on the basis of the declination and paleolatitude of 42° S observed in Yunnan, China. Paleolongitudes are indeterminate so that the equatorial position of South China may have been more to the east or west with respect to Gondwana. Figure from Fang, Van der Voo & Liang (1989).

Table 7.10. *Declinations and paleolatitudes for Korea at 37°N, 128°E from the reference poles of Table 7.6*

Age interval	Siberia		N. China	
	D	Lat.	D	Lat.
Tl (37–66)	13,	44		
Ku (67–97)	13,	53	9,	44
Kl (98–144)	25,	43	16,	37
Ju, uJm (145–176)	30,	44	23,	33
Same**	16,	51		
lJm, Jl (177–195)	1,	60		
Tru/Jl (196–215)	349,	66		
Tru, uTrm (216–232)	17,	76		
Trl/m, Trl (233–245)	40,	70	311,	16

Age abbreviations as in Table 5.1. The reference poles are taken from Table 7.6. Declination (D) and paleolatitude (Lat.) values are calculated for a common location in Korea (37° N, 128° E)
**For the Late Jurassic there are two options for the Siberian reference pole (see Table 5.1).

In summary, the paleomagnetic data appear to indicate a Gondwana affinity for Carboniferous and older Paleozoic times (e.g., Figure 7.62), a northward movement with respect to Gondwana and South China well underway during the Early Triassic (Figure 7.61), and a subsequent southward movement with respect to South China in the Tertiary. As such, Sibumasu fits the definition (and the timing of motion) of a typical Cimmerian continental element (Şengör, 1987), as already indicated by Figures 7.2 and 7.27.

Korea

Before descending farther towards eastern Indonesia and New Guinea, it is worth looking briefly at the data from Korea. The area is treated separately, despite its common inclusion with North China in the Sino-Korea Block, because the paleomagnetic data indicate complexities that argue against all of Korea being a (stable) part of the North China Block. Korea is not, strictly speaking, a Tethyan terrane either, but it logically belongs in this section as a link between the Tethyan–Cathaysian blocks and Circum-Pacific terranes farther east and north in Asia.

The available paleopoles (Table A10) and predicted declinations and paleolatitudes for a common reference point at 37° N, 128° E (Table 7.10) have been used to construct the plots of Figures 7.63 and 7.64. The observed paleolatitudes generally agree with those predicted from North China, although in the Late Jurassic they range from 30° to 50° N. There is no systematic difference between paleolatitude results from Korea north of the Jurassic–Cretaceous Okchŏn mobile belt (see Figure 7.65) and results from the Okchŏn belt and areas farther south (symbols marked by * in Figure 7.63).

The Late Jurassic and younger declinations for all of Korea, with the exception of those from the Okchŏn belt, are slightly more easterly than those predicted from either North China or Siberia. The pre-Late Jurassic declinations from the results with Q ≥ 3 deviate from those of North China, indicating significant relative clockwise rotations. On the other hand, the Late Jurassic to Early Cretaceous declinations from the Okchŏn belt indicate counterclockwise rotations. The map of Figure 7.65 shows the extent of the deviating declinations observed in several widespread localities of granitic plutons and sedimentary rocks.

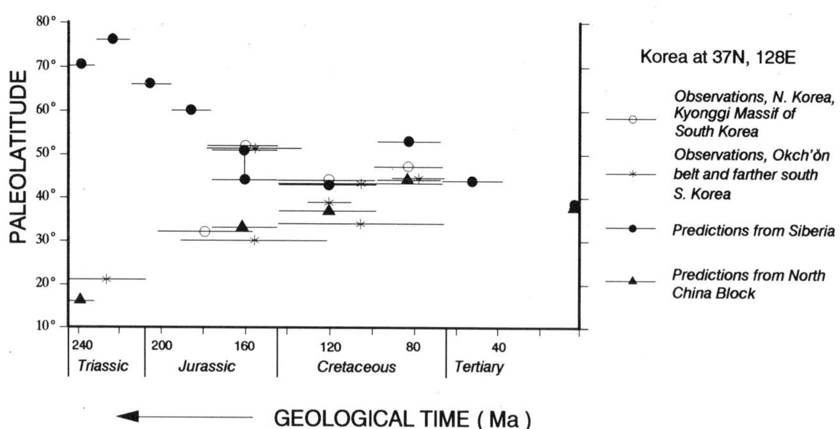

Figure 7.63. Paleolatitudes versus time for Korea, together with the values expected for the common reference site at 37° N, 128° E as extrapolated from the Siberia paleopoles (solid circles) and the North China paleopoles (triangles) according to Tables 7.6 and 7.10. Symbols marked by '*' represent results from South Korea south of the Kyonggi Massif, that is, the Okchŏn Belt, Ryongnam Massif and Gyeongsang Basin.

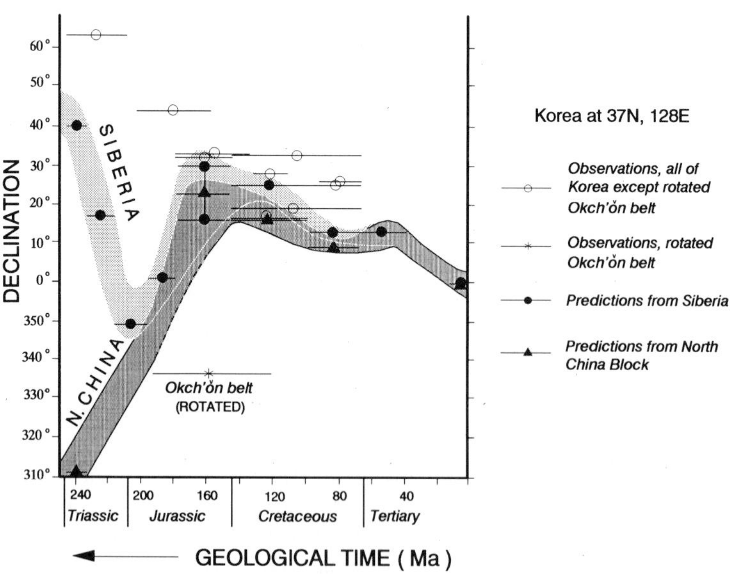

Figure 7.64. Declinations versus time for Korea, together with the values expected for the common reference site at 37° N, 128° E as extrapolated from the Siberian paleopoles (solid circles) and the North China results (triangles) according to Tables 7.6 and 7.10. The symbol marked by an asterisk represents the combined result from the (counterclockwise rotated) Okchŏn Belt.

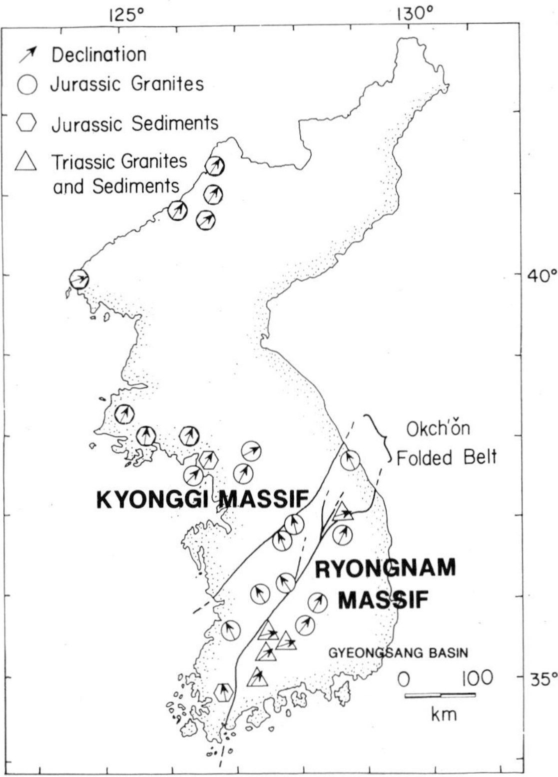

Figure 7.65. Map of Korea showing Late Jurassic paleomagnetic declinations, which illustrate the counterclockwise rotations of a large part of the Okch'ŏn Belt with respect to areas both north and south of the belt. Figure from Kim & Van der Voo (1990).

The rotation patterns seem fairly systematic, but cannot be directly related to North China and South China Block rotations (e.g., Figure 5.22) which suggested to Kim & Van der Voo (1990) that they are only regional in nature and possibly related to sinistral displacements along the western paleomargin of the Pacific Ocean.

Results from Japan have not been compiled for this book and will not be discussed other than to note that Mizutani (1987) has proposed that geological aspects of some terranes from central Japan match those for tectonic elements along the Asian coast. He locates these parts of Japan in pre-Neogene time between Korea and the Nadanhada terrane in northeastern China and Sikhote Alin in Russia (Figure 7.66). In Figure 5.19 the northward displacement of Sikhote Alin with respect to Siberia is in good agreement with that of Figure 7.63. It is worth investigating further whether South Korea, parts of Japan and Sikhote Alin could have formed a single terrane.

In summary, the paleolatitudes of Korea show affinities with those predicted from North China, supporting the notion that 'Sino-Korea' was a single paleogeographic element in early Mesozoic and probably also Paleozoic times, as suggested by faunal and other similarities (Lee, 1987). However, it seems wrong to speak of a Sino-Korea Block, because of the different rotations of most of Korea with respect to North China. The eastern boundary of the stable part of North China could be located within China (e.g., the Tan-Lu fault) or in Korea near the Okchŏn Belt. Paleopoles from Shandong and East Laoning Provinces to the east of the Tan-Lu

Fault (Lin, 1984), which are coherent with other results from the North China Block, indicate that a boundary within Korea is most likely.

Borneo

A good paleomagnetic data set is available from the western part of the island of Borneo (Kalimantan and Sarawak) owing to the work of Haile, McElhinny & McDougall (1977) and Schmidtke *et al.* (1990). The observed paleolatitudes are all very close to the equator, suggesting no significant northward or southward displacements with respect to the rotation axis or to South China (Table 7.11).

Predicted and observed declinations are illustrated in Figure 7.67. When paleolatitudes are equatorial and large rotations are a possibility, the polarity ambiguity in paleomagnetic data always begins to play a role. An easterly declination can be taken as deviating clockwise from a northerly reference direction if the polarity was normal, but if the polarity was reversed the north-seeking declination must have been westerly and the subsequent rotation was counterclockwise. The ambiguity, however, can be remedied by studying a sequence of formations that recorded the rotation in progress. This appears to have been done successfully in western Borneo where the declinations show a progressively larger counterclockwise deviation with increasing Tertiary to Cretaceous ages (Figure 7.67).

The rest of Borneo did not necessarily share this rotation. R. McCabe (pers. comm., 1990) has obtained normal and reversed Eocene to Pleistocene declinations based on a positive fold test in southeastern Borneo that are north–south, indicating no significant rotation. The paleolatitudes from southeastern Borneo are equatorial and in agreement with those from Sarawak. Thus the rotations suggested by Schmidtke *et al.* (1990) may have been of local nature only and are not necessarily

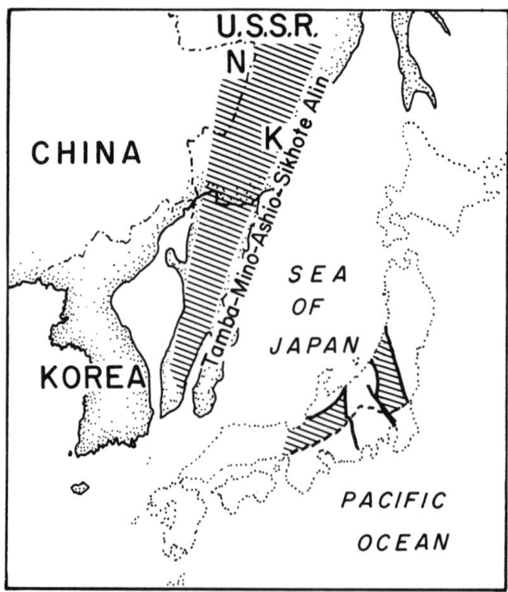

Figure 7.66. Model of Mizutani that proposes a pre-Neogene location for parts of Japan (shaded terranes) between Korea and the Nadanhada Range in China (N) and the Krasnorechensk of Sikhote Alin (K). Figure redrawn after Mizutani (1987) and reproduced by permission of the author.

Table 7.11. *Declinations and paleolatitudes for Borneo at 0° N, 110° E from the reference paleopoles of Table 7.6*

Age interval	Australia	S. China
	D Lat.	D Lat.
Tl (37–66)	354, −28	
Ku (67–97)	342, −30	6, −1
Kl (98–144)	320, −30	14, 0
Ju, uJm (145–176)	322, −20	

Age abbreviations as in Table 5.1. The declinations (D) and paleolatitude (Lat.) are calculated for the common site at 0° N, 110° E) from the reference poles given in Table 7.6.

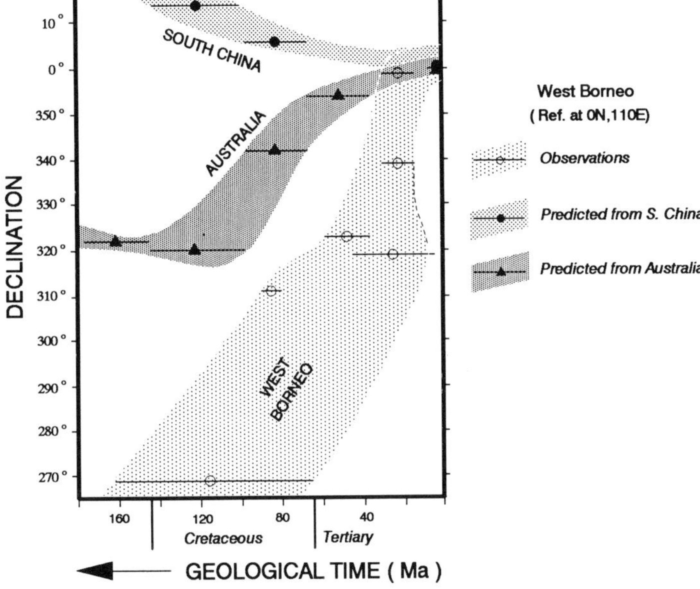

Figure 7.67. Declinations versus time for West Borneo, together with the values expected for the common reference site at 0° N, 110° E as extrapolated from the South China results (solid circles) and Australian paleopoles (triangles) according to Tables 7.6 and 7.11, indicating counterclockwise rotations for West Borneo.

in conflict with the extrusion model of Tapponnier *et al.* (1982) which predicts a very small clockwise rotation if any.

East Indonesia

The invaluable descriptions of the geology of Indonesia by Van Bemmelen (1949) and the tectonic synthesis of Hamilton (1979) are publications cited by virtually everyone working in Indonesia. Even so, much of the geological setting remains unknown, providing both a challenge and an impediment to paleomagnetic studies

Table 7.12. *Declinations and paleolatitudes for East Indonesia at 5° S, 130° E from the reference paleopoles of Table 7.6*

Age interval	Australia D	Australia Lat.
Tl (37–66)	5,	−34
Ku (67–97)	354,	−39
Kl (98–144)	329,	−46
Ju, uJm (145–176)	329,	−36
lJm, Jl (177–195)	319,	−25
Tru/Jl (196–215)	330,	−24
Tru, uTrm (216–232)	311,	−44
Trl/m, Trl (233–245)	327,	−43
Pu (246–266)	348,	−51
Pl (267–281)	8,	−48

Age abbreviations as in Table 5.1. The declinations (D) and paleolatitudes (Lat.) are calculated for the common site at (5° S, 130° E) from the reference poles given in Table 7.6.

of this complicated area. The available paleopoles, compiled in Table A10, part V, are of variable quality; as in the rest of this book, only results with $Q \geq 3$ are being considered seriously. The paleolatitudes of these results can be compared with those predicted from Australia (see Table 7.12) in Figure 7.68.

Paleolatitude analysis here is hampered by the same equatorial ambiguity as already discussed for Borneo, because in the eastern Indonesian archipelago large rotations are very likely as well. I have taken all pre-Neogene paleolatitudes to be southerly ones; in most cases the declinations, falling in the northern quadrants with the normal polarity so implied, show only small rotations. However, for two results (the Lower Cretaceous Wai Bua Formation of Timor and the Upper Cretaceous South Coast Volcanics and Dikes from Sumba) the implied rotations are very large (>140°). Younger results from Timor and Sumba do not show systematic declination deviations and thus do not provide resolution about the possible sense of rotation or the polarity of the older results, so I must emphasize a caveat about these two results. Be that as it may, the pattern of Figure 7.68 shows a fairly coherent trend which follows that predicted from Australia. The observed paleolatitudes are in almost all cases slightly more northerly than those predicted, suggesting a subsequent convergence between Australia and the East Indonesian islands. This is perhaps not surprising for Sumba, where the convergence seems significant (Figure 7.68), but for Misool and southern Timor this is unexpected. The maps of Figure 7.69 and 7.70 illustrate the locations of these islands as well as two of the possible models proposed for the evolution of the area.

In one model (Audley-Charles, 1988), the islands of Timor, Seram, Misool and the Bird's Head of New Guinea remain located on the margin of the Australian continent (Figure 7.69). Although some east–west fault zones disect the area, large north–south displacements of parts of this margin are not to be expected. In the other model (Pigram & Panggabean, 1984), Misool, Seram and the Bird's Head of

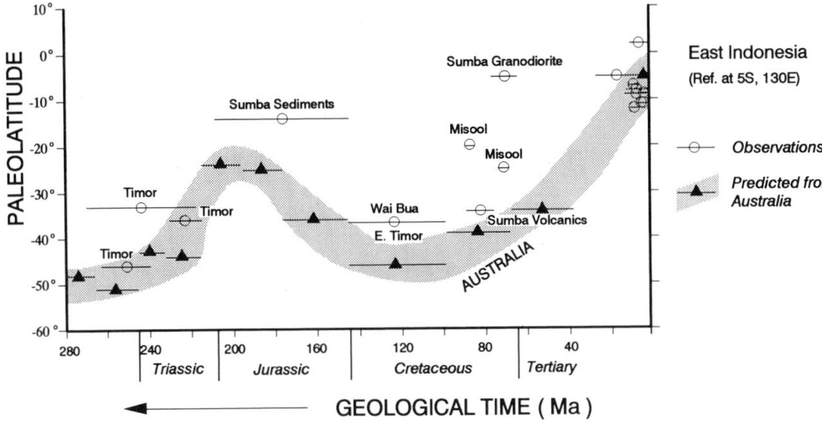

Figure 7.68. Paleolatitudes versus time for East Indonesia, together with the values expected for the common reference site at 5° S, 130° E as extrapolated from the Australian paleopoles (triangles) according to Tables 7.6 and 7.12. Deviations from the (Australian) predicted pattern imply relative north–south movements of the Indonesian island involved. For a location map see Figures 7.69 and 7.70.

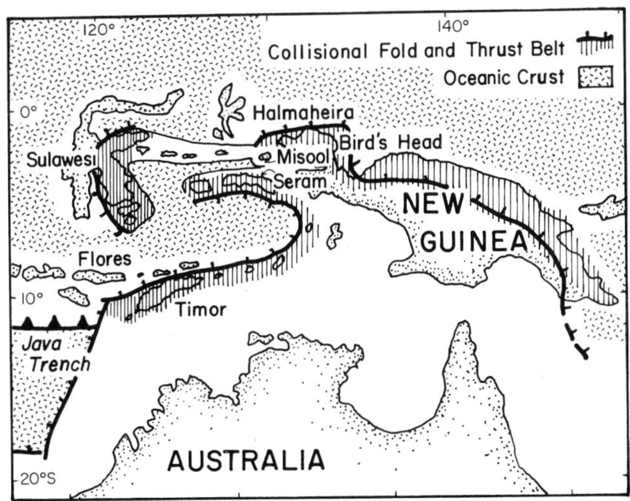

Figure 7.69. Location map of some of the East Indonesian islands and a continental margin outline of Australia hypothesized by M. G. Audley-Charles, who proposed that Timor, Seram, Misool, and the Bird's Head of New Guinea remained with Australia even after Jurassic rifting of other terranes. Figure redrawn after Audley-Charles (1988). Published with permission from the author and the Geological Society Publishing House.

New Guinea form one of a set of microcontinents, which are hypothesized to have separated from Australia (Figure 7.70).

The excellent paleomagnetic results from Misool, obtained by Wensink, Hartosukohardjo & Suryana (1989), clearly favor the second model because their observed paleolatitudes are lower by about 10° to 20° than those predicted if Misool had remained coherent with Australia (Figure 7.68). The paleolatitude results from Timor are only slightly lower than those from Australia and probably within the margin of error; this agrees with the ideas of most workers (summarized by McCabe

& Cole, 1986) that Timor has always been a part of Australia, albeit strongly affected by the somewhat chaotic structural deformation associated with the collision of Timor and the Banda Arc (e.g., Flores, Leti, etc. in Figure 7.70). Results from Seram and Sulawesi are of too low reliability to draw meaningful conclusions.

In summary, the notion that several of the islands of eastern Indonesia (e.g., Sumba, Misool) may have formed part of microcontinents which underwent displacements with respect to Australia seems supported by the available paleopoles, although more results of good reliability are badly needed. Paleomagnetism has contributed less to our understanding of the complex neotectonic setting of the area, in part because sampling has often been in allochthonous terranes, such that rotations cannot be extrapolated to larger areas, whereas Neogene north–south displacements appear to have been too small to be detected.

The Philippines

A discussion of the paleopoles from this archipelago located between Indonesia and South China is included in this Tethyan chapter because of that location, although one could argue that the Philippines are not exactly the prototype of a Tethyan terrane.

A large data set, listed in two parts in Table A10, parts VI and VII, is available. The first part concerns the northern main island of Luzon and its neighbor Marinduque, the second includes most of the smaller islands south of Luzon in the west-central Philippines. For a location map, I refer to Figure 7.71 (after Faure, Marchadier & Rangin, 1989).

Paleolatitudes for the Philippines range from 8° S to 15° N, spanning all of the Tertiary. Two older results from Cebu and Mindoro are also equatorial. Thus, the paleomagnetic data indicate only minor and generally northward displacements of

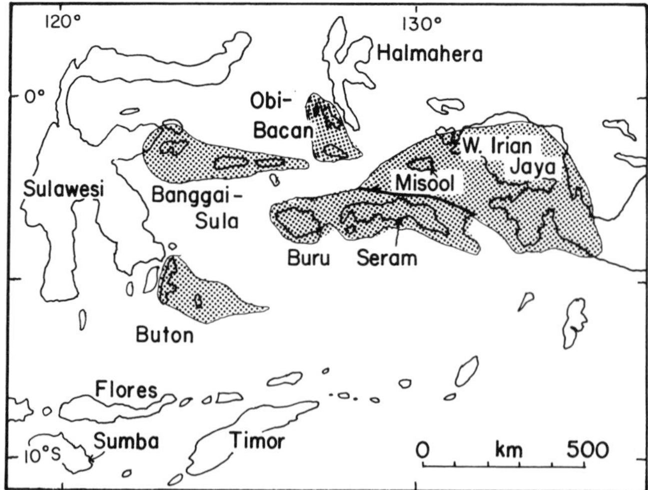

Figure 7.70. Location map of some of the East Indonesian islands showing a model of Pigram & Panggabean (1984) in which Seram, Misool, and the Bird's Head of New Guinea (West Irian Jaya) constituted one of the microcontinents which rifted away from Australia. Figure redrawn after Audley-Charles (1988). Published with permission from the author and the Geological Society Publishing House.

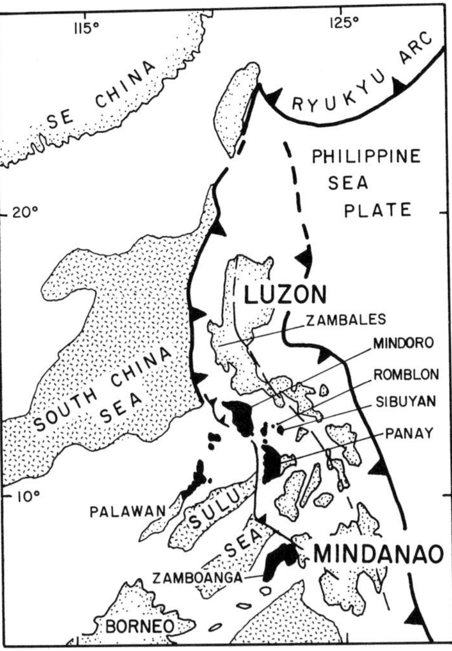

Figure 7.71. Location map of the Philippine Archipelago with the islands that formed part of the West Philippine (= North Palawan) microcontinent in black. The two main islands of the Philippines (Luzon, Mindanao) are labeled separately. The pattern in the marginal seas represents oceanic crust. Figure redrawn after Faure, Marchadier & Rangin (1989), and reproduced by permission of the authors.

the islands. The declinations, on the other hand, show large variations suggesting complicated rotation patterns. Again the equatorial paleolatitudes make it difficult to decide on the sense of rotations; both clockwise and counterclockwise rotations have been favored by the previous workers, who, moreover, sometimes changed their mind on this issue. A choice depends critically on the reliability of some Oligocene to Early Miocene results from Luzon.

To illustrate the two possibilities, Figure 7.72 presents the declination data from Luzon and Marinduque in two different versions. One (Figure 7.72A) uses all results with $Q \geq 3$ and interprets the pre-Late Miocene results as due to ccw rotations. This was previously also the interpretation of McCabe and Fuller and co-workers in their pre-1985 publications. Two results from the Coto Mine area (McCabe *et al.*, 1987; Fuller, Haston & Almasco, 1989) have declinations around 235° in this option and would thus show ccw rotations of some 120° with respect to the rotation axis and the South China Block (Figure 7.72A).

The plot of Figure 7.72B, on the other hand, omits many of the Early Miocene and Oligocene data as 'unreliable' (McCabe & Cole, 1986) and plots the older results with easterly declinations. The two results from the Coto Mine area now display a declination of about 55° (Figure 7.72B), and the island of Luzon in this interpretation has rotated clockwise. The results omitted are those with declinations in Table A10 between 302° and 342°.

The choice between these two options is difficult to make; moreover, it is also possible that Luzon rotated first one way and then the other, whereas differential rotations between various parts of the island may also have occurred. Although the omission of the Oligocene to Early Miocene results as unreliable in the interpretation of a clockwise rotation (Figure 7.72B) bothers me as a rather *ad hoc* 'solution', it also seems to me that this problem must remain unresolved for the time being.

The results from the west-central Philippines (the North Palawan Block of Figure 7.71) are fewer and indicate clockwise rotations, as do data from the Philippine Sea Plate to the east (including some results from Palau listed in Table A10, part VIII). Thus, the main ambiguity in the paleomagnetic data set is the lack of resolution about the sense of rotation of Luzon, as such forming a good example of the limits of paleomagnetism in equatorial terranes where rotations are large, undoubtedly rapid, and possibly unsystematic. In their reviews of the then available Luzon data, Fuller (1985) and McCabe & Cole (1986) mention reasons why the sense of rotation is pertinent to an understanding of the evolutionary history of the islands. This importance has to do with the sense of rotation of the Philippine Sea Plate (cw) and the possible coupling of Luzon as might be indicated by a cw rotation, or the lack thereof (if Luzon rotated ccw), with this oceanic plate to the east during the Paleogene and Early Miocene.

As already mentioned, the paleolatitudes do not provide sufficient resolution either. Although the relative displacements between Luzon and the North Palawan Block are larger with the clockwise-rotation model, nothing indicates that that could not be the case.

In summary, it appears to me that we have not yet learned enough about paleomagnetism of the Philippine Archipelago. Thus, the evolutionary history of the islands, e.g., in terms of Pacific versus East Asian provenance, remains rather hazy from the perspective of paleomagnetism.

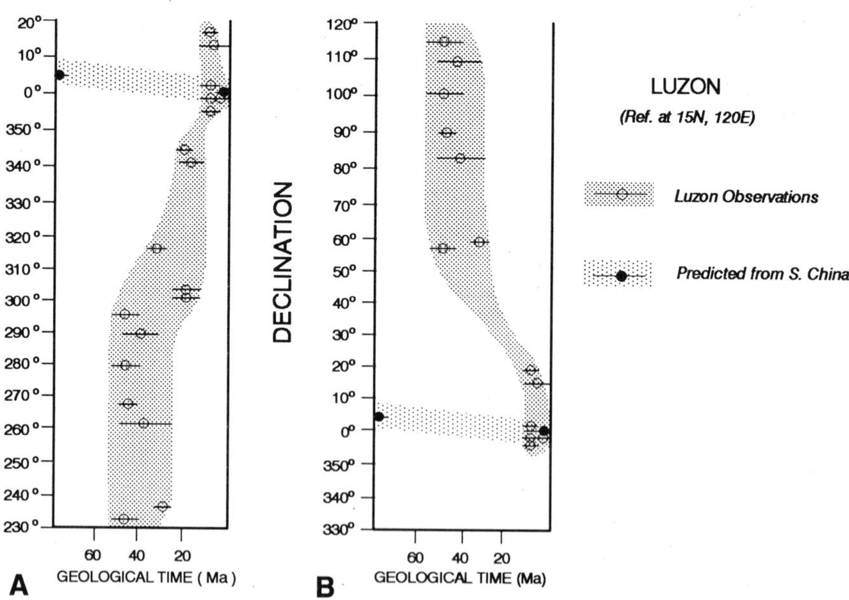

Figure 7.72. Declinations versus time for Luzon, the Philippines, together with the values expected for the common reference site at 15° N, 120° E as extrapolated from the South China results (lightly shaded envelope) according to Table 7.6, in two different plots which illustrate either (A) relative counterclockwise or (B) clockwise rotations of Luzon, depending on the choice of polarity for the observations. In B five observations with declinations between 300° and 350° have not been used. See text for further discussion and explanation.

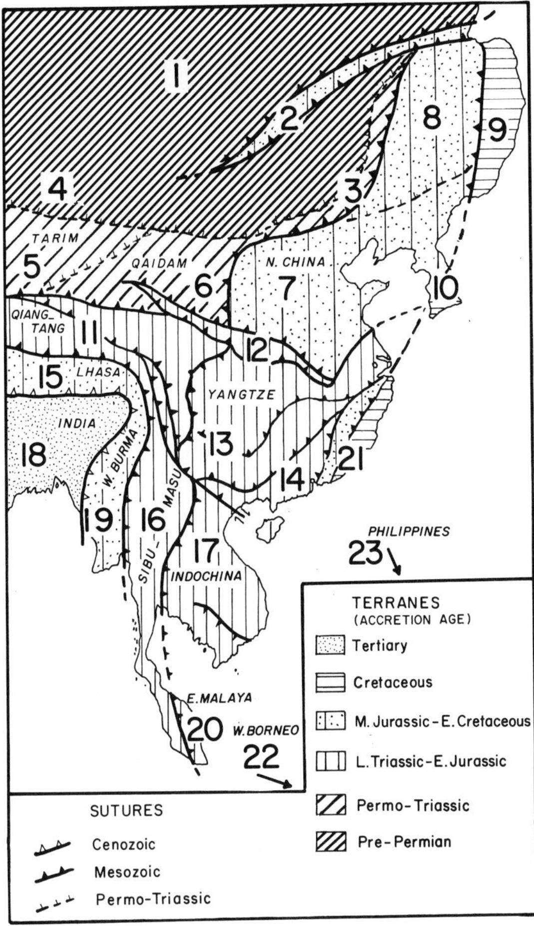

Figure 7.73. Map of Southeast and East Asia showing the different blocks and terranes with numbers tied to Table 7.13, in which the paleomagnetic information for these terranes and blocks is detailed. Map redrawn after Klimetz (1987).

Summary

This chapter is the longest one of this book, not by design but because of the complexity of the tectonics and the relative abundance of the paleomagnetic data for the Alpine–Himalayan–Indonesian realm. The data from the western Mediterranean, Iran, the Himalayas, Tarim, Sibumasu and some of the East Indonesian islands are examples of the potential provided by paleomagnetism. In contrast, for the eastern Mediterranean, Indo-China, West Indonesia and the Philippines, paleomagnetism has yet to accomplish what it potentially could reveal. And clearly, even in the areas of 'success' there is still much to be learned.

The paleomagnetic successes imply no more than the determination of rotations (sense and magnitude) and north–south displacements; with a bit of luck (and the availability of formations suitable for paleomagnetic study) also the timing of such motions could be determined. However, the results can become especially relevant when they allow us to distinguish between different models. Models for the opening/

Table 7.13. *Constraints and information provided by paleomagnetism on the continental blocks and terranes of Figure 7.73*

No.	Name	Paleomagnetic information
1	Siberia	Reference plate for many plate motions of other blocks. Very good data from Eurasia for Permian and Mesozoic, M. Paleozoic data are poor.
2	Mongolo-Okhotsk Suture Zone	No paleopoles from zone itself, but results from Tarim, Junggar. Mongolia and N. China suggest post-Permian (Jurassic?) rotation of the southern area w.r.t. Siberia.
3	Inner Mongolia	Paleopoles suggest movement together with N. China since L. Permian; w.r.t. Siberia northward movement until they joined in Jurassic.
4	Junggar	Rotation with respect to Siberia in the Permian.
5	Tarim	Movement independent from Siberia until Triassic, but may have been adjacent to Siberia since E. Permian collision with Junggar; w.r.t. N. China independent movement until Triassic–Jurassic.
6	Qaidam	No paleomagnetic information. Shown as separate element by Klimetz (1987), but taken by many other workers to be part of North China.
7	N. China	Movement (ccw rotation) independent from Siberia (and Tarim) until Jurassic.
8	Songliao-Bureya	No paleomagnetic information. Shown as a Mesozoic part of North China by Klimetz (1987), but as an element separate from N. China in the Paleozoic.
9	Sikhote Alin	Northward movement w.r.t. Siberia after M.–L. Mesozoic. May have been contiguous with parts of Japan and Korea.
10	Korea	Probably remained adjacent to N. China since Precambrian, but different parts underwent relative rotations (cw and ccw) during L. Mesozoic–E. Tertiary.

11	Qiangtang-Songpan Ganzi	Qiangtang is likely to have been with Gondwana until after the E. Permian; Songpan-Ganzi/Kunlun/N. Qiangtang elements may have drifted northward to join Tarim and Siberia by the E. Triassic; further northward movement w.r.t. Siberia in the Tertiary of Qiangtang.
12	Qinling-Datang	No paleomagnetic data from zone itself, but paleopoles from N. and S. China indicate relative movements (rotations) as late as Jurassic.
13	Yangtze (S. China)	May have been adjacent to Gondwana in E. Paleozoic but was independent block in L. Paleozoic and E. Mesozoic. Collided with N. China in Jurassic.
14	Cathaysia	No paleomagnetic information.
15	Lhasa Block	No paleomagnetic data about affinity to Gondwana; had collided with Qiangtang by Cretaceous time.
16	Sibumasu	Together with Gondwana until M. Permian; likely to have collided with equatorial S. China in the Triassic; fits Cimmeria model; possible L. Tertiary southward movement w.r.t. S. China needs substantiation.
17	Indo-China	No paleomagnetic information for pre-Tertiary times. L. Tertiary cw rotation w.r.t. S. China.
18	India	Part of Gondwana until Jurassic, rapid northward drift in L. Cretaceous/E. Paleocene and collision with Asian margin soon after. Further northward movement w.r.t. China and Siberia during most of Tertiary.
19	West Burma	No paleomagnetic information.
20	East Malaya	Appears to have been together with Sibumasu since the Cretaceous if not earlier.
21	Eastern S. China	No paleomagnetic information.
22	West Borneo	Remained on the equator since the L. Cretaceous and did not share apparent southward Tertiary drift of Malaya; large ccw rotation during Tertiary, which is not necessarily valid for the rest of Borneo.
23	Philippines	No clear information about sense of M. Tertiary rotations of about 90°; in equatorial position throughout Tertiary, with slight northward drift in the Miocene.

Abbreviations: w.r.t. = with respect to; N. = North, S. = South; E. = Early (early), M. = Middle (middle), L. = Late (late); cw = clockwise, ccw = counterclockwise.

extension of the Bay of Biscay, the Ligurian and Tyrrhenian basins and the formation of the Calabrian arc (orocline) owe much to paleomagnetic results from the surrounding rotated continental elements; farther east the Cimmerian continent hypothesis derives significant support from the paleomagnetically determined northward displacements of several terranes.

Towards eastern Asia, the situation becomes ever more complex. I can imagine that someone reading through the long descriptions and compilations of the preceding tens of pages is by now ready to throw this book in the nearest corner (if not the wastepaper basket). Thus – and in aid to those who complain justifiably about the trees obscuring the forest – I have constructed a summary table listing the contributions of paleomagnetism for each of the major continental elements in eastern Asia (Table 7.13). This table is tied to a map (Figure 7.73) constructed by Klimetz (1987) with corresponding numbers denoting each terrane or block. As worthwhile as many tectonic models, ideas or hypotheses may be, they have been ignored in this book in order to keep its size within reasonable bounds, if there are no paleopoles of relevance to them. On the other hand, there have been major and significant contributions from paleomagnetism to the solution of several Tethyan problems. It may seem to some that these have received too much attention, whereas others may complain that they did not receive enough. If so, this chapter may have been just right. And if not – as with the Tethyan terranes – nothing mandates that the truth or the terranes must lie in the middle anyway!

Me pellegrina ed orfana
Lungi dal patrio nido
Un fato inesorabile
Sospinge a stranio lido

(Giuseppe Verdi, in his opera La Forza del Destino, *which premiered in 1862 in St. Petersburg.)*

So sings Leonora in her lament: 'As a wanderer and orphan, far from my native land, an inexorable fate drives me on towards a foreign shore'. Leonora's foreign shores were along the Mediterranean; a similar fate seems to have driven the Tethyan terranes towards the shores in Europe and Asia in general, whereas the native land of these terranes often was near Gondwana.

In the next chapter, the shores and native lands of the Iapetus terranes are described. They seem more exotic and the drift histories more incredible, because the setting is in even more remote times. But that these orphaned terranes traveled is not in doubt. Iapetus and Tethyan terranes generally seemed to move northward; is there a regularity in this, I wonder?

8 The terranes, blocks and adjacent continents of the Iapetus Ocean

Iapetus is the name given by Harland & Gayer (1972) to a Paleozoic ocean that was the precursor to the modern Atlantic Ocean. The name Proto-Atlantic has also been used. No coherent remnants of Iapetus Ocean floor exist, other than dismembered ophiolites emplaced upon continental crust by obduction. The name of Iapetus, as the father of Atlas, is appropriate in other ways as well: the birth and early history are very uncertain and obscured by latest Precambrian and early Paleozoic events, and even for the later stages in the life of this ocean, as for that of the mythological figure, few aspects are known with any certainty.

Before the seminal paper of J. Tuzo Wilson (1966), who queried whether – before the opening of the present-day Atlantic – another ocean existed that closed in late Paleozoic time, it was generally and often tacitly assumed that Pangea (see Chapter 5) was valid for all of pre-Mesozoic time. Wilson based his arguments to the contrary on faunal dissimilarities in terranes now juxtaposed; in the northern Appalachians, for instance, the American craton with its early Paleozoic 'Pacific' fauna contrasts sharply with the coastal strip of eastern New England and the Canadian Maritime Provinces which yielded the so-called Acado-Baltic ('Atlantic') fauna. Today, 25 years later, faunal arguments still play an important role in discussions about Paleozoic paleogeography, as do latitudinal indications provided by sedimentary facies (glacial relicts, clastics versus carbonates, coal, evaporites, etc.).

But paleomagnetism also has made, and undoubtedly will continue to make, contributions to our knowledge about the positions of the major continents and the smaller displaced continental elements in between them. That progress has been slow is due to the many problems and ambiguities that exist for interpretations of paleomagnetic results from older rocks; not just the indeterminancy of longitude and the hemispherical ambiguity caused by polarity changes, but also the effects of later orogenies or rock-fluid interactions that may have caused pervasive remagnetization, the effects of rock deformation, the imperfections in isotopic age dating, and the incompleteness of the rock record, all cause severe problems that make the Paleozoic motions in Iapetus much more difficult to determine than the Mesozoic–Cenozoic movements of the Tethyan and Atlantic-bordering blocks. For these and additional reasons, the applications of paleomagnetism to Precambrian continental configurations are even more hazardous, but that issue lies outside the scope of this book.

For paleomagnetism to provide valuable constraints on Paleozoic paleogeography, ideally a complete APWP would have to be constructed for each craton and displaced terrane, in which local rotations would be accounted and corrected for.

This situation, however, may not exist for the displaced terranes for many decades to come, and probably will not even prevail soon for the major continents. Thus, we must make do, at least for the rest of this century, with a very incomplete record best interpreted in terms of paleolatitudinal contrasts rather than on the basis of paleopole differences or APWPs. It is worth recalling that stratified rocks can yield the paleolatitude directly, regardless of any subsequent rotations or coherent deformations. This assumes that the magnetization predates any tilting, that deformation has not included penetrative strain, that the geomagnetic field was predominantly that of a dipole, and that the paleohorizontal can be adequately determined. Thus rotations, which strongly influence the location of a paleopole, have had negligible effect on many of the paleomagnetically determined paleolatitudes. For the smaller displaced terranes, paleolatitudes derived from individual studies are therefore used as the primary data for the analyses of this chapter, whereas the mean paleopoles of the APWPs have been used for the major continents.

The APWPs of the major continents have already been discussed in Chapter 5. The continents that undoubtedly bordered Iapetus at one time or another were Laurentia (North America, Greenland and northern Scotland), Baltica (the Baltic Shield and Russian Platform) and Gondwana. It is possible, of course, that Siberia or the China Blocks also bordered Iapetus in some period of the Paleozoic, but at this time there is little evidence for this.

Since its rapid adoption after 1972, the name Iapetus has taken on a certain 'generic' aspect in that it is the name generally used for most of the oceanic domain that separated Laurentia and Baltica, but also that which separated Laurentia from Gondwana. This is also the use followed here, but with the realization that a single ocean may not be the correct choice for the changing and complicated Paleozoic paleogeographies. The initial use of the name Iapetus was restricted to that Paleozoic oceanic domain which separated Norway, the continental crust underlying the Barents Sea, Spitsbergen and Greenland.

In subsequent years, many scientists, recognizing the broader usage as somewhat imprecise, have invoked other names for oceanic domains whether separate from or contiguous with Iapetus. Thus, the names Medio-European Ocean, Tornquist Sea and Rheic Ocean have been used for the ocean between Baltica and southern-Central Europe, whereas the name Theic ocean (but also Rheic, to add to the confusion!) has been used for the domain between Gondwana and southern Europe. There were enough Titans (see Chapter 1) to provide a source for other names in the future, although the mythological hierarchy gets complicated if one recognizes that Tethys was really the mother of all oceans!

A further complexity was introduced when it became clear that island arcs and back-arc basins may have existed within Iapetus, dividing the ocean into separate parts of different ages; thus, (instead of 'Paleo-' and 'Neo-Iapetus', which fortunately never made their appearance) names like Iapetus I and Iapetus II have been proposed. This usage runs the inherent danger that with more and more details becoming known, we may eventually face Iapetus XIV. It is difficult enough, in my opinion, to keep the French kings before Louis XIV apart in the discipline of history, than to have to face such genealogy also in geology.

Few of the many smaller, and potentially displaced terranes in the Paleozoic Iapetus have been adequately identified, thus allowing enormous leeway in the construction of terrane models or maps. Regrettably, or fortunately, depending on point of view, even fewer Paleozoic paleomagnetic data are available outside the cratons. Exceptions are formed by four main areas: (1) Britain, south of the inferred Iapetus suture, (2) the Armorican Massif of France and the Iberian Meseta of Spain and Portugal, (3) the Avalon Basement terranes of eastern North America, and (4)

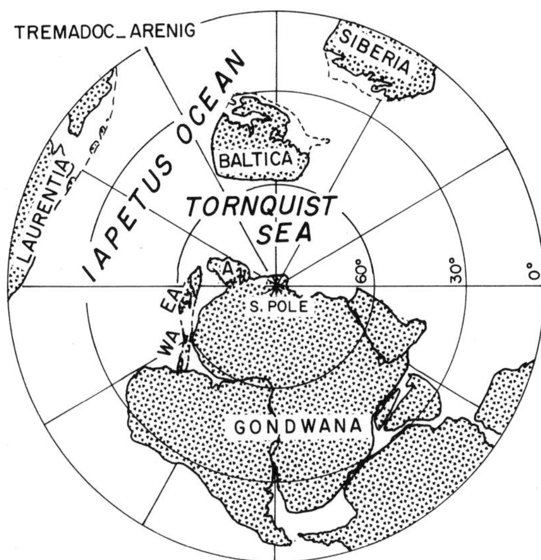

Figure 8.1. Early Ordovician paleogeographic map of the Iapetus-bordering continents of Laurentia, Baltica and Gondwana (Lawver & Scotese fit), centered on the south pole. During the Middle Ordovician these continents remain in the same paleolatitudes, although Baltica is thought to be rotated slightly counterclockwise. Peri-Gondwanide locations of Armorica (A), Western Avalon (WA) and Southern Britain (EA = Eastern Avalonia) are also shown. Figure redrawn after Torsvik & Trench (1991a). Published with permission from the Geological Society Publishing House.

an aggregate of smaller terranes now located between Avalon and the Laurentian craton. For these areas, therefore, some important constraints exist when choosing appropriate paleogeographic models. The data from each of these areas are compiled in Table A11 and will be discussed separately in sections below. No Precambrian results are included in Table A11, although in some cases these may be mentioned in the text where helpful in context. As mentioned earlier, I have generally not listed results based on undated remagnetizations, nor results that are without demagnetization or that have been superseded, unless these had been included in recent syntheses and deserve mention as 'not having been overlooked'.

The major continents and the closure of Iapetus

The pre-Permian paleopoles (recapitulated in Table 8.1) of the three major continents (Baltica, Laurentia and Gondwana) have already been discussed in Chapter 5, so here only a brief summary will be given.

Baltica and Laurentia collided in Silurian to earliest Devonian time, judging from the geology of the Caledonian orogenic belt in Scandinavia and Greenland (e.g., Roberts, 1988; Hatcher, 1988; Ziegler, 1988). The paleopoles of the two continents, when juxtaposed with the Atlantic closed, are certainly in agreement by earliest Devonian time and, in fact, appear to be similar already during the Middle Silurian (see Table 5.1). However, few Silurian paleopoles are available for Laurentia and Baltica, so this 'agreement' is in need of further confirmation.

Two excellent results have just become available for the Early to Middle Ordovician of Sweden (Torsvik & Trench, 1991b; Perroud, Robardet & Bruton, 1992) indicating paleolatitudes at that time of 30° to 60° South for Baltica (Figure 8.1),

Table 8.1. *Pre-Permian reference paleopoles for the Iapetus-bordering continents*

Age interval	Laurentia NAM coords		Laurentia EUR coords		Baltica EUR coords		West Gondwana NW-AFR coords		All Gondwana NW AFR coords	
	Lat.	Long.	Lat.	Long.	Lat.	Long.	Lat.	Long.	Lat.	Long.
Cu/Pl, Cu (282–308)	41,	129	40,	167	40,	167	33,	223	38,	231
Cm, Cl, Du/Cl (309–365)	28,	129	27,	167	27,	167	21,	227	34,	207
Du, Dm/Du (366–378)	28,	111	27,	149	27,	149	4,	200	11,	191
Dm, Dl (379–397)	23,	110	22,	148	22,	148	7,	177	30,	176
Su/Dl (398–414)	3,	97	3,	135	3,	135				
Su, Sm (415–429)	19,	124	18,	162	18,	162	43,	189	31,	177
Ou/Sl, Ou, Om (430–467)	16,	144	15,	182			−35,	158	−43,	172
Ol/m, Ol (468–505)	17,	158	16,	196	−24,	230	−30,	181	−25,	176
Єu, Єm, Є (506–542)	9,	158	9,	196			−67,	173	−55,	176
Єl (543–575)	5,	170	5,	208	11,	231	−17,	90	−9,	138

Age abbreviations as in Table 5.1. The reference poles are given in the coordinates indicated and are taken from Tables 5.1 (pre-Middle Silurian for Baltica), 5.3 (for all of Gondwana), 5.7 (for Baltica for Middle Silurian and younger time and for Laurentia) and 5.8 (for West Gondwana alone). Lat. and Long. are latitude (positive when north, negative when south) and east longitude, respectively. NAM is North America, EUR is Europe, NW AFR is northwest Africa.

but little other paleomagnetic information about its early Paleozoic paleogeography is available. Noteworthy about the Middle Ordovician position of Baltica is that it could have been positioned to the north of eastern Africa (Torsvik *et al.*, 1990b) and that its orientation makes the Uralian margin face the south pole. However, paleolongitudes remain a matter of conjecture throughout early Paleozoic time, so the paleolocation of Figure 8.1 may be modified with impunity by moving Baltica east or west.

The faunal characteristics of Baltica indicate a minor separation from Laurentia during the Early Silurian in contrast to a large separation in the Early Ordovician (Cocks & Fortey, 1990), which is in good agreement with the scant paleomagnetic evidence. The evolution of the sedimentary facies of Baltica suggests steady northward drift during the Ordovician with carbonate rocks becoming gradually more and more abundant culminating in Silurian tropical reefs (Manten, 1971). In the Cambrian and late Precambrian, on the other hand, the sedimentary record reveals mostly detrital rocks including latest Proterozoic tillites of the Varanger sequence (see, e.g., Schwab, Nystuen & Gunderson, 1988 for a good summary), suggesting higher paleolatitudes at that time.

The paleolatitudes of Laurentia are generally equatorial to subtropical throughout the Paleozoic, but with more southerly latitudes during the Late Silurian to Early Devonian (see Figure 2.10). The Iapetus (Appalachian) margin of Laurentia faced first south (Late Cambrian to Early Ordovician) and then southeast during the rest of the Paleozoic. The sedimentary facies (e.g., Schwab *et al.*, 1988; Witzke, 1990) are in excellent agreement with the paleomagnetic information. Early Cambrian and latest Precambrian paleopoles also indicate equatorial paleolatitudes, although a new study of Symons & Chiasson (1991) seems to indicate much higher (southerly?) paleolatitudes at the Precambrian–Cambrian boundary than previously thought. If this is substantiated, the Early Cambrian paleopoles thus far available (Tables A1 and 8.1) must be interpreted as later remagnetizations (a not unfamiliar development for Laurentia's Paleozoic APWP, which has earlier seen similar changes in interpretation for middle Paleozoic times). Tilloids and glacial relicts of latest Precambrian age (e.g., the Moraenesø Formation in Greenland) would then be in better agreement with the paleomagnetic information; however, many late Precambrian glacial relicts yield near-equatorial latitudes worldwide, which is one of the more baffling problems remaining in Proterozoic paleomagnetic research. I will return at the end of this chapter to the late Precambrian and its paleogeography, when discussing possible scenarios for the early opening of Iapetus.

The APWP of Gondwana, as already discussed in Chapter 5, is relatively(!) well characterized for Ordovician and late Paleozoic times, but includes large swings in the middle Paleozoic segment that are questioned by several non-paleomagnetists. The Ordovician location of Gondwana is not contentious, with the mean paleopole located in or near northwest Africa. Nevertheless, it is possible that during the Ordovician the paleopole and Africa moved somewhat with respect to each other; the details of such possible movements are not known. At any rate, Gondwana's margin along the north coasts of South America, Africa and Arabia was located in high southerly latitudes and faced north, as illustrated in Figure 8.1.

For late Paleozoic times there is little disagreement about the gradual northward movement of Gondwana resulting in a collision between it and the combined Laurentia–Baltica continental mass (Laurussia) no later than the Late Carboniferous (e.g., Ziegler, 1988). This continent–continent collision produced the Hercynian-Alleghenian orogenic belt, running east–west in the vicinity of the equator in Pangea A-type configurations (e.g., Figures 5.24–26). Sacks & Secor (1990) have recently re-emphasized that the effects of this collision lasted until well into the Early Permian.

The discussion so far, based on aspects of Paleozoic paleogeography that are not very controversial, provides the broader framework in which many details must be filled in. The more controversial issues for the major continents emerge for middle Paleozoic time, specifically because of the controversial APWP of Gondwana for the Siluro-Devonian interval.

The main questions that we ultimately need to ask in this context, are: (1) what has been the role and paleogeographical evolution of all the displaced and exotic terranes that were incorporated in Pangea, but did not form part of the major cratonic blocks during the earlier Paleozoic, (2) given the explanations in terms of continental collisions for the Caledonian belt in Norway and Greenland and the Late Carboniferous Hercynian-Alleghenian belt at the juncture between Gondwana and Laurussia, can similar explanations be proffered for the Grampian, Finnmarkian, Penobscotian, Taconic, Acadian, Ligerian, Bretonic and other orogenic phases that mark Paleozoic history at regular intervals like clockwork, (3) do plutonic, volcanic and metamorphic products of the Paleozoic orogenic belts support the possible plate tectonic interpretations based on paleomagnetic and other relevant data, and (4) do post-Ordovician faunal similarities and dissimilarities, and sedimentary facies contribute to this knowledge? Entire books and symposium volumes have been devoted to these subjects. The difficulties are rather overwhelming and a consensus does not appear to be imminent. To imagine that this chapter will resolve the problems and arrive at an acceptable synthesis is unrealistic; moreover, for many areas or time intervals there simply are no, or only grossly inadequate, paleomagnetic data. Such areas and time intervals and the controversial issues they entail, however interesting, must thus be glossed over in this book. However, for those areas where a suitable database exists, paleomagnetism may provide some constraints.

As outlined in Chapter 5, the APWP of Gondwana for the Silurian and Devonian makes two large swings, i.e., three tracks. This can be seen in the data from West Gondwana alone, but also in the mean paleopoles for all of Gondwana (Figure 8.2; Table 8.1). The first track (I) has the south pole with respect to a fixed Gondwana move from northwest Africa to the vicinity of Chile, whereas the second track (II) involves a polar movement away from the Chilean location to west-central Africa. Track III then heads back south towards the late Paleozoic paleopoles in and near Antarctica. With respect to a fixed pole, these tracks imply a northward movement of northern West Gondwana during the Late Ordovician to Early Silurian, a clockwise rotation and southward movement of northern West Gondwana during the later Devonian and finally a northward movement during the Carboniferous Period. With the large (and rapid) motions with respect to the pole that are implied by these APWP loops (see Chapter 5), it could well be possible that Siluro-Devonian collisions occurred between Gondwana and other continents, but the paleolongitude indeterminancy prevents us from knowing this with any degree of certainty. Kent & Van der Voo (1990), for instance, have speculated that the Silurian to Early Devonian (Caledonian-Acadian) orogenies on the Iapetus margin of North America could be related to the Ordovician–Silurian northward drift of Gondwana, in agreement with earlier models proposed by McKerrow & Ziegler (1972). Before getting into the details of the middle Paleozoic, let us first consider the alternatives that exist when some of these APWP swings are ignored.

If the first loop (tracks I and II) is ignored, the pole moves slowly from a location near northwest Africa in the Ordovician to a west-central African location in the Devonian (APWP similar, but not identical to pole path X of Figure 5.15a). No continent–continent collisions involving Gondwana are envisioned in this alterna-

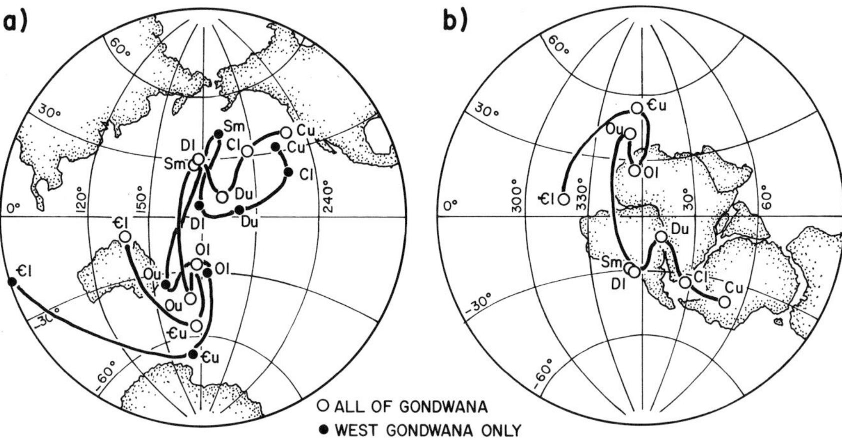

Figure 8.2. Pre-Permian apparent polar wander paths, as north poles in a), south poles in b), averaged for West Gondwana only (solid circles) and averaged for all of Gondwana (open circles), according to Tables 5.8 and 5.3 (see also Table 8.1). Abbreviations are C = Carboniferous, D = Devonian, S = Silurian, O = Ordovician, Ꞓ = Cambrian, l = Lower/Early, m = Middle and u = Upper/Late. In a) the background continents are in their present-day coordinates, whereas in b) Africa, as part of Gondwana, is the only continent in present-day coordinates.

tive, as it remains separated by a wide ocean from Laurentia and Baltica until after the Devonian. Cold to cool temperate conditions remain prevalent in West Gondwana, in agreement with some faunal characteristics, e.g., the Malvinokaffric Realm (Boucot, 1990). The Silurian *Clarkeia* fauna in this alternative does not present the problem noted by Van der Voo (1988; see also Cocks & Fortey, 1990) that this supposedly cold water brachiopod would have ranged from the south pole to equatorial latitudes implied for France in the more complicated Gondwana movement required by the Siluro-Devonian loop. On the other hand, the problem with the Devonian warm water facies in Morocco (see Chapter 5) remains unresolved, because Late Devonian paleolatitudes for Morocco would be higher than 45° S. In this option (labeled A), the Silurian Aïr pole from Africa is ignored and the Silurian to Early Devonian paleopoles from the East Australian fold belts would not be representative of cratonic Gondwana, thereby implying post-Early Devonian terrane displacements and rotations for these terranes, although these are not supported by the local geological evidence. In Figure 8.3, a simplified set of paleogeographic maps of only Gondwana and Laurentia illustrate option A. This APWP option does not result in a mid-Paleozoic collision between Gondwana and Laurentia.

If, on the other hand, the first track is accepted, but the second loop (tracks II and III) is omitted (option B), one obtains an APWP for Gondwana similar to that of Scotese & Barrett (1990), shown previously in Figure 5.15e. The pole now moves from northwest Africa in the Ordovician to northern Chile in the Silurian to Early Devonian (track I) and then along the Cape of South Africa during the Late Devonian and Carboniferous. In Figure 8.4 a simplified set of paleogeographic maps is shown for this model. This option does not pose the Devonian warm water facies problem for Morocco, but suggests rather low paleolatitudes for the earliest Carboniferous glacial relicts in Niger (Lang *et al.*, 1991). This interpretation would require, moreover, that the Late Devonian paleopoles from Australia (Canning

Basin, Worange Point, Hervey Limestones) and Africa (e.g., Ben Zireg, Bokkeveld) are not used for paleogeographic analysis. In Option B, Gondwana and Laurentia appear to collide during the mid-Paleozoic and form a Pangea assembly which remains more or less permanent throughout the rest of the Paleozoic and Early Mesozoic.

My own preference is Option C, which accepts both loops and all three tracks, because I consider the paleopoles (especially the five Late Devonian ones mentioned above) sufficiently reliable that we can use them for estimating the paleogeography (APWP of Figure 8.2 for all of Gondwana, which is similar, but not identical to Path Y of Figure 5.15a or those of 5.15b–d, f). As already noted in Chapter 5, the problems with option C are that the velocity of Gondwana with respect to the pole seems very high at times, whereas the Late Devonian facies in northern Africa has been interpreted in favor of warmer climates (e.g., Wendt, 1985) than those suggested by the temperate paleolatitudes indicated by the paleopoles. I might note, however, that the APWP for all of Gondwana is more smoothed than that of West Gondwana alone (Figure 8.2) and thus implies slightly lower velocities and less extreme paleolatitudes. This path seems therefore preferable. In Option C, Gondwana and Laurentia may have collided during the Late Silurian to Early Devonian, but Pangea did not begin to assemble until the later Carboniferous. Figure 8.5 illustrates the positions of Laurentia and Gondwana in Option C in the same fashion as the preceding figures; in a later part of this chapter, paleogeographic maps will be shown which also include Baltica and some of the displaced terranes.

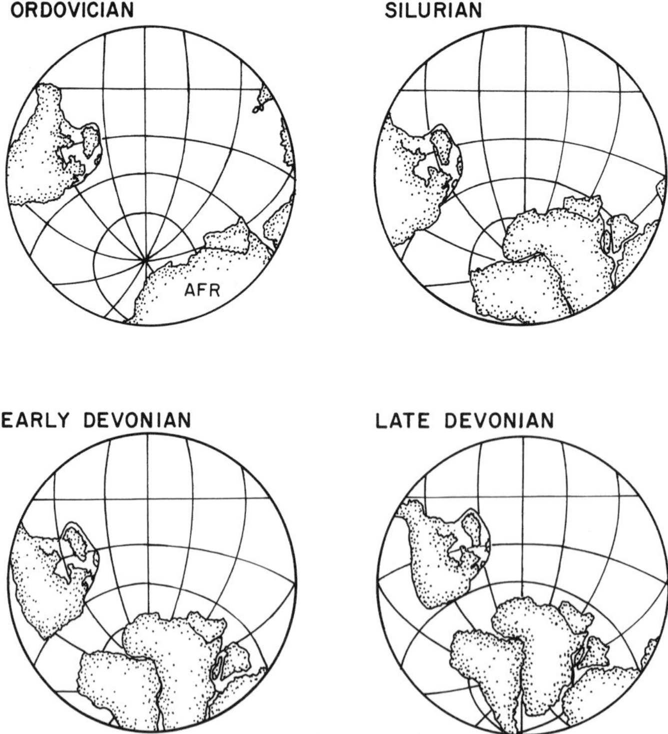

Figure 8.3. Paleogeographic maps (Option A) showing the relationships for Ordovician through Devonian time between Gondwana and Laurentia with an apparent polar wander path similar to Path X for Gondwana of Figure 5.15a (see text for discussion).

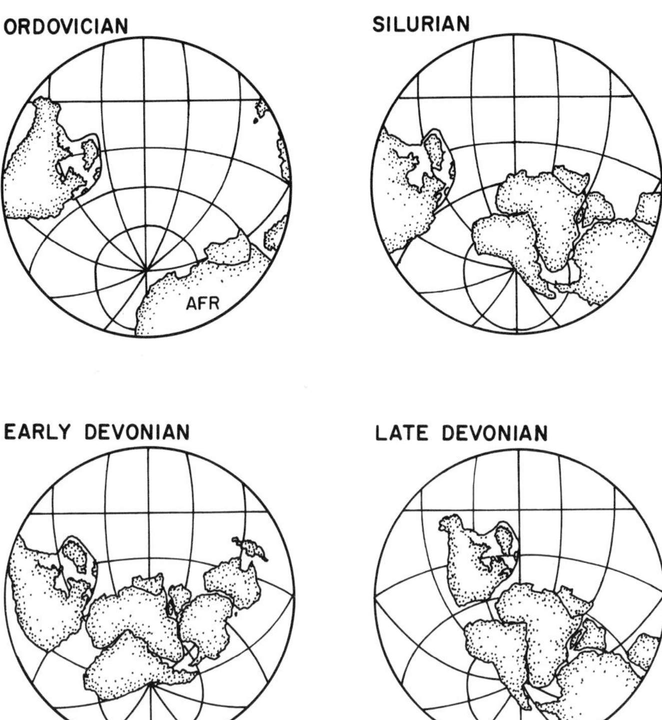

Figure 8.4. Paleogeographic maps (Option B) showing the relationships for Ordovician through Devonian time between Gondwana and Laurentia with an apparent polar wander path similar to that of Scotese & Barrett (1990) for Gondwana as shown in Figure 5.15e (see text for discussion).

In summary, we have seen that Laurentia and Baltica collided during the Silurian (Caledonian Orogeny), forming what has been called the Old Red Continent or Laurussia. This does not imply that after the Silurian there was no further deformation in the North Atlantic-bordering Caledonides; Ziegler (1988) has provided a detailed account of locally intense Devonian to Early Carboniferous disturbances. However, the available paleomagnetic data indicate that any relative post-Middle Silurian displacements between Baltica and Laurentia were minor and within the errors associated with the mean paleopoles. Gondwana may have collided with Laurussia also during the Late Silurian to Early Devonian (Acadian Orogeny), but this did not result in a lasting supercontinent configuration, because the Devonian paleopoles of Gondwana indicate a 'retreat' away from the Old Red Continent and the widening of the new ocean between them. This Devonian ocean is thought to have closed during the Carboniferous, resulting in the Hercynian-Alleghenian Orogeny and the formation of Pangea. In the following sections, I will examine how the movements and paleolatitudes of the continental elements of western and southern Europe and the displaced terranes of the Appalachian Belt fit within this larger framework.

Southern Great Britain

As discussed in Chapter 5, paleomagnetism as applied to global tectonics had an early and strong start in Great Britain in the 1950s. It is not surprising, therefore, to see the British tradition of pioneering paleomagnetic work continue in the early 1970s with Paleozoic paleomagnetic investigations designed to elucidate the movements that resulted in the Caledonian Orogeny. The geology of Great Britain is ideal for such studies because a traverse of the island(s), from north to south, crosses from the Laurentian margin in the Lewisian foreland of northern Scotland (e.g., the Hebrides) through the metamorphic Scottish Highlands and the central mobile part of the Caledonides that contains the Iapetus suture near the England–Scotland border (e.g., Phillips, Stillman & Murphy, 1976) and terminates in a southern continental element adjacent to the Welsh Basins. The basement of this southern continent is only rarely exposed, but its existence in the subsurface is not in doubt; it has been called the Midland Craton, or, also, the Brabant–London Massif which includes the northwestern part of Belgium. For excellent descriptions of the Paleozoic geology of Great Britain, I might refer to Rayner (1967), Owen (1976) and Harris, Holland & Leake (1979).

A major addition to the pre-Permian paleopoles of Britain occurred with a special issue of the Geophysical Journal of the Royal Astronomical Society devoted entirely to the paleomagnetic results obtained by the team then at the University of Leeds (e.g., Briden & Morris, 1973; Morris *et al.*, 1973; Piper & Briden, 1973; Sallomy

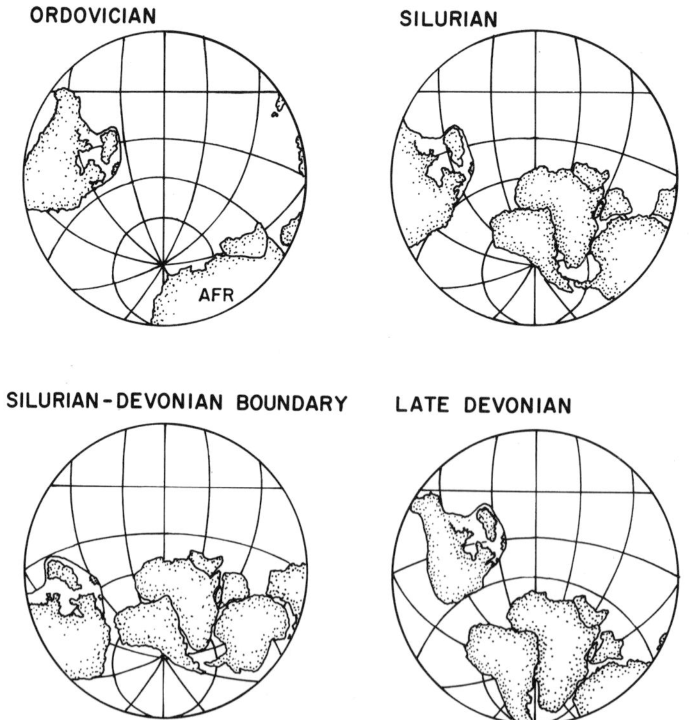

Figure 8.5. Paleogeographic maps (Option C) showing the relationships for Ordovician through Devonian time between Gondwana and Laurentia with the apparent polar wander path (APWP) of Table 8.1 (approximately equivalent to the APWPs of Figure 5.15b–d, f; see text for discussion). Option C has been adopted for the further discussions of this chapter.

& Piper, 1973). Subsequently, major reviews of Caledonian tectonics and paleomagnetism in Britain have been published by Piper (1978a, 1979a, 1987), Briden & Duff (1981), Briden, Turnell & Watts (1984), Briden *et al.* (1988), Torsvik *et al.* (1990a), Torsvik & Trench (1991a) and Trench & Torsvik (1991). In the intervening two decades a gradual change has occurred in interpretation of the movements that led to the Caledonian collision, not least because the analysis and laboratory techniques have improved and have resulted in a much more reliable data base. The change is most noticeable if one tracks the estimate of the Ordovician distance between the Laurentian and Midland cratons. Initially thought to have been negligible, this separation was proposed to be about 1000 km in 1978 (e.g., Piper, 1978a), but most recently more than 3000 km for the Middle Ordovician and about 5000 km for the Early Ordovician have been suggested (McCabe & Channell, 1989; Trench & Torsvik, 1991; see also Figure 8.1).

The Ordovician paleopoles, especially for southern Britain, have changed in this period. Paleopoles from Wales for Ordovician intrusions and the Cadr Idris Basalts (Table A11) were initially labeled anomalous by their authors (Thomas & Briden, 1976), because they gave much higher paleolatitudes than expected; although these results already then failed to meet stringent reliability criteria, the fact that more such high paleolatitudes are now being observed through modern laboratory analysis suggests that these results may perhaps not be so 'anomalous' after all.

Early Ordovician paleolatitudes of 55° S or higher imply that Southern Britain, that is Britain south of the Iapetus suture (referred to as Eastern Avalonia in Figure 8.1), was in the proximity of northwest Africa at that time. Geochemical and geological affinities between Cambrian and Ordovician rocks from the Brabant–London Massif, the Ardennes and Armorica (André, Hertogen & Deutsch, 1986; Von Hoegen, Kramm & Walter, 1991), also suggest peri-Gondwanide locations in the Early Ordovician. The paleolatitudes for Late Ordovician and younger time, however, indicate northward drift of Southern Britain (Figure 8.6) and the question has been whether this drift was together with that of Gondwana and other peri-Gondwanide terranes (as tentatively suggested by Van der Voo, 1988) or whether Southern Britain drifted independently towards Laurentia in advance of the later (Silurian?) northward drift of Armorica and Gondwana itself. The comparison of Figure 8.6 suggests that the Middle and Late Ordovician paleolatitudes for Southern Britain are significantly lower than those predicted if it had remained adjacent to Gondwana, whereas the few available Early Silurian results do not show this difference (compare also Tables 8.2 and A11). Recalling the uncertainties in the Gondwana APWP for the middle Paleozoic and the sizeable cones of confidence about Gondwana's Ordovician mean paleopoles, a definitive conclusion about the paleomagnetic affinities of Southern Britain and Gondwana seems impossible at this time. Faunal and facies comparisons (e.g., Cocks & Fortey, 1990) as well as the aggregate of presently available paleomagnetic data for the Middle and Late Ordovician (Figure 8.6) appear to favor an independent earlier northward drift of Southern Britain. On the other hand, a paleomagnetic re-examination of the Builth volcanic rocks (McCabe, Channell & Woodcock, 1991), although not yet fully published, appears to indicate paleolatitudes of about 50° S for the Middle Ordovician, which is higher by more than 15° than indicated by previous studies. A systematic re-evaluation of other previously studied Late Ordovician rock units may lead to similar revisions, in which case the northward drift of Southern Britain may turn out to be less fast and somewhat later than stated above.

Previously, adequate paleomagnetic results for the Ordovician of Baltica were unavailable, thus rendering comparisons between Southern Britain and Baltica very speculative. Many publications assumed a coherence between these two areas during

Table 8.2. *Declinations and paleolatitudes for Britain at 54° N, 4° W, predicted from the reference poles of Table 8.1*

Age interval	Laurentia		Baltica		All Gondwana	
	D,	Lat.	D,	Lat.	D,	Lat.
Cu/Pl, Cu (282–308)	7,	4	7,	4	318,	13
Cm, Cl, Du/Cl (309–365)	8,	−9	8,	−9	335,	2
Du, Dm/Du (366–378)	24,	−6	24,	−6	344,	−24
Dm, Dl (379–397)	26,	−10	26,	−10	360,	−6
Su/Dl (398–414)	46,	−24	46,	−24		
Su, Sm (415–429)	14,	−17	14,	−17	359,	−5
Ou/Sl, Ou, Om (430–467)	354,	−21			15,	−79
Ol/m, Ol (468–505)	340,	−18	285,	−40	360,	−61
€u, €m, € (506–542)	338,	−25			360,	−89
€l (543–575)	324,	−25	305,	−10	49,	−36

Age abbreviations as in Table 5.1. D is declination, Lat. is paleolatitude. Declination and paleolatitude values are calculated for a common location in Britain (54° N, 4° W) from the reference poles (Table 8.1) for Laurentia and Baltica in European coordinates and for all of Gondwana in northwest African coordinates.

the Middle and Late Ordovician on the basis of faunal similarities, resulting in paleogeographic maps that showed Southern Britain as a permanent Paleozoic, southwesterly appendage of Baltica jutting out into Iapetus. Now that excellent paleopoles for Baltica have become available, this appendage-like situation is no longer supported by the paleopoles. Instead, it can be seen (Figures 8.1, 8.6) that Southern Britain and Baltica had somewhat different paleolatitudinal positions during the Early Ordovician, but that they had very different orientations with respect to the paleomeridians; this orientation difference, recorded in the declinations, pertains to the Early Ordovician, but also to the Middle and Late Ordovician (Torsvik, pers. comm., 1991, on the basis of results in progress). The post-Ordovician convergence between Southern Britain and Baltica implied by these different positions could well explain the North German–Polish Caledonide fold belt that stretches, albeit mostly deeply buried, from the North Sea to the east–southeast along the present-day southwestern margin of the Baltic Shield to Poland (e.g., Zwart & Dornsiepen, 1978; Ziegler, 1988).

In Table A11 and Figure 8.6, the paleomagnetic results from the Lake District in northern England and those from southern England and Wales have been differentiated, but the paleolatitudes and paleopoles do not show any appreciable difference thus far, supporting the suggestion that all of Britain south of the Iapetus Suture drifted together as one block during the Ordovician. Two results from Scottish ophiolites of (Cambro-?) Ordovician age (Table A11, part I, not plotted in Figure 8.6) show paleolatitudes slightly more southerly than those predicted from Laurentia, suggesting a small latitudinal separation of about 5° to 10° from the Laurentian margin.

By Late Silurian and Early Devonian time, on the other hand, the results from Southern Britain are in excellent agreement with those from Laurentia as well as Baltica (Chapter 5, see also Figure 8.6), indicating that by then the suturing between these elements had been completed (see also Soper & Woodcock, 1990). Thus, the

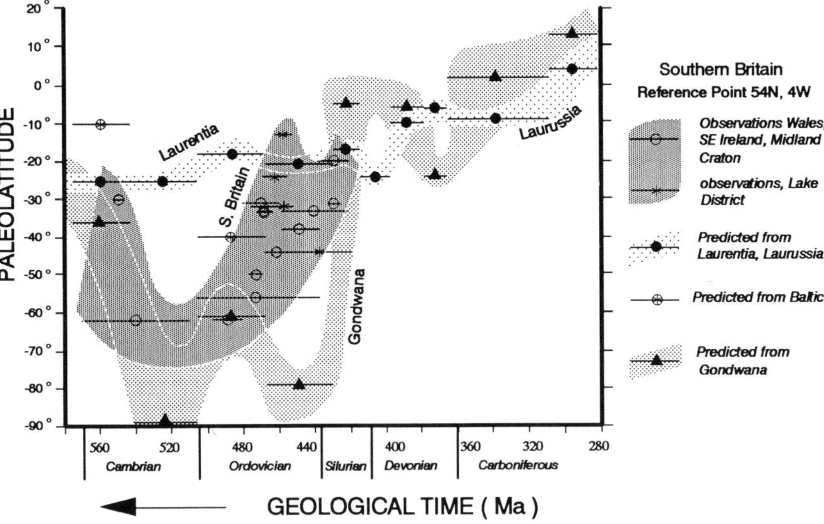

Figure 8.6. Paleolatitude versus Paleozoic time plot of the observations and predictions for a reference point of 54° N, 4° W in Southern Britain, according to the data of Tables 8.2 and A11, parts II and III ($Q \geq 3$). Symbols marked by '*' are observations from the Lake District, and it can be seen that their distribution coincides with that from Wales and the Welsh Borderlands.

parts of Iapetus or other oceans that previously separated Laurentia (i.e., northern Scotland, Greenland), Southern Britain and Baltica had disappeared by subduction during or before the Middle Silurian.

Armorica

As discussed in Chapter 5, paleomagnetic results from the Hercynian belt are valid for Stable Europe (i.e., Baltica in the Paleozoic) only for latest Carboniferous or younger time (< 300 Ma). Before the termination of the Hercynian orogeny, the different Paleozoic blocks and mobile belts of France, Germany, Iberia or Czechoslovakia may well have been separate tectonic elements in Iapetus and other Paleozoic oceans (e.g., Matte, 1991). However, the Precambrian and Cambrian geological history of these elements shows more similarities than differences, suggesting a certain paleogeographic coherence, and there is some paleomagnetic evidence for lumping several of the Hercynian areas of Europe together into one Paleozoic paleogeographic element called the Armorica plate by Van der Voo (1982, 1988).

Hercynian Europe today reveals topographic uplands (Rutten, 1969) where the Paleozoic and Precambrian basement can be found, separated by lowlands that expose post-Hercynian sedimentary cover. The uplands or massifs (Figure 8.7) are: Iberian Meseta, Armorican Massif, Massif Central, Vosges, Black Forest, Ardennes-Rhenish Slate Mountains, Harz and Bohemian Massif (ignoring less important occurrences as well as those in the Alpine belts). Of these, the Iberian, Armorican and Bohemian Massifs are characterized by a pronounced Late Precambrian to Cambrian unconformity associated with the Cadomian orogeny. The presence of this orogeny, contemporaneous with the Pan-African orogeny in Gondwana, could suggest that these areas were possibly in proximity to or contiguous with Gondwana.

This idea has been supported by paleomagnetic results (e.g., Hagstrum *et al.*, 1980; Perigo *et al.*, 1983; Ruffet, Perroud & Feraud, 1990) obtained principally from the Armorican Massif. Many of these results are for Late Precambrian rocks (e.g., Figure 8.8), and have not been included in Table A11, but in Figure 8.9 the APWPs of Hagstrum *et al.* (1980) for the Armorican Massif and Gondwana can be compared. The apparent polar wander for both was significant during the 150 million years involved, and the shapes and approximate ages of the APWPs are rather similar. For the Bohemian Massif, the results are of low quality but are not contradicting the suggestion that it, too, belonged to the same 'plate'. Hagstrum, Perigo and co-workers furthermore observed that the paleolatitudes (and many of the paleopoles) for this interval from Southern Britain were also in agreement with those from the Armorican Massif and from Gondwana (see also Piper, 1987, p. 237).

Subsequently, the paleopole similarities between Armorica and Gondwana have also been observed for Ordovician rocks from the Armorican Massif, the Iberian Meseta, and Poland (e.g., Figure 8.10); the Armorican paleolatitudes are cool-temperate to polar and would place this area in the immediate vicinity of northwest Africa (Armorica = A in Figure 8.1). Preliminary results from southern Germany also indicate high Ordovician paleolatitudes (Bachtadse, Soffel & Böhm, 1991). The paleolatitudes predicted from the reference poles of Baltica are lower, indicating an Ordovician separation between Baltica and Armorica as illustrated in Figures 8.1 and 8.10. Ordovician paleolatitudes observed in Czechoslovakia are also lower, but suspicions about remagnetizations persist (Piper, 1987, p. 236).

The Early Ordovician paleolatitudes from the Armorican Massif and the Iberian Meseta are comparable to those from Southern Britain (e.g., compare Figures 8.6 and 8.10), but they are clearly higher than those predicted from Baltica. While this

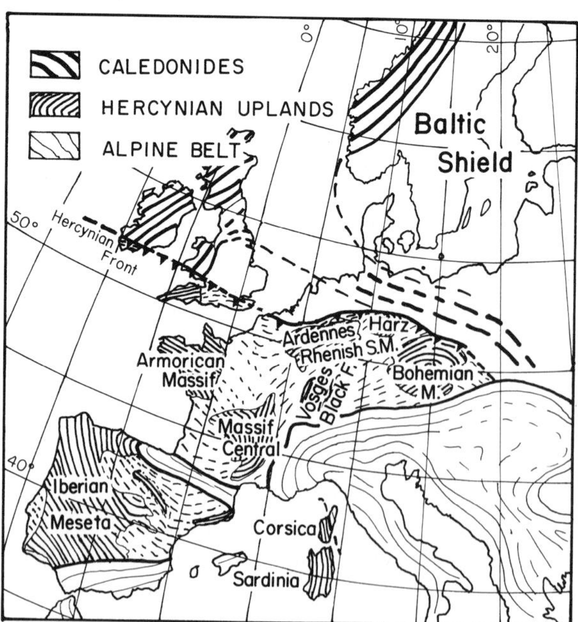

Figure 8.7. Map of Hercynian Europe, showing uplands (basement blocks discussed in the text) and intervening basins filled with post-Hercynian sediments. Figure redrawn and modified after Rutten (1969). Published with permission from Elsevier Science Publishers.

Figure 8.8. Photograph of a pillow in late Precambrian volcanic rocks of the Armorican Massif (Paimpol spilites) studied by Hagstrum *et al.* (1980). Bedding attitude and bedding tops can be determined because of the characteristic pillow elongation and shape.

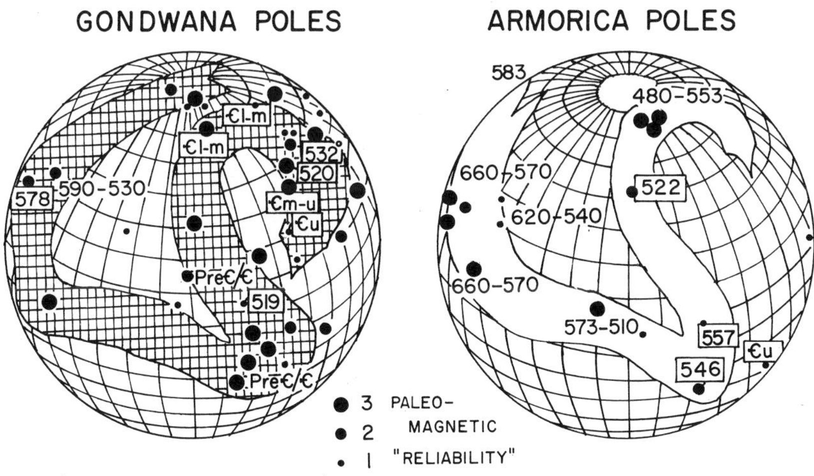

Figure 8.9. Paleopoles and apparent polar wander paths for Gondwana (left) and the Armorican Massif (right) in African and Armorican present-day coordinates for the interval of 650 to 500 Ma (redrawn after Hagstrum *et al.*, 1980). Increased symbol size indicates increased relative reliability of the paleopole, whereas the most reliable age determinations are placed in a box (in Ma for radiometrically dated rocks or as age symbol for stratigraphically dated rocks). The two paths show rapid apparent polar wander and reveal very similar shapes and ages with respect to their present-day coordinates, suggesting that Armorica and Gondwana drifted together during the 650–500 Ma interval. Published with permission from the Royal Astronomical Society.

supports a common peri-Gondwanide location for Southern Britain and Armorica in the Early Ordovician (Figure 8.1), the Late Ordovician paleolatitudes in Armorica are much higher than those from Britain, suggesting that the latter had begun drifting northward away from Armorica, as well as Gondwana.

There are only two Silurian results with Q ≥ 3 from Armorica (Perroud & Bonhommet, 1984; Perroud, Calza & Khattach, 1991) and their interpretation is ambiguous. The Almadén Volcanics of Spain have east–northeasterly declinations; in Table A11 this magnetization has been assigned a normal polarity, making the paleolatitude southerly and near-equatorial. However, when allowance is made for the inferred post-Silurian rotations of this part of the Iberian Meseta (Perroud, Calza & Khattach, 1991), the declination becomes southeasterly, suggesting a reversed polarity and a near-equatorial northerly paleolatitude. This last option has been used by Perroud *et al.* (1991) to infer coherence between this part of Iberia (and by inference, Armorica) with Gondwana during the rapid Silurian northward drift of the latter discussed in the previous section (see Figure 8.11). The other paleopole, from the Siluro-Devonian San Pedro red beds of northern Spain, resembles those for Carboniferous rocks; even though its magnetization is clearly pre-folding in age, it could well represent an early Hercynian remagnetization. Alternatively, the magnetization of the San Pedro Formation may be primary and interpreted to indicate that Armorica was together with Britain, Baltica and Laurentia already by Early Devonian time.

A series of Paleozoic (including Silurian) paleomagnetic results by Seguin (1983) from the Vendée Province, in the southern part of the Armorican Massif, has not been included in Table A11, because these have been discredited by the author's collaborators who helped in the sampling (see Robardet & Paris, 1985).

The Late Devonian results from Germany (Bachtadse, Heller & Kröner, 1983) and the Massif Central in France (Edel, 1987) are equally conflicting in that they indicate paleolatitudes of 13° S and 39° S, respectively (Table A11). Table 8.3

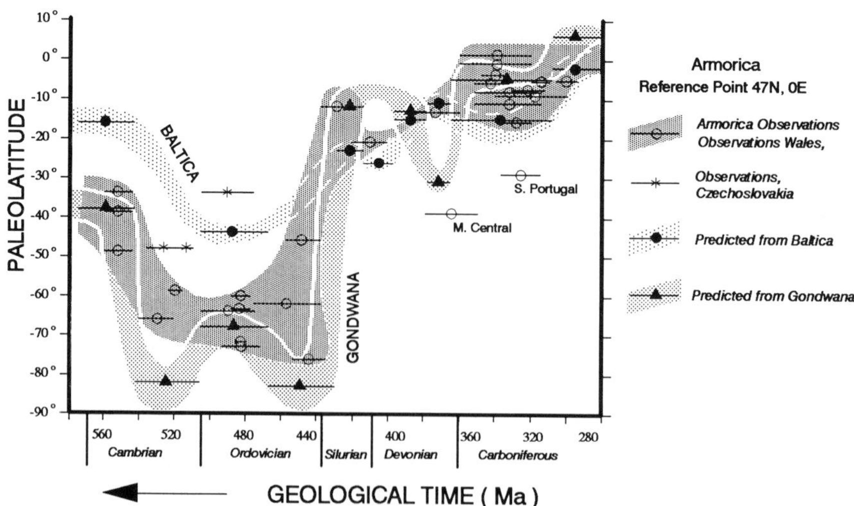

Figure 8.10. Paleolatitude versus Paleozoic time plot of the observations and predictions for a reference point of 47° N, 0° E in Hercynian Europe (Armorica), according to the data of Tables 8.3 and A11, part IV (Q ≥ 3). Symbols marked by an asterisk are observations from Czechoslovakia, which give lower paleolatitudes than those from France, Germany, Iberia and Poland for the Ordovician and Cambrian.

Table 8.3. *Declinations and paleolatitudes for Armorica at 47° N, 0° E, predicted from the reference poles of Table 8.1*

Age interval	Laurentia		Baltica		All Gondwana	
	D,	Lat.	D,	Lat.	D,	Lat.
Cu/Pl, Cu (282–308)	10,	−2	10,	−2	322,	6
Cm, Cl, Du/Cl (309–365)	12,	−15	12,	−15	338,	−5
Du, Dm/Du (366–378)	28,	−11	28,	−11	347,	−31
Dm, Dl (379–397)	31,	−15	31,	−15	4,	−13
Su/Dl (398–414)	52,	−26	52,	−26		
Su, Sm (415–429)	19,	−23	19,	−23	3,	−12
Ou/Sl, Ou, Om (430–467)	358,	−28			58,	−83
Ol/m, Ol (468–505)	343,	−25	282,	−44	10,	−68
Єu, Єm, Є (506–542)	341,	−32			164,	−82
Єl (543–575)	326,	−32	307,	−16	57,	−38

Age abbreviations as in Table 5.1. D is declination, Lat. is paleolatitude. Declination and paleolatitude values are calculated for a common location in Armorica (47° N, 0° E) from the reference poles (Table 8.1) for Laurentia and Baltica in European coordinates and for all of Gondwana in northwest African coordinates.

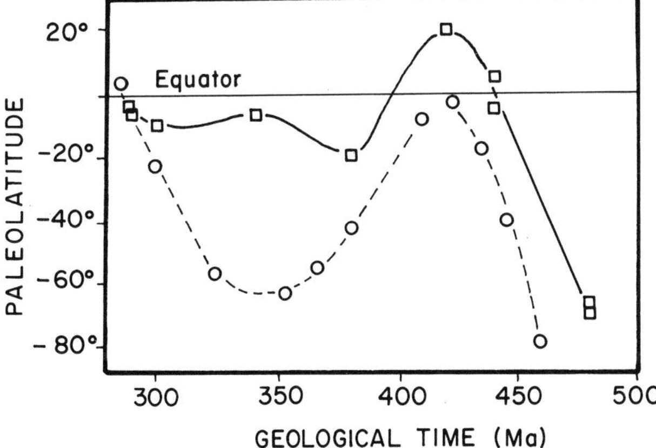

Figure 8.11. Paleolatitude versus time plot for northern Spain (solid line) and the northern margin of Africa (dashed line, based on apparent polar wander path from Bachtadse & Briden, 1990, shown in Figure 5.15), from Ordovician to Carboniferous. Figure redrawn after Perroud, Calza & Kattach (1991). Published with permission from the authors.

reveals that such Late Devonian paleolatitudes would mean a 'belonging' to Baltica-Laurentia or Gondwana, respectively. My preference for a choice between these two poles is to put stronger weight on the results from the stratified rocks of Germany than on the poorly defined magnetization component D of the French Metamorphics, but I stress that more Devonian results from Europe are clearly needed before we can state without hesitation that Armorica had parted from Gondwana and joined Baltica already in the Early Devonian. With my co-authors, I have postulated the Devonian unity of Armorica and Baltica in previous papers (e.g., Perroud, Van der Voo & Bonhommet, 1984) at a time that a remagnetization of the Montmartin red bed results (Jones, Van der Voo & Bonhommet, 1979) had not yet been documented by Perroud *et al.* (1985). Noting the paleogeographic arguments of Paris & Robardet (1990) for Devonian continuity between Armorica and Gondwana, I am no longer so sure. As an intermediate possibility, Ziegler (1988) has proposed a scheme for accretion of Avalon and various parts of Hercynian Europe at different times during the Devonian; this accretion took place in a compressional Acadian/Ligerian/late Caledonian regime, but it was accompanied, intermittently, according to Ziegler, by Devonian back-arc extension in the Rhenohercynian domain of southernmost England, Belgium, northern France and central Germany.

There are thus few unambiguous paleomagnetic indications about the paleolatitude and paleogeography of Armorica during the Silurian and Devonian periods, which are critical for understanding the evolution and history of Hercynian Europe. Carboniferous results are more plentiful (although fairly scattered; Figure 8.10), but by this time Gondwana and Laurussia paleopoles predict similar paleolatitudes for Armorica, thereby disallowing the use of them in a paleomagnetic test of the Devonian affinity of Armorica to either Gondwana or Baltica.

When the available Carboniferous declination results from Germany, France and Iberia, including those ascribed to Hercynian remagnetizations (not compiled in Table A11), are plotted on a map, a pattern of systematic rotations can be observed, which correlates with the regional structural trends (Figure 8.12). This has been noted previously by Perroud (1986) and Bachtadse & Van der Voo (1986). It appears that the Hercynian orogenesis, in response to the Gondwana–Laurussia collision, caused these block rotations perhaps by indentation and pervasive strike – slip movements. The pre-Carboniferous geographical outline of Armorica was therefore certainly very different from that of today (the use of today's outline in the paleogeographic maps of this chapter is only for convenient recognition).

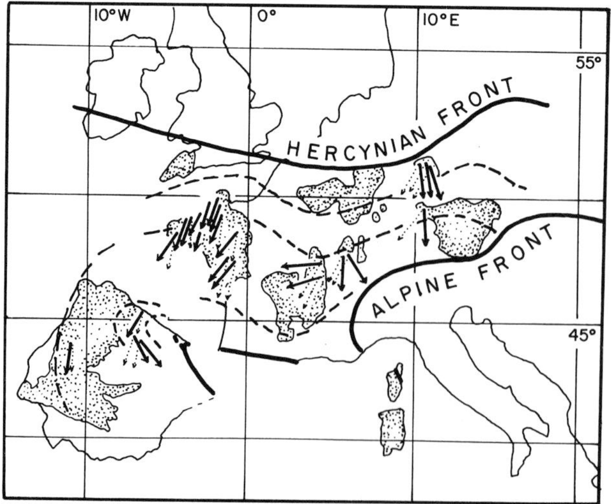

Figure 8.12. Paleomagnetic declinations of Devonian and Carboniferous age for Hercynian Europe. Solid lines are the observed values, and dotted lines are the expected ones, illustrating regional rotations. The Hercynian and Alpine fronts are also shown for reference. Figure redrawn after Perroud (1986) and published with permission from the author.

In summary, Armorica, which is thought to include the Armorican Massif, Iberian Meseta, parts of Germany and possibly the Bohemian Massif and southern Poland, was in near-polar southerly latitudes and adjacent to the northwest African margin of Gondwana during the Ordovician. During the latest Precambrian and Cambrian a similar situation prevailed according to the paleopoles. The history of Armorica for Silurian and Devonian times, however, needs further elucidation and it is therefore unclear whether the rapid Siluro-Devonian drift postulated for Gondwana is also seen in Armorica, although some workers think so (e.g., Perroud, Calza & Khattach, 1991). By Carboniferous time, the Hercynian orogeny was imminent or underway as Gondwana began to collide with Laurussia, thereby causing the Hercynian deformation, extensive metamorphism and plutonism (e.g., Finger & Steyrer, 1990), the pre-Permian rotations in the Hercynian belt, and pervasive Late Carboniferous remagnetizations and overprints.

Appalachian displaced terranes

The Avalon Terrane has its type locality in the Avalon Peninsula of eastern Newfoundland, but its name is also used for a coastal strip in eastern New Brunswick and Nova Scotia and for the Boston Basin in eastern Massachusetts (see Figure 8.13). Several workers have also suggested that the eastern parts of the Piedmont Province south of New York belonged to Avalon. Because the usage of Eastern Avalonia has become popular to denote Southern Britain, the American part of this terrane is now sometimes called Western Avalonia. I will stick to the use of simply Avalon for the American part.

The Avalon Terrane shares many geological characteristics with Armorica, including aspects of the coastal landscape in Newfoundland and Brittany, the occurrence of a Pan-African to Cadomian-type orogeny during the latest Precambrian, an epicontinental sedimentary facies during much of the Cambrian, and coherent geophysical (crustal) anomalies in both areas (e.g., Van der Voo, 1982; Lefort, Max & Roussel, 1988; Rast, Sturt & Harris, 1988; Skehan, 1988). Its paleomagnetic signatures are also similar: high Ordovician paleolatitudes are probably the most diagnostic, but there is also good agreement between predicted paleolatitudes from Gondwana and observations for the Cambrian and late Precambrian (e.g., Johnson & Van der Voo, 1986). In contrast, the pre-Silurian paleolatitudes of Laurentia and Avalon are very different, indicating that these two elements were separated by the Iapetus Ocean (e.g., Figure 8.1).

The paleolatitude of the latest Ordovician to Early Silurian Dunn Point Formation of Nova Scotia illustrates that by that time, Avalon had been drifting northward. Whether this means that it had separated from Gondwana depends critically on the (uncertain) timing of Gondwana's own rapid northward drift, leaving the affinity of Avalon with Gondwana during the Silurian just as unresolved as the similarly ambiguous affinity of Armorica with Gondwana. What seems clear, however, is that by Late Silurian to Early Devonian time, Avalon and Laurentia had joined; the earlier deeper-water marine sedimentation was replaced by continental red beds and volcanic rocks in Newfoundland and paleopoles for Avalon are in agreement with those from the North American craton for Late Silurian and younger time (see Table A11, Mascarene A–B component paleopole, and Figure 8.14).

Two results from plutons in the Avalon Terrane (Roy, Anderson & Lapointe, 1979) have been marked by an asterisk in Table A11 because they give unusually high paleolatitudes, compared to what would be expected for their Siluro-Devonian age. The lack of structural control may play a role in causing these anomalies.

Between Avalon and the North American craton, the northern Appalachians in Newfoundland and farther south include many other Ordovician terranes surrounded by Siluro-Devonian sediments in what has been called the Central Mobile Belt (van der Pluijm & Van Staal, 1988). This belt had previously been subdivided into the Dunnage Zone and the more easterly Gander Zone, with the former thought to represent the vestiges of Iapetus and the latter the continental rise sediments adjacent to Avalon, although there is no geological link between the Gander and Avalon zones (e.g., Williams, 1978, 1979; Williams, Colman-Sadd & Swinden, 1988). Faunal characteristics of these Ordovician terranes led Neuman

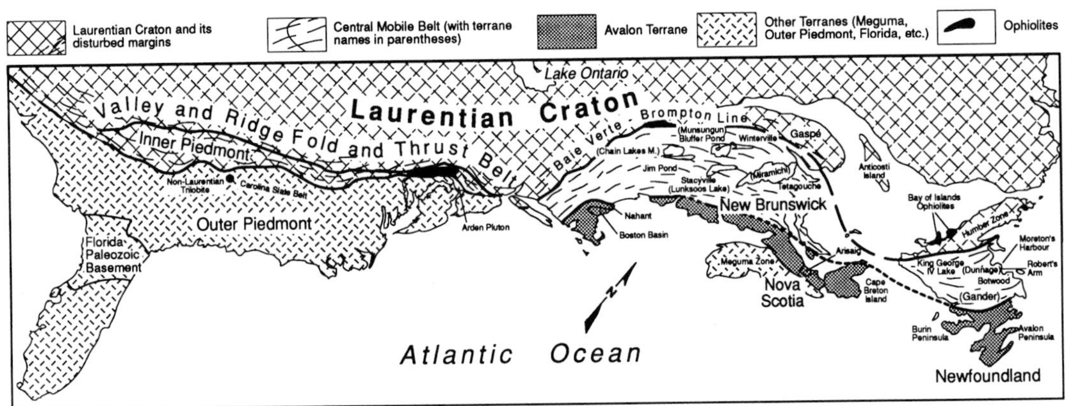

Figure 8.13. Map of the Appalachians, showing various terranes discussed in the text.

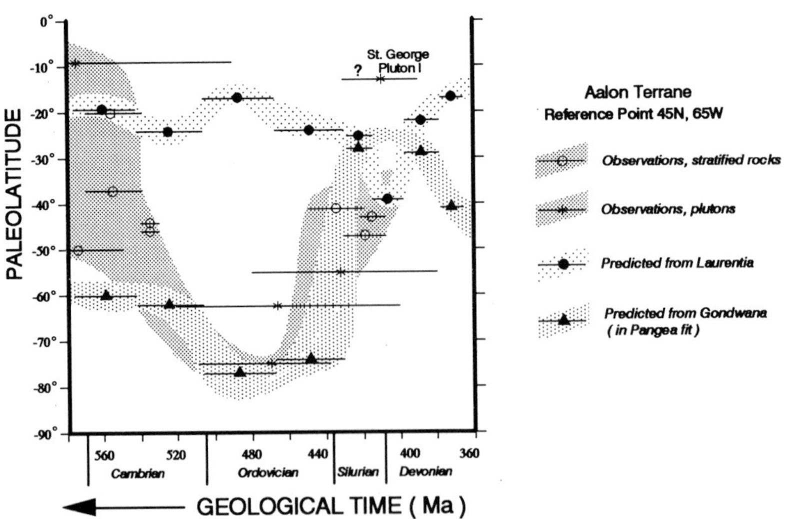

Figure 8.14. Paleolatitude versus Paleozoic time plot of the observations from Avalon and the predictions for a reference point of 45° N, 65° W in the Northern Appalachians, according to the data of Tables 8.4 and A11, part V ($Q \geq 3$). Symbols marked by an asterisk are observations from plutons, which may be incorrect because of the lack of tilt correction.

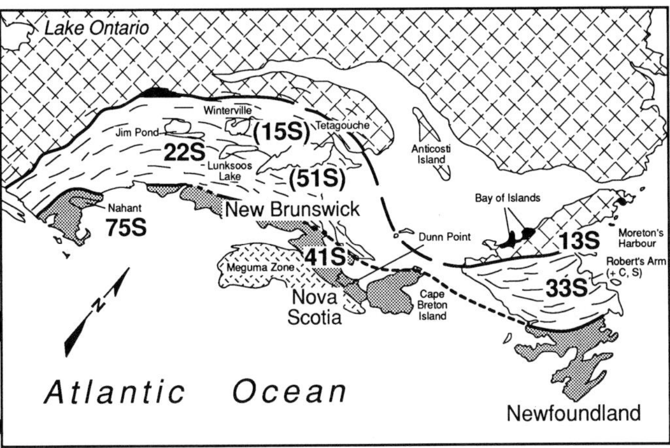

Figure 8.15. Map of the northern Appalachians (as in Figure 8.13), showing Ordovician paleolatitude determinations for Avalon and other terranes discussed in the text.

(1984) to postulate that several of them may have formed islands within Iapetus, which is supported by the geochemistry of the Ordovician extrusives that has been interpreted in several cases as indicating island-arc and ocean-island settings for their formation (Strong & Payne, 1973; Kean & Strong, 1975; Williams & Payne, 1975; Jacobi & Wasowski, 1985; Swinden, 1987).

A systematic study of the paleomagnetism of the stratified Ordovician volcanic rocks has revealed paleolatitudes ranging from about 15° S to more than 30° S or even higher (Deutsch & Rao, 1977; Wellensiek *et al.,* 1990; Johnson, van der Pluijm & Van der Voo, 1991; Liss, van der Pluijm & Van der Voo, 1991; Potts, Van der Voo & van der Pluijm, 1991; Van der Voo *et al.*, 1991). Published results for the Central Mobile Belt have been included in Table A11, part VI, together with a few results from the Piedmont Zone south of New York. As the distance from the craton increases within the present-day Central Mobile Belt, so does in general the observed paleolatitude, as can be seen in the map of Figure 8.15. Preliminary (unpublished) results (Potts *et al.*, 1991; Liss *et al.*, 1991) are placed in parentheses on this map. Figure 8.16 presents the fully published results in the usual paleolatitude versus time framework, with paleolatitudes predicted from Laurentia and Gondwana (in Pangea fit) plotted for comparison. Many of the Ordovician paleolatitudes are intermediate between the Laurentia and Gondwana predictions, indicating paleopositions for the rocks in the middle of Iapetus. In the Silurian, a convergence can be recognized between the terrane paleolatitudes and those of Laurentia, indicating that accretion to Laurentia is in progress.

The Early Silurian results nevertheless present an ambiguity reminiscent of the Silurian Almadén results of Spain discussed earlier. The King George IV Lake results (Buchan & Hodych, 1989) yield a near-equatorial paleolatitude, whereas the paleolatitude predicted from Laurentia would be approximately 25° S. On the other hand, the Botwood results (Wigwam and Lawrenceton red beds and volcanics) of the same age (Lapointe, 1979a; Gales, van der Pluijm & Van der Voo, 1989) yield paleolatitudes in agreement with those from Laurentia (Table A11). Does this mean that the King George IV Lake area moved southward with respect to Laurentia before final accretion, whereas the Botwood sampling region did not, or do the results represent an internal inconsistency? All indications are that they represent the same terrane, so I suspect the latter. Even so, both these Silurian paleomagnetic results appear to be of excellent quality, passing conglomerate and fold tests and revealing reversals. If the results from the King George IV Lake area prove to be accurate, they imply that this area in the Central Mobile Belt of Newfoundland was moving southward as it was being emplaced by dextral strike slip. Given that Gondwana was moving southward with respect to Laurentia during post-Middle Silurian time (e.g., Table 8.4), it may well be that parts of the Central Mobile Belt were moving with it.

Besides Avalon and the Central Mobile Belt, a third Appalachian region, namely the Piedmont Province, is allochthonous, that is, likely to have been displaced. In the Carolina Slate Belt, for instance, Middle Cambrian trilobites show strong affinities with contemporaneous cool water faunas from Armorica and Baltica, whereas they are very different from those of Laurentia (Samson *et al.*, 1990). The story has it that Sara Samson found these fossils in South Carolina when she tied her shoelace on the outcrop. The few paleomagnetic results from the Carolinas (Vick, Channell & Opdyke, 1987; Noel, Spariosu & Dallmeyer, 1988), however, show very shallow Cambrian and Ordovician inclinations, which do not match those expected from Avalon, suggesting that these results either represent younger remagnetizations or that the area was a terrane much different from Avalon. On the other hand, two results from Delaware and surroundings give very high paleolatitudes (Rao & Van der Voo, 1980; Brown & Van der Voo, 1983), which have puzzled many geologists familiar with the area. The rocks studied are thought to have been metamorphosed during a Late Ordovician event, ascribed to the Taconic Orogeny, which is a typical eastern North American phenomenon. Now, if the magnetizations are reset during the Taconic metamorphism of eastern Laurentia, why do they show paleolatitudes that are much higher than expected for Laurentia? The lack of structural correction can be invoked to call these results less reliable. Alternatively, the rocks are part of a displaced terrane that underwent a Late Ordovician metamorphism elsewhere. At this time, it is not possible to decide what these magnetizations from the Delaware area mean.

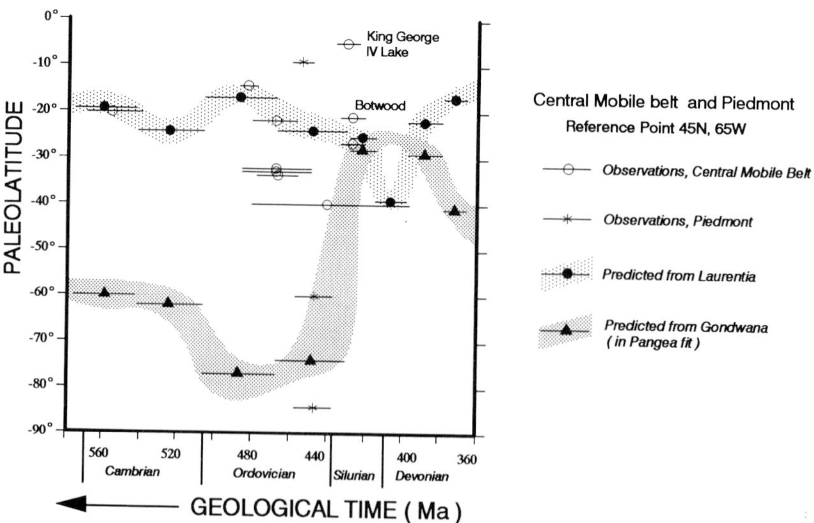

Figure 8.16. Paleolatitude versus Paleozoic time plot of the observations from the Central Mobile Belt and Piedmont Province and the predictions for a reference point of 45° N, 65° W in the Northern Appalachians, according to the data of Tables 8.4 and A11, part VI (Q ≥ 3). Symbols marked by '*' are observations from the Piedmont.

Table 8.4. *Declinations and paleolatitudes for the Appalachian terranes at 45° N, 65° W, predicted from the reference poles of Table 8.1*

Age interval	Laurentia	All Gondwana
	D, Lat.	D, Lat.
Cu/Pl, Cu (282–308)	350, −3	353, 3
Cm, Cl, Du/Cl (309–365)	347, −16	4, −14
Du, Dm/Du (366–378)	4, −17	1, −41
Dm, Dl (379–397)	5, −22	27, −29
Su/Dl (398–414)	24, −39	
Su, Sm (415–429)	351, −25	26, −28
Ou/Sl, Ou, Om (430–467)	329, −24	249, −74
Ol/m, Ol (468–505)	317, −17	326, −77
Ꞓu, Ꞓm, Ꞓ (506–542)	313, −24	239, −62
Ꞓl (543–575)	300, −20	93, −60

Age abbreviations as in Table 5.1. D is declination, Lat. is paleolatitude. Declination and paleolatitude values are calculated for a common location in the Northern Appalachians (45° N, 65° W) from the reference poles (Table 8.1) for Laurentia and for all of Gondwana in North American coordinates, using the rotation parameters of Lottes & Rowley (1990).

Farther south, subsurface cores from Florida with Ordovician–Silurian ages of sedimentation have yielded a paleolatitude of 50° (Opdyke *et al.*, 1987b). This value is compatible with the widely accepted idea that Florida drifted with West Gondwana in the Paleozoic.

In summary, the results from the Avalon Terrane in the Northern Appalachians present a rather clear picture, resembling that of Armorica and Southern Britain in its high Ordovician paleolatitudes and indicating post-Arenig displacements with respect to Laurentia of more than 4000 km. The resemblance also exists for the uncertainties in the Silurian results. Results from the Central Mobile Belt between Avalon and Laurentia show variable post-Middle Ordovician displacements, illustrating that this Belt is made up of separate terranes that had their origin in Iapetus at different paleolatitudes.

The opening of Iapetus and possible break-up of a Late Proterozoic supercontinent

A chapter on the Iapetus Ocean would not be complete without a discussion of the initial rifting of this ocean and the pre-Iapetus continental configuration. The opening and closing of Iapetus is the classical example of the Wilson Cycle, named after J. Tuzo Wilson because he was the first to postulate the existence of pre-Mesozoic oceans. However, the locations of the continents during the Late Proterozoic and Early Cambrian are very poorly known and the sketchy situation of the few paleopole data that do exist is full of uncertainties.

What seems clear – if any supercontinent existed before the rifting of Iapetus – is that the continents did not come together at the end of the Paleozoic Wilson cycle in the same configuration from which they rifted apart at its beginning. Everything else about a possible Precambrian supercontinent seems to me at this time to be very speculative. Let us first examine the proposals for latest Proterozoic supercontinental configurations and then discuss their paleomagnetic support or the lack thereof.

A longstanding model for a Precambrian supercontinent is that of Piper (1976a, 1987), who has argued for a relatively stable and unchanging configuration involving all the major cratons for the interval between the late Archean and about 1000 Ma. After about 1000 Ma, the relative position of Baltica changed with respect to Laurentia, and the East Gondwana cratons regrouped themselves with respect to West Gondwana; details of the late Proterozoic supercontinental configurations of Piper are shown in Figure 8.17. In this model, the southwest coast of Laurentia

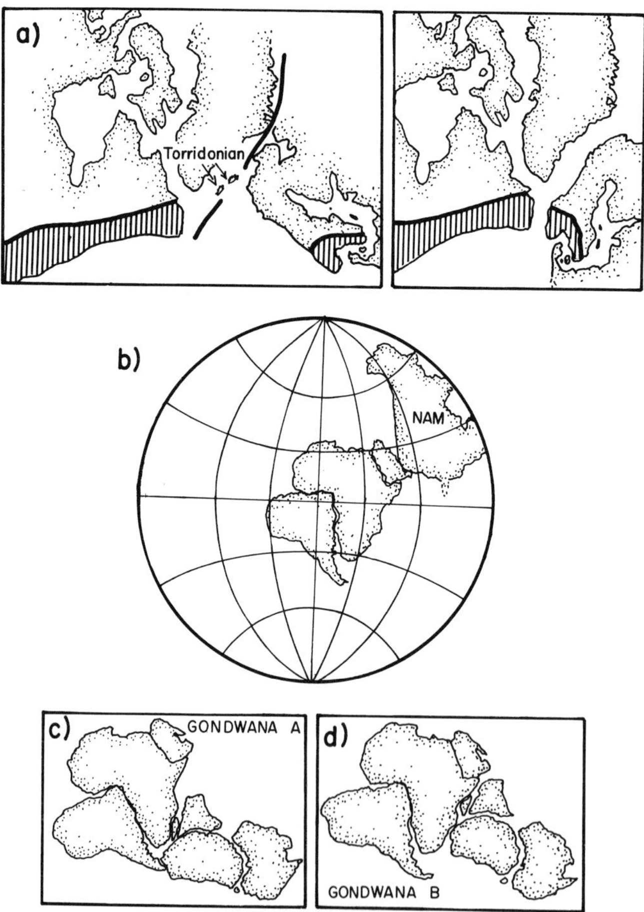

Figure 8.17. Details of the (Late) Precambrian supercontinent configuration of Piper (1987) and its proposed evolution in Precambrian – Cambrian time. a) The relationships between Laurentia (including Greenland) and Baltica before (left) and after (right) about 1000 Ma. The Grenville Belt of Laurentia and the coeval Sveconorwegian Belt of Baltica are shaded. Piper, in this model, has linked the timing of the Torridonian sedimentation in northern Scotland to the plate readjustments in this configuration. b) The relationships between Laurentia and West Gondwana, proposed by Piper as valid from Archean until Early Cambrian time. c) The configuration of the Gondwana continents for the interval of 1000–570 Ma. d) The configuration of the Gondwana continents of Smith & Hallam (1970) as used by Piper for Paleozoic to Early Mesozoic time.

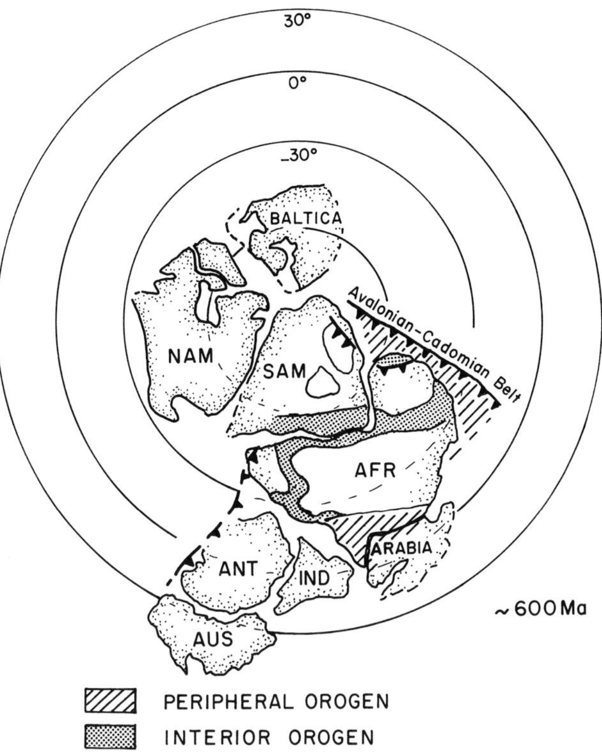

Figure 8.18. Late Precambrian supercontinent reconstruction, after Bond, Nickeson & Kominz (1984) and Murphy & Nance (1991). This fit places the Appalachian Margin of Laurentia against the Andean Margin of South America. The Pan African Belts (dense stipple) are thought to have been ensimatic (see text for discussion), but the collisions of the blocks are argued to have occurred before the time of this configuration (about 600 Ma). Note the locations of the Avalon and Armorican (= Cadomian) displaced terranes in this reconstruction, which agree with those shown in, or implied by, Figures 8.1 and 8.9 at locations peripheral to Gondwana. Published with permission from the authors.

(Arizona, California) is fitted against the Arabian (Persian Gulf) margin of Gondwanaland. The Pan-African deformation belts of Gondwana are ascribed to ensialic orogenies, such that the late Proterozoic Gondwana assembly (Figure 8.17c) is thought to have been a rather permanent entity from about 1000 Ma until the Cambrian, whereupon East Gondwana shifted northwards with respect to West Gondwana (e.g., Figure 8.17d). In Piper's model, Laurentia rifted away from Gondwana in the Early to Middle Cambrian. Thus, paleopoles from Laurentia and Gondwana for the interval of 1000–540 Ma should be in agreement with each other in this fit (called here the Piper fit), if it is to derive support from paleomagnetism.

In a different configuration (Bond, Nickeson & Kominz, 1984), Gondwana is also fitted against Laurentia, but such that the Appalachian margin of the latter is placed against the Andean margin of South America (Figure 8.18, from Murphy & Nance, 1991). Bond & co-workers argued that the latest Precambrian to Early Paleozoic sedimentary sequences on the margins of Laurentia show that formation of new ocean floor most likely occurred during the Early Cambrian, while initial break-up may have started at about 615 Ma. The Pan-African belts in this model may well have been ensimatic (i.e., accompanied by ocean floor spreading and subduction) and can thus be ascribed to relative movements between the different nuclei of Africa and South America, but the collisions of these blocks are thought to have been completed by 650–600 Ma (Murphy & Nance, 1991). The location of Laurentia in this Bond model is about 120° removed from that of the Piper model and its orientation is such that reliable paleopole sets for Gondwana and Laurentia with ages of about 650–550 Ma would be able to distinguish between the Bond and Piper models.

A more recent set of models (Moores, 1991; Dalziel, 1991), recognizing Grenville-aged orogenic belts in Antarctica, as well as Laurentia (and Baltica), places these continents in each other's proximity in the late Proterozoic (Figure 8.19), while retaining the Bond-fit between the east coast of Laurentia and South America. Stump (1992) has argued that it is likely that the breakup between Laurentia and Antartica is marked by bimodal volcanism at 750 Ma, although actual separation may have started as late as shortly before the Early Cambrian. If this Moores-Dalziel model remains valid throughout the later Proterozoic until about 550 Ma, it could be readily distinguished by paleomagnetic techniques from the Bond and Piper fits, again provided that reliable paleopole sets are available for Laurentia, West and East Gondwana.

A fourth model (Hoffman, 1991) differs from the previous two in its treatment of West Gondwana; the older nuclei of Africa and South America are reconstructed independently of each other (Figure 8.20), implying significant post-700 Ma relative motions and collisions during the latest Precambrian that resulted in the Pan-African Belts surrounding these cratons.

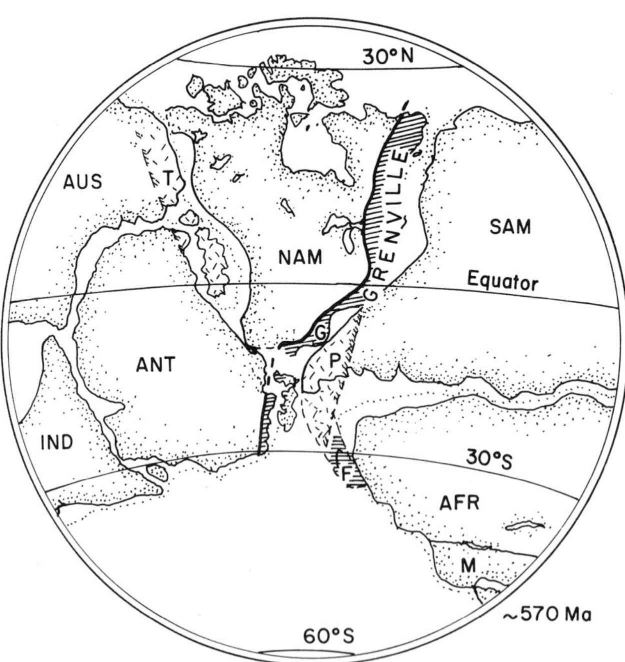

Figure 8.19. Reconstruction of the Gondwana and Laurentia cratons at the beginning of the Cambrian (about 570 Ma) in the Moores-Dalziel model, with paleolatitude lines based on McWilliams (1981). In this configuration, Grenville belts of Laurentia and East Antarctica are brought into alignment. F = Falkland Islands, G = Grenville Belt in Laurentia, M = Pan-African Mozambique Belt in Africa, P = Patagonia in South America, T = Tasman orogenic belt in Australia/Antarctica. Figure redrawn and modified after Dalziel (1991) and published with permission from the author.

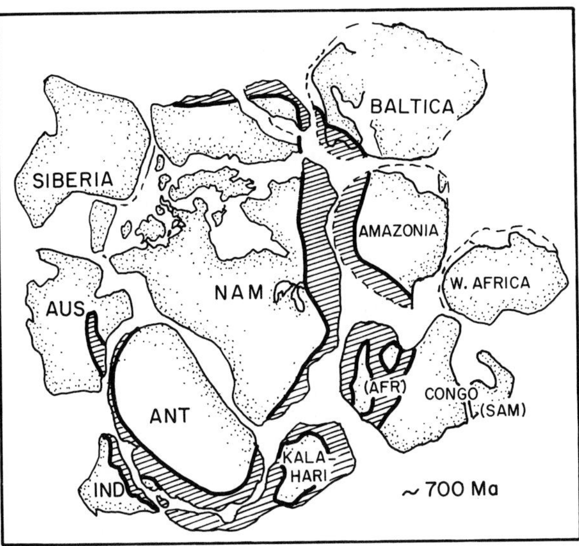

Figure 8.20. Reconstruction of the Gondwana and Laurentia cratons at the end of the Precambrian (about 700 Ma) of Hoffman (1991). In this configuration, Grenville-aged belts of Laurentia, Baltica, western South America, Antarctica and within the separate elements of Africa are brought into alignment. AFR = Africa, ANT = Antarctica, AUS = Australia, IND = India, NAM = North America, SAM = South America. Figure redrawn and modified after Hoffman (1991; Copyright 1991 by the AAAS), and published with permission of the American Association for the Advancement of Science.

A different class of models also ascribes the Pan-African Belts to ensimatic orogenies and relative plate movements, but recognizes that the block collisions may have occurred as late as the Cambrian (such models are similar, but not identical to that of McWilliams, 1981). If true, this precludes a testing of any rifting models between Laurentia and Gondwana as a whole. Testing of a Baltica–Laurentia rifting or of the drifting apart of Laurentia from a single Gondwana nucleus (e.g., Guyana or northwest Africa in Figure 8.18) remains a possibility, but the paleopole datasets have to be rather complete and robust for such testing to be definitive. As we will see shortly, this is definitely not the case for the Gondwana cratons.

Cambrian paleopoles for Laurentia (see Table A1) were briefly discussed in Chapter 5 (see also Figure 5.4) as well as earlier in this chapter. The most reliable Cambrian paleopoles and also those few for the late Precambrian (700–570 Ma, not listed in Table A1) generally fall near the equator at about 170° E. The available Early Cambrian poles from the Tapeats Sandstone, Bradore Sandstone, the Unicoi Basalts and Buckingham flows (Table A1) form a coherent set. The existence of a Middle to Late Cambrian APWP loop (Watts *et al.*, 1980a,b) is still very speculative, but with or without such a loop the mean Early and latest Cambrian paleopoles agree with each other and with the just-mentioned equatorial location. These paleopoles imply equatorial paleolatitudes for Laurentia throughout the interval 700–540 Ma. However, work in progress (e.g., Symons & Chiasson, 1991) suggests much higher paleolatitudes; because the results of isotopic age determinations on the rocks in this study are unpublished, it is very difficult to be definitive about the Early Cambrian paleolatitudes of Laurentia at this time.

The situation is exacerbated by an examination of the paleopoles of Africa for the late Proterozoic and Cambrian. An evaluation of the available results (Van der Voo

& Meert, 1991) led us to state that the data set is of such low reliability that paleomagnetic conclusions about late Precambrian tectonics and relative movements of Africa as a whole or of African cratonic nuclei individually are virtually precluded at this time. The few well-dated (Q ≥ 3) results that do exist for the three African cratons (northwest Africa, Congo and Kalahari) are for different time intervals for each of the three blocks, rendering a paleomagnetic analysis of their late Precambrian movements impossible. This is all the more noteworthy if one realizes that the late Proterozoic data for the other Gondwana continents are even more scarce and less reliable in terms of age of magnetization and paleopole location (e.g., Idnurm & Giddings, 1988).

If Africa is taken to be one piece, despite such ambiguities, its latest Proterozoic APWP (700–570 Ma) is characterized by rapid apparent polar wander, as is the case for the rest of Gondwana (e.g., McWilliams, 1981; Piper, 1987, pp. 221–6). This can also be seen in Figure 8.9. In contrast, the Laurentian APWP for this interval has generally been characterized by a near-stationary APWP segment. So, unless further work by Dave Symons drastically changes this APWP segment – or if one includes unreliable paleopoles or those based on undated remagnetizations (as done by Piper, 1987) – there is no paleomagnetic support for a relatively long-lived supercontinent configuration of Laurentia and Gondwana, because a match between a short and a very long APWP segment is impossible. A short-lived configuration, on the other hand, remains possible but is untestable by paleomagnetic means at this time; 'short-lived' here would refer to an interval of 35 Ma or less.

There are additional complications that matter greatly, but these are sometimes ignored by those advocating one fit or another. One well-known problem is that for the Cambrian Period or earlier, we may not have identified the correct polarity of a paleopole, leaving the option of selecting north versus south poles wide open. A second problem is that there is no consensus on the exact isotopic age of the Precambrian–Cambrian boundary, with estimates ranging from 590 to 530 Ma (e.g., Luo Huilin *et al.*, 1984; Palmer, 1983; Harland *et al.*, 1982; Odin *et al.*, 1983; see also Cowie & Johnson, 1985). This latter issue plays a role in comparisons between radiometrically dated results and paleopoles obtained from stratigraphically dated sequences. A third problem, mentioned earlier, is that glacial relicts and paleopoles of latest Proterozoic age either are in serious discord in terms of the paleolatitudes predicted for many continental blocks (e.g., Embleton & Williams, 1986), or are suggesting that the typical uniformitarian approach to this issue is incorrect and that the usual climatic, obliquity/orbital eccentricity or geomagnetic dipole models do not apply to this interval (e.g., Williams, 1975).

Is there then no hope for a paleomagnetic contribution to the problem of determining from what configuration Iapetus opened? On the contrary, there is hope, as suggested by the Laurentia–Baltica situation to be discussed shortly, but at this time there is a complete lack of certainty about the Gondwana–Laurentia paleogeographic relationships in the latest Proterozoic.

The Laurentia–Baltica situation for Grenville and younger Precambrian times (about 1100–570 Ma) illustrates that some late Precambrian paleogeographic configurations can be successfully worked out. Figure 8.17a (right-hand figure) illustrates Piper's fit for these two blocks; this fit agrees reasonably well with the available paleopoles for Laurentia (Piper, 1987, p. 220) and Baltica (Pesonen *et al.*, 1989) for the interval of 1000–850 Ma, as shown in Figure 8.21. For the interval of 850–600 Ma there are insufficient well-dated paleopoles available for a comparison. When Precambrian paleopoles eventually become as reliable and available for other continental elements as they are for Baltica and Laurentia, then it may well be that progress can be made in determining pre-Iapetus continental reconstructions.

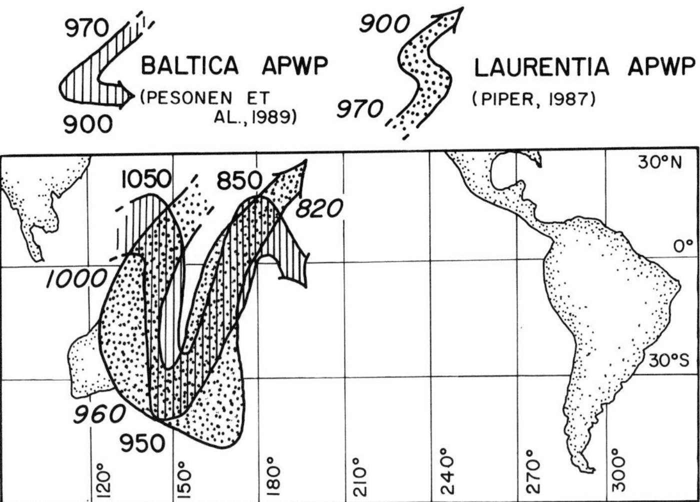

Figure 8.21. Late Proterozoic paleopoles for the interval 1000–850 Ma from Laurentia and Baltica in the reconstruction of these cratons by Piper (1987).

In summary of this section, I conclude that the existence of a late Proterozoic supercontinent is likely, but that most of its configuration is not yet clear from paleomagnetic data alone, whereas the non-paleomagnetic evidence for specific models is even more ambiguous. The question whether Pan-African belts are ensialic or ensimatic is entirely unresolved from a paleomagnetic viewpoint. The paucity of the available data renders paleomagnetic testing of models for any break-up configuration, related to the opening of Iapetus, nearly impossible. The most reliable paleomagnetic datasets are for Europe and North America and these support Piper's (1987) model for the pre-rifting configuration of Baltica and the Greenland margin of Laurentia and, presumably, for the opening of that part of Iapetus that was located between them after about 600 Ma. Late Precambrian paleopoles from Gondwana are very poorly documented; as reliable paleopoles are essential for making progress in determining the early history of the part of Iapetus that was located between Gondwana and the Appalachian, southern or western margins of Laurentia, good late Precambrian and Cambrian paleomagnetic data from Gondwana are urgently needed.

Summary

This chapter on the Paleozoic Iapetus Ocean has been less definitive than many earlier chapters, and the uncertainties in the paleomagnetic data have been greater (by up to 30°) than the typical error limits involved in latest Paleozoic, Mesozoic and Early Tertiary mean paleopoles. These uncertainties play a role, especially, in the discussions about the latest Precambrian–Cambrian paleogeographies and the initial rifting of Iapetus, but also for the movements of Gondwana during middle Paleozoic times. In the following, an attempt will be made to synthesize our current paleomagnetic knowledge about the Paleozoic history of Iapetus with the aid of paleogeographic maps for the Late Cambrian through Early Carboniferous interval. The Iapetus-bordering continents are positioned according to the mean paleopoles of Table 8.1 and the paleolatitudes for the smaller terranes are taken from Table

A11. Readers should keep in mind that the data base sometimes is very sparse and that for some continents (e.g., Baltica) several half-periods lack adequate paleopoles entirely.

The Late Cambrian (Figure 8.22)

Laurentia is located on the equator, Baltica is omitted because of a lack of reliable paleopoles, and Gondwana is 'upside down' in the opposite hemisphere from Laurentia; Gondwana is taken to be in one piece and is rapidly moving with respect to the pole at this time, so its location is a snapshot and very imprecise (A_{95} = 42°, Table 5.3). Southern Britain and Avalon lack representative paleopoles, but are assumed to be in a peri-Gondwanide position, as is Armorica, by analogy with earlier and later times. Siberia is shown according to its mean paleopole of Table 5.4 and the China Blocks are provisionally located near the equator, adjacent to East Gondwana in accordance with Figure 5.23 and the discussion of Chapter 5.

The Early Ordovician (Figure 8.23)

Laurentia remains in roughly the same position, whereas Baltica is in intermediate to high southerly paleolatitudes (see also Fig. 8.1). Gondwana is located with northwestern Africa close to the south pole (Table 8.1), but this paleopole location is again not very precisely known (A_{95} = 22°, Table 5.3). Southern Britain, Armorica and Avalon are shown adjacent to northwest Africa according to the observed Early Ordovician paleolatitudes of these terranes. Siberia is shown according to its mean paleopole of Table 5.4 and the China Blocks are located near East Gondwana in accordance with the mean paleopoles of Table 5.5, the paleomagnetic paleolatitudes

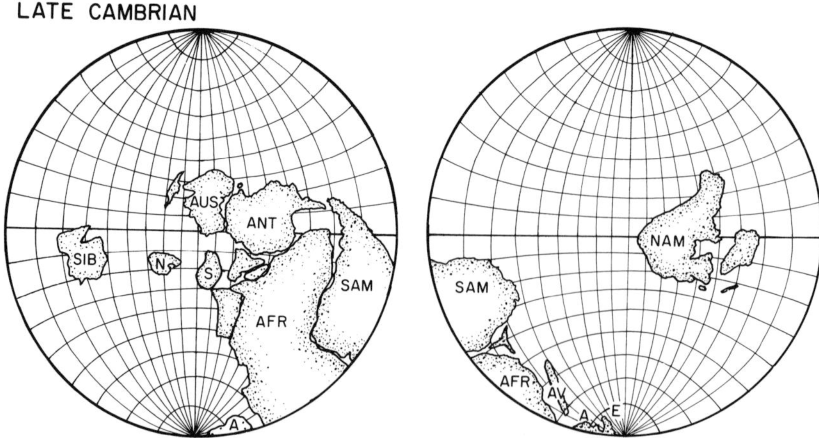

Figure 8.22. Late Cambrian paleogeographic map. See text for discussion. In this and following figures, the major continents are: AFR = Africa, ANT = Antarctica, AUS = Australia, BAL = Baltica, NAM = North America, SAM = South America, SIB = Siberia; smaller blocks: A = Armorica, AV = Avalon, N = North China Block, S = South China Block, E = Southern Britain. The major continents are positioned according to the mean paleopoles of Tables 8.1 (for NAM, BAL and all of Gondwana), 5.4 (SIB) and 5.5 (for N, S). Note that paleolongitudes are arbitrary and that continental blocks can be moved east- or westward with impunity.

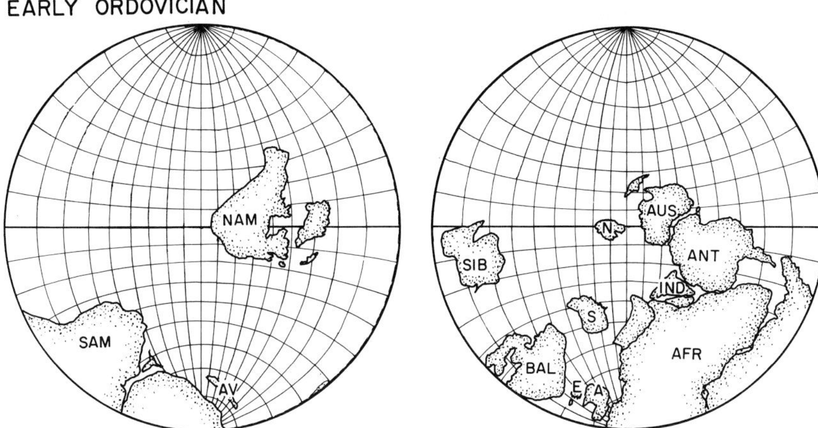

Figure 8.23. Early Ordovician paleogeographic map. See Figure 8.22 and text for explanation and discussion.

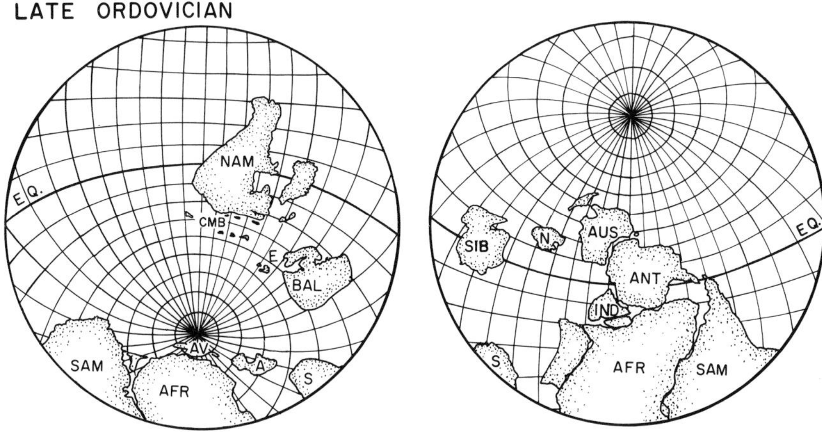

Figure 8.24. Late Ordovician paleogeographic map. See Figure 8.22 and text for explanation and discussion. CMB = parts of the Central Mobile Belt of the northern Appalachians.

of Figure 5.23 and the discussions of Chapters 5 and 7. Recall that paleolongitudes are unconstrained and that continents can be moved along lines of latitude without violating the paleomagnetic data.

The Late Ordovician (Figure 8.24)

Laurentia remains on the equator but has rotated slightly ccw, whereas Baltica is moving northwards and also rotating ccw. The Late Ordovician paleopole of Gondwana is not very precisely known, as it is based on only four studies (Table 5.2B). Southern Britain is shown as moving away from Gondwana towards Laurentia, whereas Avalon and Armorica are still close to northwest Africa. Some of the Central Mobile Belt terranes are shown in the middle of Iapetus, while others at this time are already close to the Laurentian margin in post-accretionary positions presumably related to their collision with Laurentia during the Taconic Orogeny. Siberia is shown according to its mean Middle Ordovician paleopole of Table 5.4 and the China Blocks are located adjacent to East Gondwana in positions similar to those of Figure 8.23.

The Middle Silurian (Figure 8.25)

Laurentia is continuing its ccw rotation and is colliding with Baltica by about this time, leading to the formation of the Old Red Continent of latest Siluro-Devonian times. Gondwana has completed its rapid northward drift, so that northwest Africa is now in low latitudes. The locations of Southern Britain, Armorica and Avalon are rather uncertain; their locations on the maps may give the impression that we know better where they were than we actually do. Siberia and North China have not been shown for lack of reliable data and the South China Block is positioned according to its mean pole of Table 5.5.

The Siluro-Devonian Boundary (Figure 8.26)

At this time, Laurentia is in its most southerly position, and Gondwana in a relatively far northerly position, according to a paleopole interpolated between those for the Middle Silurian and the Early Devonian. The Acadian and Caledonian orogenies are underway or in their waning stages, requiring that Laurentia, Baltica and Gondwana are in juxtaposition. The uncertainties continue about the paleolocation of Armorica, which is therefore not shown, but Avalon and Southern Britain are likely to have arrived at the positions with respect to Laurentia and Baltica that they still occupy today. Siberia and North China are again omitted and South China is shown in interpolated position.

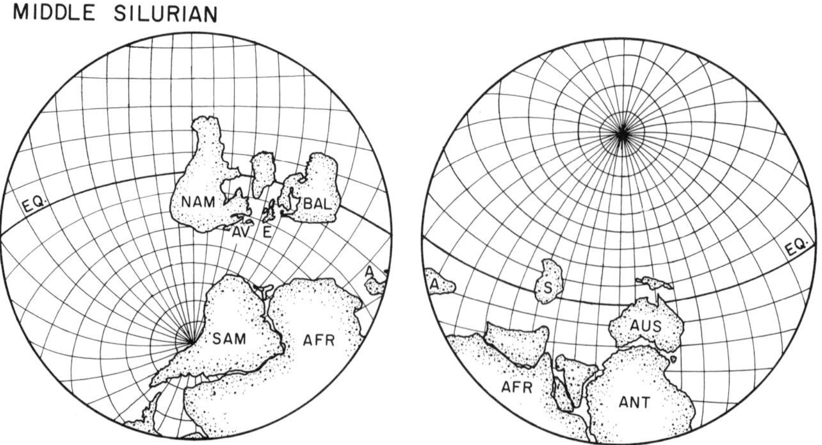

Figure 8.25. Middle Silurian paleogeographic map. See Figure 8.22 and text for explanation and discussion. N and SIB are not plotted for lack of reliable paleomagnetic data.

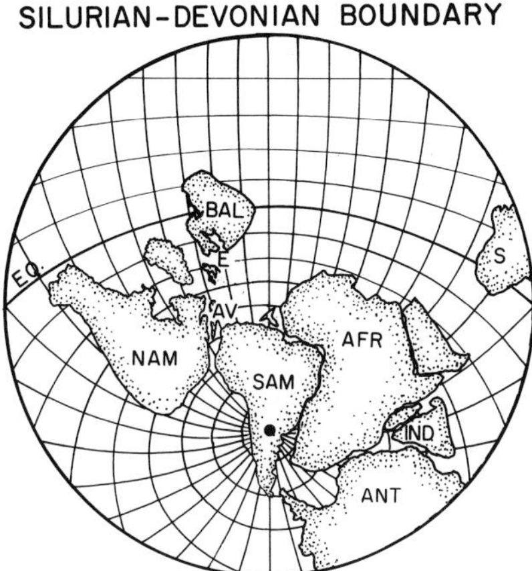

Figure 8.26. Paleogeographic map for Siluro-Devonian boundary time. See Figure 8.22 and text for explanation and discussion. A, N and SIB are not plotted for lack of reliable paleomagnetic data.

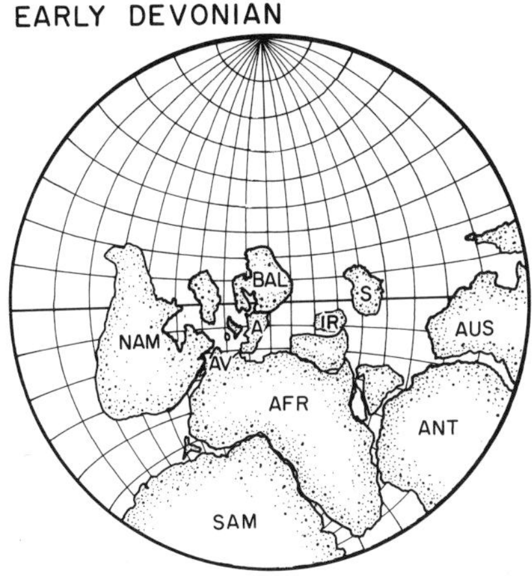

Figure 8.27. Early Devonian paleogeographic map. See Figure 8.22 and text for explanation and discussion. IR = Iran. N and SIB are not plotted for lack of reliable paleomagnetic data.

The Early Devonian (Figure 8.27)

The Old Red Continent (Laurentia–Baltica) has been moving northward with respect to a quasi-stationary Gondwana. The position of Siberia is not shown for lack of data, whereas South China is shown in an equatorial position. Iran is placed adjacent to Arabia, in accordance with the data discussed in Chapter 7.

The Late Devonian (Figure 8.28)

Gondwana is retreating from the Old Red Continent accompanied by the widening of the ocean between them. The location of South China shows that it is no longer adjacent to Gondwana, whereas Iran still is. Siberia's location is tentative and so is that of Armorica. Sibumasu is shown adjacent to Australia's west coast, as discussed in Chapter 7.

The Early Carboniferous (Figure 8.29)

Gondwana and the Old Red Continent (Laurussia) are in their final collisional approach which will result in the Alleghenian–Hercynian orogeny and the formation of Pangea. Siberia will begin moving towards Baltica, although the Ural orogeny will not reach its peak until the Permian. The Kunlun area and the Sibumasu Block are plotted adjacent to Gondwana, in agreement with the discussion of Chapter 7.

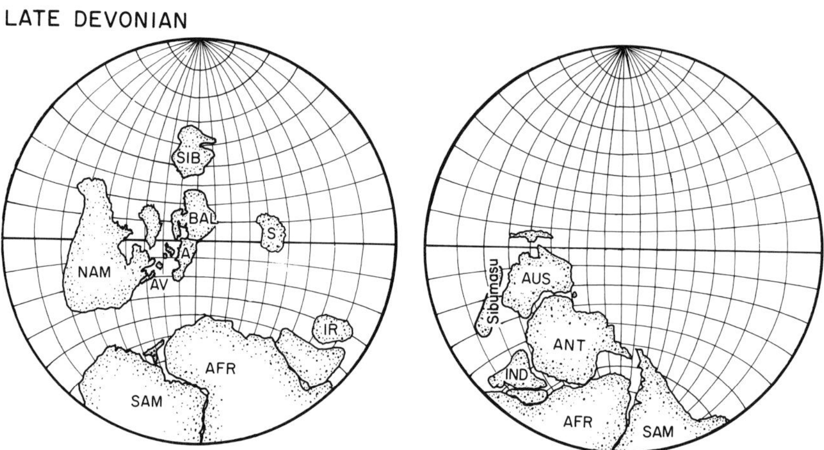

Figure 8.28. Late Devonian paleogeographic map. See Figure 8.22 and text for explanation and discussion.

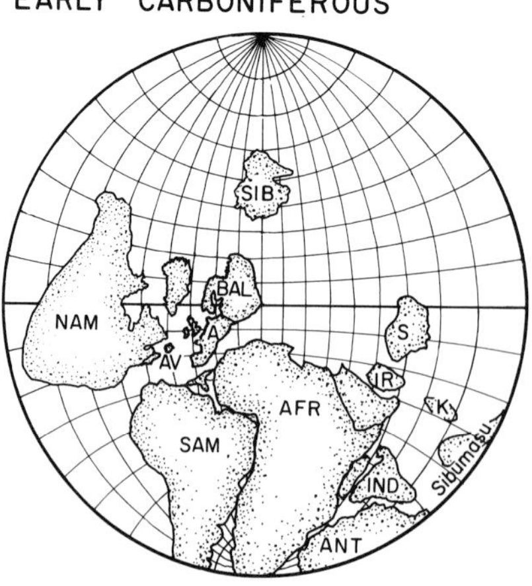

Figure 8.29. Early Carboniferous paleogeographic map. See Figure 8.22 and text for explanation and discussion. K = Kunlun. Note that paleolongitudes are arbitrary and that Gondwana, SIB or S can be moved east-or westward as long as they do not significantly overlap with other continents.

In closing, much remains to be learned about the Paleozoic paleogeography of the major continents, let alone the smaller blocks and displaced terranes. However, with careful paleomagnetic laboratory work and judicious choices of the rocktypes to be analysed, much progress can be made even in the mobile belts, as many recent studies in the Appalachians and Britain have shown. To those who wish to argue that not much of global significance remains to be done in paleomagnetic research, the above summary and the uncertainties enumerated clearly demonstrate how wrong they are.

A middle-aged man with a theory to which he has long been attached . . . grows less interested in whether it *is* true and more obsessed that it be *accepted* as true, and that he be honored for it in his lifetime

(From Joseph Heller's 1988 book Picture This; *italics in the quote are mine for emphasis.)*

After the first quote, I wrote that sometimes these quotes give me a chance to give a personal reaction to the material. Allow me a personal reaction, as my age is middle and my theories are attached to me and I to them. I, however, am still very much interested in whether my theories are true. I do not care so much whether they are accepted; it would be nice, but not necessary for me to continue thinking about them, studying new rocks and finding new ways to test them. I hope that my fellow paleomagnetists will do the same. As the preceding chapter shows, there is still an immense amount of work to be done, and theories to be tested. Let's leave no stone unturned (unspun)!

9
Epilogue

At this time of writing, it is July 1991, my sabbatical is nearly over, Barcelona is filling up with tourists, and suddenly I am suffering from a bit of writer's block. What *does* one put into a concluding chapter of a book like this, I wonder, when most sections of the preceding four chapters were already syntheses by themselves? How does one reach a conclusion on the basis of a few hundred pages of 'conclusions'?

Clearly, it is necessary to reflect back on the goals of the book as outlined in the first chapter – but these goals were reader-oriented and I am at a loss to predict what the average reader will feel when she or he has arrived at this point: disgust, puzzlement, delight, frustration, a modicum of appreciation, or just a severe case of snake-bite poisoning?

A few colleagues of fame who are experts in paleomagnetic analysis have volunteered their critiques and praise of some or all of the preceding chapters, pleasingly more praise than criticisms. However, this book, I recall, was not *only* written for the expert; my intent was to guide the tectonician, the structural geologist or the biogeographer through the maze of the extensive paleomagnetic data base and to help them recognize, if not appreciate fully, the forest more than the shrubs and trees.

What then have we learned? I would venture, foremost, that we have learned that the paleomagnetic data are overwhelmingly in support of a mobilistic Earth perspective; no matter whether the process is called the drift of the major continents (Chapter 5), plate tectonics (neo- and paleo- as in Chapters 6 and 8, respectively), microplates or displaced terranes (Chapters 7 and 8), the surface of the Earth is seen to consist of continental and oceanic elements that are, and have been, in constant relative motion. These relative movements have velocities ranging from a few centimeters per year, as in the post-Jurassic opening of the Atlantic Ocean, to some 10 cm per year, as in the rapid dispersal and northward drift of some Gondwana fragments (e.g., India in the Late Cretaceous), to perhaps several tens of centimeters per year if we can believe the mid-Paleozoic paleopoles for Gondwana (Chapters 5 and 8).

We have also learned, I believe, that robust paleomagnetic data sets exist for some continental blocks (e.g., Laurentia, Stable Europe, Iberia, Adria, Iran, etc.) which are in good agreement with prevailing theories and hypotheses for their movements as derived from other earth science fields.

While we have observed this sometimes excellent paleomagnetic support for various continental reconstructions and drift histories, we have also had a chance to reflect on the occasionally disappointing lack of accuracy and the low reliability of parts of the data base. I repeat, once again, that there are good data and bad data and that it would be nice if we could stick accurate labels on the poles stating which are which. My attempts to do this, on the basis of the criteria of Chapter 4, are neither a disastrous failure nor an overwhelming success in my opinion, proving the difficulty of the 'labelling' process with which many paleomagnetic colleagues have struggled before me. However, I feel compelled to point to the entries with $Q < 3$ of Tables A1–11 and to recommend the (tedious) excercise of compiling mean paleopoles that include these low-Q results. In doing so, one will see a deterioration of the statistical confidence limits and, hence, a diminished resolution with which paleomagnetic data can be put to use.

At times, I have optimistic dreams about an era (some decades away) when analyses based on results with Q of 5 or greater will be possible, as well as meaningful, for many if not all of the areas and issues treated in this book. As it stands now, such a severely filtered database would be so small that perhaps only about 15% of my conclusions about Tethyan displaced terranes would survive. This indicates that much paleomagnetic work remains to be done, provided, however, that it will be done well – once again, only fold-, contact-, and conglomerate tests, extensive rock magnetic experiments, detailed demagnetization, elaborate searches for reversals, precise age determinations, and careful attention to structural control, will achieve this. In short 'the WORKS', complete and unabridged (i.e., the man-bites-snake approach of chapter 1), are necessary for all paleomagnetic work from now on. That this will take much time and money is self-evident, but it is not merely a matter of meaningful employment for self-serving paleomagnetists; with full attention to reliability criteria, paleomagnetism can also reach its potential as a rewarding and uniquely quantifying technique for unravelling the dynamic evolution of our planet. It has been remarked that 'plate tectonics is so simple and clear in the ocean basins, where the magnetization maps it all out for us, but upon reaching the continents, it seems to totally fall apart' (Morley, 1991). I am not convinced that plate tectonics falls apart on the continents, but it certainly becomes more complicated. This is one of the reasons that many paleomagnetic data sets must become robust, before significant conclusions can be drawn, because there are not many other data that can document plate motions quantitatively. Hypotheses based on just one or two paleopoles may be correct but remain rather speculative until confirmed by more results.

The true frontiers of paleomagnetism as applied to tectonics, I believe, are still those related to topics and areas where pre-Mesozoic plate tectonics can be documented; the Paleozoic and Precambrian may yet provide us with surprising and non-uniformitarian aspects of geodynamics. Not only unusually high plate velocities of large continents, but also unsuspected rates of true polar wander, different (ensialic?) regimes of mountain building such as suspected for some of the Pan-African mobile belts, or presently baffling occurrences of glacial relicts in low Precambrian paleolatitudes, all provide the future paleomagnetist with a plethora of research opportunities.

The Tethyan and Circum-Pacific domains, moreover, are not to be dismissed lightly either as largely solved problems or issues of only regional importance. The Mesozoic drift histories of many Tethyan terranes in Asia are still largely uncharted as we have seen in Chapter 7. And even Tertiary motions for areas such as the eastern Mediterranean, central Asia, Indo-China, Indonesia and the Philippines remain rather uncertain for lack of reliable data. In short, paleomagnetism is by far not yet reduced to the level of quadrangle mapping (as useful as that activity, for which I have much respect, may be).

Now that we are talking nitty-gritty, I might also mention the surprising lack of resolution of the late Paleozoic to early Mesozoic paleopoles from the Gondwana continents, which results in lingering ambiguities about Pangea configurations and

Table 9.1. *The worst cases: typical uncertainties associated with some of the more ambiguous paleopositions of the major continents where data are available*

Age (Ma)	Uncertainty (degrees)	Description
800–540	>60	The positions of Laurentia for some intervals of the Late Precambrian and Cambrian
505–375	40	Ordovician paleopoles for Gondwana (A_{95}s range from 22° to 33°), and the Silurian to Early Devonian paleolocation for Gondwana
505–280	35	Paleozoic positions of the South China Block
400–360	25	Siluro-Devonian positions of Siberia
300–207	20	Pangea configurations, and the fit between East and West Gondwana
245–65	20	Mesozoic positions of the China Blocks
175–144	15	Middle to Late Jurassic paleopoles for the Atlantic-bordering continents
65–40	13	Early Tertiary paleopoles for the South Atlantic-bordering continents

about amounts of continental margin extension before the break-up of South America, Africa, and the East Gondwana continents (Chapter 5).

It is perhaps illustrative to reflect briefly on the scale (in degrees) of some currently existing major uncertainties for some of the larger and better studied continents, as discussed earlier in this book, and to consider how the magnitude of these uncertainties correlates with increasing age of the rocks. Table 9.1 gives a summary of examples ranging from the latest Precambrian of Laurentia (uncertainties perhaps >60°), via the Ordovician–Silurian of Gondwana (about 40°), the Permian of East versus West Gondwana versus Laurentia (about 15° to 25°), the Jurassic of the Atlantic-bordering continents (about 10° to 15°) and on down to the level of what I consider the minimum uncertainty in 1990-style paleomagnetic data bases no matter how robust they are (about 5°). For some of the displaced terranes in the Tethys and Iapetus, where few data are available for some geological intervals, these uncertainties are even larger. It is all too tempting to be a bit disdainful about the larger ambiguities in this list, ascribing them to an intrinsic weakness of the paleomagnetic method. Some geoscientists are critical of our field, but I stress that it is not the paleomagnetic method that is flawed; instead, the limits on data quality imposed by field situations or poor investigator planning, the deceptive situations posed by the possibilities of unrecognized remagnetization, and the faulty interpretations of the data that result from low-reliability results are more often to be blamed than the measurements! I must emphasize that these 'worst cases' of Table 9.1 are for continents and periods for which paleomagnetic data do exist. Obviously, the uncertainties are not quantifiable when paleopole data are simply not available.

To counter the negative, I hasten to add a Table 9.2 which highlights some of the more remarkable achievements of paleomagnetism, again as a function of geological time between 1000 and 2 Ma. This list is limited by what I have been able to discuss in this book and is certainly subjective. Table 9.2 is meant only as a counterweight to Table 9.1 and probably excludes several other examples that my paleomagnetic colleagues might value highly.

From both Tables 9.1 and 9.2 it is clear that older geological periods usually

Table 9.2. *The best cases: typical uncertainties associated with some of the better known paleopositions of continents and blocks*

Age (Ma)	Uncertainty (degrees)	Description
1000–800	30	The relative Late Precambrian positions of Laurentia and Baltica
505–400	20	Ordovician–Silurian paleoposition differences between Southern Britain and Laurentia
400–65	15	The paleolatitudinal drift history of Iran with respect to Gondwana and Eurasia
300–180	5	Stable Europe–North America Pangea fits
300–180	8	Iberia's rotated position with respect to Stable Europe
300–45	8	Coherence between the paleolatitudes of India and Himalayan areas
300–45	9	Coherence between Adria and West Gondwana
144–98	5	Early Cretaceous paleopoles of the Atlantic-bordering continents
15–2	8	Rotations in the Hellenic Arc of Greece, Crete and southern Turkey

are associated with greater uncertainties. Extrapolating these uncertainties to the Precambrian, although interesting as an excercise, is not completely justified. Nevertheless, I wish to note that the contributions made by paleomagnetism to Precambrian continental reconstructions and tectonics are still extremely limited and controversial, with previously proposed models spanning the complete spectrum of possibilities. At the one extreme of the spectrum is the Precambrian supercontinent model of Piper (1987), as discussed in Chapter 8. On the other is a view of the Precambrian Earth that is very similar to that of modern-day plate tectonics (e.g., Hoffman, 1988). I predict that in a few decades, the Precambrian data base will have expanded sufficiently in quantity and quality to test some of these models for some intervals of Proterozoic time in a definitive manner. At present, I believe, the data are neither abundant nor precisely enough dated, nor of sufficiently high quality in terms of other paleomagnetic parameters to do this for more than just a few examples and for time intervals of very limited duration. This is one of the reasons that the Precambrian has not been treated in this book. While the Laurentian Proterozoic APWP is rather well determined (e.g., Piper, 1987), and the Baltica path is also becoming more and more reliable (e.g., Pesonen *et al.*, 1989), the problem lies with the APWPs from the other continental nuclei, especially those in Asia and the Gondwana continents (e.g., Idnurm & Giddings, 1988; Van der Voo & Meert, 1991). I stress, however, that this is a very personal viewpoint and that, surely, John Piper and some other paleomagneticians will not agree with the cautious statements above.

My writing about our time travel to the now extinct Iapetus and Tethys oceans is coming to an end, as I am flying back to the United States across the ocean named after Atlas. From my window seat, I see that majestic mountain range of the Pyrenees in which I did my first graduate field work and, as always, it reminds me of the struggle between the Titans and their son/nephew Zeus. The latter, as we know, has won the battle but not until he conquered one after the other of the huge mountain ranges that the angry Titans erected. The names of these mythological mountains (Ossa, Pelion, Oeta) do not mean much to us geologists who are more used to conquering the Rocky Mountains, the Tien-Shan, the Andes or the Himalayas (Conquer? Well, not all of us go to the top). Since his victory, Zeus remained enthroned high on Mount Olympus unthreatened and unchallenged by his unsympathetic family. The geologists and paleomagnetists of today have yet to conquer many mountain belts, traverse many terranes, or defeat the multitude of snakes and other beasts that are hiding remagnetizations from us, but wouldn't you, dear reader, agree with me that we have made progress with our quest?

Appendix

Table A1. *Phanerozoic pole positions for the North American craton*

Rock unit, location	Age		Paleopole		(Paleopole)			Reliability								Reference
	Lo	Hi	Lat.	Long.	(unrotated)	k	α_{95}	1	2	3	4	5	6	7	Q	
Cenozoic (Paleocene, Eocene) results																
Spanish Peaks Dikes, CO, NM	024	058	81,	211		6	13					x			1	Larson & Strangway (1969)
Mistastin Lake Volc., Labrador	032	046	86,	118		—	2	x	x			x			3	Currie & Larochelle (1969)
Beaverhead Flows, MT	030	065	66,	239		15	6		x					x	2	Hanna (1967)
Flagstaff Intrusives, CO	030	065	68,	189		8	14				x			x	2	McMahon & Strangway (1968)
East Fork Washakia Basin, WY	038	050	84,	144		40	8	x	x	x		x	x		5	Flynn (1986)
Monterey Intrusives, VA	042	049	88,	46		52	9	x	x	x		x	x		5	Løvlie & Opdyke (1974)
Monterey Intrusives, VA, WV	042	049	86,	244		21	10	x	x	x		x			4	Ressetar & Martin (1980)
Absaroka Flows, WY	045	050	84,	177		15	8	x	x	x		x	x		5	Shive & Pruss (1977)
Rattlesnake Hills Volc., WY	042	050	79,	146		14	10	x	x	x		x			4	Sheriff & Shive (1980)
Green River Fm., CO	045	049	19,	135		25	6	x	x			x		x	*	Strangway & McMahon (1973)
Combined Eocene Intrusives, MT	047	054	82,	170		19	4	x	x	x		x	x	x	6	Diehl *et al.* (1983)
Robinson Anticline Intr., MT	048	053	77,	146		46	4	x	x	x			x	x	5	Harlan *et al.* (1988)
Black Peaks Fm., TX	053	062	76,	86		25	6	x	x	x		x	x	x	6	Rapp *et al.* (1983)
Cape Dyer Flows, Baffin I., NWT	056	062	83,	305		155	6	x				x	x		3	Deutsch *et al.* (1971)
Fort Union Fm., MT	056	062	69,	133		11	18	x		x		x		x	4	Butler *et al.* (1987)
Eureka Sound Fm., Ellesmere, NWT	058	066	62,	215		17	31	x		x	x	x	x		5	Tauxe & Clarke (1987)
Alkalic Intrusives, MT	058	066	81,	185		19	6	x	x			x	x	x	5	Jacobson *et al.* (1980)
Golden Basalt Flow 2, CO	054	066	81,	64		9	11	x				x			2	Larson *et al.* (1969)
Nacimiento Fm., NM	060	065	76,	148		22	3	x	x	x		x	x	x	6	Butler & Taylor (1978)
Edmonton Group, Alberta	063	064	72,	183		24	7	x	x	x				x	4	Lerbeckmo & Coulter (1985)
Gringo Gulch Volcanics, AZ	063	067	77,	201		678	1	x	x	x				x	4	Barnes & Butler (1980)
Combined Paleocene Intr., MT	060	069	82,	181		20	5	x	x	x		x	x	x	6	Diehl *et al.* (1983)
Mesozoic results																
Roskruge Volcanics, AZ	068	076	70,	177	(74, 176)	86	6	x	x	x			x	x	5	Vugteveen *et al.* (1981)
Boulder Batholith, MT	070	080	73,	249		17	7	x	x				x	x	*	Hanna (1973)
Elkhorn Volcanics, MT	076	080	69,	189		60	5	x	x		x		x	x	5	Hanna (1967; 1973)
Mesaverde Group, WY, UT	074	084	65,	198		4	13	x				x		x	3	Kilbourne (1969)

Niobrara Fm., WY, CO, KS	069	089	64,	188		170	4	x	x		x	x		x	5	Shive & Frerichs (1974)
Maudlow Fm. Welded Tuffs, MT	074	086	70,	208		22	9	x	x	x	x		x	x	6	Swenson & McWilliams (1989)
Caryn Seamount, W. Atl.	066	124	74,	178		—	—					x		x	2	Irving *et al.* (1976b: 13.44)
Magnet Cove & other Intr., AR	098	102	74,	193		34	6	x	x	x		x		x	5	Globerman & Irving (1988)
Isachsen Diabase, NWT	104	114	69,	180		20	8	x				x		x	3	Larochelle *et al.* (1965)
Sverdrup Basin Intr., NWT	088	140	67,	203		—	7		x	x	x	x	x	x	6	Jackson & Halls (1988)
Tatnic Gabbro, ME	116	124	62,	209		136	3	x		x		x		x	4	Hurley & Shearer (1981)
Rehoboth, Gillis Sea Mts, W. Atl.	098	144	71,	171		48	18	x				x		x	3	Mayhew (1986)
Isachsen, Strand Fj. Fms., NWT	112	132	65,	221		7	8	x		x	x		x	x	*	Wynne *et al.* (1988)
Monteregian Hills Intr., Québec	118	136	72,	191		49	2	x	x	x		x	x	x	6	Foster & Symons (1979)
Notre Dame Bay Dikes, Newfoundl.	115	144	67,	212		180	4	x	x	x		x		x	5	Lapointe (1979b)
Mt. Ascutney Gabbro, VT	125	135	64,	187		335	14	x				x		x	3	Opdyke & Wensink (1966)
Upper Morrison Fm., CO	144	150	64,	168	(68, 162)	663	4	x	x			x	x	x	5	Steiner & Helsley (1975)
Lower Morrison Fm., CO	147	152	57,	148	(61, 142)	549	5	x	x			x	x	x	5	Steiner & Helsley (1975)
Stump Fm., WY	152	156	64,	170		18	15	x	x	x					*	Schwartz & Van der Voo (1984)
Canelo Hills Volcanics, AZ	149	153	59,	139	(63, 132)	33	6	x	x	x	x		x	x	6	Kluth *et al.* (1982)
Kelvin, other Seamounts, W. Atl.	147	158	74,	147		75	11	x	x			x		x	4	Mayhew (1986)
White Mountain Intrus., VT, NH	140	180	79,	167		—	—	x	x	x		x	x		*	Van Alstine (1979)
Oldest Seamounts, W. Atl.	160	170	69,	98		66	11	x	x			x		x	4	Mayhew (1986)
Summerville Sandstone, UT	156	174	64,	121	(67, 110)	51	4	x		x		x	x	x	5	Steiner (1978)
Moat Volcanics, NH	163	175	79,	90		49	7	x	x	x					3	Van Fossen & Kent (1990)
Corral Canyon Rocks, AZ	166	178	59,	125	(62, 116)	50	6	x	x	x			x	x	5	May *et al.* (1986)
Twin Creek Limestone, WY	163	176	68,	145		32	9	x	x	x					*	McCabe *et al.* (1982)
Navajo Sandstone, UT	146	208	59,	94	(60, 84)	4	16					x		x	2	Johnson (1976)
Diabase Dikes, Anticosti, Québec	168	191	76,	85		—	1	x				x		x	3	Larochelle (1971)
Newark Volcanics II, CT–MD	176	182	65,	103		92	1	x	x			x		x	4	Smith & Noltimier (1979)
Combined Dikes, NB, NS, NFL	186	195	73,	89		132	11	x	x	x		x		x	5	Hodych & Hayatsu (1988)
Hartford, Newark Basins, CT, NJ	185	194	67,	94		47	8	x	x	x		x		x	5	Prévot & McWilliams (1989)
Picton Dike, Ontario	180	208	76,	346		242	3	x		x	x	x		x	*	Barnett *et al.* (1984)
Watchung Basalts, NJ, PA	180	208	63,	90		26	2	x	x	x		x		x	5	McIntosh *et al.* (1985)
New Jersey Volcanics, NJ	180	208	63,	108		49	4	x	x			x		x	4	Opdyke (1961)
Sil Nakya Fm., AZ	180	208	73,	98	(74, 80)	37	8	x		x				x	3	Kluger-Cohen *et al.* (1986)
Conn. Valley Volc., CT, MA	180	208	65,	87		31	11	x	x			x		x	4	De Boer (1968)
Piedmont dikes, GA, SC, NC	182	205	66,	86		49	3	x	x	x		x		x	5	Dooley & Smith (1982)
Newark Volcanics I, CT–MD	191	199	63,	83		56	2	x	x			x		x	4	Smith & Noltimier (1979)

Table A1. *Phanerozoic pole positions for the North American craton – continued*

Rock unit, location	Age		Paleopole		(Paleopole)	k	α_{95}	Reliability								Reference
	Lo	Hi	Lat.	Long.	(unrotated)			1	2	3	4	5	6	7	Q	
Kayenta Fm., UT	193	200	60,	94	(61, 83)	51	10	x	x			x		x	4	Johnson & Nairn (1972)
Kayenta Fm., UT	193	200	62,	86	(62, 75)	81	7	x	x			x	x	x	5	Steiner & Helsley (1974)
North Mountain Basalt, NB	186	208	67,	69		24	11	x	x	x	x	x		x	6	Hodych & Hayatsu (1988)
Newark Basin N-Polar., NJ, PA	198	225	62,	91		8	3	x		x		x		x	4	McIntosh *et al.* (1985)
Moenave Fm., AZ, UT	196	208	60,	62	(58, 52)	45	5	x	x	x		x	x	x	6	Ekstrand & Butler (1989)
Piedmont dikes, NC	180	214	62,	55		14	8		x	x		x	x	x	5	Smith (1987)
Wingate Fm., UT	198	204	60,	73	(59, 63)	—	8	x	x			x		x	4	Reeve (1975)
Diabases, PA	190	210	62,	105		118	2	x	x			x		x	4	Beck (1972a)
Holyoke, Granby Flows, MA	184	208	55,	88		41	11	x				x		x	3	Irving & Banks (1961)
Hettangian Newark Red Beds, NJ	204	208	55,	95		72	5	x	x	x		x		x	5	Witte & Kent (1990)
Newark Basin, West Limb, NJ	200	225	51,	49		58	13	x	x	x	x			x	5	Van Fossen *et al.* (1986)
Manicouagan Structure, Québec	205	225	60,	89		58	5	x	x			x		x	4	Irving *et al.* (1976b: 8.135)
Newark Basin, East Limb, NJ	200	225	62,	115		41	11	x	x	x		x			4	Van Fossen *et al.* (1986)
Chinle Fm., UT	208	220	62,	75	(61, 64)	892	4	x	x		x	x	x	x	6	Reeve (1975)
Chinle Fm., NM	208	220	58,	79		62	4	x	x	x		x	x	x	6	Reeve & Helsley (1972)
Newark Basin, Lower Redbeds, PA	208	230	54,	102		50	5	x	x	x		x	x	x	6	Witte & Kent (1989)
Newark Basin R-Polar., NJ, PA	208	230	49,	89		6	5	x		x		x		x	4	McIntosh *et al.* (1985)
Luning Fm., NV	208	230	74,	71	(73, 54)	504	11	x	x	x				x	*	Kluger-Cohen *et al.* (1986)
L. Fundy Grp., N. Scotia, NB.	208	235	45,	97		21	7	x	x	x	x	x	x	x	7	Symons *et al.* (1989)
Abbott Pluton, ME	213	229	48,	92		242	4	x	x	x				x	4	Fang & Van der Voo (1988)
Shinarump Memb., Chinle Fm., NM	225	228	60,	99	(61, 89)	130	3	x	x	x	x	x	x	x	7	Molina *et al.* (1991)
Popo Agie Fm., WY	220	240	56,	96		14	14	x		x		x			3	Van der Voo & Grubbs (1977)
Agamenticus Pluton, ME	223	233	48,	99		258	3	x	x	x	x		x	x	6	Fang & Van der Voo (1988)
Moenkopi Fm., NM	230	240	56,	109	(58, 100)	74	3	x	x	x	x	x	x	x	7	Molina *et al.* (1991)
Upper Maroon Fm., CO	210	270	58,	112	(60, 102)	26	9		x		x	x	x		4	McMahon & Strangway (1968)
Moenkopi Fm., CO	235	245	56,	99	(57, 89)	85	5	x	x			x	x	x	5	Helsley (1969)
Upper Moenkopi Fm., CO	235	245	53,	111	(55, 103)	102	3	x	x			x	x	x	5	Helsley & Steiner (1974)
Upper Moenkopi drillcore, CO	235	245	55,	109	(57, 100)	368	6	x	x			x	x	x	5	Baag & Helsley (1974)
Moenkopi Fm., AZ	235	245	56,	110	(58, 101)	10	2	x	x	x	x	x	x	x	7	Purucker *et al.* (1980)
Ankareh Fm., WY	221	245	51,	105		143	4		x	x		x		x	4	Grubbs & Van der Voo (1976)

Upper Red Peak Fm., WY	230	240	49,	105		731	5	x	x	x		x	x	x	6	Van der Voo & Grubbs (1977)
Lower Red Peak Fm., WY	237	245	46,	121		170	7	x	x	x		x		x	5	Van der Voo & Grubbs (1977)
Chugwater Fm., WY	230	245	47,	114		999	2	x	x	x		x	x	x	6	Shive *et al.* (1984)
Chugwater Fm., WY	230	245	45,	115		47	4	x	x	x		x	x	x	6	Herrero-Bervera & Helsley (1983)
Paleozoic results																
Basic Sill, Pr. Edward I.	245	248	51,	111		—	—	x		x		x		x	4	Larochelle (1967)
Ochoan Redbeds, NM, OK	245	253	55,	119		39	15	x	x			x	x		4	Peterson & Nairn (1971)
Bernal Fm., NM	245	258	50,	120	(53, 112)	238	8	x		x		x		x	4	Molina *et al.* (1991)
Guadalupian Redbeds, NM, OK	253	263	51,	125		214	5	x	x			x	x	x	5	Peterson & Nairn (1971)
Dewey Lake Formation, TX	245	255	51,	126		40	5	x	x	x		x	x	x	6	Molina-Garza *et al.* (1989)
Hoskinnini Tongue, AZ	245	279	47,	128	(50, 121)	23	7					x		x	2	Farrell & May (1969)
Pegmatite dikes, CT	257	259	35,	126		63	10	x	x					x	*	De Boer & Brookins (1972)
Esayoo Volcs., Ellesmere, NWT	263	286	48,	159		105	4	x	x	x	x		x	x	*	Wynne *et al.* (1983)
Toroweap Fm., AZ	263	286	52,	125	(55, 117)	77	10	x				x		x	3	Peterson & Nairn (1971)
Elephant Canyon F., UT	263	286	41,	126	(44, 120)	27	5	x	x			x		x	4	Gose & Helsley (1972)
Ingelside Fm., CO	263	286	43,	128	(46, 122)	147	2	x	x	x		x		x	5	Diehl & Shive (1979)
Cutler Fm., Lisbon Valley, UT	263	286	41,	124	(44, 118)	—	—	x	x	x		x		x	5	Reynolds *et al.* (1985)
Cutler Fm., CO	263	286	40,	128	(43, 122)	25	6	x	x			x		x	4	Helsley (1971)
Cutler Fm., UT	263	286	41,	122	(44, 116)	21	2	x	x			x		x	4	Gose & Helsley (1972)
Fountain & Lykins Fms., CO	263	286	45,	126	(48, 119)	6	13	x			x	x		x	4	McMahon & Strangway (1968)
Minturn & Maroon Fms., CO	263	286	40,	121	(43, 115)	19	3	x	x	x		x		x	5	Miller & Opdyke (1985)
Leonardian Subset, OK only	263	286	52,	119		761	5	x				x		x	3	Peterson & Nairn (1971)
Casper Fm., WY	263	286	51,	123		129	2	x	x	x		x		x	5	Diehl & Shive (1981)
Abo Formation, NM	263	286	47,	125	(50, 118)	55	2	x	x	x		x		x	5	Steiner (1988)
Wolfcampian Redbeds, UT, AZ, NM	263	286	38,	124	(41, 118)	46	11	x	x			x		x	*	Peterson & Nairn (1971)
Helderberg Synfolding, NY–VA	245	286	49,	115		64	—		x	x	x	x		x	5	Scotese (1985)
Mauch Chunk Synfolding, PA	245	286	51,	118		35	5		x	x	x	x		x	5	Kent & Opdyke (1985)
Pictou Red Beds, Pr. Edward I.	263	296	42,	126		88	3	x	x	x	x	x	x	x	7	Symons (1990)
Laborcita Formation, NM	268	296	42,	132		43	2	x	x	x		x		x	5	Steiner (1988)
Churchland Pluton, NC	276	288	34,	126		4	16	x		x				x	3	Barton & Brown (1983)
Dunkard Fm., VA	280	296	44,	123		162	5	x	x			x	x	x	5	Helsley (1965)
Casper Fm., WY	286	300	46,	129		40	2	x	x	x		x		x	5	Diehl & Shive (1981)
Mason Creek, IL	296	302	37,	131		45	8	x	x	x		x		x	5	Scotese (1985)
Buffalo Siltstone, PA	290	315	27,	123		13	6	x		x		x		x	*	Payne *et al.* (1981)

Table A1. *Phanerozoic pole positions for the North American craton – continued*

Rock unit, location	Age		Paleopole		(Paleopole)			Reliability								Reference
	Lo	Hi	Lat.	Long.	(unrotated)	k	α_{95}	1	2	3	4	5	6	7	Q	
Brush Creek Limestone, PA	290	315	36,	124		13	4	x		x		x		x	4	Payne *et al.* (1981)
Wescogame Fm. (Supai), AZ	286	296	44,	125	(47, 119)	42	3	x	x	x		x		x	5	Steiner (1988)
Piedmont Mafic Intrusions, SC	258	320	39,	121		27	10		x	x				x	3	Dooley (1983)
Pr. Edward I. Redbeds	268	308	42,	133		36	6	x	x			x		x	3	Roy (1966)
Pr. Edward I. Redbeds	268	308	41,	126		433	6	x	x			x		x	4	Black (1964)
Tormentine Fm., Pr. Edward I.	286	308	41,	132		107	4	x				x		x	3	Roy (1966)
Hurley Creek Fm., N. Brunswick	290	310	39,	125		58	10	x				x		x	3	Roy *et al.* (1968)
Bonaventure Fm., N. Brunswick	290	320	38,	133		20	10	x				x		x	3	Roy (1966)
Cumberland Group, N. Scotia	296	315	36,	125		85	5	x	x		x	x		x	5	Roy (1969)
Morien Group, N. Scotia	286	308	40,	131		32	6	x	x	x	x	x		x	6	Scotese *et al.* (1984)
Riversdale Group, N. Scotia	308	315	36,	122		91	6	x	x			x	x	x	5	Roy (1977)
Pomquet & Lismore Fms, N. Scotia	315	333	29,	122		60	33	x		x			x	x	4	Scotese *et al.* (1984)
Barachois Grp., W. Newfoundl.	308	333	34,	143		107	7	x	x	x	x	x		x	6	Murthy (1985)
New Brunswick Volcanics I	308	324	21,	135		49	10	x	x				x	x	4	Seguin *et al.* (1985)
New Brunswick Volcanics II	320	330	36,	136		78	6	x	x	x			x	x	5	Seguin & Fyffe (1986)
Gaspé Dikes & Contacts, Québec	308	380	11,	148		43	10		x	x	x		x	x	5	Seguin (1987; 1986)
Minudie Point, N. Scotia	308	315	36,	122		91	6	x	x				x	x	4	Roy (1977)
Shepody Fm., N. Scotia	315	325	36,	124		13	5	x	x	x	x	x	x	x	7	DiVenere & Opdyke (1990)
Hopewell Group, N. Brunswick	315	343	34,	118		32	7	x	x		x		x	x	5	Roy & Park (1969)
Maringouin Fm., N. Scotia	325	333	32,	121		12	4	x	x	x	x	x	x	x	7	DiVenere & Opdyke (1990)
Lilesville Pluton, NC	299	353	40,	134		41	3	x	x	x					*	Barton & Brown (1983)
Chèverie Fm., N. Scotia	352	360	24,	152		59	6	x	x	x	x			x	*	Spariosu *et al.* (1984)
Maringouin & Shepody, N. Scotia	319	343	36,	122		89	3	x	x		x		x	x	5	Roy & Park (1974)
Mauch Chunk Fm., south, PA	320	340	24,	141		18	10	x	x	x	x		x	x	6	Kent (1988)
Mauch Chunk Fm., north, PA	320	340	26,	143	(34, 119)	19	9	x	x	x	x		x	x	6	Kent & Opdyke (1985)
Deer Lake Fm., W. Newfoundl.	320	352	22,	122		40	9	x	x	x	x	x	x	x	7	Irving & Strong (1984)
Fisset Br. Synfold., N. Scotia	325	356	20,	139		91	5		x	x	x	x	x	x	6	Johnson & Van der Voo (1989)
Jeffreys Village Mbr., W. Newfoundl.	333	345	27,	131		54	8	x	x	x		x	x	x	6	Murthy (1985)
Windsor Group, N. Scotia	333	352	36,	137		118	6	x	x	x	x			x	5	Scotese *et al.* (1984)
Spout Falls Fm., W. Newfoundl.	352	357	29,	140		61	7	x	x	x		x		x	5	Murthy (1985)
Horton Group, N. Scotia	352	360	32,	136		87	8	x	x	x				x	4	Scotese *et al.* (1984)
St. Lawrence Alaskite, Newfoundl.	354	365	12,	120		14	10	x		x		x		x	4	Irving & Strong (1985)

Peekskill Granite, NY	350	370	23,	117		—	16	x		x		x		x	4	Miller & Kent (1989b)
Terrenceville Fm., E. Newfoundl.	352	380	27,	124		26	10	x	x	x			x	x	5	Kent (1982)
Metamorphics, MA	360	380	23,	126		11	10	x	x					x	3	Schutts *et al.* (1976)
Catskill Redbeds, south, PA	360	380	26,	124		16	16	x		x			x		3	Miller & Kent (1986a)
Catskill Redbeds, north, PA	360	380	33,	117	(33, 90)	165	7	x	x	x	x			x	5	Miller & Kent (1986b)
McAras Brook B, N. Scotia	350	387	32,	109		27	24			x			x	x	3	Stearns & Van der Voo (1988)
McAras Brook A, N. Scotia	360	387	35,	76		821	4	x		x			x	x	4	Stearns & Van der Voo (1988)
Compton Metasediments, Québec	360	380	28,	77		30	7	x	x	x	x		x	x	6	Seguin *et al.* (1982)
Dockendorff, ME	350	387	24,	87		11	7		x	x				x	*	Brown & Kelly (1980)
Gander Zone Dikes, E.–C. Newfoundl.	335	400	−47,	105		28	6		x					x	2	Murthy (1983a, b)
Traveler Felsite, ME	360	401	29,	82		16	11	x	x	x			x	x	5	Spariosu & Kent (1983)
Gaspé Sediments C Comp., Québec	350	410	16,	112		68	15			x		x		x	3	Recalc. from Seguin (1986)
Eastport Fm., ME	350	421	24,	114		19	9		x	x	x		x	x	5	Kent & Opdyke (1980)
Hersey Fm., ME	350	421	20,	129		36	6		x	x	x		x	x	5	Kent & Opdyke (1980)
Peel Sound MDL pole, NWT	350	421	25,	99		66	9		x	x		x	x	x	5	Dankers (1982)
Pleasant Lake Granite, ME	390	410	2,	95		—	17	x		x				x	3	Miller & Kent (1989b)
Dalhousie Volcanics, N. Brunswick	394	421	1,	88		236	7	x		x				x	3	Seguin & Gahé (1985)
Peel Sound E pole, NWT	380	421	1,	91		18	10	x		x		x		x	4	Dankers (1982)
Andreas Redbeds, PA	380	421	13,	105		13	9	x		x	x			x	4	Miller & Kent (1988)
Bloomsburg Fm., south, PA–VA	408	421	31,	117		36	9	x	x	x	x		x		*	Kent (1988)
Bloomsburg Fm., north, PA	408	421	29,	129	(33, 103)	57	7	x	x	x	x				*	Kent (1988)
Rose Hill Fm., PA–VA	414	428	19,	129		18	6	x	x	x	x		x	x	6	French & Van der Voo (1979)
Wabash Reef Ls., IN	414	428	17,	125		74	5	x	x	x	x	x	x	x	7	McCabe *et al.* (1985)
Beemerville Complex, NJ	415	455	35,	126		23	20	x							1	Proko & Hargraves (1973)
Ringgold Gap sediments, GA	428	448	28,	142		62	7	x	x	x			x		4	Morrison & Ellwood (1986)
Anticosti I., B Compt., Québec	428	448	19,	129		—	18	x		x		x	x	x	5	Seguin & Petryk (1986)
Cordova Secondary, Ontario	445	450	31,	102		18	11		x	x		x	x		*	Dunlop & Stirling (1985)
Juniata north, PA	438	458	17,	126	(21, 105)	29	13	x	x	x	x			x	5	Miller & Kent (1989a)
Juniata south, PA	438	458	19,	128		14	21	x		x	x			x	4	Miller & Kent (1989a)
Juniata south, PA–VA	438	458	32,	114		53	5	x	x	x	x		x		*	Van der Voo & French (1977)
Steel Mtn. Secondary, W. Newfoundl.	416	486	23,	139		22	13		x			x			2	Murthy & Rao (1976)
Chapman Ridge Fm., TN	448	478	27,	112		38	15	x	x	x	x				*	Watts & Van der Voo (1979)

Table A1. *Phanerozoic pole positions for the North American craton – continued*

Rock unit, location	Age		Paleopole		(Paleopole)			Reliability								Reference
	Lo	Hi	Lat.	Long.	(unrotated)	k	α_{95}	1	2	3	4	5	6	7	Q	
Martinsburg Shale, PA	448	478	36,	160		5	14	x					x	x	*	Hower (1979)
Moccasin-Bays Fms., TN	448	478	33,	147		135	6	x	x	x	x		x	x	6	Watts & Van der Voo (1979)
Granby & other volcs., Québec	448	478	6,	191		—	19	x					x	x	3	Seguin (1977)
Table Head Grp., W. Newfoundl.	438	478	11,	142		72	11			x	x	x		x	4	Hodych (1989)
Table Head Grp., W. Newfoundl.	438	478	15,	156		95	10		x	x		x		x	4	Hall & Evans (1988)
Table Head Grp., W. Newfoundl.	438	478	13,	149		394	5			x		x		x	3	Deutsch & Prasad (1987)
St. George Grp., W. Newfoundl.	438	505	21,	138		13	7		x	x		x			3	Beaubouef *et al.* (1990)
St. George Grp., W. Newfoundl.	438	505	18,	152		144	9		x	x		x		x	4	Deutsch & Prasad (1987)
Oneota Dolomite, IA, MN, WI	478	505	10,	166		18	12	x	x	x		x	x	x	6	Jackson & Van der Voo (1985)
Buckingham Dike, Québec	483	511	−3,	123		—	7		x	x	x	x		x	*	Dankers & Lapointe (1981)
Black, Unaweap Canyon dikes, CO	481	513	35,	108	(37, 102)	—	9	x	x				x		*	Larson *et al.* (1985)
Kingston Dike, Ontario	440	570	20,	197		30	11					x		x	2	Park & Irving (1972)
NE Frontenac Dikes, Ontario	440	570	−5,	130		32	22					x		x	2	Park & Irving (1972)
Royer Dolomite, OK	505	520	13,	157		33	4	x	x	x	x	x		x	6	Nick & Elmore (1990)
Lamotte Fm. (recalculated), MO	505	523	1,	168		—	—	x				x		x	*	Van der Voo *et al.* (1976b)
Pt. Peak Mbr. (Wilberns Fm.) TX	505	523	6,	159		171	5	x	x	x		x		x	5	Van der Voo *et al.* (1976b)
Taum Sauk Lst. (Bonneterre), MO	505	523	−4,	176		12	7	x	x	x		x		x	5	Dunn & Elmore (1985)
Lion Mtn. Mbr. (Riley Fm.) TX	505	523	27,	146		13	10	x		x		x			*	Watts *et al.* (1980b)
Cap Mtn. Mbr. (Riley Fm.) TX	505	523	33,	140		8	8	x		x		x	x		*	Watts *et al.* (1980b)
Morgan Cr.-Welge, Wilberns, TX	505	523	24,	151		20	4	x	x	x		x		x	*	Watts *et al.* (1980b)
Morgan Creek Memb., Wilberns, TX	505	523	11,	158		29	10	x	x	x		x		x	5	Loucks & Elmore (1986)
Covey Hill Fm., Potsdam, Québec	505	523	46,	149		111	5	x	x	x		x		x	*	Seguin *et al.* (1981)
Châteaugay Fm., Potsdam, Québec	505	523	37,	156		78	5	x	x	x		x	x	x	*	Seguin et al. (1981)
Hickory Mbr. (Riley Fm.) TX	523	540	34,	145		12	4	x	x	x		x		x	*	Watts *et al.* (1980b)
Wichita Granites, OK	520	560	4,	164		3	18	x						x	2	Spall (1968; 1970)
McClure Mountain Complex, CO	530	540	14,	146	(18, 142)	—	—	x	x					x	3	Lynnes & Van der Voo (1984)
Colorado Intrusives I	485	704	11,	146	(15, 142)	22	10		x	x			x	x	4	French *et al.* (1977)
Colorado Intrusives II	485	704	1,	177	(5, 174)	39	8			x			x	x	3	French *et al.* (1977)
Colorado Intrusives III	485	704	46,	114	(48, 107)	16	15			x			x		2	French *et al.* (1977)
Sept Iles Intrusion, Québec	528	562	20,	141		34	8	x	x	x		x	x	x	6	Tanczyk *et al.* (1987)

Rock unit	Age		Lat., Long.		Unrotated	k	α_{95}	1	2	3	4	5	6	7	Q	Reference
Rome-Waynesboro Fms., TN–MD	523	540	36,	150		21	10	x	x	x	x		x	x	*	Watts *et al.* (1980a)
Grenville Front Dikes, Ontario	510	570	10,	138		64	—					x		x	2	Palmer *et al.* (1977)
Tapeats Sandstone, AZ	540	570	1,	161	(5, 158)	15	3	x	x			x	x	x	5	Elston & Bressler (1977)
Bradore Sandstone, W. Newfoundl.	540	570	29,	167		109	5	x	x			x		x	4	Rao & Deutsch (1976)
Unicoi Basalts, VA, TN	555	570	0,	178		24	14	x		x			x	x	4	Brown & Van der Voo (1982)
Buckingham flows I, Québec	541	605	6,	154		76	14	x		x		x			3	Dankers & Lapointe (1981)
Buckingham flows II, Québec	541	605	1,	173		159	6	x	x	x		x		x	5	Dankers & Lapointe (1981)
Buckingham flows III, Québec	541	605	−10,	186		405	13	x		x		x		x	4	Dankers & Lapointe (1981)

The entries in this Table are for the North American craton, including its thrust belt margins in so far as no suspicions exist about relative displacements; obviously rotations can occur there and this mandates that reliability criterion 5 can not be checked as being met. The craton outline used is that of Figure 3.1, while for Devonian and younger time New England and the Canadian Maritime provinces have also been included.

For the Colorado Plateau, paleopoles are rotated according to the parameters of Bryan & Gordon (1988), Euler pole at 37°N, 103°W, angle 5.4° counterclockwise. For the northern limb of the Pennsylvania Salient, paleopoles have been rotated about an Euler pole at 41°N, 77°W, angle 23° counterclockwise (Kent, 1988). Unrotated paleopoles are given in parentheses.

The reliability criteria (1–7) and Q are explained in the text. A '*' is placed in the Q column when the result is not to be used for apparent polar wander path construction, as explained below; a 'x' mark under any of the seven reliability columns means that this particular criterion is being met.

Abbreviations: Lat. = latitude, Long. = longitude, k and α_{95} are the precision parameter and radius of the cone of 95% confidence, respectively (Fisher, 1953).

*Poles that would have $Q \geq 3$ but are excluded from mean pole determinations and other compilations, are listed below with the reasons in parentheses: Eocene Green River Formation (magnetization thought to be transitional or composite); Upper Cretaceous Boulder Batholith (no structural control, tilting suspected); Lower Cretaceous Isachsen and Strand Fjord formations (local structural rotations); Upper Jurassic Stump Formation (remagnetized); Upper Jurassic White Mountain Intrusives (contaminated by present-day field overprint); Middle Jurassic Twin Creek Limestone (remagnetized); Lower Jurassic Picton Dike (secular variation not averaged); Upper Triassic Luning Formation (local rotations suspected); Upper Permian Pegmatite Dikes (contaminated by overprint); Lower Permian Esayoo Volcanics (local rotations); Lower Permian Wolfcampian Red Beds (superseded); Upper Carboniferous Buffalo Siltstone (contaminated by overprint); Lower Carboniferous Lilesville Pluton (remagnetized); Lower Carboniferous Chèverie Formation (local rotations); Upper Devonian Dockendorff (remagnetized); Upper Silurian Bloomsburg Formation (directions affected by strain); Upper Ordovician Cordova (age date of secondary magnetization is suspect); Upper Ordovician Juniata South of PA–VA (superseded); Upper Ordovician Chapman Ridge Formation (remagnetized); Upper Ordovician Martinsburg Shale (local rotations); Lower Ordovician Buckingham Dike (secular variation not averaged, anisotropy problem); Black and Unaweap Canyon Dikes (secular variation not averaged or remagnetized); Upper Cambrian Lamotte Recalculated (too few samples); Upper Cambrian Lion Mountain, Cap Mountain and Morgan Creek-Welge members, Covey Hill, Châteaugay formations and Hickory member (possibly remagnetized; problem of the 'Cambrian loop'); Middle Cambrian Rome and Waynesboro formations (possibly remagnetized).

Table A2. *Phanerozoic pole positions for 'Stable' Europe*

Rock unit, location	Age		Pole position				Reliability								Reference
	Lo	Hi	Lat.	Long.	k	α_{95}	1	2	3	4	5	6	7	Q	
Cenozoic (Paleocene, Eocene) results															
Sorlifjell Lavas, Spitsbergen	002	024	75,	235	110	4		x					x	2	Sandal & Halvorsen (1973)
Halvdanpiggen Volc., Spitsbergen	002	045	40,	180	—	—							x	1	Halvorsen (1972; pers. comm. 1990)
Faeroe Upper Lavas	000	010	84,	218	54	4		x	x		x			*	Løvlie & Kvingedal (1975)
Dellen Impact Structure, Sweden	037	066	68,	165	135	3					x		x	2	Bylund (1974)
Kragerø Dikes, Norway	037	066	79,	147	63	9					x		x	2	Storetvedt (1968)
Dikes, Baked Contacts, Scotland	034	057	74,	197	21	4	x	x		x	x	x	x	6	Smith (1966)
Volcanics, Germany	039	065	80,	143	16	14	x	x			x	x		4	Nairn & Negendank (1973)
Lundy Island Dikes, Wales	045	054	83,	155	33	4	x	x			x	x	x	5	Mussett *et al.* (1976)
Mull Dikes at Loch Ba, Scotland	049	059	75,	298	33	10	x				x		x	3	Ade-Hall (1974)
British Dikes	040	060	77,	211	11	9	x	x		x	x		x	5	Dagley (1969)
Fishnish Dikes, Scotland	044	060	74,	140	15	10	x	x			x	x	x	5	Dagley & Mussett (1978)
Sleat Dikes, Skye, Scotland	044	058	82,	162	—	2	x	x				x	x	4	Wilson *et al.* (1982)
Ardnamurchan Gabbro, Scotland	044	058	69,	165	15	16	x				x		x	3	Khan (1960)
Faeroe Basalts	050	060	77,	161	253	2	x	x			x	x	x	5	Tarling (1970)
Faeroe Weathered/Baking Lava	050	060	62,	182	258	2	x		x	x	x		x	5	Løvlie & Kvingedal (1975)
Arran Dikes, Scotland	054	060	82,	180	37	1	x	x			x	x	x	5	Dagley *et al.* (1978)
Skye Lavas & Dikes, Scotland	054	061	88,	103	128	7	x	x	x		x	x		5	Dagley *et al.* (1990)
Cleveland-A. Dike, Mull, Scotland	052	065	75,	240	145	4	x	x	x	x	x		x	6	Giddings *et al.* (1974)
Vaternish Dikes, Scotland	052	065	76,	160	10	5	x	x			x	x	x	5	Wilson *et al.* (1974)
Skye Lavas, Scotland	050	065	72,	165	36	3	x	x					x	3	Wilson *et al.* (1972)
Rhum/Canna Volcanics, Scotland	058	060	77,	173	68	9	x	x	x		x	x	x	6	Dagley & Mussett (1981)
Ardnamurchan Ign. Complex, UK	056	063	77,	175	—	3	x	x			x		x	4	Dagley *et al.* (1984)
Muck and Eigg Volcs., UK	053	063	76,	174	116	7	x	x			x	x	x	5	Dagley & Mussett (1986)
Mull Dike Swarm, Scotland	058	063	78,	187	17	3	x	x	x	x	x	x	x	7	Ade-Hall *et al.* (1972)
Mull Lavas, Scotland	061	063	72,	168	32	3	x	x	x		x		x	5	Hall *et al.* (1977)
Mesozoic results															
Triassic Redbeds, Czech.	000	010	87,	198	27	3		x			x			*	Krs (1967)
Cretaceous Sandstones, Czech.	000	010	87,	347	—	—					x			*	Andreeva *et al.* (1965)

Malm delta Limest., Germany	001	040	79,	124	—	5		x	x		x	x		*	Heller (1978)
Jura Tan Limestones, Switzerland	001	007	88,	211	43	6		x	x	x	x	x		*	Johnson *et al.* (1984)
Antrim Basalts, Ireland	060	080	71,	126	12	10	x	x	x		x		x	5	Løvlie *et al.* (1972)
Antrim Basalts, Ireland	060	080	70,	163	25	5	x	x			x		x	4	Wilson (1970b)
Aix-en-Provence Seds., France	064	084	73,	156	33	8	x	x	x	x	x	x	x	7	Westphal & Durand (1989)
Münsterland Turonian, Germany	088	091	68,	149	—	5	x		x		x		x	4	Kerth (1985)
Münster Basin Ls., Germany	088	098	76,	181	276	4	x	x	x	x	x			5	Heller & Channell (1979)
Comb. Dikes, Sills, Spitsbergen	100	115	62,	223	73	5	x	x	x		x	x	x	6	Vincenz & Jelenska (1985)
Vectis Fm., Isle of Wight, UK	116	122	61,	189	148	21	x		x			x		3	Kerth & Hailwood (1988)
Berriasian Stratotype, France	138	144	75,	179	16	3	x	x	x		x	x	x	6	Galbrun (1985)
Sussex Iron Grit, UK	098	131	84,	11	—	—	x				x	x		*	Edwards (1965)
North Caucasus Carb., Georgia	113	144	76,	155	150	8	x	x			x	x		4	Khramov (pole 4.45, 1984)
Hinlopenstr. Dol., Spitsbergen	132	166	66,	200	118	4	x	x	x		x	x	x	6	Halvorsen (1989)
Isfjorden Doler., Spitsbergen	130	170	58,	179	—	9	x				x		x	3	Krumsiek *et al.* (1968)
Malm alpha, beta Ls., Germany	152	163	68,	130	—	5	x	x	x		x			*	Heller (1978)
Malm gamma Ls., Germany	144	158	62,	122	—	8	x	x	x		x			*	Heller (1978)
Jura Blue Limest., Switzerland	144	169	78,	148	26	6	x	x	x	x		x		5*	Johnson *et al.* (1984)
Kujawi Sediments, Poland	156	163	70,	147	288	4	x	x	x					*	Kruczyk & Kadzialko-Hofmokl (1988)
Krakow Upland Sediments, Poland	160	169	74,	200	665	4	x	x	x			x	x	5	Ogg *et al.* (1991)
Krakow Upland Seds., Poland	156	169	72,	150	58	7	x	x	x			x		4*	Kadzialko-Hofmokl & Kruczyk (1987)
Subtatric Nappe Seds., Poland	152	176	72,	132	198	4	x	x	x			x		4*	Kadzialko-Hofmokl & Kruczyk (1987)
Bathonian Clays Donbass, Ukraine	169	176	63,	117	137	2	x	x			x	x		4*	Rusakov (in Khramov, 1971)
Scania Basalts, Sweden	149	216	64,	96	—	—					x		x	2	Bylund (1981)
Alsace Bajocian Seds., France	176	180	63,	120	92	6	x	x	x	x	x	x	x	7	Kadzialko-Hofmokl *et al.* (1988)
Dorset-Yorkshire Seds., UK	176	187	50,	37	27	6	x	x			x	x	x	*	Sallomy & Briden (1975)
Toarcian Stratotype, France	187	193	73,	105	30	4	x	x	x		x	x	x	6	Galbrun *et al.* (1988)
Pliensbachian Sediments, UK	193	198	77,	135	37	2	x	x			x		x	4	Hijab & Tarling (1982)
Liassic Volcanics, France	180	208	65,	143	70	7	x	x		x			x	4	Girdler (1968)
Keuper Volcanics, France	208	230	62,	114	69	7	x	x		x			x	4	Girdler (1968)
Keuper Redbeds, France	208	230	56,	148	5	32	x				x	x		3	Biquand (1967)
Keuper Marls, UK	208	230	44,	134	—	—	x				x	x	x	4	Creer (1959)
Yushatyr & Bukobai Urals, Russia	208	230	48,	157	14	7	x	x			x		x	4	Khramov (pole 6.02, 1975)
Sunnhordland Dikes, Norway	164	274	46,	117	—	—			x		x		x	*	Løvlie (1981)
Rochechouart Impact, France	160	275	44,	110	310	4		x	x		x		x	4	Pohl & Soffel (1971)

Table A2. *Phanerozoic pole positions for 'Stable' Europe – continued*

Rock unit, location	Age		Pole position				Reliability								Reference
	Lo	Hi	Lat.	Long.	k	α_{95}	1	2	3	4	5	6	7	Q	
St. Bee Sandstone, UK	208	245	43,	134	—	—			x		x	x	x	*	Turner & Ixer (1977), Turner (1981)
Bunter & Musschelk. Germany	230	245	49,	146	14	15	x	x			x	x	x	5	Rother (1971)
Tuffs Central Urals, Russia	230	245	48,	153	37	13	x	x			x	x	x	5	Karmanova (in Khramov, 1971)
Duncansby Neck, A Comp., UK	210	270	60,	127	155	2		x	x				x	3	Storetvedt *et al.* (1978b)
Kingscourt Redbeds, Ireland	239	245	59,	146	44	7	x		x		x		x	4	Mulder (1972)
Brumunddal Lavas, Norway	210	270	48,	138	41	—			x		x	x	x	4	Storetvedt *et al.* (1978a)
Volga Clays & Sandst., Russia	239	245	52,	167	6	10	x				x	x	x	4	Khramov (pole 6.10, 1975)
Blyumenthal Grp, S. Urals, Russia	239	245	49,	159	6	11	x				x	x	x	4	Khramov (pole 6.16C, 1975)
Romaskka, other Redbeds, Russia	239	245	52,	165	46	4	x	x			x	x	x	5	Khramov (pole 6.46, 1975)
Induan & Bashk. Stages, Ukraine	239	245	52,	145	20	3	x	x			x		x	4	Khramov (pole 6.09, 1975)
Serebryansk, Donbass, Ukraine	239	245	56,	146	110	12	x	x			x	x	x	5	Khramov (pole 6.08, 1975)
Varieg. Suite, Urals, Russia	239	245	51,	151	106	1	x	x			x		x	4	Irving *et al.* (1976b: 8.189)
Upper Buntsandstein, France	239	245	43,	146	14	5	x	x	x		x	x	x	6	Biquand (1977)
Vetluga Red Clays, Russia	239	245	53,	158	31	4	x	x			x	x	x	5	Khramov (pole 6.43B, 1975)
Paleozoic results															
Stornoway Fm., Scotland, UK	208	286	50,	138	30	9					x	x		2	Storetvedt & Steel (1977)
Maures Uplift, France	245	258	51,	161	140	4	x	x	x	x		x	x	6	Merabet & Daly (1986)
Estérel Volcs. & Seds, France	245	268	50,	143	112	4	x	x	x	x	x			5	Zijderveld (1975)
Dôme de Barrot Redbeds, France	245	268	52,	144	41	3	x	x	x	x	x			5	Van den Ende (1977)
W. Pyrenees Flows, Dikes, Spain	245	268	40,	126	25	7	x	x	x	x			x	*	Van der Voo & Boessenkool (1973)
Upper Tatarian Redbeds, Russia	245	250	51,	166	183	4	x	x			x	x	x	5	Irving *et al.* (1976a: 7.288)
Lower Tatarian Redbeds, Russia	248	253	43,	169	948	2	x	x			x		x	4	Irving *et al.* (1976a: 7.289)
Kazanian Stage Combined, Russia	253	258	47,	171	86	6	x	x			x		x	4	Irving *et al.* (1976a: 7.246)
Vindodden Lst., Spitsbergen	245	296	31,	140	12	14			x				x	2	Jelenska (1987)
Upper Lodève Sandstones, France	258	268	47,	156	33	5	x	x			x			3	Kruseman (1962)
Lodève Saxonian, France	258	268	49,	154	2096	1	x	x	x		x		x	5	Merabet & Guillaume (1988)
Skaane Melaphyre Dikes, Sweden	258	286	54,	172	26	11	x	x			x		x	4	Bylund (1974)
Nideck-Donon Upp. Volc., France	258	286	47,	168	134	4	x	x			x		x	4	Roche *et al.* (1962)
Bohemian Redbeds, Czechoslovakia	258	286	40,	164	238	4	x	x	x		x		x	5	Krs (1968)
Cupriferous Ss. Comb., Ukraine	258	286	40,	166	185	4	x	x			x	x	x	5	Irving *et al.* (1976: 7.236)

Exeter Lavas, UK	258	286	48,	163	29	10	x	x		x	x			4	Cornwell (1967)
Exeter Lavas, UK	258	286	50,	149	211	4	x	x	x	x	x			5	Zijderveld (1967)
Nahe Volcanics, Germany	258	286	46,	167	54	13	x	x			x			3	Nijenhuis (1961)
Thüringer F. Seds., Germany	258	286	42,	160	10	6	x		x		x		x	4	Mauritsch & Rother (1983)
Thüringer F. Volcs., Germany	258	286	37,	170	10	7	x		x		x		x	4	Mauritsch & Rother (1983)
Oslo Volcanics, Norway	258	286	47,	157	21	1	x	x	x		x		x	5	Van Everdingen (1960)
Krakow Volcanics, Poland	258	286	43,	165	35	8	x	x			x		x	4	Birkenmajer & Nairn (1964)
Lower Silesia Volcanics, Poland	258	286	44,	167	30	10	x	x			x			3	Birkenmajer *et al.* (1968)
Rhomb Porphyry Dikes, Sweden	258	286	46,	165	72	7	x	x	x		x		x	5	Thorning & Abrahamsen (1980)
Porphyry Dolerites, Sweden	258	286	57,	175	36	21	x		x		x			3	Thorning & Abrahamsen (1980)
Lodève Red bed drillcore, France	258	286	53,	151	122	2	x	x	x				x	4	Evans & Maillol (1986)
Lodève Autunian, France	268	286	42,	169	517	2	x	x	x		x		x	5	Merabet & Guillaume (1988)
Lower Lodève Sandstones, France	268	286	44,	170	28	8	x				x		x	3	Kruseman (1962)
Lodève Autunian, France	268	286	49,	161	53	17	x		x		x		x	4	Cogné *et al.* (1990)
St. Affrique (rotated), France	268	286	43,	193	37	21	x		x				x	*	Cogné *et al.* (1990)
Dolerites, SW Swedeb	258	286	56,	162	8	29	x		x		x			3	Thorning & Abrahamsen (1980)
Spitsbergen Sediments	245	323	36,	141	85	10		x	x	x			x	*	Vincenz & Jelenska (1985)
Moissey Volcanics, France	258	320	41,	172	123	5		x	x			x	x	4	Thompson *et al.* (1986)
Queensferry Sill, UK	268	320	38,	174	241	5			x				x	2	Torsvik *et al.* (1989)
Arendal Diabase, Norway	275	296	43,	160	166	7	x		x		x			3	Halvorsen (1972)
Nideck-Donon L. Volc., France	275	296	42,	168	44	19	x		x		x			3	Westphal (1972)
Saar-Nahe Volcanics, Germany	275	296	41,	169	9	16	x				x		x	3	Berthold *et al.* (1975)
Black Forest Volc., Germany	275	296	48,	174	35	6	x	x			x		x	4	Konrad & Nairn (1972)
Ny Hellesund Sills, Norway	275	296	39,	161	284	3	x	x	x		x		x	5	Halvorsen (1970)
Mt. Billinger Sill, Sweden	278	288	31,	174	529	2	x	x	x		x		x	5	Mulder (1971)
Mt. Hunneberg Sill, Sweden	278	288	38,	166	660	5	x	x	x		x		x	5	Mulder (1971)
Skaane Dolerite Dikes, Sweden	278	288	37,	174	60	7	x	x	x		x		x	5	Mulder (1971)
Peterhead Dike, Scotland, UK	256	323	41,	162	184	1		x	x		x		x	4	Torsvik (1985a)
Stabben Island Syenite, Norway	289	305	32,	174	—	2	x	x	x				x	4	Sturt & Torsvik (1987)
Skaane Quartz-dolerites, Sweden	286	320	38,	168	40	7	x	x	x		x		x	5	Bylund (1974)
Gzhelian Stage Seds., Russia	286	295	43,	170	196	5	x	x			x		x	4	Irving *et al.* (1976a: 6.245)
Donbass Arauc. & Avilov, Ukraine	286	295	46,	176	24	4	x	x			x		x	4	Irving *et al.* (1976a: 6.246)
Westph./Steph. Redbeds, Czech.	286	315	39,	163	56	9	x	x	x		x		x	5	Krs (1968)
Westph./Steph. Volcanics, Poland	286	315	43,	174	17	14	x	x		x	x			4	Birkenmajer *et al.* (1968)
Sarna Alkaline Intrusion, Sweden	273	301	38,	164	25	7	x		x		x		x	4	Smith & Piper (1979)
Sudetic Mnt. Granitoids, Poland	275	305	42,	166	29	17	x		x	x	x	x	x	6	Halvorsen *et al.* (1989)
Great Whin Sill, UK	289	301	44,	159	259	5	x	x	x		x			4	Storetvedt & Gidskehaug (1969)

Table A2. *Phanerozoic pole positions for 'Stable' Europe – continued*

Rock unit, location	Age		Pole position				Reliability								Reference
	Lo	Hi	Lat.	Long.	k	α_{95}	1	2	3	4	5	6	7	Q	
Wackerfield Dike, UK	298	308	49,	169	174	3	x				x		x	3	Tarling *et al.* (1973)
Sudetic Mnts., Poland	300	305	39,	181	27	11	x	x	x		x	x	x	6	Westphal *et al.* (1987)
Argyllshire Dikes, UK	316	324	35,	175	61	5	x	x	x		x		x	5	Esang & Piper (1984a)
Kinghorn Lavas, UK	320	360	18,	161	17	8	x	x	x	x	x	x	x	7	Wilson & Everitt (1963)
Kinghorn, Burntisl. Lavas, UK	320	360	14,	152	31	7	x	x	x		x	x	x	6	Torsvik *et al.* (1989)
Billefjorden Grp., Spitsbergen	320	360	27,	153	31	12	x		x			x	x	4	Watts (1985a)
Portishead Redbeds, UK	348	380	32,	158	23	10	x	x		x	x			4	Morris *et al.* (1973)
Zilair Suite, S. Urals, Russia	360	374	20,	141	102	8	x				x	x	x	4	Irving *et al.* (1976a: 6.254)
Leningrad Redbeds Comb., Russia	360	374	32,	161	108	7	x	x			x	x		4	Irving *et al.* (1976a: 5.120)
Koltubansk, S. Urals, Russia	360	374	33,	170	22	3	x	x			x		x	4	Irving *et al.* (1976a: 5.153)
Orkney Lavas, UK	365	385	24,	150	45	9	x				x		x	3	Storetvedt & Petersen (1972)
Esha Ness, Shetland, UK	365	385	21,	134	58	3	x	x	x		x	x	x	6	Storetvedt & Torsvik (1985)
Ulutau Suite, S. Urals, Russia	374	380	31,	156	93	10	x	x			x			3	Irving *et al.* (1976a: 5.126)
Foyers Old Red SS., UK	374	387	30,	147	8	29	x				x		x	3	Kneen (1973)
Lower Caithness Redbeds, UK	374	387	27,	149	20	6	x		x		x	x		4	Storetvedt & Torsvik (1983)
Holy Cross Mntn. Ss., Poland	374	394	24,	142	39	6	x				x		x	3	Lewandowski *et al.* (1987)
Sarclet Sandstone, UK	380	400	24,	161	14	6	x	x	x		x	x		5	Storhaug & Storetvedt (1985)
Hendre and Blodwell Intr., UK	374	394	32,	166	21	15	x	x						2	Piper (1978b)
Saerv Nappe Dolerites, Sweden	350	420	21,	170	11	12		x	x	x	x			*	Bylund & Zellman (1981)
Hoy Lavas, UK	387	406	14,	154	66	4	x		x		x		x	4	Storetvedt & Meland (1985)
Wood Bay Fm., Spitsbergen	390	408	12,	135	44	10	x	x	x			x	x	5	Jelenska & Lewandowski (1986)
Eday Sandstone, UK	390	415	8,	167	8	8	x		x	x	x		x	5	Robinson (1985)
Orkney Lavas, UK	390	415	−2,	147	18	14	x	x			x	x	x	5	Morris *et al.* (1973)
Cheviot Hills Combined, UK	384	412	−4,	141	11	13	x	x			x	x	x	5	Thorning (1974)
Midland Valley Lavas, UK	394	411	4,	140	13	6	x	x		x	x	x	x	6	Sallomy & Piper (1973)
Zmeinny Island Redbeds, Ukraine	387	421	6,	128	25	6					x		x	2	Irving *et al.* (1976a: 4.55)
Moine Metasediments, UK	387	421	6,	133	74	6		x	x			x	x	4	Watts (1982)
Garabal Hill-Glen Fyne, UK	388	422	5,	146	11	24	x		x				x	3	Briden (1970)
Strathmore Lavas, Midland, UK	394	421	−2,	138	18	3	x	x	x	x	x	x	x	7	Torsvik (1985b)
ORS, Anglo-Welsh Cuvette, UK	401	414	−3,	118	5	13	x		x	x	x		x	5	Chamalaun & Creer (1964)
Dniester Sediments, Ukraine	401	414	4,	128	4	18	x				x	x	x	4	Irving *et al.* (1976a: 4.50)
Comrie Intrusion, UK	403	413	6,	107	19	6	x	x	x				x	4	Turnell (1985)

Lorne Plateau Lavas, UK	396	422	−2,	141	43	6	x	x		x	x		x	5	Latham & Briden (1975)
Arrochar Ign. Complex, UK	396	422	8,	144	167	5	x	x	x	x			x	5	Briden (1970)
Salrock Fm., Ireland	400	428	2,	108	44	4	x	x	x	x		x	x	6	Smethurst & Briden (1988)
Ringerike Old Red, Norway	408	428	19,	164	15	9	x	x	x	x	x	x	x	7	Douglass (1988)
Gotland and Skaane Ls., Sweden	414	428	21,	166	313	5	x	x	x		x		x	5	Claesson (1979 in Piper, 1987)
Mjøsa Ls., Oslo Graben, Norway	448	455	−4,	143	14	25	x		x	x	x			*	Perroud *et al.* (1992)
Orthoceras Limestones, Sweden	468	488	−18,	226	176	5	x	x	x	x	x	x	x	7	Torsvik & Trench (1991b)
Orthoceras Limestones, Sweden	478	490	−30,	235	94	9	x	x	x	x	x		x	6	Perroud *et al.* (1992)
N. Serginski Intr., Russia	458	505	16,	184	—	11		x					x	2	Irving *et al.* (1976a: 3.60)
Asha Series, S. Urals, Russia	540	570	8,	189	44	12	x	x			x	x	x	5	Irving *et al.* (1976a: 2.108)
Nexø Sandstone, Denmark	520	590	−38,	314	12	13		x			x			2	Prasad & Sharma (1978)
Alnön Complex, Sweden	547	559	8,	272	13	9	x	x	x		x	x	x	6	Piper (1981a)
Fen/Rödberg Complex, Norway	540	630	−63,	322	138	3			x					1	Poorter (1972)
Fen/Tinguaites Intr., Norway	575	595	−50,	324	27	6	x	x	x		x	x		5	Piper (1988b)

The entries in this Table are for the Stable European craton, including its thrust belt margins in so far as no suspicions exist about relative displacements; obviously rotations can occur there and this mandates that reliability criterion 5 can not be checked as being met. Stable Europe includes the Scandinavian cratonic nucleus and the Russian Platform, Caledonian Europe for post-Silurian time (414 Ma and younger) and Hercynian Europe for post-Middle Carboniferous time (299 Ma and younger), conforming to Figure 5.5.
The reliability criteria (1–7) and Q are explained in the text. A '*' is placed in the Q column when the result is not to be used for apparent polar wander path construction, as explained below; a 'x' mark under any of the seven reliability columns means that this particular criterion is being met.
Abbreviations: Lat. = latitude, Long. = longitude, k and α_{95} are the precision parameter and radius of the cone of 95% confidence, respectively (Fisher, 1953).

Poles that would have $Q \geq 3$ but are excluded from mean pole determinations and other compilations, are listed below with the reasons in parentheses: Lower Tertiary Faeroe Upper Lavas (thought to be remagnetized according to the authors); Triassic Red Beds, Cretaceous Sandstones, Malm Delta and Jura Tan Limestones (all thought to be remagnetized); Lower Cretaceous Sussex Iron Grit (remagnetized); Upper Jurassic Limestone poles, e.g., Malm alpha, Malm gamma, Kujawi (probably remagnetized, as discussed in Chapter 6); four Late Jurassic Limestone poles marked by Q are discussed also in Chapter 6; Lower Middle Jurassic Dorset-Yorkshire Sediments (contaminated by overprint); Jurassic (to Permian?) Sunnhordland Dikes (too poorly dated); Lower Triassic St. Bee Sandstone (contaminated by overprint); Upper Permian West Pyrenees flows (in rotated block); Lower Permian St. Affrique (rotated); Permo-Carboniferous Sediments from Spitsbergen (poorly dated, contaminated by overprint); Devonian Saerv Nappe dolerites (rotated, poorly dated); Upper Ordovician Mjøsa Limestone (remagnetized).

Table A2 continued. *Pre-latest Silurian (Phanerozoic) paleopoles from Caledonian Europe*

Rock unit, location	Age		Pole position				Reliability								Reference
	Lo	Hi	Lat.	Long.	k	α_{95}	1	2	3	4	5	6	7	Q	
Great Britain, North of Iapetus Suture															
Aberdeenshire Gabbros III, UK	398	438	5,	151	621	5			x				x	2	Watts & Briden (1984)
Ratagan Intrusion, UK	410	420	15,	167	30	8	x	x	x	x		x	x	6	Turnell (1985)
Clare Island, Ireland	421	428	28,	154	27	10	x	x	x	x			x	5	Smethurst & Briden (1988)
Peterhead Granite, UK	414	438	21,	178	19	5		x	x				x	3	Torsvik (1985a)
Foyers Plutonic Complex, UK	400	438	16,	168	6	19				x		x		2	Kneen (1973)
Foyers Granite, UK	400	438	27,	167	15	6		x	x				x	3	Torsvik (1984)
Helmsdale Granite, UK	400	438	23,	175	32	6		x	x				x	3	Torsvik *et al.* (1983)
Borrolan Syenite, UK	408	438	18,	147	11	8		x	x				x	3	Turnell & Briden (1983)
N. Highlands Zone 5, UK	380	438	16,	168	14	6		x	x			x	x	4	Esang & Piper (1984b)
Loch Ailsh, UK	408	438	16,	141	14	8		x	x				x	3	Turnell & Briden (1983)
Assynt Dikes Combined, UK	408	438	16,	148	32	5		x	x				x	3	Turnell & Briden (1983)
Knockhaven Grp., Ireland	424	430	−3,	143	10	31	x							1	Morris *et al.* (1973)
Strontian Granite, UK	408	460	21,	164	18	5		x	x				x	3	Torsvik (1984)
Borrolan Ledmorite, UK	420	440	10,	189	13	7	x	x	x			x	x	5	Turnell & Briden (1983)
Caledonian Dolerites, UK	420	450	15,	168	10	10	x	x	x			x	x	5	Esang & Piper (1984b)
Caledonian Microdiorites, UK	420	450	17,	166	14	6	x	x	x			x	x	5	Esang & Piper (1984b)
Tourmakeady, Glensaul, Ireland	408	458	31,	169	38	4		x	x					2	Deutsch & Storetvedt (1988)

Borrolan Pseudoleucite, UK	428	458	24,	178	16	8			x			x	x	3	Turnell & Briden (1983)
Ben Loyal, UK	428	458	4,	178	7	16			x			x	x	3	Turnell & Briden (1983)
Aberdeenshire Gabbros II, UK	446	448	8,	178	26	14	x	x	x				x	4	Watts & Briden (1984)
Barrovian of Angus, UK	438	478	3,	202	47	7		x	x	x			x	4	Watts (1985b)
Port Askaig Tillite Overp., UK	438	505	28,	198	4	12			x				x	2	Stupavsky *et al.* (1982)
Ballantrae Intr. Combined, UK	438	478	12,	207	24	8		x	x				x	3	Piper (1978a)
Aberdeenshire Gabbros I, UK	458	505	16,	212	121	8		x	x				x	3	Watts & Briden (1984)
Northern British Caledonide displaced, rotated terranes															
Killary Harbour Intr.I, Ireland	409	424	42,	248	10	11	x			x			x	3	Morris (1976)
Killary Harbour Intr.II, Ireland	409	424	36,	264	15	16	x						x	2	Morris (1976)
Mweelrea Ignimbrites, Ireland	458	468	9,	214	23	20	x			x			x	3	Deutsch & Somayajulu (1970)
Slockenray Fm., Scotland, UK	468	488	−12,	120	—	4	x	x	x	x				4	Trench *et al.* (1988)
Connemara Gabbros, Ireland	480	510	−6,	122	5	16	x		x	x				3	Robertson (1988)
Scandinavian Caledonide possibly displaced or rotated terranes															
West Finnmark, Compt. A, Norway	385	438	−5,	155	100	8		x	x	x			x	4	Torsvik *et al.* (1990b)
Sulitjelma Gabbro, Norway	380	450	14,	180	6	28						x	x	2	Piper (1974, 1975a)
Fongen-Hyllingen, P 1, Norway	390	430	19,	225	33	10		x					x	2	Abrahamsen *et al.* (1979)
Fongen-Hyllingen, P 2, Norway	390	430	19,	128	24	12		x					x	2	Abrahamsen *et al.* (1979)
West Finnmark, Compt.C, Norway	480	575	−31,	266	22	11		x	x				x	3	Torsvik *et al.* (1990b)

These paleopoles are listed separately to highlight that they are not representative of Stable Europe, but of a former part of Laurentia, now incorporated into Europe. For explanation, see above part of Table A2.

Table A3. *Phanerozoic pole positions for Greenland*

Rock unit, location	Age		Paleopole		(Paleopole)			Reliability								Reference
	Lo	Hi	Lat.	Long.	(unrotated)	k	α_{95}	1	2	3	4	5	6	7	Q	
Cenozoic results																
Angmagssalik Dikes, E. Greenland	023	100	75,	133	(73, 140)	45	8		x	x	x			x	4	Beckman (1982)
Skaergaard Intr., E. Greenland	054	058	63,	161	(61, 165)	40	4	x		x				x	3	Schwarz *et al.* (1979)
Scoresby Sund Lavas, E. Greenland	050	060	65,	171	(63, 174)	42	15	x	x			x		x	4	Tarling (1967)
Scoresby Sund Lavas, E. Greenland	050	060	72,	179	(70, 181)	88	4	x	x			x		x	4	Hailwood (1977)
Kangerdlugssuak Lavas, E. Greenland	045	060	65,	183	(63, 185)	105	9	x	x			x		x	4	Faller (1975)
K. Irminger/THOL 1 Dikes, E. Greenland	053	060	65,	178	(63, 180)	58	6	x	x	x				x	4	Faller & Soper (1979)
Jacobsen Fjord Dikes, E. Greenland	053	060	70,	176	(68, 178)	69	4	x	x	x		x	x	x	6	Faller & Soper (1979)
Jacobsen Fjord Lavas, E. Greenland	053	060	58,	180	(56, 183)	—	6	x	x	x		x		x	5	Faller & Soper (1979)
Ubekendt Ejland Lavas, W. Greenland	057	070	57,	206	(55, 209)	36	11	x	x			x		x	4	Tarling & Otulana (1972)
Mesozoic results																
S. Disko Island Volcs., W. Greenland	066	074	65,	184	(62, 191)	18	6	x	x			x		x	4	Deutsch & Kristjansson (1974)
N. Disko Island Volcs., W. Greenland	066	074	74,	204	(72, 206)	308	7	x	x	x		x	x	x	6	Athavale & Sharma (1975)
Coast-Parallel Dikes, W. Greenland	157	173	80,	176	(72, 191)	90	5	x	x			x	x		4	Piper (1975b)

Coast-Parallel Dikes, S. Greenland	161 176	64,	145	(56, 168)	—	6	x	x			x	x	x	5	Fahrig & Freda (1975)
Fleming Fjord Fm., E. Greenland	215 240	35,	80	(34, 103)	25	5	x	x				x	x	4	Reeve *et al.* (1974)
Pingodal Fm., E. Greenland	240 245	43,	166	(35, 184)	—	7	x				x		x	*	Athavale & Sharma (1974)
Gipsdalen Fm., E. Greenland	240 245	56,	135	(49, 158)	—	10	x				x		x	*	Athavale & Sharma (1974)
Paleozoic results															
Kap Stanton Fm., N. Greenland	360 440	2,	104	(−1, 122)	16	11		x	x	x				3	Stearns *et al.* (1989)
Pseudotachylites, Ikertoq, W. Greenland	540 620	17,	205	(10, 222)	23	13		x	x		x	x	x	5	Piper (1981b)
Kimberl. & Lamprophyres I, W. Greenland	570 590	10,	197	(3, 215)	47	5	x	x			x		x	4	Piper (1981b)

The entries in this Table are for the Greenland craton, including its thrust belt margins in so far as no suspicions exist about relative displacements; obviously rotations can occur there and this mandates that reliability criterion 5 can not be checked as being met. All of Greenland is included, and poles are given in unrotated coordinates (in parentheses), as well as in North American coordinates; rotations have been performed with the parameters of Bullard *et al.* (1965), Euler pole 65.8°, 92.4°W, angle 19.7 clockwise for times before 144 Ma, and with variable parameters from Olivet *et al.* (1984) for Late Cretaceous and younger times.

The reliability criteria (1–7) and Q are explained in the text. A '*' is placed in the Q column when the result is not to be used for apparent polar wander path construction, as explained below; a 'x' mark under any of the seven reliability columns means that this particular criterion is being met.

Abbreviations: Lat = latitude, Long = longitude, k and α_{95} are the precision parameter and radius of the cone of 95% confidence, respectively (Fisher, 1953).

*Poles that would have $Q \geq 3$ but are excluded from mean pole determinations and other compilations, are listed below with the reasons in parentheses: Lower Triassic Pingodal and Gipsdalen formations (internally discordant results, possibly contaminated by overprint).

Table A4. *Phanerozoic pole positions for cratonic West Gondwana, rotated to West African coordinates*

Rock unit, location	Age		Paleopole		(Paleopole)			Reliability								Reference
	Lo	Hi	Lat.	Long.	(unrotated)	k	α_{95}	1	2	3	4	5	6	7	Q	
I. AFRICA																
Cenozoic (Paleocene, Eocene) results																
Jebel Nefousa Basalt, Libya	001	005	86,	152		84	3		x			x	x	x	4	Schult & Soffel (1973)
Qatrani and Quseir Flows, Egypt	024	032	68,	102		35	10	x	x	x		x		x	5	Ressetar *et al.* (1981)
Ethiopian Flood Basalts	005	037	81,	168		35	6		x			x		x	3	Brock *et al.* (1970)
Ethiopian Traps	005	037	87,	254		17	11					x		x	2	Grasty (1964)
Afar Volcanics, Ethiopia	005	037	76,	135		—	9					x	x	x	3	Pouchan & Roche (1971)
Tororo Ring Complex, Uganda	025	058	76,	196		157	10					x		x	2	Raja & Vise (1973)
Jebel Nefousa, Lybia	033	053	86,	152		84	3	x	x			x	x		4	Schult & Soffel (1973)
Baharia Iron Ores, Egypt	037	050	84,	139		65	6	x	x	x		x	x		5	Schult *et al.* (1981)
Wadi Abu Tereifiya volcs., Egypt	037	058	69,	188		137	6	x	x					x	3	Hussain *et al.* (1979)
Gebel Gifata Ls, Sh., Egypt	060	070	82,	225		50	9	x	x			x	x		4	Saradeth *et al.* (1989)
Mesozoic results																
Fuerteventura Lavas, Canary Isl.	038	098	68,	224		47	3		x	x		x	x	x	5	Storetvedt *et al.* (1979)
North Sudan Volcanic Field	061	082	56,	278		36	11	x	x	x		x		x	5	Saradeth *et al.* (1989)
Melilite Basalts, S. Africa	064	077	85,	243		28	—	x				x			2	Duncan *et al.* (1978)
Usfan Fm., Saudi Arabia	065	074	65,	217	63, 231	18	13	x	x			x		x	4	Yousif & Beckman (1981)
Quseir Trachytes, Egypt	063	092	63,	252		134	3	x	x	x		x		x	5	Ressetar *et al.* (1981)
East Maio Volcs., Cape Verde	040	120	57,	235		24	6		x					x	2	Storetvedt & Løvlie (1983)
East El Oweinat Volc., Egypt	065	097	68,	269		16	13	x	x	x				x	4	Hussain & Aziz (1983)
Basalt Series I, Gran Canaria	065	097	72,	251		31	8	x	x			x	x	x	5	Storetvedt *et al.* (1978c)
Gebel El Kahfa Ring C., Egypt	074	095	61,	238		148	8	x		x		x		x	4	Ressetar *et al.* (1981)
Abu Khrug Ring Complex, Egypt	087	091	59,	266		28	13	x	x	x		x		x	5	Ressetar *et al.* (1981)
E. Aswan Nubian ss., Egypt	045	135	75,	203		23	11					x		x	2	El Shazly & Krs (1973)
Wadi N. Nubian Ss. Comb., Egypt	045	135	82,	223		84	3		x	x		x	x		*	Schult *et al.* (1978; 1981)
Group I Kimberlites, S. Africa	081	100	64,	226		59	5	x	x	x		x		x	5	Hargraves (1989)
Wadi Natash volcs., Egypt	086	100	69,	258		21	9	x	x	x		x		x	5	Schult *et al.* (1981)
Wadi Natash ss. & volcs., Egypt	078	111	64,	218		5	6	x				x		x	3	El Shazly & Krs (1973)
Wadi Natash Intrusions, Egypt	078	111	76,	228		18	18	x		x		x			3	Ressetar *et al.* (1981)

Kasr es Souk Red beds, Morocco	065	145	66,	227		16	16					x		x	2	Martin *et al.* (1978)
Lupata Volcanics, Mozambique	109	113	64,	257	62, 260	339	3	x	x			x		x	4	Gough & Opdyke (1963)
Kaoko Basalts, Namibia	120	140	56,	264	48, 267	40	3	x	x			x	x	x	5	Gidskehaug *et al.* (1975)
Beni Mellal Intrusives, Morocco	110	130	46,	258		—	15	x	x				x	x	4	Westphal *et al.* (1979)
Infra-Cenomanian seds., Morocco	098	144	75,	227		117	7	x	x						2	Hailwood (1975)
Mlanje Syenite, Malawi	119	131	64,	258	60, 262	36	9	x				x		x	3	Briden (1967)
Group II Kimberlites, S. Africa	113	145	56,	268	48, 270	26	10	x	x	x		x	x	x	6	Hargraves (1989)
Wadi Abu Shihat Dikes, Egypt	065	200	51,	272	45, 273	6	43					x	x	x	3	Hussain *et al.* (1979)
Central Atlas intr. Morocco	123	162	53,	262		10	22	x						x	2	Hailwood & Mitchell (1971)
Swartruggens Kimberlite, S. Africa	146	165	39,	277	31, 278	—	6	x	x	x		x	x	x	6	Hargraves & Onstott (1980)
Djebel Oust Seds., Tunisia	144	169	65,	200		121	6	x	x	x		x	x	x	6	Nairn *et al.* (1981)
Intrusives, Nigeria	140	180	63,	242		13	19	x				x		x	3	Marton & Marton (1976)
Hoachanas Basalts, Namibia	160	175	68,	246	61, 254	19	7	x				x		x	3	Gidskehaug *et al.* (1975)
Mateke Hills Intr., Zimbabwe	153	183	63,	267	55, 270	29	11	x	x			x	x	x	5	Gough *et al.* (1964)
Beni Mellal Basalts, Morocco	170	177	45,	251		27	9	x	x		x	x	x	x	6	Bardon *et al.* (1973)
Beni Mellal Basalts, Morocco	170	177	45,	248		—	11	x	x				x	x	4	Westphal *et al.* (1979)
Sahara Dolerites, Morocco	165	195	69,	255		24	14	x	x			x		x	4	Bardon *et al.* (1973)
Draa V. & other volcs, Morocco	165	195	62,	245		26	6	x	x			x		x	4	Hailwood & Mitchell (1971)
Diabase dikes, Liberia	173	192	69,	242		25	4	x	x			x	x	x	5	Dalrymple *et al.* (1975)
Freetown Ign. Complex, S. Leone	164	202	81,	225		43	7	x				x	x		3	Briden *et al.* (1971b)
Hank, N. Mauritania	182	192	69,	232		99	4	x	x			x		x	4	Sichler *et al.* (1980)
Hodh, S. Mauritania	182	192	71,	240		47	6	x	x			x		x	4	Sichler *et al.* (1980)
Marangudzi Complex, Zimbabwe	172	206	78,	289	70, 285	41	9	x	x			x	x	x	5	Brock (1968)
Nusab el Balgum Complex, Egypt	157	221	69,	221	64, 231	92	5		x			x			2	Saradeth *et al.* (1989)
Stormberg Lavas Comb., Lesotho	183	203	79,	263	71, 269	68	15	x	x			x	x		4	Van Zijl *et al.* (1962a)
Karroo Dolerites Comb., S. Africa	183	203	72,	245	65, 255	16	13	x	x		x	x	x		5	McElhinny & Jones (1965)
Karroo Lavas Combined, S. Africa	183	203	65,	259	57, 264	41	8	x	x			x			3	McElhinny *et al.* (1968)
Uvinza Dike, Tanzania	187	201	60,	237	54, 245	26	13	x				x			2	Piper (1972)
Midelt Limestones, Morocco	187	203	62,	210		48	7	x						x	2	Martin *et al.* (1978)
Atlas & Meseta Volcs., Morocco	175	217	71,	216		17	7	x	x					x	3	Bardon *et al.* (1973)
Zambia Red Sandstone	187	225	72,	211	68, 230	208	5	x	x			x			3	Opdyke (1964)
Shawa Ijolite, Zimbabwe	193	225	72,	261	64, 266	40	12	x				x			2	Gough & Brock (1964)
Argana Red beds, Morocco	208	230	51,	251		8	12	x					x	x	3	Martin *et al.* (1978)

Table A4. *Phanerozoic pole positions for cratonic West Gondwana, rotated to West African coordinates – continued*

Rock unit, location	Age Lo	Age Hi	Paleopole Lat.	Paleopole Long.	(Paleopole) (unrotated)	k	α_{95}	1	2	3	4	5	6	7	Q	Reference
Issaldain Lavas, Morocco	208	245	77,	141		88	7		x	x					2	Hailwood (1975)
Jerada Limestones, Morocco	230	240	79,	276		156	4	x							1	Daly & Pozzi (1976)
Jerada Red Marls, Morocco	230	240	72,	229		201	2	x							1	Daly & Pozzi (1976)
Combined Basalts, Morocco	230	240	65,	254		39	20	x							1	Daly & Pozzi (1976)
Ait-Aadel Lavas, Morocco	230	240	72,	255		27	8	x	x						2	Hailwood (1975)
Titchka Red beds, Morocco	230	240	66,	211		19	22	x				x			2	Hailwood (1975)
Cassanje Series, Angola	230	240	61,	254	54, 259	20	8	x				x	x		3	Valencio *et al.* (1978)
Maji-ya-Chumvi Beds, Kenya	235	245	75,	264	67, 269	14	17	x				x			2	McElhinny & Brock (1975)
Paleozoic results																
Permian Sediments, Tunisia	245	270	49,	200		18	22							x	1	Abou-deeb & Tarling (1984)
Djebel Tarhat Red beds, Morocco	268	286	24,	244		35	8	x	x					x	3	Martin *et al.* (1978)
Chougrane Red beds, Morocco	268	286	32,	244		20	5	x	x		x	x		x	5	Daly & Pozzi (1976)
Taztot Trachyandesites, Morocco	268	286	39,	237		80	5	x		x				x	3	Daly & Pozzi (1976)
Abadla Red beds, Morocco	268	286	29,	240		59	6	x	x	x		x		x	5	Morel *et al.* (1981)
Songwe-K. (K3) Red beds, Tanzania	268	286	35,	267	27, 269	60	12	x	x					x	3	Opdyke (1964)
Galula (K3) Redbeds, Tanzania	268	296	50,	211	46, 220	227	5	x	x					x	3	Opdyke (1964)
Mechra, Chougrane Vcs., Morocco	268	296	36,	238		23	20	x		x				x	3	Westphal *et al.* (1979)
Dwyka Varves, S. Central Africa	286	320	30,	202	27, 207	52	11	x	x		x	x	x	x	6	McElhinny & Opdyke (1968)
Algerian Sediments & Volcs.	279	333	28,	238		39	13		x					x	2	Abou-deeb & Tarling (1984)
Sabaloka Ring Complex, Sudan	319	406 ?	47,	231	42, 236	21	8		x	x		x	x		*	Soffel *et al.* (1990)
Ain-ech-Chebbi Fm., Algeria	315	333	25,	235		—	4	x	x	x				x	4	Conrad & Westphal (1975)
Ain-ech-Chebbi Fm., Algeria	307	333	25,	235		128	4	x	x	x				x	4	Daly & Irving (1983)
Volcs., Redbeds & Ls., Morocco	330	352	8,	237		—	6	x	x		x			x	4	Martin *et al.* (1978)
Ben Zireg Synfolding, Algeria	352	360	25,	201		8600	1		x	x	x	x		x	5	Aïfa *et al.* (1990)
Msissi Norite, Morocco	136	408	1,	205		29	12		x			x		x	*	Hailwood (1974)

Ben Zireg Griotte, Algeria	360	367	19,	200		24	4	x	x	x	x	x		x	6	Aïfa *et al.* (1990)
Bokkeveld Grp., S. Africa	374	387	−9,	195	−10, 195	33	7	x				x	x	x	4	Bachtadse *et al.* (1987b)
Gneiguira-Dikkel, Mauritania	380	428	35,	224		89	5		x	x					2	Kent *et al.* (1984)
Aïr Ring Complexes, Niger	418	438	43,	189		50	6	x	x	x		x		x	5	Hargraves *et al.* (1987)
Pakhuis/Cedarberg, S. Africa	421	458	−28,	165	−25, 163	9	18	x		x		x	x	x	5	Bachtadse *et al.* (1987b)
Salala Ring Compl., Sudan	434	478	−42,	153	−40, 150	44	9	x	x	x		x	x	x	6	Bachtadse & Briden (1989)
Blaubeker Fm. (NBX), Namibia	445	497	−52,	180	−51, 172	45	20					x		x	2	Kröner *et al.* (1980)
Graafwater Fm., S. Africa	478	505	−27,	197	−28, 194	25	9	x	x	x		x	x	x	6	Bachtadse *et al.* (1987b)
Hasi Messaud Seds., Algeria	476	523	−53,	206		—	3	x				x		x	3	Bucur (1971)
Ntonya Ring Structure, Malawi	481	526	−31,	168	−28, 165	1045	2	x	x	x		x		x	5	Briden (1968b)
Um Rus Dikes, Egypt	464	550	−84,	301	−86, 13	48	6		x	x			x		3	Davies *et al.* (1980)
Quena-Safaga Dikes, Egypt	480	550	−86,	233	−87, 124	36	5		x	x			x		3	Davies *et al.* (1980)
Esh El Mellaha Dikes, Egypt	445	580	−84,	269	−89, 322	101	12		x	x			x		3	Abdullah *et al.* (1984)
Hook Intrusives, Zambia	499	535	−18,	157	−14, 156	13	34	x				x		x	3	Brock (1967)
Nama Grp. Combined (N3), Namibia	480	580	−1,	153	3, 154	26	10		x	x			x	x	4	Kröner *et al.* (1980)
Adoud. Andesites, Seds., Morocco	514	550	−47,	222		182	7	x	x					x	*	Daly & Pozzi (1977b)
Sidi-Said-Maachou Volc., Morocco	523	540	14,	231		70	6	x	x	x	x			x	*	Khattach *et al.* (1989)
Mulden Grp., Namibia	480	605	−21,	89	−13, 90	18	16			x		x	x	x	4	McWilliams & Kröner (1981)
Fish River, Nama (N2), Namibia	520	620	−13,	90	−5, 91	41	14			x		x	x	x	4	Kröner *et al.* (1980)
Série Lie-de-Vin, Morocco	536	610	−29,	63		8	21								0	Martin *et al.* (1978)
Sijarira Grp., Zimbabwe	540	620	−4,	171	−2, 172	9	18				x	x			2	Reid (1968)
II. SOUTH AMERICA																
Cenozoic (Paleocene, Eocene) results																
Abrolhos Islands, Brazil	047	052	69,	73	59, 54	—	6	x				x	x	x	4	Pacca & Valencio (1972)
Rio de los Molinos dikes, Argentina	058	066	67,	231	77, 198	50	6	x				x		x	3	Linares & Valencio (1975)
Los Elquinos Fm., Chile	037	094	73,	254	84, 245	32	12			x			x		2	Palmer *et al.* (1980)
Mesozoic results																
Andacollo Remagn., Argentina	049	089	63,	260	74, 245	—	5		x			x		x	3	Vilas & Valencio (1978a)
Patagonian Plat. Bas., Argentina	064	079	69,	226	79, 178	32	6	x	x	x		x	x	x	6	Butler *et al.* (1991)
Poços de Caldas, Brazil	066	088	85,	231	81, 53	32	11	x	x			x	x		4	Opdyke & MacDonald (1973)

Table A4. *Phanerozoic pole positions for cratonic West Gondwana, rotated to West African coordinates – continued*

Rock unit, location	Age		Paleopole		(Paleopole)			Reliability								Reference
	Lo	Hi	Lat.	Long.	(unrotated)	k	α_{95}	1	2	3	4	5	6	7	Q	
San Luis & Cordoba Volcs., Argentina	066	098	56,	253	70, 225	14	12	x		x			x	x	4	Valencio *et al.* (1983)
C. Sto. Agostinho Volcs., Brazil	085	099	72,	243	88, 135	114	5	x	x			x			3	Schult & Guerreiro (1980)
Vinita Fm., Chile	066	113	78,	227	82, 72	32	8	x	x	x					*	Palmer *et al.* (1980)
Pirgua Basalts, Red beds, Argentina	077	114	74,	248	85, 42	27	8	x	x		x	x			4	Valencio *et al.* (1977a)
Sierra los Condores Grp., Argentina	098	119	58,	265	84, 271	130	5	x	x			x			3	Vilas (1976)
Quebrada Marquesa Fm., Chile	098	131	72,	271	77, 13	66	6	x	x	x	x			x	*	Palmer *et al.* (1980)
East Maranhao Volcanics, Brazil	112	124	56,	264	84, 261	122	3	x	x			x		x	4	Schult & Guerreiro (1979)
Cerro Colorado Fm., Argentina	118	124	51,	251	81, 194	42	10	x	x			x			3	Valencio (1972)
Serra Geral Fm., Brazil	114	144	59,	264	85, 295	40	4	x	x			x	x		4	Pacca & Hiodo (1976)
El Salto-Almafuerte Lavas, Argentina	119	127	42,	252	72, 205	43	6	x	x			x	x	x	5	Mendia (1978)
Rumipalla Fm., Argentina	098	121	62,	260	88, 326	—	9	x	x			x			3	Valencio & Vilas (1976)
Arqueros Fm., Chile	119	131	65,	272	81, 345	41	11	x		x					2	Palmer *et al.* (1980)
Older Rio Molinos Dikes, Argentina	129	150	50,	248	79, 188	50	11	x				x	x		3	Linares & Valencio (1975)
Rio Gr. do Norte Dikes, Brazil	125	167	63,	237	81, 95	50	9		x	x		x	x		4	Bücker *et al.* (1986)
West Maranhao Volcanics, Brazil	146	170	63,	248	85, 83	18	9	x	x			x			3	Schult & Guerreiro (1979)
Camaraca Fm., Africa, Chile	153	161	56,	221	71, 190	19	6	x	x	x			x	x	*	Palmer *et al.* (1980)
Chon Aike Lavas, Argentina	161	171	65,	259	85, 17	10	6	x	x			x	x		4	Vilas (1974)
Lago San Martin Volcs., Argentina	144	208	68,	232	77, 77	13	13						x		1	Burns *et al.* (1980)
Pescadero Porphr., Venezuela	187	199	61,	252	88, 97	23	11	x							1	MacDonald & Opdyke (1974)
Guacamayas Grp., Venezuela	188	199	56,	221	71, 121	17	19	x				x			2	MacDonald & Opdyke (1974)
Bolivar Dikes, Venezuela	194	204	74,	206	67, 66	246	5	x	x			x		x	4	MacDonald & Opdyke (1974)

Paramillos, Cach-Uspallata, Argentina	201	207	67,	223	74, 86	—	13	x	x			x	x		4	Creer *et al.* (1970)
Quebrada Alumbrera, Argentina	208	232	58,	211	67, 110	11	10	x					x	x	3	Valencio & Vilas (1985)
Ischigualasto Volcs., Argentina	219	229	70,	243	79, 59	14	14	x	x				x		3	Valencio *et al.* (1975a)
Surinam Dolerites	209	245	57,	242	82, 140	25	10	x	x	x		x			4	Veldkamp *et al.* (1971)
Puesto Viejo Fm., Argentina	228	236	72,	239	76, 56	27	13	x	x			x	x		4	Valencio *et al.* (1975a)
Giron Fm., Colombia	208	245	69,	233	77, 74	2	30					x			1	Creer (1970)
La Quinta Tuffs, Venezuela	208	245	76,	232	72, 53	38	46					x			1	Hargraves (1975)
Dolerites, British Guyana	208	245	85,	217	64, 42	21	14		x			x		x	*	Hargraves (1975)
Rio Chasquil, U. Paganzo, Argentina	170	270	56,	260	85, 244	—	8					x			1	Thompson (1972)
Motuca Ss., Brazil	208	245	60,	238	81, 116	7	10					x	x		2	Creer (1970)
Villa Unión, U. Paganzo, Argentina	204	258	69,	279	77, 355	—	6					x		x	2	Creer *et al.* (1970)
Cochico Fm., Mendoza, Argentina	204	258	66,	282	77, 343	18	5		x		x	x	x	x	5	Valencio & Vilas (1985)
Cuesta los Terneros, Argentina	204	258	66,	218	72, 88	91	8		x			x	x		3	Vilas (1969; recalculated)
Amana, U. Paganzo, Argentina	230	245	43,	222	63, 149	4	12	x				x	x	x	4	Valencio *et al.* (1977b)
Corumbatai Fm., Brazil	230	258	60,	248	86, 114	17	11					x	x		2	Valencio *et al.* (1975b)
Paleozoic results																
S. de la Ventana, Pillah., Argentina	245	268	72,	253	78, 39	—	14	x			x	x			3	Creer *et al.* (1970)
Irati Fm., Tubarao Grp., Brazil	245	268	53,	260	83, 235	225	5	x				x			2	Pascholati *et al.* (1976)
Upper Los Color's, Paganzo, Argentina	245	268	54,	246	82, 169	—	4	x				x			2	Thompson (1972)
Quebrada del Pimiento, Argentina	258	268	55,	232	76, 136	64	12	x	x			x	x		4	Creer *et al.* (1971)
La Colina Fm., M. Paganzo, Argentina	261	271	56,	241	81, 147	—	3	x	x			x			3	Thompson (1972)
La Colina Fm., M. Paganzo, Argentina	261	271	67,	234	78, 77	15	3	x	x			x			3	Valencio *et al.* (1977b)
M. Los Colorados, Paganzo, Argentina	245	286	54,	228	74, 133	—	5		x			x		x	3	Thompson (1972)
Mitu Fm., Peru	245	290	68,	259	82, 26	—	8						x		1	Creer (1970)
Lowest Middle Paganzo, Argentina	268	286	40,	232	65, 167	—	4	x	x			x		x	4	Creer (1965)
Lower Los Color's, Paganzo, Argentina	268	286	32,	235	59, 178	—	3	x	x			x		x	4	Embleton (1970)
Huaco Gorge, M. Paganzo, Argentina	268	286	36,	236	63, 176	—	8	x				x		x	3	Embleton (1970)

Table A4. *Phanerozoic pole positions for cratonic West Gondwana, rotated to West African coordinates – continued*

Rock unit, location	Age		Paleopole		(Paleopole)			Reliability								Reference
	Lo	Hi	Lat.	Long.	(unrotated)	k	α_{95}	1	2	3	4	5	6	7	Q	
Port. del Cenizo Fm., Argentina	256	303	62,	70	32, 31	41	8		x					x	2	Valencio & Vilas (1985)
Itarare, Tubarao Fm., Brazil	268	296	31,	233	57, 177	14	11	x				x		x	3	Valencio *et al.* (1975b)
La Colina Basalt, Paganzo, Argentina	281	308	41,	233	66, 168	—	5	x				x		x	3	Thompson (1972)
Tubarao Fm., Brazil	286	308	29,	229	54, 171	—	6	x				x		x	3	Pascholati & Pacca (1976)
La Tabla Fm., Chile	245	360	27,	225	51, 167	72	6		x	x	x			x	4	Jesinkey *et al.* (1987)
L. Paganzo (Lagares/Bum-B), Argentina	286	320	57,	248	85, 155	—	5					x	x		2	Thompson (1972)
La Colina Fm., M. Paganzo, Argentina	286	320	27,	221	49, 163	16	5	x	x	x		x	x	x	6	Sinito *et al.* (1979)
Taiguati Fm., Bolivia	286	320	14,	198	28, 146	—	5	x	x			x	x	x	5	Creer (1970)
Upper Piaui Fm., Brazil	286	320	30,	229	55, 171	—	12	x				x		x	3	Creer (1970)
Lower Piaui Fm., Brazil	286	320	40,	232	65, 167	—	9	x				x		x	3	Creer (1965)
Pilar & Cas Fms., Chile	280	333	32,	229	57, 170	—	10	x	x	x	x		x	x	6	Jesinkey *et al.* (1987)
Yalgnarez Fm., Mendoza, Argentina	303	333	79,	359	56, 15	25	7	x	x					x	*	Valencio & Vilas (1985)
Guandacol Fm., Argentina	320	360	75,	260	75, 30	11	9	x	x			x	x		*	Valencio & Vilas (1985)
Picos & Passagem Series, Brazil	380	401	21,	189	30, 133	2	26	x				x	x	x	4	Creer (1970)
Red beds, Bolivia	380	401	−8,	165	−7, 127	2	19	x				x		x	3	Creer (1970)
Salta & Jujuy Red beds, Argentina	360	460	−24,	192	−9, 158	—	8					x		x	2	Creer (1965; 1970)
Urucum Fm., Argentina	408	438	−49,	182	−34, 164	—	9					x		x	2	Creer (1965)
Acampamento Velho Fm. B1, Brazil	300	600	39,	225	61, 158	—	17			x			x		2	D'Agrelho-Filho & Pacca (1988)
Alcaparrosa Fm, Argentina	420	478	26,	255	56, 213	12	16					x		x	2	Vilas & Valencio (1978b)
Red beds, Bolivia	438	505	−2,	162	−4, 122	4	22				x	x		x	3	Creer (1970)
Salta Red Beds, Argentina	438	505	−58,	220	−31, 193	—	45					x		x	2	Thompson (1973)
Jujuy Red beds, Argentina	438	505	−24,	186	−11, 153	2	9					x		x	2	Creer (1970)
Suri Fm., Argentina	458	505	−18,	227	9, 186	21	6		x					x	2	Valencio *et al.* (1980)
Campo Alegre Fm., Itajai, Brazil	430	585	−85,	317	−57, 223	56	10		x	x			x	x	4	D'Agrelho-Filho & Pacca (1988)
Castro Group, Brazil	425	600	−47,	66	−76, 4	37	12			x					1	D'Agrelho-Filho & Pacca (1988)

Acampamento Velho Fm. B2, Brazil	504	530	−57,	172	−44, 165	—	42	x		x			x		3	D'Agrelho-Filho & Pacca (1988)
Vargas Fm., Bom Jardin, Brazil	495	560	−21,	18	−34, 322	5	17								0	D'Agrelho-Filho & Pacca (1988)
Hilario Fm., Bom Jardin, Brazil	494	560	−15,	270	14, 228	97	12							x	1	D'Agrelho-Filho & Pacca (1988)
Purmamarca, Argentina	505	570	−35,	262	−5, 219	—	15					x		x	2	Thompson (1973)
Purmamarca Village, Argentina	505	570	−45,	118	−61, 113	—	26					x			1	Thompson (1973)
Abra de Cajas, Argentina	505	570	−32,	249	−2, 208	—	50					x		x	2	Thompson (1973)
North Tilcara, Argentina	505	570	−78,	223	−49, 204	—	23					x		x	2	Thompson (1973)
South Tilcara, Argentina	505	570	−81,	226	−52, 207	—	23					x	x	x	*	Thompson (1973)
Argentina Sediments, Combined	505	570	−66,	226	−38, 199	4	43					x	x	x	3	Irving *et al.* (1976a; 2.134)

The entries in this Table are for cratonic West Gondwana, including its thrust belt margins in so far as no suspicions exist about relative displacements; obviously rotations can occur there and this mandates that reliability criterion 5 can not be checked as being met. West Gondwana includes the cratonic nuclei of Africa and South America. Pole coordinates have been rotated; they are given for each cratonic element in its own (unrotated) geographic coordinate grid of the present-day as well as in the coordinates of the West African craton. Rotation parameters for times preceding 130 Ma are as follows: southern Africa to West Africa, Euler pole at 9.34°N, 5.7°E, angle 7.82 clockwise; northeast Africa to southern Africa Euler pole at 16.3°S, 41.7°E, angle 2.5 counterclockwise; northeast Africa to West Africa, Euler pole at 19.24°N, 352.61°E, angle 6.26 clockwise; Arabia to southern Africa, Euler pole at 45.23°S, 196.28°E, angle 7.12 counterclockwise; South America to West Africa, Euler pole at 53.0°N, 325.0°E, angle 51.01 counterclockwise (all from Lottes & Rowley, 1990). For times between 130 and 100 Ma paleopoles for the African blocks have been partially rotated and for times after 100 Ma they have not been rotated. For South America, for times between 116 and 80 Ma, rotation parameters have been used from Klitgord & Schouten (1986) and for times after 80 Ma from Cande *et al.* (1988).

The reliability criteria (1–7) and Q are explained in the text. An asterisk is placed in the Q column when the result is not to be used for apparent polar wander path construction, as explained below; a 'x' mark under any of the seven reliability columns means that this particular criterion is being met.

Abbreviations: Lat. = latitude, Long. = longitude, k and α_{95} are the precision parameter and radius of the cone of 95% confidence, respectively (Fisher, 1953).

*Poles that would have $Q \geq 3$ but are excluded from mean pole determinations and other compilations, are listed below with the reasons in parentheses: *Africa.* Upper Cretaceous Wadi Natash Nubian Sandstone (remagnetized); Devono-Carboniferous Sabaloka Ring Complex (poorly dated, possibly remagnetized according to authors); Upper Devonian Msissi Norite (erroneous age and pole according to Salmon *et al.*, 1986); Middle Cambrian Adoudounian Andesites (contaminated by overprint according to Khattach *et al.*, 1989); Middle Cambrian Sidi-Said-Maachou Volcanics (possibly displaced terrane). *South America:* Upper Cretaceous Vinita Formation (rotated); Lower Cretaceous Quebrada Marquesa Formation (rotated); Upper Jurassic Camaraca Formation (rotated); Triassic dolerites of British Guyana (contaminated by overprint); Lower Carboniferous Yalgnarez and Guandacol formations (rotated or displaced terrane); Cambrian South Tilcara (contaminated by overprint).

Table A5. *Phanerozoic pole positions for East Gondwana*

Rock unit, location	Age		Paleopole				Reliability								Reference
	Lo	Hi	Lat.	Long.	k	α_{95}	1	2	3	4	5	6	7	Q	
I. WEST ANTARCTICA *(rotations possible)*															
Cenozoic (Paleocene, Eocene) results															
Livingston Isl. Tonalite, dikes	005	058	82,	210	—	3							x	1	Dalziel *et al.* (1973)
Ezcurra & Pt. Hennequin Volcs.	042	058	79,	228	42	5	x	x	x			x	x	5	Watts *et al.* (1984)
Snow Island Sills	044	064	61,	243	117	5	x	x	x			x	x	5	Watts *et al.* (1984)
Mesozoic results															
25 de Mayo Island Flows	025	120	85,	12	28	9						x		1	Valencio *et al.* (1979)
Lassiter Coast Intrusions	066	098	87,	49	10	12	x	x						2	Kellogg & Reynolds (1978)
Layered Intrus., Greenwich Isl.	066	098	66,	261	343	5	x		x				x	3	Watts *et al.* (1984)
Lepley Nun., Jones Mnts. Intrusions	086	092	80,	341	117	3	x	x	x				x	4	Grunow *et al.* (1991)
Jones Mnts. Granites (Remagn.?)	080	245?	72,	276	105	12								0	Scharnberger & Scharon (1972)
Byers Peninsula, Livingston Isl.	075	111	77,	312	57	8	x	x	x				x	4	Watts *et al.* (1984)
Cape Spring Intr., Graham Land	088	107	86,	297	77	9	x							1	Valencio *et al.* (1979)
Marie Byrd Land Basic Intrusions	090	110	61,	242	137	8	x	x					x	3	Scharnberger & Scharon (1972)
Marie Byrd Land Basic Intrusions	090	110	66,	61	40	5	x	x	x				x	4	Grindley & Oliver (1983)
Wide Open, Etna, Danger Isl.	075	130	75,	291	38	20			x					1	Watts *et al.* (1984)
Dustin & McNamara Isl. Intrusions	105	113	73,	30	64	8	x	x	x				x	4	Grunow *et al.* (1991)
Orville Coast Plutons, Dikes	093	120	71,	15	25	6	x	x					x	3	Kellogg (1980)
Marie Byrd Land Granites	116	118	36,	296	17	31	x						x	2	Scharnberger & Scharon (1972)
Belknap Nunatak Intrusions	120	124	49,	52	233	8	x	x	x				x	4	Grunow *et al.* (1987, 1991)
Landfall Pk., Mt. Simpson Intrusions	143	154	65,	326	123	7	x	x	x				x	4	Grunow *et al.* (1991)
Nash Hill-Pagano Nunatak Intrusions	167	183	41,	55	110	5	x	x	x			x	x	5	Grunow *et al.* (1987)
Mt. Dowling Volcanoclastics	164	182	67,	289	371	2	x		x				x	3	Grunow *et al.* (1991)
Cole, Jason Penins. Intrusives	170	180	49,	61	45	4	x	x	x				x	4	Longshaw & Griffiths (1983)
Ferrar Dol., Mt. Schopf, Ohio V.	174	184	58,	51	60	7	x		x				x	3	Kellogg (1988)
Argentine Island Lavas	144	208	44,	217	20	35							x	1	Evans (1965)
Mt. Bramhall Granodiorite	223	233	61,	297	41	19	x		x			x		3	Grunow *et al.* (1991)
Paleozoic results															
U. Heritage Grp., Ellsworth M.	505	523	4,	296	52	11	x	x		x			x	4	Watts & Bramall (1981)
Nash Hills Metasediments, Ells	505	570	7	292	28	8			x				x	2	Grunow *et al.* (1987)

II. EAST ANTARCTICA															
Mesozoic results															
Ferrar Dolerites Combined	147	163	49,	42	34	3	x	x		x	x		x	5	Irving *et al.* (1976b; 9.174)
Ferrar Dolerite, McMurdo Sound	150	166	51,	31	82	10	x	x	x		x		x	5	Funaki (1984)
Ferrar Dolerite, Wright Valley	147	163	45,	28	—	2	x	x	x		x		x	5	Funaki (1984)
Ferrar Dolerite, Mt. Cerberus	174	184	58,	44	245	3	x		x				x	3	Kellogg (1988)
Vestfjella Dikes Flows Qu. Maud	152	176	51,	23	39	4	x	x	x			x	x	5	Løvlie (1979)
Dufek Intrusion, Pensacola Mts.	163	167	57,	12	12	5	x	x				x	x	4	Beck (1972b)
Storm Peak Lavas, Qu. Alexandra	182	208	44,	52	—	7	x	x			x		x	4	Ostrander (1971)
Mt. Falla Lavas, Qu. Alexandra	182	208	54,	43	—	4	x	x			x		x	4	Ostrander (1971)
Mt. Falla Intr., Qu. Alexandra	182	208	54,	40	—	5	x	x			x		x	4	Ostrander (1971)
Vestfjella Flows Dikes Qu. Maud	160	230	42,	47	47	4		x	x		x	x	x	5	Løvlie (1979)
Paleozoic results															
Gneisses, Ongul Island	360	505	12,	207	56	17					x		x	2	Irving *et al.* (1976a: 2.73)
Lamprophyres, Taylor Valley	450	490	9,	207	75	11	x		x		x		x	4	Manzoni & Nanni (1977)
Vande Porphyry, Wright Valley	463	480	3,	204	—	5	x	x	x		x		x	5	Funaki (1984)
Sør Rondane Intrusives	460	510	28,	190	—	5	x		x		x			3	Zijderveld (1968)
Wright Valley Granitic Rocks	457	543	5,	199	—	5			x	x	x		x	4	Funaki (1984)
Mirnyi St. Charnockitic Rocks	478	526	2,	209	25	15	x	x	x		x		x	5	McQueen *et al.* (1972)
III. AUSTRALIA															
Cenozoic (Paleocene, Eocene) results															
Nerriga Province, NSW	044	048	36,	295	211	3	x	x			x		x	*	McElhinny *et al.* (1974a)
Barrington Volcano, NSW	051	053	70,	306	48	4	x	x			x	x	x	5	Wellman *et al.* (1969)
Basalts, NSW	025	060	63,	317	16	4	x	x			x		x	4	Irving *et al.* (1961)
Woy Woy S. Basalts, NSW	045	048	54,	273	—	4	x		x		x		x	4	Embleton *et al.* (1985)
Tasmanian Basalts	025	090	73,	305	29	—					x		x	2	Green & Irving (1958)
Older Volcanics, Victoria	025	060	63,	320	—	9	x	x			x	x	x	5	Mumme (1963)
Peat's Ridge Basalt, NSW	048	050	66,	278	333	2	x		x		x		x	4	Robertson (1979)
Mogo Hill Basalt, NSW	056	059	41,	310	121	5	x	x	x		x		x	5	Schmidt & Embleton (1981)
Morney Profile, Queensland	038	084	60,	299	36	4		x			x	x	x	4	Idnurm & Senior (1978)
Mesozoic results															
Mt. Dromedary Complex, NSW	083	103	56,	318	47	5	x	x		x	x	x	x	6	Robertson (1963)
Bunbury Basalt, Western Aus.	084	116	49,	341	142	6	x	x			x	x	x	5	Schmidt (1976)
Cygnet Alk. Complex, Tasmania	084	124	50,	338	776	5	x	x		x	x		x	5	Robertson & Hastie (1962)
Otway Group, Victoria	110	115	49,	329	97	4	x	x	x		x		x	5	Idnurm (1985)

Table A5. *Phanerozoic pole positions for East Gondwana – continued*

Rock unit, location	Age		Paleopole				Reliability								Reference
	Lo	Hi	Lat.	Long.	k	α_{95}	1	2	3	4	5	6	7	Q	
Noosa Heads Intr., Queensland	137	142	36,	312	48	13	x	x		x	x	x	x	6	Robertson (1963)
Dundas Breccia Pipe, Sydney B.	130	150	31,	15	182	9	x		x		x		x	4	Schmidt (1982)
Erskine Park Sill, NSW	100	160	23,	331	81	5			x		x		x	3	Robertson (1979)
Bendigo Dikes, Victoria	142	159	47,	315	19	22	x				x	x	x	4	Schmidt (1976)
N. Bondi Volcanic Neck, NSW	148	154	55,	360	—	7	x		x		x		x	4	Embleton *et al.* (1985)
Hornsby Breccia, Sydney Basin	146	178	29,	346	75	7	x	x	x		x		x	5	Schmidt & Embleton (1981)
Kangaroo Island Basalt, S. Aus.	159	179	39,	3	22	11	x						x	2	Schmidt (1976)
Tasmanian Dolerites, West Decl.	162	179	51,	355	120	3	x	x		x	x		x	5	Schmidt & McDougall (1977)
Tasmanian Dolerites, East Decl.	162	179	48,	304	66	5	x	x			x		x	*	Schmidt & McDougall (1977)
Luddenham Dike, Sydney, NSW	160	185	47,	353	30	12	x		x		x		x	4	Manwaring (1963)
Gingenbullen Dolerite, NSW	168	181	53,	324	23	8					x			1	Boesen *et al.* (1961)
Prospect & Other Intr., NSW	168	181	51,	6	38	11	x	x			x	x	x	5	Schmidt (1976)
Sydney Basin Dikes, NSW	170	190	52,	358	324	5	x	x	x		x	x	x	6	Robertson (1979)
Western Victoria Basalt	167	223	47,	358	14	18					x	x	x	3	Schmidt (1976)
Garrawilla & Nombi Extr., NSW	183	203	46,	355	52	10	x	x			x		x	4	Schmidt (1976)
Springfield Basin Seds., S. Aus.	208	230	32,	350	26	24	x				x		x	3	Schmidt *et al.* (1976)
Brisbane Tuff, Queensland	225	240	57,	323	145	6	x				x			2	Robertson (1963)
Chocolate Sh., Narrabeen, NSW	230	245	49,	340	85	10	x	x			x			*	Irving (1963)
Patonga Clayst., Narrab., NSW	230	245	30,	327	231	4	x	x	x		x	x	x	6	Embleton & McDonnell (1981)
Milton Monzonite, NSW	220	260	32,	350	14	7	x				x		x	3	Robertson (1964)
Paleozoic results															
Dundee Rhyodacite, New England	244	268	26,	316	—	11	x	x	x				x	4	Lackie (1988)
Dundee Ignimbrite, New England	244	268	37,	335	—	6	x	x	x			x		4	Lackie (1988)
Upper Marine Latites, NSW	244	252	44,	312	26	6	x	x			x		x	4	Irving & Parry (1963)
Goono Goono Secondary, NSW	230	296	32,	10	—	—					x			1	Irving (1966)
Dotswood Secondary, Queensland	230	296	46,	316	9	28					x			1	Chamalaun (1968)
New Council Quarry Dike, NSW	230	296	12,	307	118	23					x		x	2	Irving (1966)
Moonbi Lamprophyre, NSW	230	296	35,	308	100	11					x		x	2	Irving (1966)
Lower Marine Basalt, NSW	263	286	46,	302	170	5	x				x		x	3	Irving & Parry (1963)
Upper Kuttung Combined, NSW	286	320	50,	320	15	9	x	x	x	x	x			5	Irving (1966)
Paterson Toscanite, NSW	283	313	73,	327	300	3	x				x		x	3	Irving (1966)
Percy Creek Volcs., Queensland	286	333	23,	317	20	18	x						x	2	Chamalaun (1968)

Yetholme Adamellite, NSW	309	318	84,	141	57	10	x	x		x				3	Facer (1977)
Isismurra & Gilmore, NSW	320	360	73,	34	6	21	x					x		2	Luck (1973)
Alice Springs secondary overpr.	320	408	48,	293	11	8		x	x			x		*	Klootwijk (1980)
Canning Basin Limest., W. Aus.	360	374	49,	218	62	8	x	x	x		x	x	x	6	Hurley & Van der Voo (1987)
Ross River Secondary, Amadeus B.	320	420	60,	248	42	5			x				x	*	Kirschvink (1978)
Mereenie Sandstone, Amadeus B.	320	434	42,	221	15	11					x	x	x	3	Embleton (1972)
Stairway Sandstone, Amadeus B.	450	478	2,	231	25	10	x				x		x	3	Embleton (1972)
Flinders R. secondary, S. Aus.	418	525	−26,	255	107	7		x	x			x		*	Klootwijk (1980)
Kangaroo Island secondary, S. Aus.	418	525	−17,	256	8	8			x			x		2	Klootwijk (1980)
Amadeus Basin older secondary	418	525	−37,	249	6	5			x			x		2	Klootwijk (1980)
Tumblagooda Sandstone, W. Aus.	418	525	30,	211	16	9					x	x	x	3	Embleton & Giddings (1974)
Jinduckin Fm., N. Territory	478	505	13,	205	8	13	x				x	x	x	4	Luck (1972)
Upper Lake Frome Group, S. Aus.	438	576	38,	206	18	12						x	x	2	Embleton & Giddings (1974)
Ross River Section, Amadeus B.	505	523	−15,	215	—	6	x						x	2	Klootwijk (1980)
Lower Lake Frome Group, S. Aus.	523	540	8,	205	10	15	x					x	x	3	Embleton & Giddings (1974)
Areyonga Section, Amadeus Basin	523	540	33,	192	574	5	x	x	x		x	x	x	6	Klootwijk (1980)
Basal Lake Frome Grp., S. Aus.	523	540	29,	206	15	13	x	x	x			x	x	5	Klootwijk (1980)
Hudson Fm., W. Aus./N. Terr.	523	540	−18,	199	7	14	x				x		x	3	Luck (1972)
Pantapinna Fm., S. Australia	505	570	36,	209	32	17			x			x	x	3	Klootwijk (1980)
Hugh River Shale, Amadeus B.	505	570	−11,	217	44	9					x	x	x	3	Embleton (1972)
Billy Crk./Aroona Crk., S. Aus.	523	555	37,	200	11	14	x	x	x	x		x	x	6	Klootwijk (1980)
Kangaroo Isl. Red beds, S. Aus.	540	570	34,	195	9	12	x		x			x		3	Klootwijk (1980)
Aroona Dam Sediments, S. Aus.	540	570	36,	213	23	17	x					x		2	Embleton & Giddings (1974)
Antrim Pl. Basalts, W. Aus./NT	540	570	9,	160	11	13	x	x			x	x	x	5	McElhinny & Luck (1970)
Todd R. & other fms., Amadeus B.	540	570	43,	160	15	7	x	x	x	x	x	x	x	7	Kirschvink (1978)
Hawker Grp., South Australia	540	570	21,	195	12	11	x	x	x	x		x		5	Klootwijk (1980)
IV. INDIA															
Cenozoic (Paleocene, Eocene) results															
Basal Murree Fm., Jammu, India	038	050	35,	145	40	4	x	x	x				x	*	Klootwijk *et al.* (1986a)
Mysore Dikes, S. India	020	076	43,	281	17	23					x		x	2	Hasnain & Qureshy (1971)
Sanjawi Ls., Pakistan	037	066	55,	304	14	15	x		x				x	3	Klootwijk *et al.* (1981)
Basal Kalakot Fm., Jammu, India	050	056	31,	319	29	9	x		x				x	3	Klootwijk *et al.* (1986a)
Dhak, Chichali Comb., Pakistan	050	065	54,	253	18	22	x		x			x	x	4	Klootwijk *et al.* (1986b)
Sonhat Coal Basin Sill, C. India	057	063	37,	285	395	2	x	x	x				x	4	Klootwijk (1974)
Upper Brewery Ls., Kirthar, Pak.	058	066	53,	299	9	22	x		x			x	x	4	Klootwijk *et al.* (1981)

Table A5. *Phanerozoic pole positions for East Gondwana – continued*

Rock unit, location	Age		Paleopole				Reliability								Reference
	Lo	Hi	Lat.	Long.	k	α_{95}	1	2	3	4	5	6	7	Q	
Mount Pavagarh Deccan Traps	058	066	39,	286	38	5	x	x			x		x	4	Verma & Mital (1974)
Mahabaleshwar Deccan Traps	059	067	38,	263	16	7	x	x			x	x	x	5	Kono *et al.* (1972)
U. Deccan Traps Comb'd, W. Ghats	059	067	39,	283	920	3	x	x	x		x	x	x	6	Wensink & Klootwijk (1971)
L. Deccan Traps Comb'd, W. Ghats	062	070	33,	275	288	5	x	x	x		x		x	5	Wensink & Klootwijk (1971)
Gondwana Dikes, Damodar Valley	016	104	33,	291	37	7		x					x	2	Athavale & Verma (1970)
Deccan Trap Superpole	061	074	37,	281	22	2	x	x	x		x	x	x	6	Vandame *et al.* (1991)
Mesozoic results															
L. Deccan Traps Comb'd, Nagpur	064	072	28,	289	36	8	x	x	x		x		x	5	Wensink (1973)
Satyavedu Sandstone, SE India	050	110	26,	293	80	4					x		x	2	Mital *et al.* (1970)
Fort Munro Fm., Rakhi Gaj, Pak.	066	075	28,	295	13	27	x		x			x	x	4	Klootwijk *et al.* (1981)
Tirupati Sandstone, Godavary V.	050	110	28,	287	—	4		x			x	x	x	4	Verma & Pullaiah (1967)
Rajmahendri Traps, SE India	040	150	22,	312	—	4		x			x		x	3	Athavale *et al.* (1970)
Mughal Kot Fm., Rakhi Gaj, Pak.	066	084	4,	290	50	3	x	x	x				x	4	Klootwijk *et al.* (1981)
Goru Fm.-Parh Ls., Brewery, Pak.	075	119	15,	295	17	4	x	x	x				x	4	Klootwijk *et al.* (1981)
Goru Fm.-Parh Ls., Sanjawi, Pak.	075	119	18,	297	33	3	x	x	x	x			x	5	Klootwijk *et al.* (1981)
Rajmahal Traps Combined, Bihar	113	121	8,	297	85	3	x	x	x		x	x	x	6	Klootwijk (1971)
Sylhet Traps, E. India	075	170	16,	300	—	7		x			x		x	3	Athavale *et al.* (1963)
Chiltan Ls., Kirthar R., Pak.	152	163	1,	312	17	10	x		x			x	x	4	Klootwijk *et al.* (1981)
Loralai Ls., Sanjawi, Pakistan	163	187	2,	308	9	8	x		x				x	3	Klootwijk *et al.* (1981)
Pachmarhi Red beds, Satpura	193	230	10,	310	33	5	x	x	x		x		x	5	Wensink (1968)
Parsora Sandstone, East C. India	190	240	30,	305	200	6		x			x			2	Bhalla & Verma (1969)
Mangli Beds, Wardha/Godavary V.	230	245	−7,	304	43	5	x		x		x		x	4	Wensink (1968)
Panchet Clay Beds, Damodar V.	230	260	−8,	301	49	6	x	x	x		x		x	5	Klootwijk (1974)
Wardha Valley Red Beds	230	260	−4,	309	24	7	x	x	x		x	x	x	6	Klootwijk (1975)
Paleozoic results															
Kamthi Ss., Wun, Wardha/Godavary	245	263	−21,	310	63	2	x	x	x		x		x	*	Wensink (1968)
Kamthi Ss., Tadoba, Godavary V.	245	263	4,	283	411	2	x				x		x	3	Wensink (1968)
Chhidru Fm., Lunda G., Pakistan	235	266	24,	281	10	20			x			x		2	Klootwijk *et al.* (1986b)
Speckled (Warccha) Ss., Salt R.	245	320	−13,	317	50	7		x	x				x	*	Wensink (1975)
Alozai Fm., Quetta, Pakistan	245	360	−18,	291	6	12			x		x		x	3	Klootwijk *et al.* (1981)
Talchir Series, Nagpur area	286	320	−32,	314	109	3	x				x		x	*	Wensink & Klootwijk (1968)

Jutana Fm., Saidu V., Pakistan	530	554	21,	231	9	11	x		x			x	x	4	Klootwijk *et al.* (1986b)
Salt Pseudom. (Baghanw.) Pak.	523	570	27,	214	176	5	x	x				x	x	4	Wensink (1972)
Purple (Khewra) Ss., Salt Range	540	570	28,	212	15	12	x					x	x	3	McElhinny (1970)
Upper Bhander Sandstone, MP	505	620	32,	199	200	6					x		x	2	Athavale *et al.* (1972)
Upper Bhander Ss., Gwalior	505	620	49,	214	138	6		x	x		x		x	4	Klootwijk (1973)
Upper Bhander Red beds, MP	505	620	51,	223	37	11		x			x		x	3	McElhinny *et al.* (1978)
V. MADAGASCAR															
Mesozoic results															
Androy Volcanics	066	076	64,	243	78	8	x	x			x		x	4	Andriamirado (1971)
Southeast Coast Volcanics	070	075	66,	216	82	4	x	x			x		x	4	Andriamirado (1971)
Tamatave Dolerites	070	082	60,	212	286	3	x	x					x	3	Andriamirado (1971)
Mangoky-Onilahy Volcanics	084	090	74,	253	28	9	x	x			x		x	4	Andriamirado (1971)
Mailaka Volcanics	088	092	70,	243	49	7	x	x			x		x	4	Andriamirado (1971)
Antanimena Volcanics	088	092	66,	230	79	5	x	x			x		x	4	Andriamirado (1971)
Isalo Group sediments	163	240	74,	277	78	6		x			x		x	3	Embleton & McElhinny (1975)
Sakamena Group sediments	228	263	66,	292	96	6		x			x	x		3	McElhinny *et al.* (1976)
Paleozoic Results															
Sakoa Group sediments	263	320	53,	261	94	8		x			x			2	McElhinny *et al.* (1976)

The entries in this Table are for cratonic eastern Gondwana, including its thrust belt margins in so far as no suspicions exist about relative displacements; obviously rotations can occur there and this mandates that reliability criterion 5 can not be checked as being met. East Gondwana includes the cratonic nuclei of Madagascar, India, Australia and Antarctica and the East Australian foldbelts for post-Devonian time. Pole coordinates have not been rotated; they are given for each cratonic element in its own geographic coordinate grid of the present-day.

The reliability criteria (1–7) and Q are explained in the text. An asterisk is placed in the Q column when the result is not to be used for apparent polar wander path construction, as explained below; a 'x' mark under any of the seven reliability columns means that this particular criterion is being met.

Abbreviations: Lat. = latitude, Long. = longitude, k and α_{95} are the precision parameter and radius of the cone of 95% confidence, respectively (Fisher, 1953).

*Poles that would have $Q \geq 3$ but are excluded from mean pole determinations and other compilations, are listed below with the reasons in parentheses: *Australia*. Eocene Nerriga Province (secular variation not averaged); Middle Jurassic Tasmanian dolerites (inconsistent result with other dolerites); Lower Triassic Chocolate Shale (contaminated by overprint); Devono-Carboniferous Alice Springs, Ross River, Ordovician Flinders Range secondary components (remagnetized, poorly dated). *India*. Eocene Lower Murree Formation (anomalous result); Upper Permian Kamthi Sandstone at Wun, Permo-Carboniferous Speckled Sandstone and Talchir Series (remagnetized or contaminated by overprint).

Table A5. (Continued.) *Pre-Carboniferous pole positions for the Eastern Australian foldbelt*

Rock unit, location	Age		Paleopole				Reliability								Reference
	Lo	Hi	Lat.	Long.	k	α_{95}	1	2	3	4	5	6	7	Q	
Worange Point Fm., NSW	360	370	69,	208	11	6	x	x	x			x	x	5	Thrupp *et al.* (1991)
Hervey Group, NSW	352	387	54,	204	14	15	x	x	x			x	x	5	Li *et al.* (1988)
Mulga Downs Grp., NSW	320	387	54,	276	18	8						x		1	Embleton (1977)
Nethercote Basalts, NSW	363	377	65,	160	29	14	x					x	x	3	Green (1961)
Comerong Volcanics, NSW	367	387	77,	151	42	7	x	x	x	x		x	x	6	Schmidt *et al.* (1986)
Catombal Grp., NSW	360	394	49,	132	33	10		x					x	2	Williamson & Robertson (1976)
Lochiel Formation, NSW	360	394	28,	92	18	18							x	1	Embleton & Shepherd (1977)
Wellington-Cowra area, NSW	374	408	44,	73	11	29	x					x	x	3	Goleby (1981)
Snowy River Volcanics, Victoria	387	408	74,	43	26	10	x	x	x	x		x	x	6	Schmidt *et al.* (1987)
Ainslie Volcanics, ACT	360	420	72,	161	33	11							x	1	Luck (1973); Goleby (1981)
Bowning Group, NSW	387	408	64,	225	37	10	x	x		x				3	Luck (1973)
Mugga Mugga Porphyry, ACT	408	425	63,	188	29	7	x						x	2	Briden (1966); Goleby (1981)
Cowra Trough/Molong H., NSW	408	428	47,	177	29	9	x	x		x		x	x	5	Goleby (1981)
Yass Area Combined, NSW	408	428	54,	91	18	10	x	x				x	x	4	Luck (1973)
Cowra Trough/Molong H., NSW	414	438	38,	215	38	8	x	x		x		x	x	5	Goleby (1981)
Molong High fms., NSW	438	468	−6,	199	57	6	x	x		x		x	x	5	Goleby (1981)
Walli & Mt. Pleasant And., NSW	478	505	11,	183	—	—	x					x	x	3	Goleby (1981)
Dundas Grp., Tasmania	505	523	23,	193	—	12	x			x			x	3	Giddings & Embleton (1974)

These paleopoles are listed separately to highlight that they may not be representative of cratonic Gondwana, but may be indicative of displaced terranes in the eastern fold belts of Australia. For other explanations see above part of Table A5.

Table A6. *Phanerozoic pole positions for the North and South China blocks*

Rock unit, location	Age		Paleopole				Reliability								Reference
	Lo	Hi	Lat.	Long.	k	α_{95}	1	2	3	4	5	6	7	Q	
I. NORTH CHINA BLOCK, INCLUDING MONGOLIA AFTER EARLY JURASSIC TIME															
Linju County Basalts, Shandong	013	018	76,	178	230	6	x	x				x	x	4	Lin (1984); Lin *et al.* (1985)
Pingzhuang, Lining Bas. Inner Mong.	014	025	89,	29	26	8	x	x	x	x	x	x	x	7	Zhao *et al.* (1990)
Zuoyun Fm., Datong, Shanxi	067	098	80,	170	474	4	x	x	x		x	x		5	Zheng *et al.* (1991)
Outer Mongolia Basalts, Seds.	065	144	83,	222	19	4		x	x					2	Pruner (1987)
Early Cretac. Sediments, Ordos Basin	098	144	85,	310	—	4	x	x						2	Cheng *et al.* (1988)
Pingzhuang Tuffs, Dalai Lake Volcs.	100	144	83,	250	138	6	x	x	x	x	x	x	x	7	Zhao *et al.* (1990)
Qingshan Fm. Andesites, Shandong	120	144	69,	201	40	12	x	x	x			x	x	5	Lin (1984); Lin *et al.* (1985)
Laiyang Fm., Shandong	144	163	71,	226	38	8	x	x	x				x	4	Lin (1984); Lin *et al.* (1985)
Manzhouli Andes., Tuffs, Inner M.	139	176	62,	225	67	5	x	x	x	x	x	x	x	7	Zhao *et al.* (1990)
Hua-an, Zalantum And., Seds., Inner M.	139	176	73,	255	51	8	x	x	x	x	x	x	x	7	Zhao *et al.* (1990)
Balinzuo Qi Tuffs, Inner Mongolia	139	176	74,	238	39	10	x		x		x	x	x	5	Zhao *et al.* (1990)
Shizhuangfang Granite, Inner Mong.	144	187	63,	221	656	3			x		x		x	3	Zhao *et al.* (1990)
Mengying Fm. Sandst., Shandong	144	163	74,	216	57	4	x						x	2	quoted in Lin (1984)
Santai Fm., Shandong	144	187	72,	202	194	7		x	x				x	3	Lin (1984); Lin *et al.* (1985)
Middle Jurassic Seds., Ordos Basin	163	187	80,	249	113	12	x	x						2	Cheng *et al.* (1988)
Wulian Sandstone, Shandong	163	186	82,	253	41	10	x							1	quoted in Lin (1984)
Yungan Fm., Datong, Shanxi	163	187	76,	200	443	6	x		x		x			3	Zheng *et al.* (1991)
Yanniou Valley, Qiliang, Qinghai	208	245	22,	41	12	19			x				x	2	Li & Zhang (1986)
Middle Triassic Seds., Ordos Basin	230	240	49,	13	—	4	x	x			x		x	4	Cheng *et al.* (1988)
Shiqianfeng Fm. Red Beds, Liaoning	240	245	35,	39	15	18	x					x	x	3	Lin (1984); Lin *et al.* (1985)
Late Permian Sediments, Ordos Basin	240	258	42,	14	103	12	x	x			x	x	x	5	Cheng *et al.* (1988)
Shihezi Fm., Shanxi & Hebei	245	258	57,	355	35	21	x		x	x	x	x	x	6	Zhao & Coe (1987)
U. Shihezi Fm., Shanxi	245	258	38,	6	—	—	x		x				x	3	Lin (1984); Lin *et al.* (1985)
U. Shihezi Fm., Shanxi	245	258	49,	359	—	4	x	x			x		x	4	Zhang (1984)
U. Shihezi & Shiqianfen Fms., Shanxi	245	258	44,	358	40	7	x	x	x		x	x	x	6	McElhinny *et al.* (1981)
Yanniou Valley, Qiliang, Qinghai	360	374	35,	222	291	4	x		x				x	*	Li & Zhang (1986)
Majiagou Fm., Limestone, Liaoning	458	478	43,	333	53	11	x					x	x	3	Lin (1984); Lin *et al.* (1985)
Yeli Fm. Limestone, Shandong	478	505	26,	17	49	11	x						x	2	Lin (1984)
Carbonates, Liaoning & Shandong	505	570	15,	299	38	10		x					x	2	Lin (1984); Lin *et al.* (1985)

Table A6. *Phanerozoic pole positions for the North and South China blocks – continued*

Rock unit, location	Age		Paleopole				Reliability								Reference
	Lo	Hi	Lat.	Long.	k	α_{95}	1	2	3	4	5	6	7	Q	
II. SOUTH CHINA BLOCK, TAIWAN															
Fung Mt. Basalts, Nanjing, Jiangsu	000	002	81,	330	—	6	x						x	2	Liu Chun *et al.* (1975)
Lin Yuen Mntn. Basalts, Nanjing	000	005	77,	193	—	14	x						x	2	Liu Chun *et al.* (1975)
Plio-pleistocene Volcanics, Taiwan	001	005	82,	256	144	8	x	x				x	x	4	Hsu I-Chi *et al.* (1966)
Tuluanshan Fm., Taiwan Coastal Range	003	022	43,	161	11	17	x					x	x	*	Yang *et al.* (1982)
Lanhsu Island Lavas, Taiwan	005	021	75,	292	40	7	x	x				x	x	4	Yang *et al.* (1982)
Lutao Island Pyroclastics, Taiwan	005	024	81,	188	230	5	x						x	2	Yang *et al.* (1982)
Chiapanshan Area Basalts, Taiwan	005	024	82,	294	—	13	x	x				x	x	4	Hsu I-Chi *et al.* (1966)
Shiukuran River Andesites, Taiwan	005	024	80,	293	—	16	x						x	2	Hsu I-Chi *et al.* (1966)
Tertiary Basalts & Red Beds Combined	002	058	74,	37	148	8		x				x	x	3	Lin (1984); Lin *et al.* (1985)
Sec'ry, Pz. Seds., Jiangsu, Hubei	005	025?	75,	41	495	11		x	x				x	3	Kent *et al.* (1987)
Lunan and Kunming Basins, Yunnan	032	060	84,	326	—	—						x	x	2	Liang *et al.* (1986)
Xiaoba & Leidashu Fmns., Sichuan	055	098	81,	297	27	9			x	x	x	x		4	Kent *et al.* (1986)
Puko & Yanzijing Fmns., Nanjing	065	098	76,	173	42	8	x	x	x	x	x		x	6	Kent *et al.* (1986)
Ya'an (rotated area?), Sichuan	065	144	76,	275	22	11		x	x	x		x	x	*	Otofuji *et al.* (1990)
South China Red Beds Combined	065	144	59,	159	82	7					x		x	2	Chen *et al.* (1965)
Sec'ry, Pz. Seds., Jiangsu, Hubei	065	144?	73,	180	38	7		x	x			x		3	Kent *et al.* (1987)
Tuffs & Red Beds, Zhejiang, Xinchang	098	144	77,	228	120	6	x	x	x			x	x	5	Lin (1984)
Ignimbrites & Sandstones, Zhejiang	144	187	73,	214	29	13		x	x					2	Lin (1984); Lin *et al.* (1985)
Ziliujing Fm., Guizhou	163	187	68,	185	25	12	x					x		2	Lin (1984); Lin *et al.* (1985)
Ls., Siltst., Nanjing, Jiangsu	208	245	45,	224	12	9			x		x	x	x	4	Opdyke *et al.* (1986)
Huachi Fm., Guizhou	225	240	55,	209	—	6	x		x	x	x	x	x	6	Chan *et al.* (1984)
Dulaying Dolomite, Guizhou	230	240	30,	183	—	3	x		x				x	3	Lin (1984)
Guiyang Carbonates, Guizhou	230	240	55,	210	95	6	x						x	2	Lin (1984); Lin *et al.* (1985)
Yelang Fm., Sichuan & Guizhou	240	245	46,	219	23	11	x	x	x	x	x	x	x	7	Opdyke *et al.* (1986)
Feixianguan Fm., Sichuan	240	245	33,	230	—	19	x		x		x		x	4	Chan *et al.* (1984)
Hechuan HPT, HT Sec. Seds., Sichuan	240	245	39,	211	30	4	x	x	x	x		x	x	6	Steiner *et al.* (1989)
Changxing Fm., Limestones, Zhejiang	230	258	51,	205	38	7		x	x				x	3	Lin (1984); Lin *et al.* (1985)
Xuanwei Fm., Sichuan	230	260	52,	243	15	—			x		x		x	3	McElhinny *et al.* (1981)
Sichuan Limestones	235	255	48,	225	—	5	x	x	x		x	x	x	6	Heller *et al.* (1988)
Shangsi GPW Section Seds., Sichuan	242	255	47,	226	33	3	x	x	x		x	x	x	6	Steiner *et al.* (1989)
Chongging LP Section Seds., Sichuan	244	251	46,	225	28	5	x	x	x		x	x	x	6	Steiner *et al.* (1989)

Emeishan Basalts, N polar., Yunnan	245	258	50,	241	46	6	x	x	x	x	x		x	6	Fang *et al.* (1990a)
Emeishan Basalts, R polar., Yunnan	245	258	23,	206	—	—	x		x		x		x	*	Fang *et al.* (1990a)
Emeishan Basalts, N, Sichuan	245	258	51,	248	27	3	x	x			x		x	4	Zhang (1984)
Emeishan Basalts, N+R, Daqiao Section	245	258	54,	242	37	10	x	x	x		x	x	x	6	Huang *et al.* (1986)
Emeishan Basalts, N+R, Dapinzi	245	258	52,	226	6	25	x		x		x	x	x	5	Huang *et al.* (1986)
Emeishan Basalts, R, Lijiao Section	245	258	25,	204	26	25	x		x		x		x	4	Huang *et al.* (1986)
Emeishan Basalts, N, Ls., Sichuan	245	258	54,	242	—	20	x		x		x		x	4	Chan *et al.* (1984)
Emeishan Basalts, N, Sichuan	245	258	53,	253	27	6	x	x	x	x	x		x	6	McElhinny *et al.* (1981)
Emeishan Basalts, N, Meigu, Sichuan	245	258	42,	229	—	30	x				x		x	3	Zhou *et al.* (1986)
Emeishan Basalts, R, Miyi Section	245	258	24,	221	201	9	x				x		x	3	Zhou *et al.* (1986)
Emeishan Basalts, N, Sichuan	245	258	49,	251	39	4	x	x	x		x		x	5	Zhao & Coe (1987)
Emeishan Basalts, R, Zhijin, Guizhou	245	258	29,	235	41	12	x						x	2	Lin (1984); Lin *et al.* (1985)
Chuanshan Fm., Limestones, Zhejiang	286	296	22,	225	28	18	x		x				x	3	Lin (1984); Lin *et al.* (1985)
Zhuzhangwu Fm., Red Beds, Zhejiang	320	360	25,	221	12	28	x		x					2	Lin (1984); Lin *et al.* (1985)
Devonian Limestones, Yunnan	360	408	−9,	190	13	7		x	x		x	x	x	5	Fang *et al.* (1989)
Cuifengshan Fm., Qujing, Yunnan	387	408	48,	228	—	21	x				x			2	Liu & Liang (1984)
Yulongshi Fm., Qujing, Yunnan	408	421	51,	229	—	20	x				x			2	Liu & Liang (1984)
Silurian Sediments, Sichuan, Yunnan	408	438	5,	195	41	6		x	x		x	x		4	Opdyke *et al.* (1987a)
Hongshiya Fm., Yunnan	478	500	−39,	236	21	17	x		x	x	x		x	5	Fang *et al.* (1990b)
Hetang Fm., Zhejiang	540	570	3,	196	29	17	x		x		x			3	Lin *et al.* (1985)
Tianheban Fm., Hubei	540	570	7,	190	29	23	x		x		x			3	Lin *et al.* (1985)
Shuijingtuo Fm., Shennongjia, Sichuan	540	570	59,	218	—	14	x	x			x			3	Zhang & Zhang (1985)
Shuijingtuo Fm., Zhejiang	554	594	32,	180	227	6	x		x		x		x	*	Lin *et al.* (1985)
Meishucun Fm., Yunnan	560	615	−9,	211	17	10		x			x			2	Lin *et al.* (1985)
Meishucun Section, Yunnan	560	615	69,	271	9	7			x		x	x	x	4	Fang *et al.* (1988/9)

The entries in this Table are for the stable blocks ('cratons') of North China and South China, including their disturbed margins in so far as no suspicions exist about relative displacements; obviously rotations can occur there and this mandates that reliability criterion 5 cannot be checked as being met. The North China Block (NCB) includes Mongolia for times after the Early Jurassic, whereas results from Korea (traditionally included in the NCB) have been listed separately. The reliability criteria are explained in the text (see also Table 4.1). A '*' is placed in the Q column when the result is not to be used for apparent polar wander path construction. For other explanations see Table A1.

*Poles that would have Q ≥ 3 but are excluded from mean pole determinations and other compilations, are listed below with the reasons in parentheses. *North China Block*: Upper Devonian Yanniou Valley (possibly rotated or displaced). *South China Block*: Upper Tertiary Tuluanshan Formation (rotated); Upper Cretaceous Ya'an (rotated); Upper Permian Emeishan Basalts, reversed results of Fang *et al.* (too few samples); Lower Cambrian Shujingtao Formation (too few samples, possibly contaminated by overprint).

Table A7. *Phanerozoic pole positions for the western Mediterranean area*

Rock unit, location	Age		Pole position				D,	Lat.	Reliability								Reference
	Lo	Hi	Lat.	Long.	k	α_{95}	40°N,	4°W	1	2	3	4	5	6	7	Q	
I. IBERIAN MESETA, LATE CARBONIFEROUS AND YOUNGER																	
Lisbon Volcs. Overprint, Portugal	030	076	50,	215	18	7	336,	6			x		x		x	*	Storetvedt *et al.* (1987)
East Ebro Basin N-polarity	030	039	69,	159	—	—	6,	20	x		x		x		x	4	Pascual & Parès (1990)
East Ebro Basin R-polarity	030	039	65,	121	—	—	22,	23	x		x		x		x	4	Pascual & Parès (1990)
Ebro-Basin, NE Spain	036	061	70,	113	114	6	20,	29	x	x	x			x	x	5	Parès *et al.* (1988a)
Burgau Intrusion, Portugal	020	080?	50,	216	545	2	335,	7			x				x	2	Storetvedt *et al.* (1990)
Salema Intrusion, Portugal	062	070	69,	228	94	2	342,	26	x		x				x	3	Storetvedt *et al.* (1990)
Sines Complex, Portugal	061	081	52,	91	92	9	44,	28	x	x	x				x	*	Storetvedt *et al.* (1987)
Monchique Intrusion, Dikes, Portugal	070	074	77,	163	39	5	3,	27	x	x	x		x		x	5	Storetvedt *et al.* (1990)
Monchique Syenite, Portugal	070	074	73,	166	73	7	3,	23	x		x		x		x	4	Van der Voo (1969)
Lisbon Volcanics, Portugal	069	076	73,	197	66	3	353,	24	x	x	x	x	x		x	6	Van der Voo & Zijderveld (1971)
Sintra Complex, mafics only, Portugal	074	099	65,	176	180	5	0,	15	x	x	x				x	4	Storetvedt *et al.* (1987)
Sintra Complex Granites, Portugal	074	099	77,	174	46	8	1,	27	x	x	x			x	x	5	Van der Voo (1969)
Aptian Sediments, Portugal	113	119	77,	226	32	2	348,	31	x		x		x		x	4	Galdeano *et al.* (1989)
Wealden, Vega de Pas, Cantabria	113	124	67,	202	—	1	350,	19	x	x	x				x	4	VandenBerg (1980)
Hauterivian-Barremian Seds., Portugal	119	131	57,	275	100	4	318,	37	x		x	x	x		x	5	Galdeano *et al.* (1989)
Wealden Series, N. Central Spain	131	156	41,	237	186	5	318,	8	x	x	x	x		x	x	6	Schott & Peres (1987a)
Tazones Red Beds, Danois Block	144	163	86,	78	—	3	5,	40	x		x					2	VandenBerg (1980)
Zona de Enlace, Catalunya	144	187	79,	175	—	8	0,	29				x				1	Parès *et al.* (1988b)
Aguilón Section, N. Central Spain	156	160	55,	255	240	6	321,	26	x	x	x	x		x	x	6	Steiner *et al.* (1985)
Messejana Dike, W. Spain & Portugal	159	201	71,	236	36	7	341,	29		x	x		x		x	4	Schott *et al.* (1981)
Alcázar Red Beds, Central Spain	208	233	63,	178	47	3	359,	13	x	x	x		x		x	5	Van der Voo (1967)
Catalunya Red Beds, NE Spain	230	240	51,	181	205	6	357,	1	x	x	x	x		x	x	6	Parès *et al.* (1988a)
Catalunya L & M. Triassic, NE Spain	230	245	57,	176	32	17	0,	7	x		x				x	3	Parès *et al.* (1988b)
Villaviciosa Red Beds, Danois Block	230	260	55,	152	—	4	14,	7		x	x			x	x	*	VandenBerg (1980)
West Cantabrian Red Beds	230	263	53,	226	405	6	332,	13		x	x				x	3	VandenBerg (1980)

Cantabrian Red Beds, NW Spain	243	258	47,	215	107	7	335,	4	x	x	x	x		x	x	6	Schott & Perès (1987b)
Asturias Red Beds, NW Spain	243	258	51,	202	70	9	344,	4	x	x	x				x	4	Schott & Perès (1987b)
Coastal Catalan Dikes, NE Spain	263	286	49,	202	83	9	343,	2	x	x	x	x			x	5	Parès (1988)
Rio de Gallo Section, C. Spain	263	286	49,	193	—	9	349,	0	x		x				x	3	Turner *et al.* (1989)
Viar Intrus., Red beds, S. Meseta	263	296	43,	211	110	6	335,	−1	x		x		x		x	4	Van der Voo (1969)
Buçaco Red Beds, N. Portugal	268	291	36,	211	332	7	332,	−8	x		x	x	x		x	5	Van der Voo (1969)
Atienza Andesites, Central Spain	275	299	42,	208	25	14	337,	−3	x	x	x		x		x	5	Hernando *et al.* (1980)
Atienza Andesites, Central Spain	275	299	36,	203	31	12	338,	−10	x	x	x		x		x	5	Van der Voo (1967)
II. MESETA, EARLY CARBONIFEROUS OR OLDER, AND IBERIAN PYRENEES (internal rotations possible)																	
Deva/Zumaya Flysch, Bilbao Area	098	109	73,	179	—	3	359,	23	x	x	x				x	4	VandenBerg (1980)
Guernica Carniolas, Bilbao Area	187	225	39,	116	—	2	43,	6			x				x	2	VandenBerg (1980)
Garralda Red Beds, Navarra	208	245	55,	196	—	—	349,	7			x	x		x	x	4	Van der Voo (1969)
Anayet Andesites & Red Beds, Huesca	208	263	51,	207	—	10	341,	5							x	1	Van der Lingen (1960)
Rio Aragón Andesites, Huesca	208	263	51,	229	65	4	329,	12							x	1	Schwarz (1963)
Nogueras Sheet 7 only, Rotated	230	260	53,	151	—	12	15,	6			x	x		x	x	4	Bates (1989)
Seo de Urgel Sediments Comb., E. Pyr.	230	286	54,	218	22	8	336,	11			x			x	x	3	Van Dongen (1967)
Pyrenean Red Beds I, European Side?	230	245	50,	145	48	18	19,	4	x		x				x	3	Schott & Peres (1988)
Pyrenean Red Beds II, Rotated	230	245	18,	99	42	9	68,	2	x	x	x	x		x	x	6	Schott & Peres (1988)
Pyrenean Red Beds III, Rotated	230	258	−14,	62	41	20	116,	8			x				x	2	Schott & Peres (1988)
Cincovillas Massif, Flows, Dikes	245	268	40,	126	25	7	36,	2	x	x	x	x			x	5	Van der Voo & Boessenkool (1973)
Sierra del Cadi Andesites, E. Pyr.	256	286	49,	197	63	6	346,	1	x	x	x	x			x	5	Van Dongen (1967)
Col du Somport (Strain rem'd), W. Pyr.	268	286	42,	211	28	4	335,	−2	x	x	x	x			x	5	Cogné (1987)
Beja Gabbro, S. Portugal	315	337	7,	149	15	16	34,	−37	x		x			x	x	4	Ruffet (1990)
Alba Griotte (S. Emiliano), Asturias	315	352	4,	249	53	2	284,	−10	x	x	x	x			x	5	Bonhommet *et al.* (1981)
Alba & Candas Red Beds, Asturias	315	352	35,	175	39	9	1,	−15	x	x	x	x			x	5	Perroud (1983)
San Pedro Red Beds, Asturias	360	421	3,	235	30	10	293,	−21		x	x	x			x	4	Perroud & Bonhommet (1984)
Almadén Volcanics, S. Meseta	421	438	8,	118	12	14	62,	−18	x	x	x					3	Perroud *et al.* (1991)
Cabo de Peñas Volcs., Asturias	438	476	−22,	165	50	6	31,	−70	x		x				x	3	Perroud (1983)
Buçaco Red Beds, N. Portugal	458	523	−23,	182	83	6	342,	−72			x				x	2	Perroud & Bonhommet (1981)

Table A7. *Phanerozoic pole positions for the western Mediterranean area – continued*

Rock unit, location	Age Lo	Age Hi	Pole position Lat.	Pole position Long.	k	α_{95}	D, 40°N,	Lat. 4°W	Reliability 1	2	3	4	5	6	7	Q	Reference
III. BETIC CORDILLERA AND BALEARES (internal rotations possible)																	
Agost Section, C29N, Alicante	065	066	77,	122	17	8	12,	32	x		x				x	3	Groot *et al.* (1989)
Agost Section, C29R, Alicante	066	067	36,	100	72	4	54,	13	x		x				x	*	Groot *et al.* (1989)
Agost Section, C30N, Alicante	066	067	81,	159	61	5	3,	31	x		x					2	Groot *et al.* (1989)
Agost Red Limestones, Alicante	077	082	66,	269	—	5	329,	37	x	x	x				x	4	VandenBerg (1980)
Cehegin Section VCY, Murcia	132	142	64,	101	52	4	29,	29	x	x	x			x	x	5	Ogg *et al.* (1988)
Cehegin Section VCZ, Murcia	132	142	30,	74	39	4	73,	27	x	x	x			x	x	5	Ogg *et al.* (1988)
Sierra de Lugar, Murcia	139	158	2,	54	13	3	110,	25	x	x	x			x	x	5	Mazaud *et al.* (1986)
Sierra Gorda Tithonian Carbonates	144	152	53,	244	79	2	324,	20	x	x	x			x	x	5	Ogg *et al.* (1984)
Carcabuey Tithonian Section	144	152	61,	105	160	2	31,	26	x	x				x	x	4	Ogg *et al.* (1984)
Sierra Gorda Kimmeridgian Section	152	156	50,	242	25	4	322,	17	x	x	x			x	x	5	Ogg *et al.* (1984)
Carcabuey Kimmeridgian Section	152	156	63,	121	56	3	24,	22	x	x				x	x	4	Ogg *et al.* (1984)
Subbetic Limest. & Volcs., Granada	138	187	52,	107	16	17	38,	20			x	x		x	x	4	Osete *et al.* (1988)
Mid-Upper Jurassic Limest., Mallorca	144	187	55,	98	—	7	39,	26		x	x	x			x	4	Freeman *et al.* (1989)
Carcabuey-S. Harana Subbetic Sections	172	183	37,	81	85	13	62,	26	x	x	x	x		x	x	6	Steiner *et al.* (1987)
Iznaloz Toarcian Sec., Jaen Subbet.	179	188	20,	296	12	5	267,	35	x	x	x			x	x	5	Galbrun *et al.* (1989)
Subbetic Sediments, Cieza-Cehegin	225	235	32,	241	7	9	310,	4	x		x	x		x	x	5	Mäkel *et al.* (1984)
Malaguide Red Sst., Aledo Area	230	245	33,	234	8	10	315,	1	x		x				x	3	Mäkel *et al.* (1984)
Buntsandstein, Serra Tram., Mallorca	235	245	55,	165	—	6	6,	6	x						x	2	Parès *et al.* (1988b)

Rock unit, location	Age Lo	Age Hi	Pole position Lat.	Pole position Long.	k	α_{95}	D, 40°N,	Lat. 9°E	Reliability 1	2	3	4	5	6	7	Q	Reference
IV. CORSICA AND SARDINIA																	
Pre-Tort. Logudoro Basalts, Sardinia	011	030	65,	260	13	18	333,	29	x				x	x	x	4	Bobier & Coulon (1970)
Extrusiv., Ignimbr., Andes., Sardinia	018	020	81,	189	—	3	0,	31	x	x			x	x	x	5	Montigny *et al.* (1981)

Extrusiv., Ignimbr., Andes., Sardinia	005	038	67,	269	19	4	333,	33	x	x	x		x	x	x	6	Edel (1979)
St. Florent Molasse, Corsica	015	021	70,	240	99	9	343,	26	x	x	x			x	x	5	Vigliotti & Kent (1990)
Coarse Molasse, Site E, Corsica	015	021	33,	292	124	3	289,	30	x		x					2	Vigliotti & Kent (1990)
Castelsardo Volcanics, Sardinia	015	030	56,	279	16	17	319,	32	x		x		x	x	x	5	De Jong *et al.* (1973)
Mid-Tertiary Volcanics, Sardinia	020	034	64,	270	—	2	329,	32	x	x			x	x	x	5	Montigny *et al.* (1981)
Alghero Trachyandesites, Sardinia	022	037	54,	263	20	11	322,	23	x	x	x		x	x	x	6	De Jong *et al.* (1969)
Olig./Miocene Combined, Sardinia	017	037	62,	257	14	10	331,	26	x	x	x		x	x	x	6	Manzoni (1974)
Belagne Calcarenite, Corsica	044	052	63,	257	80	4	332,	26	x		x		x		x	4	Vigliotti & Kent (1990)
M. Albo/Baunei Limestones, E. Sardinia	152	169	40,	292	11	6	297,	33	x	x	x		x	x	x	6	Horner & Lowrie (1981)
M. Maiore Musschelkalk, C. Sardinia	230	240	33,	285	999	4	293,	25	x	x	x	x	x		x	6	Horner & Lowrie (1981)
Combined Corsica-Sardinia Dikes	240	286	34,	249	38	5	314,	3		x	x				x	3	Vigliotti *et al.* (1990)
Gallura Rhyolitic Ignimbr., Sardinia	240	286	38,	239	34	12	323,	0		x	x		x		x	4	Zijderveld *et al.* (1970b)
Gallura Volcanic Complex, Sardinia	240	286	37,	238	12	23	323,	−1					x		x	2	Westphal *et al.* (1976)
Gallura Volcanic Complex, Sardinia	240	286	45,	211	99	6	345,	−3		x	x		x		x	4	Storetvedt & Markhus (1978)
Porto Ferro, Cala B. Red Beds, Sardinia	245	286	20,	271	—	—	290,	7				x	x		x	3	Zijderveld *et al.* (1970b)
Cinto Volcanic Complex, Corsica	245	292	44,	209	64	10	346,	−2					x		x	2	Westphal *et al.* (1976)
Osani Flows & Dikes Combined, Corsica	245	292	55,	194	9	8	357,	7		x			x		x	3	Nairn & Westphal (1968)
Osani And./Rhyol. Volcs., Corsica	245	292	50,	154	86	7	22,	7					x		x	2	Storetvedt & Petersen (1976)
Ota Gabbrodiorite, Corsica	260	400	54,	173	56	5	9,	7							x	1	Nairn & Westphal (1967)
Nurra Volcanics, N. Sardinia	288	312	27,	253	—	—	307,	0	x		x		x		x	4	Edel *et al.* (1981a)
Barbagia-Gerrei Area Volcs., Sardinia	280	315	0,	284	27	6	267,	4	x	x	x		x		x	5	Edel *et al.* (1981a)

Table A7. *Phanerozoic pole positions for the western Mediterranean area – continued*

Rock unit, location	Age		Pole position				D,	Lat.	Reliability								Reference
	Lo	Hi	Lat.	Long.	k	α_{95}	42°N,	13°E	1	2	3	4	5	6	7	Q	
V. ALPS NORTH AND WEST OF INSUBRIC LINE, NEAR GAILTAL LINE, AUSTRIAN AND YUGOSLAVIAN ALPS (internal rotations probable)																	
C. Lepontine Metamorph., Switzerland	012	025	69,	271	58	7	335,	35	x	x	x			x	x	5	Heller (1980)
W. Lepontine Metamorph., Locarno Area	020	030	67,	260	—	7	335,	30	x						x	2	Heller (1977)
Bergell Granite, Switzerland	025	030	71,	264	48	3	338,	34	x	x				x	x	4	Heller (1973)
Sesio-Lanza Zone Dikes, Ivrea Zone	030	033	61,	235	—	10	340,	19	x	x				x	x	4	Lanza (1977)
Amphibolites, N. Italy	065	144	61,	222	19	19	345,	16							x	1	Förster *et al.* (1975)
Osterhorn Grp., NCA, Austria	144	205	38,	103	29	11	60,	24		x					x	2	Mauritsch & Frisch (1978)
Adnet Lst., Karwendel, Austria	187	208	54,	98	19	10	46,	35	x	x	x	x		x	x	6	Channell *et al.* (1990b)
Adnet Lst., Lofer, Austria	187	208	73,	148	71	7	14,	29	x	x	x			x	x	5	Channell *et al.* (1990b)
Adnet Lst., Adnet, Austria	187	208	33,	83	64	16	75,	35	x		x			x	x	4	Channell *et al.* (1990b)
Pelvoux Spilites, France	204	225	55,	235	13	27	337,	14	x					x	x	3	Westphal (1973)
Tarvisio Porphyries, Carnic Alps	225	235	60,	133	—	—	28,	23	x		x	x			x	4	Guicherit (1964)
Vorarlberg Quartz Porphyries, Austria	204	286	42,	264	8	10	313,	16				x			x	2	Soffel (1975)
Leogang Ss., N. Calcareous Alps	235	250	44,	147	415	4	31,	5	x	x					x	3	Soffel (1979)
E. Ljubljana (NE Slovenia), Yugoslavia	230	263	35,	135	7	19	44,	3			x	x			x	3	Soffel *et al.* (1983)
Grödener Sandstone, Coccau, Gailtal	245	260	45,	142	—	—	33,	8	x						x	2	Guicherit (1964)
Guil Volcanics, France	245	286	49,	225	30	4	339,	5		x					x	2	Westphal (1973)
Guil Volcanics, France	245	286	52,	242	—	7	331,	13		x					x	2	Henry (1976)
Allevard Sandst., Belledonne, France	245	286	35,	125	—	—	50,	9							x	1	Westphal (1973)
Argentéra Red Beds, France	245	263	52,	147	63	10	27,	12	x	x	x	x			x	5	Bogdanoff & Schott (1977)
Tirol/Oetztal dikes-Grp. III, Austria	267	330	29,	255	3	10	309,	1							x	1	Förster *et al.* (1975)
Pre-Permian Intrusions, N. Italy	290	363	31,	250	20	14	313,	0		x					x	2	Förster *et al.* (1975)

Monte Lauro and other flows, Sicily	003	005	86,	339	26	5	357,	45	x	x	x		x	x	x	6	Grasso *et al*. (1983)
Trubi Fm., Iblei, Sicily	003	005	84,	261	167	5	353,	40	x	x	x	x	x	x	x	7	Besse *et al*. (1984a)
Carlentini & Tellaro Fms., Sicily	006	008	82,	201	15	9	359,	34	x	x	x		x	x		5	Grasso *et al*. (1983)
Tellaro Fm., Sicily	009	014	79,	245	137	10	349,	35	x		x	x	x		x	5	Besse *et al*. (1984a)
Colli Eugani & M. Lessini, Padova	024	037	70,	134	28	6	20,	30	x	x				x	x	4	Soffel (1978)
Colli Eugani & M. Lessini, Padova	037	052	64,	213	12	8	351,	17	x	x				x	x	4	Soffel (1978)
Adamello Contact, Lombardy	024	058	78,	164	—	6	7,	31	x		x			x	x	4	Kipfer & Heller (1988)
Malo/Priabona Basalts, Vic. Alps	044	052	75,	214	—	—	354,	28	x				x		x	3	De Boer (1965)
Vizzini Sediments, Sicily	046	050	71,	219	140	2	351,	25	x	x	x				x	4	Besse *et al*. (1984a)
Capo Passero Dikes, Iblei, Sicily	069	086	63,	229	62	7	343,	19	x	x			x		x	4	Barberi *et al*. (1974)
C. Passero flows, Iblei, Sicily	071	086	69,	212	36	7	353,	22	x	x	x		x		x	5	Grasso *et al*. (1983)
C. Passero Volcs., Iblei, Sicily	069	086	62,	223	23	7	346,	17	x	x			x		x	4	Schult (1973)
Capo Passero Basalts & Dikes, Sicily	069	086	64,	236	13	10	341,	21	x	x			x		x	4	Gregor *et al*. (1975)
Bassano Area Combined, Vic. Alps	066	090	66,	230	139	6	345,	22	x	x	x	x	x		x	6	VandenBerg & Wonders (1976)
Senonian, Trentino & Vic. Alps	066	089	64,	229	34	8	344,	20	x	x	x		x		x	5	Channell & Tarling (1975)
Gargano Limestones, Apulia, S. Italy	066	091	56,	259	—	4	326,	23	x	x	x		x		x	5	VandenBerg (1983)
Istria Peninsula, Yugoslavia	066	098	59,	279	30	23	322,	33	x		x			x	x	4	Marton & Veljovic (1983)
Gargano Bauxites, Apulia, S. Italy	066	117	64,	236	46	4	341,	21		x			x		x	3	Channell & Tarling (1975)
Schio Biogenic Sediments, Vic. Alps	066	144	65,	200	36	16	357,	17					x			1	De Boer (1963)
Belluno/Bassano Sections, Vic. Alps	080	091	66,	229	305	4	345,	22	x	x	x		x	x	x	6	VandenBerg & Wonders (1980)
Cenomanian, Trentino & Vic. Alps	091	098	55,	245	70	15	332,	17	x				x	x	x	4	Channell & Tarling (1975)
Belluno/Bassano Sections, Vic. Alps	094	098	58,	237	—	5	337,	17	x	x	x		x	x	x	6	VandenBerg & Wonders (1980)
Belluno/Bassano Sections, Vic. Alps	098	105	53,	252	—	5	327,	18	x	x	x		x	x	x	6	VandenBerg & Wonders (1980)
Kvarner, Istria & Trieste, Yugoslavia	088	124	65,	245	75	6	338,	25	x	x	x	x		x	x	6	Marton *et al*. (1990a)
Gargano, Peschici, Apulia, S. Italy	080	144	61,	249	31	7	334,	23		x	x		x		x	4	Channell (1977)
Istria Peninsula, Yugoslavia	098	138	52,	270	26	18	318,	25	x		x			x	x	4	Marton & Veljovic (1983)
Gargano Limestones, Apulia, S. Italy	098	144	49,	274	81	6	313,	26	x	x	x		x		x	5	VandenBerg (1983)

Table A7. *Phanerozoic pole positions for the western Mediterranean area – continued*

Rock unit, location	Age		Pole position				D,	Lat.	Reliability								Reference
	Lo	Hi	Lat.	Long.	k	α_{95}	42°N,	13°E	1	2	3	4	5	6	7	Q	
Bassano Area Section, Vic. Alps	152	156	59,	255	—	3	330,	23	x		x	x	x		x	5	VandenBerg & Wonders (1976)
Volcanics, Sicily	150	190	38,	67	11	22	77,	48						x	x	2	Schult (1976)
Schio Biogenic Sediments, Vic. Alps	144	208	51,	175	24	19	11,	4					x		x	2	De Boer (1963)
Alpe Turati Section, Western Lombardy	180	198	63,	238	—	7	340,	21	x	x	x		x	x		5	Horner & Heller (1983)
Norian Dolostones, Vicentinian Alps	208	225	69,	151	—	—	15,	25	x				x			2	De Boer (1963)
Ladinian-Carnian Volcanics, Dolomites	227	233	48,	240	22	9	330,	9	x	x		x	x		x	5	Manzoni (1970)
Predazzo Dikes, Dolomites	220	235	61,	242	—	—	337,	21	x							1	Manzoni (1970)
Trl/m Vicentinian Alps Combined	230	245	57,	249	218	4	331,	20	x	x		x	x	x		5	De Boer (1963)
Valle di Scalve Porphyrite, Lombardy	232	238	52,	221	—	6	343,	7	x		x		x		x	4	Zijderveld & De Jong (1969)
Verrucano Lombardo Meta-ss., Lombardy	240	268	47,	237	—	6	331,	7	x		x		x	x	x	5	Kipfer & Heller (1988)
Verrucano Lombardo Red Beds, Lombardy	240	268	48,	239	—	5	331,	9	x	x	x		x		x	5	Kipfer & Heller (1988)
Grödener Sst., S. Martino, Dolomites	245	260	51,	235	—	—	334,	10	x		x		x		x	4	Guicherit (1964)
Grödener (Val Gardena) Sst., Dolom.	245	260	42,	237	26	18	329,	3	x			x	x		x	4	Manzoni (1970)
W. Ljubljana (Julian Alps), Yugoslavia	230	263	43,	235	9	19	330,	3			x	x			x	3	Soffel *et al.* (1983)
Staro & Camparmo Acid Flows, Vic. Alps	245	258	53,	241	220	6	332,	14	x		x		x	x	x	5	De Boer (1963)
Grödener Clastics, Cortiana, Vic. Alps	253	258	48,	238	117	7	331,	9	x				x		x	3	De Boer (1963)
P-Tr. Posina Acid Complex, Vic. Alps	230	286	50,	244	157	5	329,	12							x	1	De Boer (1963)
Bolzano Area Flows, Grödener, Grp. I	230	286	46,	217	9	7	343,	1					x	x	x	3	Förster *et al.* (1975)

Pre-Grödener Dikes, Vicentinian Alps	260	300	49,	252	95	6	324,	15							x	1	De Boer (1963)
Bolzano Quartz Porphyries Combined	263	286	45,	239	—	4	329,	6	x	x	x	x	x		x	6	Zijderveld *et al.* (1970a)
L. Collio & Auccia Volcanics, Lomb.	263	296	39,	252	15	20	318,	7	x		x		x		x	4	Zijderveld & De Jong (1969)
Lugano (Ganna) Porphyries	263	296	43,	243	76	10	325,	6	x	x	x	x	x		x	6	Heiniger (1979)
Auccia Volcanics, Lombardy	263	296	38,	245	89	8	321,	3	x	x	x		x		x	5	Heiniger (1979)
Arona Volcanics, Lombardy	263	296	35,	248	47	14	318,	2	x	x	x		x		x	5	Heiniger (1979)
Auernig Grp., Carnic Alps	286	296	36,	243	8	18	321,	1	x		x		x		x	4	Manzoni *et al.* (1989)
Brixen-Merano Intrusives, Dolomites	267	330	24,	225	3	9	329,	−17					x		x	2	Förster *et al.* (1975)
VII. APENNINES, CALABRIAN AND SICILIAN THRUSTBELTS (internal rotations probable)																	
Porto Empedocle Seds., Sicily	002	005	50,	98	25	7	50,	33	x		x			x	x	4	Besse *et al.* (1984a)
Ramero-Garbagna Sections, Piemont	024	030	49,	259	62	6	321,	18	x	x	x			x	x	5	Thio (1988)
Contessa Quarry Section, N. Umbria	024	037	65,	247	19	3	338,	25	x	x	x			x	x	5	Lowrie *et al.* (1982)
Piemonte Turbidites, Voltri, Liguria	030	044	65,	255	—	6	335,	27	x	x	x	x		x	x	6	VandenBerg (1979b)
Calcari di Castro (Remagn.?), Apulia	000	056	68,	123	41	6	24,	32		x	x			x		3	Tozzi *et al.* (1988)
Contessa Section, Gubbio, N. Umbria	037	058	54,	260	13	2	324,	22	x	x	x			x	x	5	Lowrie *et al.* (1982)
Cagli Paleogene, N. Umbrian Arc	037	064	63,	264	—	3	330,	29	x		x	x			x	4	VandenBerg *et al.* (1978)
Gubbio Eocene Section, N. Umbria	044	060	64,	255	69	5	334,	27	x	x	x			x	x	5	Napoleone *et al.* (1983)
M. S. Calogero (Saccense), Sicily	024	098	71,	209	—	7	354,	24			x				x	2	Channell *et al.* (1980)
M. Genuardo (Saccense Pl.), Sicily	024	098	47,	109	12	23	49,	26			x			x	x	3	Channell *et al.* (1980)
M. Barracu-M. Colomba (Sicani), Sicily	024	098	9,	77	16	20	101,	25			x			x	x	3	Channell *et al.* (1980)
M. Rose (Sicani Basin), Sicily	024	098	6,	79	—	—	102,	21			x				x	2	Channell *et al.* (1980)
M. Kumeta (Trapanese Pl.), Sicily	024	098	35,	98	8	13	65,	26			x			x	x	3	Channell *et al.* (1980)
Sagana & Cozzo d.L. (Imerese), Sicily	024	098	−11,	65	12	10	125,	18		x	x	x		x	x	5	Channell *et al.* (1980)
Panormide Platform Eocene, Sicily	037	058	−18,	64	63	5	131,	14	x		x				x	3	Channell *et al.* (1990a)
Contessa Section, Gubbio, N. Umbria	058	066	51,	262	7	9	321,	21	x		x			x	x	4	Lowrie *et al.* (1982)

Table A7. *Phanerozoic pole positions for the western Mediterranean area – continued*

Rock unit, location	Age		Pole position				D,	Lat.	Reliability								Reference
	Lo	Hi	Lat.	Long.	k	α_{95}	42°N,	13°E	1	2	3	4	5	6	7	Q	
Gubbio Section Paleocene, N. Umbria	058	066	56,	259	7	6	326,	23	x					x	x	3	Roggenthen & Napoleone (1977)
Mt. Cerviero Limburgites, Calabria	058	075	16,	297	22	17	273,	21	x						x	2	Manzoni (1975)
Maastrichtian, N. Umbrian Arc	066	075	48,	269	23	10	315,	22	x	x	x	x		x	x	6	Channell *et al.* (1978)
Gubbio Maastrichtian, N. Umbrian Arc	066	075	44,	277	10	8	308,	24	x					x	x	3	Roggenthen & Napoleone (1977)
Maastrichtian, S. Umbrian Arc	066	075	68,	216	21	11	351,	21	x	x	x			x	x	5	Channell *et al.* (1978)
Cagli Campan./Maastr., N. Umbrian Arc	066	084	62,	264	249	8	330,	29	x	x	x	x		x	x	6	VandenBerg *et al.* (1978)
Scaglia Rossa, Umbrian Arc Combined	044	091	51,	270	17	6	317,	25		x				x	x	3	Lowrie & Alvarez (1975)
Moria Section, N. Umbria	063	098	59,	255	35	9	330,	25	x	x		x		x	x	5	Alvarez & Lowrie (1978)
Gubbio Section, N. Umbrian Arc	066	092	42,	279	7	3	305,	24	x		x	x		x	x	5	Lowrie & Alvarez (1977)
Panormide, Ku Combined, Sicily	066	098	5,	85	17	17	98,	16	x	x	x	x		x	x	6	Channell *et al.* (1990a)
Trapanese, Erice, Sicily	066	098	49,	123	—	—	41,	19	x	x	x			x	x	5	Channell *et al.* (1990a)
Internal Saccense, Sambuca, Sicily	066	098	63,	147	89	4	21,	21	x		x				x	3	Channell *et al.* (1990a)
Custonaci Volcs., Panormide, Sicily	066	098	21,	100	29	17	76,	16	x						x	2	Schult (1976)
Matese Mt. Bauxites & Lst., Campania	066	117	48,	257	16	7	321,	17		x	x			x	x	4	Channell & Tarling (1975)
Tur./Coniacian, N. Umbrian Arc	087	091	48,	265	17	17	317,	20	x		x	x			x	4	Channell *et al.* (1978)
Tur./Coniacian, S. Umbrian Arc	087	091	60,	245	21	9	335,	21	x	x	x				x	4	Channell *et al.* (1978)
Cagli Cenom./Santon., N. Umbrian Arc	084	094	46,	269	60	10	314,	21	x	x	x	x			x	5	VandenBerg *et al.* (1978)
Poggio le Guaine Section, N. Umbria	098	116	40,	277	128	11	305,	22	x	x		x			x	4	Lowrie *et al.* (1980)
Valdorbia E. Albian, N. Umbrian Arc	104	113	41,	264	17	12	312,	15	x			x			x	3	Lowrie *et al.* (1980)
Valdorbia L. Aptian, N. Umbrian Arc	113	116	57,	264	6	19	325,	26	x			x		x	x	4	Lowrie *et al.* (1980)
Cagli L. Aptian, N. Umbrian Arc	113	117	40,	286	158	7	300,	28	x	x	x	x		x	x	6	VandenBerg *et al.* (1978)
Gorgo a Cerbara Section, N. Umbria	113	124	30,	292	96	13	288,	26	x	x		x		x	x	5	Lowrie *et al.* (1980)

Gorgo a Cerbara Section, N. Umbria	118	128	19,	296	14	3	276,	22	x	x	x			x	x	5	Lowrie & Alvarez (1984)
Presale Section, N. Umbria	118	125	24,	293	18	3	282,	23	x	x	x			x	x	5	Lowrie & Alvarez (1984)
Frontale Section, N. Umbria	118	130	32,	278	15	4	297,	18	x	x	x			x	x	5	Lowrie & Alvarez (1984)
Maiolica Fm., Northern Umbrian Arc	115	148	32,	282	57	5	295,	20	x	x	x	x		x	x	6	Hirt & Lowrie (1988)
Maiolica Fm., Southern Umbrian Arc	115	148	48,	264	33	7	318,	20	x	x	x	x		x	x	6	Hirt & Lowrie (1988)
Maiolica Fm., F. Giordano, N. Umbria	121	148	19,	288	149	10	281,	17	x	x	x	x		x	x	6	Cirilli *et al.* (1984)
Monte Maiella Ls., S. Umbria	119	152	57,	263	44	8	325,	25			x			x	x	3	Jackson (1990)
Cagli E. Neocomian, N. Umbrian Arc	131	144	39,	287	—	7	299,	28	x		x	x			x	4	VandenBerg *et al.* (1978)
Elba Ophiolites (Secondary?)	113	186	63,	147	19	6	20,	21		x	x	x		x	x	5	Soffel (1981)
Cagli Kimm./Tith., N. Umbrian Arc	144	156	34,	295	—	8	290,	30	x		x	x			x	4	VandenBerg *et al.* (1978)
Gorgo a Cerbara Section, Umbria	148	152	7,	305	6	7	260,	21	x		x			x	x	4	Channell *et al.* (1984)
Fonte Avellana U. Section, N. Umbria	144	163	46,	271	14	4	313,	22	x	x	x					3	Channell *et al.* (1984)
Valdorbia U. Mixed Section, N. Umbria	144	185	28,	280	9	4	293,	17			x			x	x	3	Channell *et al.* (1984)
Valdorbia Quiet Z. Section, N. Umbria	152	187	44,	276	22	3	309,	24		x	x					2	Channell *et al.* (1984)
Malvito Lavas, Calabria	144	208	−21,	332	—	15	219,	17							x	1	Manzoni & Vigliotti (1983)
Scisti Silic., L. Nappe, Lagonegro B.	144	213	46,	133	162	3	38,	13			x				x	2	Incoronato *et al.* (1985)
Scisti Silic., U. Nappe, Lagonegro B.	144	213	−29,	344	174	7	206,	14			x				x	2	Incoronato *et al.* (1985)
Panormide, Cozzo Lupo, Sicily	163	187	−10,	87	20	8	108,	5	x						x	2	Channell *et al.* (1990a)
Trapanese Jm combined, Sicily	163	187	52,	129	35	8	35,	19	x	x	x			x	x	5	Channell *et al.* (1990a)
Internal Saccense, St Maria, Sicily	163	187	56,	109	55	4	41,	30	x		x				x	3	Channell *et al.* (1990a)
External Saccense, R. Nadore, Sicily	163	187	56,	233	14	10	338,	14	x		x				x	3	Channell *et al.* (1990a)
Cagli Toarc./Aal., N. Umbrian Arc	183	193	38,	292	—	6	296,	31	x		x	x				3	VandenBerg *et al.* (1978)
Valdorbia L. Mixed Section, N. Umbria	187	191	42,	283	10	5	304,	27	x		x			x		3	Channell *et al.* (1984)
Fonte Avellana L. Section, N. Umbria	187	204	44,	275	5	7	309,	23	x		x			x		3	Channell *et al.* (1984)

Table A7. *Phanerozoic pole positions for the western Mediterranean area – continued*

Rock unit, location	Age		Pole position				D,	Lat.	Reliability								Reference
	Lo	Hi	Lat.	Long.	k	α_{95}	42°N,	13°E	1	2	3	4	5	6	7	Q	
Cingoli Section, N. Umbrian Arc	193	204	70,	269	10	6	336,	35	x	x	x			x	x	5	Channell *et al.* (1984)
Sila Massif Liassic, Calabria	193	208	15,	311	86	13	262,	31	x	x	x				x	4	Manzoni & Vigliotti (1983)
Verrucano, Siena Area, Tuscany	225	235	48,	299	47	10	304,	40	x	x	x	x				4	VandenBerg & Wonders (1976)
Monte Quoio Fm., Verrucano, Tuscany	238	245	36,	257	8	19	313,	8	x		x	x		x	x	5	VandenBerg (1979a)
Sila Nappes, Calabria	258	286	−23,	311	24	10	234,	4	x	x	x				x	4	Manzoni (1979)
Acqua Prenzano Amphibolites, Calabria	258	320	−24,	295	—	11	244,	−7			x				x	2	Manzoni & Vigliotti (1983)
Farma Fm., Verrucano, Siena, Tuscany	258	320	42,	252	—	13	320,	10			x				x	2	VandenBerg & Wonders (1976)

The entries in this Table are for the blocks and mobile belts of the Western Mediterranean; in the mobile belts rotations can occur and this mandates that reliability criterion 5 generally cannot be checked as being met. The reliability criteria are explained in the text (see also Table 4.1). A '*' is placed in the Q column when the result is not to be used for apparent polar wander path construction or paleolatitude determinations, with the reasons given below. For other explanations see Table A1.

*Poles that would have Q ≥ 3 but are excluded from mean pole determinations and other compilations, are listed below with the reasons in parentheses. *Iberian Meseta*: Tertiary overprint of the Lisbon Volcanics (poorly dated, too few samples); Upper Cretaceous Sines Complex (anomalous result, possibly rotated); Permo-Triassic Villaviciosa Red Beds (remagnetized according to Schott & Peres, 1987b). *Betic Cordillera*: Tertiary–Cretaceous Agost Section C29R (too few samples, non-antipodal reversals).

Table A8. *Phanerozoic pole positions for the eastern Mediterranean area and the Middle East*

Rock unit, location	Age Lo	Age Hi	Pole position Lat.	Pole position Long.	k	α_{95}	D, 40°N,	Lat. 22°E	Reliability 1	2	3	4	5	6	7	Q	Reference
I. GREECE, S. BULGARIA (internal rotations probable)																	
Peloponnesos Sediments, Greece	001	004	76,	148	75	9	13,	31	x	x	x			x	x	5	Laj *et al.* (1982)
Kephallinia Seds., Ionian I., Greece	001	004	82,	141	316	5	9,	36	x	x	x			x	x	5	Laj *et al.* (1982)
Volos Lavas, Pelagonian Zone, Greece	002	004	88,	320	85	8	358,	41	x	x	x				x	4	Kissel *et al.* (1986a)
Almopias, C. Macedonia, Vardar Zone	002	004	75,	132	32	16	17,	34	x					x	x	3	Kondopoulou & Lauer (1984)
Almopias, C. Macedonia, Vardar Zone	002	004	77,	70	40	10	14,	48	x						x	2	Bobier (1968)
Crete Pliocene Sediments, Greece	002	005	82,	204	62	7	360,	32	x	x	x			x	x	5	Laj *et al.* (1982)
Corfu Sediments, Ionian Isl., Greece	002	011	68,	143	35	10	21,	27	x	x	x			x	x	5	Laj *et al.* (1982)
Zakinthos Seds., Ionian Isl., Greece	001	013	71,	134	46	18	21,	31	x		x			x	x	4	Laj *et al.* (1982)
Evia Lavas & Seds., Aegean C., Greece	002	014	48,	115	56	7	49,	27	x	x	x				x	4	Kissel *et al.* (1986a)
Crete Tortonian Sediments, Greece	008	011	81,	225	52	5	356,	32	x	x	x			x	x	5	Laj *et al.* (1982)
Skyros Lavas, Aegean Sea, Greece	014	016	65,	136	82	8	26,	27	x	x	x				x	4	Kissel *et al.* (1986c)
Strymon, N. Serbomacedonia, Greece	023	028	56,	288	40	15	320,	30	x					x	x	3	Pavlides *et al.* (1988)
Mesohellenic Trough Sediments, Greece	024	036	65,	131	25	10	27,	28	x	x	x			x	x	5	Kissel & Laj (1988)
Thrace Volcs., Intr., Rhodope, Greece	025	035	76,	180	23	8	6,	27	x	x	x			x	x	5	Kissel *et al.* (1986b)
Epirus Flysch, Ionian Z., NW Greece	024	037	38,	113	30	5	59,	23	x	x	x	x		x	x	6	Kissel *et al.* (1985)
Arnea, Gomati Intr., Chalkid., Greece	030	041	50,	139	28	9	37,	16	x	x	x			x	x	5	Kondopoulou & Westphal (1986)
Sithonia Bathol., Chalkid., Greece	039	042	49,	144	22	16	35,	13	x					x	x	3	Kondopoulou & Lauer (1984)
Lefkas Island, Ionian Zone, W. Greece	037	058	15,	92	17	13	91,	25	x					x	x	3	Marton *et al.* (1990b)

Table A8. *Phanerozoic pole positions for the eastern Mediterranean area and the Middle East – continued*

Rock unit, location	Age		Pole position				D,	Lat.	Reliability								Reference
	Lo	Hi	Lat.	Long.	k	α_{95}	40°N,	22°E	1	2	3	4	5	6	7	Q	
Flysch, Pindos Zone, W. Greece	037	066	58,	257	—	—	333,	18	x					x	x	3	Marton *et al.* (1990b)
Microbreccious Ls., Ionian Z., Greece	037	066	48,	116	25	4	48,	26	x	x	x	x		x	x	6	Horner & Freeman (1983)
Senonian Reversed Andes., S. Bulgaria	066	089	55,	163	—	—	22,	11	x						x	2	Nozharov & Velevich (1974)
Pindos & W. Thessaly Ls., W. Greece	058	098	55,	111	18	30	43,	32						x	x	2	Marton *et al.* (1990b)
S. Bulgarian Andesites	066	098	66,	177	26	13	10,	18	x					x	x	3	Vollstadt *et al.* (1967)
S. & W. Bulgarian Volcanics Combined	066	098	86,	223	23	7	358,	36	x	x				x		3	McElhinny List XVI; pole 16.069
Theopetra Area, Thessalia, Greece	126	135	47,	267	—	3	320,	14	x	x	x			x	x	5	Surmont (1989)
C. Chalkidiki Plutons, Greece	120	150	87,	350	—	3	358,	42	x							1	Pavlides *et al.* (1988)
Stara Planina Zone Comb., S. Bulgaria	135	154	80,	172	582	5	6,	31	x	x	x			x		4	Surmont *et al.* (1991)
Bafi Area, Argolis, Pelagonian Zone	164	190	−4,	316	—	2	251,	15	x	x	x			x	x	5	Surmont (1989)
Polaten Fm., Stara P. Zone, S. Bulgaria	169	180	84,	269	57	3	353,	37	x	x						2	Surmont *et al.* (1991)
Korou Dere Olistolith, S. Bulgaria	163	187	51,	108	175	2	48,	32	x	x					x	3	Peybernes *et al.* (1989)
Pindos Mtns., Gabbros, Greece	144	208	53,	248	—	9	334,	11						x	x	2	Pucher *et al.* (1974)
Argolis Pen. Diabase, Pelagonian Z., Greece	144	208	−12,	70	—	17	129,	22						x	x	2	Pucher *et al.* (1974)
Argolis Ls., Pelagonian Z., Greece	150	178	27,	107	23	21	71,	21	x		x				x	3	Turnell (1988)
Atalanti Volcs., Pelagonian Z., Greece	215	245	54,	221	44	9	349,	5		x	x				x	3	Turnell (1988)

Rock unit, location	Age		Pole position				D,	Lat.	Reliability								Reference
	Lo	Hi	Lat.	Long.	k	α_{95}	38°N,	32°E	1	2	3	4	5	6	7	Q	
II. CYPRUS, TURKEY, EAST AEGEAN ISLANDS (internal rotations probable)																	
Rhodos Sediments, Aegean, Greece	002	011	82,	202	86	8	2,	30	x	x	x			x	x	5	Laj *et al.* (1982)
Antalya Basin Sediments, S. Turkey	001	015	84,	198	80	5	2,	32	x	x	x	x		x	x	6	Kissel & Poisson (1986)
Afyon Lavas, Tuffs, Menderes Massif	009	015	73,	144	—	—	18,	30	x		x			x	x	4	Kondopoulou & Lauer (1984)
Izmir-Bergama, Aegean Coast, Turkey	007	019	63,	294	20	9	329,	30	x	x	x			x	x	5	Kissel *et al.* (1987)
Cannakkale Volc., Aegean C., Turkey	010	020	83,	164	254	4	6,	33	x	x	x			x	x	5	Kissel *et al.* (1987)
Izmir-Foca Region Lavas, W. Turkey	015	019	65,	286	19	11	333,	28	x	x	x			x	x	5	Kissel *et al.* (1986c)
Lesbos Isl. Volcs., Aegean Coast	015	018	78,	146	11	15	13,	32	x	x				x	x	4	Kondopoulou & Lauer (1984)
Lesbos Isl., Aegean Coast, Greece	015	019	80,	176	25	7	7,	30	x	x	x			x	x	5	Kissel *et al.* (1986c; 1987)
South of Bergama, Aegean C., Turkey	017	018	65,	152	16	20	23,	23	x		x			x	x	4	Kissel *et al.* (1987)
Karaburun Andes., Aegean C., Turkey	017	019	48,	101	34	21	55,	40	x		x				x	3	Kondopoulou & Lauer (1984)
Karaburun Penins., Aegean C., Turkey	017	022	50,	112	9	16	50,	34	x		x			x	x	4	Kissel *et al.* (1987)
Akrokipia Fm., Cyprus	012	033	69,	321	76	9	333,	42	x	x	x				x	4	Abrahamsen & Schonharting (1987)
Bey Daglari Sediments, S. Turkey	015	069	58,	282	19	9	327,	22		x	x			x	x	4	Kissel & Poisson (1987)
Tunceli Red Beds & Volcs., Taurides	037	058	65,	294	36	6	331,	31	x		x	x		x	x	5	Van der Voo (1968)
Almacik Flake, Bolu Area, W. Turkey	037	058	−8,	6	23	16	213,	38	x		x			x	x	4	Saribudak *et al.* (1990)
Central Northern Pontides Area	037	058	64,	246	19	29	345,	16	x		x				x	3	Saribudak (1989)
Mesudiye Dikes, N. Pontides	037	058	42,	139	31	18	47,	14	x		x				x	*	Orbay & Bayburdi (1979)
Lower Eocene Pelagic Chalks, Cyprus	044	058	77,	225	—	10	357,	25	x	x	x				x	4	Clube *et al.* (1985)
Fluvial Red Beds, Sakarya C., Turkey	063	066	65,	194	16	32	8,	14	x						x	2	Evans & Hall (1990)

Table A8. *Phanerozoic pole positions for the eastern Mediterranean area and the Middle East – continued*

Rock unit, location	Age		Pole position				D,	Lat.	Reliability								Reference
	Lo	Hi	Lat.	Long.	k	α_{95}	38°N,	32°E	1	2	3	4	5	6	7	Q	
Umber, Lefkara Fms., Cyprus	038	088	52,	282	15	18	322,	19			x			x	x	3	Abrahamsen & Schonharting (1987)
Gumushane Lst. & Volcs., E. Pontides	047	084	70,	261	67	4	344,	24			x	x			x	3	Van der Voo (1968)
Gerede I, Site 16, W. Pontides	055	073	78,	273	—	—	348,	31	x						x	2	Lauer (1981a)
R-polarity Pelagic Chalk, Cyprus	058	074	52,	263	—	15	331,	10	x	x	x				x	*	Clube *et al.* (1985)
N-polarity Pelagic Chalk, Cyprus	058	074	63,	270	—	13	336,	21	x	x	x				x	4	Clube *et al.* (1985)
Central Northern Pontides Area	066	098	70,	240	27	12	350,	20	x	x	x			x	x	5	Saribudak (1989)
Sariyer, Sites 14–15, W. Pontides	066	098	63,	273	—	—	335,	22	x						x	2	Lauer (1981a)
Mesudiye Tuffs, N. Pontides	066	098	75,	275	24	11	345,	30	x	x	x				x	4	Orbay & Bayburdi (1979)
Troodos, Akamas Pillow Lavas, Cyprus	088	091	14,	318	—	12	272,	21	x	x	x				x	4	Clube *et al.* (1985)
Sheeted Dike Complex, Cyprus	080	100	26,	322	29	8	281,	31	x	x	x	x			x	5	Bonhommet *et al.* (1988)
Gumushane Sst. & Volcs., E. Pontides	075	119	61,	278	22	6	331,	22	x	x	x	x			x	5	Van der Voo (1968)
Niksar Basalts, E. Pontides, Turkey	066	144	51,	296	11	16	316,	215			x				x	2	Van der Voo (1968)
Aptian KA Site, E. Pontides	113	119	47,	207	190	5	3,	−5	x		x					2	Saribudak (1989; pers.comm.1991)
Bilecik Limestone, NW Anatolia	144	187	19,	92	—	8	92,	35		x					x	2	Evans *et al.* (1982)
Gerede III, Site 18, W. Pontides	152	156	44,	259	—	—	328,	2	x						x	2	Lauer (1981a)
Kimmeridgian Pontides Combined	150	160	42,	260	—	8	327,	1	x	x	x				x	4	Lauer (1981a)
Bayburt Volcs., Seds., E. Pontides	144	208	39,	265	51	8	322,	1			x				x	2	Van der Voo (1968)
Bademli Redbeds, W. Taurides, Turkey	187	230	−43,	330	22	7	221,	−9			x				x	2	Van der Voo & Van der Kleijn (1970)
Balya, Site 60, Sakariya, W. Turkey	187	230	44,	244	—	—	388,	−3							x	1	Lauer (1981a)
Kure Basalt, Qtzdior., Pontides	144	286	49,	93	—	5	55,	46		x	x				x	*	Guner (1982)
Kure Diabase Dike, Pontides	144	286	47,	167	7	15	29,	4			x				x	2	Guner (1982)
Kure Massive Sulphide, Pontides	144	286	18,	80	15	9	102,	44			x				x	2	Guner (1982)
Kure Dacite Dike, Pontides	144	286	1,	128	4	23	86,	−4			x				x	2	Guner (1982)
Kure Peridotite Dike, Pontides	144	286	−2,	72	15	11	128,	36			x				x	2	Guner (1982)
Kure Amphibolitized Diabase, Pontides	144	286	40,	325	15	13	295,	39			x				x	2	Guner (1982)

Rock unit, location	Age Lo	Age Hi	Pole position Lat.	Long.	k	α_{95}	D, 33°N,	Lat. 36°E	Reliability 1	2	3	4	5	6	7	Q	Reference
Denizlikoy Limestone, W. Pontides	235	240	59,	132	118	5	35,	27	x		x				x	*	Saribudak *et al.* (1989)
Gumusdere Basalt, W. Bosporus, Turkey	240	245	34,	309	187	3	295,	25	x		x				x	*	Saribudak *et al.* (1989)
Tavsancil Andesite, W. Pontides	240	245	52,	111	49	5	48,	35	x		x				x	*	Saribudak *et al.* (1989)
Amasra Red Beds, W. Pontides, Turkey	225	286	11,	282	63	18	291,	−8			x	x			x	3	Gregor & Zijderveld (1964)
Cakraz Red Beds, Amasra, W. Pontides	225	286	19,	277	30	9	300,	−7			x				x	2	Evans *et al.* (1991)
Zonguldak Fm., W. Pontides, Turkey	286	320	48,	250	9	12	336,	2	x		x					2	Evans *et al.* (1991)
Inkum Limest., Bartin, W. Pontides	360	374	39,	234	33	6	342,	−10	x		x			x	x	4	Evans *et al.* (1991)
Ls., Denizli, W. Pontides, Turkey	360	374	−5,	253	62	7	301,	−40	x		x				x	3	Evans *et al.* (1991)
Ls., Gebze Area, W. Pontides, Turkey	360	374	−5,	288	36	6	275,	−14	x		x				x	3	Evans *et al.* (1991)
Kartal Fm., Cam Dag, W. Pontides	380	387	50,	206	28	8	4,	−2	x		x					2	Evans *et al.* (1991)
Quartzite, Yumrukaya, W. Pontides	408	438	7,	281	36	6	289,	−12			x				x	2	Evans *et al.* (1991)
III. LEVANT (internal rotations probable)																	
Miocene Lower Basalt, Israel	005	024	26,	310	14	—	290,	17	x						x	2	Nur & Helsley (1971)
Neogene Basalts, Jordan	005	025	76,	99	1000	2	16,	38	x		x				x	3	Sallomy & Krs (1980)
Miocene Volcanics, Syria	015	025	75,	220	28	6	359,	18	x	x	x		x	x	x	6	Roperch & Bonhommet (1986)
Carmel Region Basalts, Israel	066	089	42,	264	52	5	326,	−3	x				x		x	3	Helsley & Nur (1970)
Cretaceous Sediments, Israel	065	144	77,	251	23	20	352,	22			x			x		2	Freund & Tarling (1979)
Ramon Anticline Basalts, Israel	098	144	53,	265	15	9	333,	6	x				x		x	3	Helsley & Nur (1970)
Essexite Laccolith, Israel	098	163	48,	266	84	14	329,	3			x				x	2	Freund & Tarling (1979)
Mt. Hermon Volcanics, Lebanon	113	119	4,	302	9	16	276,	−1	x		x		x	x	x	5	Ron (1987)
Aptian Basalts & Limest., Lebanon	113	119	38,	282	—	6	314,	4	x		x		x		x	4	Van Dongen *et al.* (1967)
Neocomian Basalts, Lebanon	119	144	25,	285	39	9	302,	−2	x	x			x	x	x	5	Gregor *et al.* (1974)
Kimmeridgian Basalts, Lebanon	153	156	−2,	294	28	11	275,	−11	x	x	x		x		x	5	Gregor *et al.* (1974)
Kimmeridgian Basalts, Lebanon	152	156	−1,	300	—	3	272,	−6	x		x		x		x	4	Van Dongen *et al.* (1967)
Jurassic Sediments, Israel	163	213	75,	249	25	16	351,	20			x			x		2	Freund & Tarling (1979)
Saharonim Fm., Israel	213	248	44,	254	2	—	334,	−6			x			x		2	Freund & Tarling (1979)
Red Beds, Jordan	476	517	37,	323	—	8	296,	32					x	x	x	3	Burek (1969)
Cambrian Dikes, Jordan	525	590	26,	161	26	9	49,	−11			x				x	2	Sallomy & Krs (1980)

Table A8. *Phanerozoic pole positions for the eastern Mediterranean area and the Middle East – continued*

Rock unit, location	Age		Pole position				D,	Lat.	Reliability								Reference
	Lo	Hi	Lat.	Long.	k	α_{95}	35°N,	60°E	1	2	3	4	5	6	7	Q	
IV. IRAN, OMAN, KOPET DAGH (TURKMENISTAN), AFGHANISTAN (internal rotations possible)																	
Combined Volcanics, Central Iran	000	005	77,	174	51	10	14,	29	x	x					x	3	Soffel & Förster (1981)
Dash-e-Lut Plio-Quatern. Volcs., Iran	001	010	83,	347	18	13	352,	37	x					x	x	3	Conrad *et al.* (1981)
Dasht-eh-Nawar Felsics, Afghanistan	002	005	78,	230	27	8	2,	23	x						x	2	Krumsiek (1976)
Savalan Volcanics, E. Caucasus, Iran	002	025	60,	170	15	16	30,	21							x	1	Alberti *et al.* (1974)
Hassan-Abad Series, Central Iran	018	019	76,	311	28	12	345,	29	x		x				x	3	Bina *et al.* (1986)
Dacite & Conglomerates, Central Iran	005	037	64,	303	14	17	335,	21	x						x	2	Soffel & Förster (1981)
Karaj Sills, Tuffs, Alborz, Iran	027	034	45,	119	13	10	59,	45	x		x				x	3	Bina *et al.* (1986)
Shindand Volcanics, Afghanistan	025	055	72,	219	7	11	7,	18	x					x	x	3	Krumsiek (1976)
Kuh-e-Kaleh-e-Karg. Series, Central Iran	040	045	34,	141	112	6	65,	25	x	x	x				x	4	Bina *et al.* (1986)
Ophiolites & Flows, Central Iran	037	066	62,	311	15	7	331,	22	x	x	x				x	4	Soffel & Förster (1981)
Dash-e-Lut Paleogene Volcs., Iran	037	066	12,	351	11	18	268,	24	x					x	x	3	Conrad *et al.* (1981)
Kuh-e-Shabadum Volcs., C. Iran	050	055	51,	129	15	16	49,	39	x		x				x	3	Bina *et al.* (1986)
Bajestan-Gonabad, Lut Block, Iran	056	063	79,	335	78	7	347,	35	x		x				x	3	Bina *et al.* (1986)
Seh-Deh Red Beds, East Iran	055	065	75,	2	59	6	343,	42	x		x				x	3	Bina *et al.* (1986)
Kopet-Dagh Ls., Marls, Turkmenistan	055	088	74,	193	526	3	13,	23	x	x	x			x	x	5	Bazhenov (1987)
Semail Ophiolite Sheeted Dikes, Oman	093	098	76,	257	6	7	356,	21	x		x			x	x	4	Luyendijk *et al.* (1982)
Semail Wadi Kadir Gabbro, Oman	093	098	61,	285	9	8	339,	13	x		x				x	3	Luyendijk *et al.* (1982)

Trachybasalt & Red Beds, Central Iran	066	144	79,	288	21	9	351,	27		x	x					2	Soffel & Förster (1981)
Dehuk Sandstone, Central Iran	091	113	58,	315	19	21	327,	22	x		x	x			x	4	Wensink (1982)
Kopet-Dagh Seds., Turkmenistan	091	119	75,	151	177	4	18,	33	x	x	x	x			x	5	Bazhenov (1987)
Chalus Fm., Volcs., C. Alborz, Iran	088	124	61,	148	22	7	34,	31	x	x	x	x		x	x	6	Wensink & Varekamp (1980)
SW Hilmend Block Volc., Afghanistan	098	144	71,	81	4	25	11,	52	x							1	Krumsiek (1976)
Garedu Red Beds, Central Iran	144	156	79,	217	45	14	5,	25	x	x	x					3	Wensink (1981, 1982)
Bidou Beds, Central Iran	138	175	23,	187	17	19	49,	−13			x					1	Wensink (1982)
Flood Basalts & Ss., Central Iran	144	208	38,	314	12	12	310,	10		x	x				x	3	Soffel & Förster (1981)
Liassic Volcanics, Alborz, Iran	187	208	50,	157	—	5	44,	22	x	x	x				x	4	Wensink (1979, 1981)
Naiband & Nakhlak Fm. Ls., C. Iran	208	245	15,	349	17	20	272,	24							x	1	Soffel & Förster (1981)
Sorkh Shales, Central Iran	243	248	22,	326	30	14	290,	9	x	x	x			x	x	5	Wensink (1982)
Abadeh & Jamal Fms., Andes., C. Iran	245	286	35,	307	6	28	311,	4							x	1	Soffel & Förster (1981)
M. Permian Volcanics, Alborz, Iran	255	265	23,	282	—	19	319,	−20	x		x				x	3	Wensink (1979)
Geirud Lavas, S. Alborz, Iran	356	367	0,	212	380	4	43,	−46	x	x	x				x	4	Wensink *et al.* (1978)
Devonian Seds. Combined, Central Iran	360	408	66,	277	10	20	345,	15			x					1	Soffel & Förster (1981)
Kerman Old Red Ss., Central Iran	387	408	51,	196	18	10	26,	4	x	x	x	x		x	x	6	Wensink (1983)

The entries in this Table are for the blocks and mobile belts of the eastern Mediterranean, Iran, Oman, Kopet Dagh and Afghanistan; in the mobile belts rotations can occur and this mandates that reliability criterion 5 generally cannot be checked as being met. The reliability criteria are explained in the text (see also Table 4.1). A '*' is placed in the Q column when the result is not to be used for apparent polar wander path construction or paleolatitude determinations, with the reasons given below. For other explanations see Table A1.

*Poles that would have Q $\geq$ 3 but are excluded from mean pole determinations and other compilations, are listed below with the reasons in parentheses. *Turkey and Cyprus:* Lower Tertiary Mesudiye Dikes (too few samples, no tilt correction); Upper Cretaceous to Lower Tertiary reversed-polarity Pelagic Chalk of Cyprus (overprint suspected because of non-antipodal reversals); Permian to Jurassic Kure Basalts (poorly dated, impossible and anomalous paleolatitude, see text Chapter 7); Triassic Denizlikoy, Gumusdere and Tavsancil rocks (impossible and anomalous paleolatitude, see text Chapter 7).

Table A9. *Phanerozoic pole positions for the Central Asian blocks and mobile belts*

Rock unit, location	Age		Pole position				D,	Lat.	Reliability								Reference
	Lo	Hi	Lat.	Long.	k	α_{95}	30°N,	80°E	1	2	3	4	5	6	7	Q	
I. HIMALAYAN SUTURE AREA (HINDU-KUSH, KARAKORUM, LADAKH, KASHMIR, KROL, KUMAON, NEPAL, S. XIZANG) (internal rotations possible)																	
Pagri Area Seds., Xizang Himalayas	001	005	82,	152	10	7	9,	32	x						x	2	Zhu *et al.* (1981)
Krol Sec'ry Component, Garhwal, India	002	025?	73,	196	97	7	16,	21		x	x			x	x	4	Klootwijk *et al.* (1982)
Ladakh Secondary C-4, Kargil, India	015	046?	77,	222	13	8	8,	19		x	x				x	3	Klootwijk *et al.* (1979)
Ladakh Secondary C-3, Kargil, India	015	046?	75,	179	12	8	17,	27		x	x				x	3	Klootwijk *et al.* (1979)
Krol Prefolding Sec'ry, Garhwal	025	065?	59,	199	45	10	28,	12		x	x			x	x	4	Klootwijk *et al.* (1982)
Kashmir Sediments Sec'ry Component	005	038?	70,	215	18	8	15,	15		x	x			x	x	4	Klootwijk *et al.* (1983)
Ladakh Intrus., C-6, Kargil, India	046	050	63,	250	53	4	5,	3	x	x	x			x	x	5	Klootwijk *et al.* (1979)
Ladakh Intrus., C-5, Kargil, India	046	050	65,	268	14	9	357,	5	x		x			x	x	4	Klootwijk *et al.* (1979)
Jhelum Section, Kashmir, Pakistan	045	055	57,	209	10	9	25,	7	x	x	x			x	x	5	Bossart *et al.* (1989)
Koli Grp. Sec'ry (Collision), Nepal	050	060?	57,	279	59	7	350,	−1		x	x				x	3	Klootwijk & Bingham (1980)
Indus Suture Sec'ry, Ladakh, India	050	060?	36,	329	88	27	311,	2			x			x	x	3	Klootwijk *et al.* (1984)
Indus M., Dras Flysch Sec'ry, Ladakh	050	060?	56,	252	34	12	5,	−4		x	x				x	3	Klootwijk *et al.* (1984)
Panjal Traps Secondary, Kashmir	060	070?	32,	282	20	10	340,	−25		x		x			x	3	McElhinny *et al.* (1978)
Tingri Area Ls., Xizang Himalayas	065	144	33,	299	13	17	327,	−17							x	1	Zhu *et al.* (1981)
Dzong Ss., Kali Gandaki, Nepal	116	119	12,	289	12	5	322,	−40	x	x	x				x	4	Klootwijk & Bingham (1980)
Kagbeni Ss., Kali Gandaki, Nepal	119	131	18,	290	339	14	325,	−34	x	x	x				x	4	Klootwijk & Bingham (1980)
Gyangze Area Slate, Xizang Himalayas	145	163	7,	329	22	11	287,	−14	x						x	2	Zhu *et al.* (1981)
Lumachelle Fm., Kali Gandaki, Nepal	163	188	12,	299	62	16	312,	−34	x	x	x					3	Klootwijk & Bingham (1980)
Jomosom Ls., Kali Gandaki, Nepal	188	208	19,	295	45	19	321,	−30	x		x					2	Klootwijk & Bingham (1980)

Jomosom Qtz., Kali Gandaki, Nepal	208	219	22,	290	11	8	327,	−30	x	x	x					3	Klootwijk & Bingham (1980)
Thinigoan Ls., Kali Gandaki, Nepal	219	238	26,	297	503	11	324,	−24	x	x	x					3	Klootwijk & Bingham (1980)
Khrew, barsu & Naubug Ls., Kashmir	208	245	−20,	305	104	9	273,	−48		x	x			x	x	4	Klootwijk *et al.* (1983)
Tulong Sandst., Xizang Himalayas	208	245	24,	308	22	6	314,	−19		x						1	Zhu *et al.* (1981)
Mukut Limestone, Manang Area, Nepal	227	230	22,	287	30	4	330,	−32	x	x	x	x				4	Appel *et al.* (1991)
Krol-A Ls., Garhwal Himalaya, India	208	286	−4,	290	8	15	307,	−51			x			x	x	3	Klootwijk *et al.* (1982)
Zewan Beds, Barsu, Kashmir	245	258	−28,	310	69	15	260,	−47	x	x	x			x	x	5	Klootwijk *et al.* (1983)
Tini Chu Fm., Kali Gandaki, Nepal	245	286	−7,	305	10	8	289,	−42		x	x				x	3	Klootwijk & Bingham (1980)
Bhimtal (Bhowali) Volcs., Kumaon Himalayas	245	286	−10,	318	25	25	278,	−33			x				x	2	Athavale *et al.* (1980)
Blaini Ls., Garhwal Himalaya, India	258	320	−27,	296	7	8	266,	−58			x			x	x	3	Klootwijk *et al.* (1982)
Lower Blaini Diamict., Garhwal, India	258	320	3,	262	7	12	356,	−57			x			x	x	3	Klootwijk *et al.* (1982)
Syringotheris Ls., Aismukum, Kashmir	356	360	29,	238	8	13	22,	−27	x		x			x		3	Klootwijk *et al.* (1983)
Kuragh Ironst., Chitral, E. Hindu-Kush	360	374	35,	306	14	17	323,	−12	x		x			x		3	Klootwijk & Conaghan (1979)
Rudraprayag Volcs., Kumaon Himalayas	360	438	−30,	348	26	15	244,	−16		x	x				x	3	Athavale *et al.* (1980)
Marhaum Fm., Kashmir	408	458	−12,	200	13	10	86,	−32			x				x	2	Klootwijk *et al.* (1983)
Trahagaum Fm., Kahsmir	458	505	−79,	173	23	4	167,	−30		x	x			x		3	Klootwijk *et al.* (1983)
Yadong Micagneiss, Xizang Himalayas	540	570	−83,	70	13	9	181,	−23	x	x						2	Zhu *et al.* (1981)

Table A9. *Phanerozoic pole positions for the Central Asian blocks and mobile belts – continued*

Rock unit, location	Age		Pole position				D,	Lat.	Reliability								Reference
	Lo	Hi	Lat.	Long.	k	α_{95}	35°N,	90°E	1	2	3	4	5	6	7	Q	
II. TIBET (LHASA, QIANGTANG, KUNLUN) (internal rotations possible)																	
Takena Secondary (LT) Comp., Tibet	005	060?	70,	279	9	20	357,	15			x			x	x	3	Achache *et al.* (1982; 1984)
Xigaze Ophiolite Secondary, Tibet	005	040?	52,	215	16	17	31,	9			x			x	x	3	Pozzi *et al.* (1984)
Dingri Sec'ry Component, Lhasa, Tibet	045	050?	62,	228	84	10	19,	13		x	x			x	x	4	Besse *et al.* (1984b)
Dingri Sediments, Lhasa Block	056	058	51,	308	76	10	337,	2	x	x	x	x		x	x	6	Besse *et al.* (1984b)
Rotated W. Gamnian Suture Zone rocks	050	144	21,	336	41	19	301,	−6			x				x	2	Otofuji *et al.* (1989)
Western Gamnian/Qiangtang rocks	050	163	62,	231	12	28	18,	12			x			x		2	Otofuji *et al.* (1989)
Fenghuoshan Grp., P 23, Qiangtang	050	065	61,	253	29	7	8,	7	x	x	x				x	4	Lin & Watts (1988)
Fenghuoshan Grp., P 24 Qiangtang	050	065	55,	201	79	4	34,	17	x	x	x				x	4	Lin & Watts (1988)
Lingzizong Fm., Volcs., Lhasa Block	047	080	71,	299	26	11	350,	18		x	x	x		x	x	5	Achache *et al.* (1984)
Lingzizong Fm., Volcs., Lhasa Block	047	080	69,	270	69	31	0,	14			x			x		2	Westphal *et al.* (1983)
Lhunzhub Red Beds, Xizang, Lhasa B.	058	097	69,	347	20	5	337,	28		x					x	2	Zhu *et al.* (1981)
Qelico Volcanics, Lhasa, Tibet	085	095	74,	318	32	17	347,	24	x		x					2	Lin & Watts (1988)
(Takena) Tibetan Red Beds	055	119	68,	283	66	5	355,	13								0	Zhu *et al.* (1977)
Zhamo Granodiorite, Xizang, Lhasa B.	065	097	64,	171	48	5	32,	35	x						x	2	Zhu *et al.* (1981)
Takena Fm., Lhasa, Tibet	065	119	68,	279	70	7	357,	13		x	x	x				3	Lin & Watts (1988)
Takena Fm., Tibet	065	119	68,	340	35	10	337,	25		x		x		x		3	Pozzi *et al.* (1982)
Nagqu Volcanics, Lhasa, Tibet	092	100	78,	282	54	6	357,	23	x	x	x					3	Lin & Watts (1988)
Basu Fm., Granites, E. Lhasa	069	127	72,	289	13	19	354,	18			x			x		2	Otofuji *et al.* (1990)
Takena Fm. Red Beds, Lhasa Block	080	119	64,	348	78	8	332,	26	x	x	x	x		x		5	Westphal *et al.* (1983)
Biru Red Beds, Xizang, Lhasa Block	065	144	59,	2	30	4	323,	30		x					x	2	Zhu *et al.* (1981)
Takena Fm., Tibet	091	119	68,	305	37	7	347,	16	x	x	x	x				4	Achache *et al.* (1982; 1984)

Rock unit, location	Age		Pole position				D,	Lat.	Reliability								Reference
	Lo	Hi	Lat.	Long.	k	α_{95}	42°N,	80°E	1	2	3	4	5	6	7	Q	
Xigaze Ophiolite & Flysch, Tibet	094	100	0,	15	18	30	261,	12	x		x	x			x	4	Pozzi *et al.* (1984)
Laoran Fm., Markam, E. Qiangtang	098	120	49,	176	77	9	48,	28	x		x					2	Otofuji *et al.* (1990)
Lhasa Ls., Xizang, Lhasa Block	144	163	59,	280	5	17	355,	4	x							1	Zhu *et al.* (1981)
Batang Group Volcanics, Qiangtang	219	225	59,	184	23	16	35,	27	x		x					2	Lin & Watts (1988)
Kunlun Dikes, Tibet	240	245	76,	237	60	5	8,	23	x	x	x					3	Lin & Watts (1988)
Dagangou Fm., P-41, Kunlun, Tibet	320	340	30,	292	12	28	340,	−22	x		x				x	3	Lin & Watts (1988)
Dagangou Fm., P-39,40, Kunlun, Tibet	320	340	22,	241	9	22	30,	−27	x		x				x	3	Lin & Watts (1988)
III. TARIM, TIAN SHAN, JUNGGAR (internal rotations possible)																	
Bashenjiqike Formation, Kuche, Tarim	065	098	66,	214	114	9	19,	24	x		x	x		x	x	5	Li *et al.* (1988a)
Yageliemu & Kelaza, Paicheng, Tarim	098	163	65,	209	62	9	21,	24		x	x	x		x	x	5	Li *et al.* (1988a)
Shanzhuquan Synfolding, W. Junggar	200	284	51,	179	58	6	44,	26		x	x	x			x	4	Zhao *et al.* (1990)
Kelamayi Granite Remagn., W. Junggar	200	284	63,	183	3206	—	31,	31		x	x				x	3	Zhao *et al.* (1990)
Kuche and Yangshi Red beds, Tarim	208	245	59,	160	26	13	41,	40		x		x		x	x	4	Li (1990)
Kuche Red Beds, Tarim	235	255	72,	188	45	6	21,	34	x	x	x			x		4	McFadden *et al.* (1988)
Tarim Dikes	245	258	66,	181	61	4	29,	34	x	x	x	x				4	Li *et al.* (1988b)
Heavenly Lake Section, Junggar	245	258	76,	193	24	13	16,	35	x	x	x	x		x		5	Nie *et al.* (1992)
Liusugou Sandstones, W. Junggar	245	268	26,	223	12	37	34,	−14	x		x				x	*	Zhao *et al.* (1990)
NE striking Keramayi Dikes, Junggar	245	320	82,	139	49	6	10,	46		x	x					2	Li *et al.* (1989)
ESE striking Keramayi Dikes, Junggar	245	320	66,	254	87	5	3,	18		x	x				x	3	Li *et al.* (1989)
Heshantou-Hoboksar Remagn., Junggar	245	333	69,	350	15	8	333,	39		x	x	x				3	Li *et al.* (1991)
Balikelike Fm. Ls., Aksu-Keping, Tarim	258	286	65,	163	128	4	33,	40	x	x	x	x				4	Sharps *et al.* (1989)
Biyoulitie Basalts, Aksu-Keping, Tarim	258	286	55,	172	66	4	43,	32	x	x	x	x				4	Sharps *et al.* (1989)

Table A9. *Phanerozoic pole positions for the Central Asian blocks and mobile belts – continued*

Rock unit, location	Age		Pole position				D,	Lat.	Reliability								Reference
	Lo	Hi	Lat.	Long.	k	α_{95}	42°N,	80°E	1	2	3	4	5	6	7	Q	
Aksu Region Extr., Sediments, Tarim	258	286	57,	190	42	5	34,	25	x	x	x					3	Bai *et al.* (1987)
Upper Carboniferous Sections, Tarim	286	320	41,	160	823	4	62,	32	x	x		x			x	4	Zhai *et al.* (1988, in: Li (1990))
Kangkelin Formation, Tarim	286	320	52,	180	79	9	43,	27	x	x	x					3	Bai *et al.* (1987)
Qijiagu-Aoertou Fms., Junggar	286	333	72,	7	28	7	335,	45		x	x	x			x	4	Li *et al.* (1991)
Bulungu-Hoboksar Volcs., Junggar	296	333	70,	299	36	6	346,	25	x	x	x	x			x	5	Li *et al.* (1991)
Keziertage & Yimugantawu Fms., Tarim	360	374	−11,	151	23	8	111,	6	x	x	x					3	Bai *et al.* (1987)
Devonian Red Beds, Tarim	360	408	17,	165	25	4	80,	15		x	x	x		x	x	5	Li *et al.* (1990)

Rock unit, location	Age		Pole position				D,	Lat.	Reliability								Reference
	Lo	Hi	Lat.	Long.	k	α_{95}	45°N,	115°E	1	2	3	4	5	6	7	Q	
IV. MONGOLIA BEFORE LATE JURASSIC TIME (internal rotations possible)																	
Triassic Sandstones	208	245	25,	216	3	26	177,	70						x	x	2	Jelen *et al.* (1966)
Wudan, Shizhangfang Andes., Seds.	245	268	52,	342	46	7	332,	15	x	x	x			x	x	5	Zhao *et al.* (1990)
Permian Volcanics, Outer Mongolia	245	286	32,	12	6	9	302,	14			x			x	x	3	Pruner (1987)
Balinzuo Qi Limest., Inner Mongolia	286	320	33,	338	39	4	325,	−3		x	x	x			x	4	Zhao *et al.* (1990)
Carboniferous Tuffs, Outer Mongolia	286	360	41,	307	23	4	351,	−3		x	x				x	3	Pruner (1987)

The entries in this Table are for the blocks and mobile belts of the Himalayas, Tibet, Tarim, Tian Shan, Junggar and Mongolia; in these mobile belts rotations can occur and this mandates that reliability criterion 5 generally cannot be checked as being met. The reliability criteria are explained in the text (see also Table 4.1). A '*' is placed in the Q column when the result is not to be used for apparent polar wander path construction or paleolatitude determinations, with the reasons given below. For other explanations see Table A1.
*Poles that would have $Q \geq 3$ but are excluded from mean pole determinations and other compilations, are listed below with the reasons in parentheses. *Junggar:* Permo-Triassic Liusugou Sandstones (three sites, statistical parameters worsen upon bedding correction).

Table A10. *Phanerozoic pole positions for the Far Eastern blocks and mobile belts*

Rock unit, location	Age		Pole position				D,	Lat.	Reliability								Reference
	Lo	Hi	Lat.	Long.	k	α_{95}	20°N,	100°E	1	2	3	4	5	6	7	Q	
I. INDO-CHINA (internal rotations possible)																	
Vietnam Basalts	000	007	86,	313	25	5	358,	17	x	x			x	x	x	5	Giang Nguyen (1982)
Rattanakiri Complex Comb., Cambodia	000	002	85,	307	—	—	358,	15	x				x	x	x	4	Lacombe & Roche (1970)
Khorat Plateau Basalts, E. Thailand	000	011	86,	171	15	5	4,	21	x	x	x		x	x	x	6	McCabe *et al.* (1988)
Chao Phraya Basin Flows, C. Thailand	000	011	66,	190	19	7	25,	18	x	x	x			x	x	5	McCabe *et al.* (1988)
Denchai Basalts, W. Thailand	005	007	71,	193	148	5	20,	18	x						x	2	Barr & MacDonald (1979)
Remagnetized Paleo-, Mesozoic, Thailand	001	200?	59,	175	25	8	33,	25		x	x			x		*	Chen Yan & Courtillot (1989)
Cretaceous, Thailand	066	144	79,	176	36	43	12,	22								0	Bunopas (1982)
M.-U. Jurassic Thai Red Sediments	144	187	80,	111	13	26	2,	30	x							1	Barr *et al.* (1978)
M.-U. Jurassic Thai Sediments	144	187	67,	177	10	14	25,	23	x	x						2	Bunopas (1982)
Khorat Group & Low Sak Tuff, Thailand	098	230	57,	188	45	9	35,	18		x	x			x		*	Maranate & Vella (1986)
Jurassic Red Beds Combined, Thailand	144	213	69,	177	20	12	22,	23								0	Haile & Tarling (1975)
Jurassic Red Beds, Thailand	144	213	43,	202	—	15	46,	5			x				x	2	Bunopas & Pitakpaivan (1978)
Lower Jurassic Sediments, Thailand	187	208	53,	180	19	9	40,	22	x	x						2	Bunopas (1982)
U. Triassic-L. Jurassic Seds., Thail.	187	230	60,	185	11	18	32,	20								0	Bunopas (1982)
Tr-J Thai Red Sediments, Thailand	188	243	56,	176	32	12	37,	24								0	Barr *et al.* (1978)
Huai Hin Lat Fm., Khorat Grp. Thailand	219	225	49,	172	106	8	44,	27	x	x	x			x		*	Achache & Courtillot (1985)

Rock unit, location	Age		Pole position				D,	Lat.	Reliability								Reference
	Lo	Hi	Lat.	Long.	k	α_{95}	18°N,	95°E	1	2	3	4	5	6	7	Q	
II. SIBUMASU, EAST MALAYA AND EAST SUMATRA (internal rotations possible)																	
Toba Tuffs, W. Sumatra	000	001	85,	183	28	5	5,	18	x	x			x	x	x	5	Yokoyama & Dharma (1982)
Kuantan Area Lavas, E. Malaysia	001	003	79,	317	6	54	353,	10	x			x			x	3	Haile *et al.* (1983)

Table A10. *Phanerozoic pole positions for the Far Eastern blocks and mobile belts – continued*

Rock unit, location	Age		Pole position				D,	Lat.	Reliability								Reference
	Lo	Hi	Lat.	Long.	k	α_{95}	18°N,	95°E	1	2	3	4	5	6	7	Q	
Granodiorite & Metabasalt, W. Sumatra	023	037	56,	190	—	12	35,	12	x						x	2	Haile (1979b)
Volcanics and Sediments, E. Sumatra	001	065	84,	130	20	10	4,	23		x				x	x	3	Sasajima *et al.* (1978)
Diorite & metavolcanics, W. Sumatra	023	066	82,	183	—	6	8,	18							x	1	Haile (1979b)
Seganat Basalts & Dikes, Malaya	062	110	44,	35	16	9	311,	34	x					x	x	3	McElhinny *et al.* (1974b)
Kuantan Area dikes, E. Malaysia	077	131	58,	52	28	6	332,	39		x				x	x	3	Haile *et al.* (1983)
Pahang and Johore Red beds, E. Malaya	098	163	71,	41	20	8	343,	28		x					x	2	Haile & Khoo Han Peng (1980)
Limestones, W. Sumatra	208	230	49,	209	—	7	37,	−1	x						x	2	Haile (1979b)
Limestones and Shales, W. Sumatra	208	245	55,	213	—	32	30,	0						x	x	2	Sasajima *et al.* (1978)
Iwanshui Fm., West Yunnan, China	240	245	35,	342	—	7	311,	−7	x		x				x	3	Chan *et al.* (1984)
Pengerang Rhyolite, W. Malaya	240	258	−57,	332	25	10	213,	−33						x	x	2	McElhinny *et al.* (1974b)
Padang Area Basalts, W. Sumatra	245	286	47,	358	—	5	317,	8							x	1	Sasajima *et al.* (1978)
Singapore Gabbro & Dikes, E. Malaya	240	500?	−72,	55	17	19	192,	−4							x	1	McElhinny *et al.* (1974b)
Sempah Cgl. & Rhyolite, W. Malaya	258	320	−55,	344	14	11	217,	−27							x	1	McElhinny *et al.* (1974b)
Woniusi Fm., Flows, Baoshan, W. Yunnan	275	300	17,	258	40	9	27,	−51	x	x	x	x			x	5	Huang & Opdyke (1991)
Woniusi Fm., Flows, Yongde, W. Yunnan	275	300	−34,	230	74	12	121,	−47	x	x	x				x	4	Huang & Opdyke (1991)
Bentong Grp., Singa Fm., W. Malaya	320	360	−57,	2	25	11	215,	−17	x						x	2	McElhinny *et al.* (1974b)
Devonian Ls., West Yunnan, China	360	408	−67,	314	21	6	198,	−35		x	x	x		x	x	5	Fang *et al.* (1989)
Upper Setul Ls., Langkawi, W. Malaya	408	421	−17,	259	21	20	89,	−75	x						x	2	Haile (1980)
Lower Setul Ls., Langkawi, W. Malaya	438	505	−46,	256	33	12	155,	−58		x					x	2	Haile (1980)

Rock unit, location	Age		Pole position				D,	Lat.	Reliability								Reference
	Lo	Hi	Lat.	Long.	k	α_{95}	37°N,	128°E	1	2	3	4	5	6	7	Q	
III. KOREA (internal rotations possible)																	
Gyeongsang Seds. & Flows South Korea	065	090	69,	191	85	8	26,	44	x	x	x		x		x	5	Otofuji *et al.* (1983)
Sandstones, North Korea	066	098	69,	182	—	4	25,	47	x	x			x		x	4	Gurariy *et al.* (1966)
Gyeongsang Group Seds., South Korea	065	144	64,	195	195	6	33,	43		x		x	x		x	4	Otofuji *et al.* (1986)
South Korean Igneous rocks	065	144	74,	222	44	10	19,	34		x			x		x	3	Kienzle & Scharon (1966)
Cretaceous Granites, South Korea	071	121	87,	186	66	8	3,	38		x			x			2	Ito & Tokeida (1980)
North Korean Cretaceous Combined	098	144	76,	183	260	3	16,	44	x	x			x		x	4	Kang & Li (1977)
Gyeongsang Extrus., Seds., S. Korea	110	130	68,	205	55	5	28,	39	x	x	x	x	x	x	x	7	Lee *et al.* (1987)
North Korean Intr., Seds., Combined	144	178	63,	174	41	9	32,	52	x	x					x	3	Kang & Li (1977)
Okchŏn Belt Intr., Seds., S. Korea	121	191	69,	25	32	11	336,	30		x	x			x	x	4	Kim & Van der Voo (1990)
Ryongnam Massif Intr., South Korea	133	178	63,	176	211	9	33,	51		x	x				x	3	Kim & Van der Voo (1990)
Kyonggi Massif Intr., Seds., S. Korea	157	202	54,	213	30	17	44,	32			x			x	x	3	Kim & Van der Voo (1990)
Bansong, Nampo Fms., Okchŏn, S. Korea	144	208	49,	39	11	16	312,	28							x	1	Otofuji *et al.* (1986)
Ryongnam & Okchŏn Intr., Seds., S.K.	208	245	34,	216	27	15	63,	21		x	x			x	x	3	Kim & Van der Voo (1990)
Nogam Fm., Okchŏn Belt, S. Korea	208	245	60,	30	20	12	326,	28		x					x	2	Otofuji *et al.* (1986)
Nogam Fm., Okchŏn Belt, S. Korea	208	245	11,	186	19	29	101,	32							x	1	Shibuya *et al.* (1988)
Gobangsan Fm., Okchŏn Belt, S. Korea	245	286	55,	2	17	13	332,	13		x					x	2	Otofuji *et al.* (1986)
Gobangsan Fm., Okchŏn Belt, S. Korea	245	286	40,	233	1000	4	50,	13							x	1	Shibuya *et al.* (1988)

Table A10. *Phanerozoic pole positions for the Far Eastern blocks and mobile belts – continued*

Rock unit, location	Age		Pole position				D,	Lat.	Reliability								Reference
	Lo	Hi	Lat.	Long.	k	α_{95}	37°N,128°E		1	2	3	4	5	6	7	Q	
Hongjeom Fm., Okchŏn Belt, S. Korea	286	360	2,	216	24	—	90,	3							x	1	Shibuya *et al.* (1988)
Middle Cambrian Sandstones, N. Korea	523	540	8,	313	40	5	353,	−45	x						x	2	Gurariy *et al.* (1966)
Lower Cambrian Sandstones, N. Korea	540	570	14,	321	250	2	344,	−38	x						x	2	Gurariy *et al.* (1966)

Rock unit, location	Age		Pole position				D,	Lat.	Reliability								Reference
	Lo	Hi	Lat.	Long.	k	α_{95}	0°N, 110°E		1	2	3	4	5	6	7	Q	
IV. BORNEO (internal rotations possible)																	
Redbeds, West Sarawak, Sarawak	005	037	88,	10	—	4	358,	0	x						x	2	Haile (1981b)
Nanga Raun Basic Sills, W. Kalimantan	005	037	85,	241	—	5	4,	−3	x						x	2	Untung *et al.* (1987)
Non-deflected Intrusions, W. Sarawak	015	030	89,	24	79	10	359,	0	x	x				x	x	4	Schmidtke *et al.* (1990)
Deflected Intrusions, W. Sarawak	015	030	69,	25	20	15	339,	2	x	x				x	x	4	Schmidtke *et al.* (1990)
Silantek Red Beds, Sarawak	005	044	88,	10	169	4	358,	0	x						x	2	Haile (1979a)
Silanted Red and Grey Beds, Sarawak	005	044	47,	39	196	9	319,	13	x					x	x	3	Schmidtke *et al.* (1990)
Kalasin Area Sediments, C. Kalimantan	037	058	53,	25	—	14	323,	3	x	x					x	3	Untung *et al.* (1987)
Schwaner Zone Igneous, W. Kalimantan	080	090	41,	21	8	8	311,	1	x					x	x	3	Haile *et al.* (1977)
Pedawan Fm. Seds., Volcs., Sarawak	065	163	−1,	12	67	15	269,	−8					x	x	x	3	Schmidtke *et al.* (1990)
Bau Limest. Fm., Sarawak	098	163	2,	16	46	5	272,	−4					x		x	2	Schmidtke *et al.* (1990)
Kedadom Fm., Limest., Sarawak	144	163	41,	18	3	23	311,	−1	x							1	Schmidtke *et al.* (1990)
Tenguwe Region Sed., W. Kalimantan	144	208	−3,	21	-	7	267,	1		x					x	2	Untung *et al.* (1987)

Rock unit, location	Age Lo	Age Hi	Pole position Lat.	Pole position Long.	k	α_{95}	D, 5°S,130°E	Lat. 5°S,130°E	Reliability 1	2	3	4	5	6	7	Q	Reference
V. EAST INDONESIA, SULAWESI, NEW GUINEA (internal rotations possible)																	
Sulawesi Lavas & Sills	000	002	81,	359	32	16	353,	−11	x					x	x	3	Haile (1978)
Baluan & Other Basalts, New Guinea	000	001	86,	330	21	6	359,	−9	x	x					x	3	Manwaring (1971)
Pinogu Volcanics, Sulawesi	000	005	83,	345	122	4	356,	−11	x		x				x	3	Otofuji *et al.* (1981b)
Lower Edie Porphyry, Papua New Guinea	002	005	63,	36	37	15	333,	−6	x						x	2	Rogerson & Moaina (1987)
Kwikla Volcanics, Papua New Guinea	002	005	34,	55	3	—	306,	9	x						x	2	Rogerson & Moaina (1987)
Oecussi Volcanics, West Timor	004	006	54,	51	—	12	325,	2	x		x			x	x	4	Chamalaun & Sunata (1982)
Neogene Volcanics, Sulawesi	000	013	84,	55	21	17	354,	−3	x						x	2	Sasajima *et al.* (1980)
Sumba Fm. Tuffs. Sumba	002	011	82,	249	—	8	7,	−9	x		x				x	3	Chamalaun & Sunata (1982)
Jawilla, Edeklara Volcs., Sumba	002	011	82,	245	—	10	7,	−8	x		x				x	3	Chamalaun & Sunata (1982)
Manamas Fm. Combined, W. Timor	006	008	40,	28	8	18	310,	−12	x		x			x	x	4	Wensink & Hartosukohardjo (1990a)
Papua New Guinea Extrusives	002	025	83,	208	14	17	7,	−3	x						x	2	Manwaring (1974)
Papua New Guinea Extrusives	002	025	6,	232	4	30	85,	−12	x						x	2	Manwaring (1974)
Kelang Lavas, Seram	005	010	16,	34	37	4	286,	−7	x	x					x	3	Haile (1981a)
Yonkie & Bismark Intr., Papua New G.	005	024	10,	219	8	25	80,	0	x						x	2	Rogerson & Moaina (1987)
Bauine Granodiorite, Papua New Guinea	005	024	55,	220	2	—	35,	−4	x						x	2	Rogerson & Moaina (1987)
Miocene of Sumba, Sumbawa & Flores	005	025	88,	231	11	16	2,	−5	x	x				x	x	4	Otofuji *et al.* (1981a)
Bilangula Volcanics, Sulawesi	005	025	40,	218	2	—	50,	−2	x		x			x	x	*	Otofuji *et al.* (1981b)
Quelicai Volcanics, East Timor	000	065	75,	264	—	16	11,	−15			x				x	2	Chamalaun & Sunata (1982)
Sapphire Creek Gabbro, Papua New G.	002	065	42,	220	3	—	48,	−3							x	1	Manwaring (1974)
Menado Lavas, Sulawesi	005	065	63,	187	21	58	23,	10			x				x	*	Otofuji *et al.* (1981b)

Table A10. *Phanerozoic pole positions for the Far Eastern blocks and mobile belts – continued*

Rock unit, location	Age		Pole position				D,	Lat.	Reliability								Reference
	Lo	Hi	Lat.	Long.	k	α_{95}	5°S,	130°	1	2	3	4	5	6	7	Q	
Tinombo Fm., Sulawesi	015	055	−8,	209	6	56	97,	12	x		x			x	x	*	Otofuji *et al.* (1981b)
Paleogene, Miocene Volcs., Sulawesi	019	063	45,	37	7	28	315,	−6						x	x	2	Sasajima *et al.* (1980)
Tanadaro Granodiorite, dikes, Sumba	063	075	48,	222	—	9	42,	−5	x		x				x	3	Chamalaun & Sunata (1982)
Fafanlap Fm., Misool	068	072	70,	321	143	5	356,	−25	x	x	x				x	4	Wensink *et al.* (1989)
South Coast Volcs., Dikes, Sumba	075	080	52,	117	—	7	351,	32	x		x				x	3	Chamalaun & Sunata (1982)
U. Cretaceous Sediments, Sulawesi	065	098	71,	347	31	22	348,	−20	x					x		2	Sasajima *et al.* (1980)
Waaf Fm., Misool	084	088	46,	16	97	5	318,	−20	x	x	x	x		x	x	6	Wensink *et al.* (1989)
Wai Bua Fm., East Timor	098	144	48,	141	34	11	9,	36	x		x	x			x	4	Chamalaun & Sunata (1982)
SW Arm Radiolar. Chert, Sulawesi	098	163	55,	32	60	7	325,	−9							x	1	Haile (1978)
SE Arm Radiolar. Chert, Sulawesi	098	163	−26,	340	109	9	222,	−47							x	1	Haile (1978)
Nakfunu Fm. I, West Timor	120	144	51,	88	13	19	333,	23	x		x			x	x	*	Wensink *et al.* (1987)
Nakfunu Fm. II, West Timor	120	144	55,	11	13	27	328,	−20	x		x			x	x	*	Wensink *et al.* (1987)
Jurassic Sediments, Sumba Island	144	208	32,	234	478	6	58,	−14		x				x	x	3	Otofuji *et al.* (1981a)
Punala River Shale, Seram	208	230	−8,	27	15	12	261,	−12	x						x	2	Haile (1981a)
Aituto Fm., West Timor	215	230	26,	2	212	6	299,	−36	x	x	x				x	4	Wensink & Hartosukohardjo (1990b)
Cribas Fm., East Timor	218	270	57,	340	14	9	341,	−33		x		x		x		3	Chamalaun (1977a)
Maubisse Fm., West Timor	240	263	4,	354	48	7	271,	−46	x	x	x	x			x	5	Wensink & Hartosukohardjo (1990b)
Maubisse Fm., East Timor	245	286	48,	5	—	13	322,	−27								0	Chamalaun (1977b)

Rock unit, location	Age		Pole position				D,	Lat.	Reliability								Reference
	Lo	Hi	Lat.	Long.	k	α_{95}	15°N,	120°E	1	2	3	4	5	6	7	Q	
VI. LUZON, MARINDUQUE (N. PHILIPPINES) (internal rotations possible)																	
Luzon Sediments, Intrusives	000	005	83,	327	79	5	357,	9	x	x				x	x	4	Fuller *et al.* (1983)
Luzon & Marinduque Volcs., Seds.	000	011	83,	329	81	4	357,	9	x	x				x	x	4	McCabe *et al.* (1982a)
Tarlac Area Plug, Zambales, Luzon	001	011	75,	238	30	11	13,	8	x		x				x	3	Fuller *et al.* (1989)
Luzon U. Miocene Extrus., Sediments	005	011	85,	272	21	12	2,	11	x	x				x	x	4	Fuller *et al.* (1983)
Late Miocene Volcanics, W-C. Luzon	005	011	74,	211	46	7	17,	14	x	x	x	x		x	x	6	McCabe *et al.* (1987)
Late Miocene Extrus., Seds., Luzon	005	011	84,	359	64	8	355,	12	x	x	x					3	McCabe *et al.* (1987)
Luzon E. Miocene Intrus., Sediments	011	024	69,	32	35	7	338,	15	x	x				x	x	4	Fuller *et al.* (1983)
Zambales Fm., Barlo Area, Luzon	011	024	33,	18	94	13	305,	−2	x		x			x	x	4	Fuller *et al.* (1989)
Marinduque Island Volcanics	011	024	33,	29	11	18	302,	7	x		x			x	x	4	McCabe *et al.* (1987)
Moriones Fm., Zambales, Luzon	015	024	72,	24	—	—	342,	12	x		x				x	3	Fuller *et al.* (1989)
Sual Area (Sec'ry?), Zambales, Luzon	005	048?	36,	16	17	12	308,	−2		x	x					2	Fuller *et al.* (1989)
Coto Mine Dikes, Zambales, Luzon	025	036	−33,	30	115	6	238,	−8	x	x	x				x	4	McCabe *et al.* (1987)
Luzon Oligocene Intrus., Sediments	025	038	44,	20	5	47	315,	3	x					x	x	3	Fuller *et al.* (1983)
Aksitero Fm., Zambales, Luzon	024	052	−6,	35	—	—	263,	3	x		x			x	x	4	Fuller *et al.* (1989)
Acoje Block, Zambales, Luzon	030	050	20,	34	21	15	289,	9	x	x	x			x	x	5	Fuller *et al.* (1989)
Barlo Chert-Spilite, Zambales, Luzon	041	047	0,	34	5	65	269,	4	x		x				x	3	Fuller *et al.* (1989)
Coto Mine Intrus., Zambales, Luzon	042	052	−32,	45	21	9	235,	4	x	x	x				x	4	Fuller *et al.* (1989)
Tarlac Area Zambales Ophiol., Luzon	038	055	10,	28	23	12	280,	1	x	x	x				x	4	Fuller *et al.* (1989)
Zambales Ophiolite, Luzon	038	055	23,	19	19	11	295,	−4	x	x				x	x	4	Fuller *et al.* (1989)
VII. CENTRAL AND SOUTHERN PHILIPPINE ISLANDS (internal rotations possible)																	
Late Neogene Lavas, Leyte	002	011	89,	13	30	7	359,	15	x	x	x			x	x	5	Cole *et al.* (1989)
Early Neogene Lavas, Leyte	011	024	67,	207	36	21	24,	15	x		x			x	x	4	Cole *et al.* (1989)
Panay, Cebu & Mindanao Extr., Seds.	011	024	66,	215	27	8	24,	12	x	x	x				x	4	McCabe *et al.* (1987)
Panay Island Basalts, Sediments	011	033	70,	209	70	8	21,	14	x	x		x		x	x	5	McCabe *et al.* (1982b)

Table A10. *Phanerozoic pole positions for the Far Eastern blocks and mobile belts – continued*

Rock unit, location	Age		Pole position				D, Lat.		Reliability								Reference
	Lo	Hi	Lat.	Long.	k	α_{95}	15°N,120°E		1	2	3	4	5	6	7	Q	
Sara Diorite, Panay	033	042	19,	206	24	12	73,	9	x						x	2	McCabe *et al.* (1982b)
Pandan Fm., Cebu	098	144	−26,	200	72	15	118,	2	x		x	x			x	4	McCabe *et al.* (1987)
Mansalay Fm., Mindoro	144	208	17,	199	17	19	76,	15			x			x	x	3	McCabe *et al.* (1987)
VIII. PALAU																	
Arakabesan Member, Palau	015	024	35,	228	20	17	52,	−5	x			x		x	x	4	Haston *et al.* (1988)
Medorm Member, Palau	025	035	23,	225	127	22	64,	−7	x					x	x	3	Haston *et al.* (1988)
Aimelik Member, Palau	030	040	−12,	231	37	43	97,	−23	x					x	x	3	Haston *et al.* (1988)
Babeldaup Fm., Palau	033	043	22,	216	56	34	67,	0	x					x	x	3	Haston *et al.* (1988)

The entries in this Table are for the blocks and mobile belts of the Far East; in these mobile belts rotations can occur and this mandates that reliability criterion 5 generally cannot be checked as being met. The reliability criteria are explained in the text (see also Table 4.1). A '*' is placed in the Q column when the result is not to be used for apparent polar wander path construction or paleolatitude determinations, with the reasons given below. For other explanations see Table A1.

*Poles that would have $Q \geq 3$ but are excluded from mean pole determinations and other compilations, are listed below with the reasons in parentheses. *Indo-China*: Paleozoic and Mesozoic sequences studied by Chen Yan & Courtillot (remagnetized); Mesozoic Khorat Group studied by Maranate & Vella and the Triassic Huai Hin Lat Formation studied by Achache & Courtillot (remagnetizated according to Chen Yan & Courtillot). *East Indonesia:* Tertiary Bilangula volcanics, Menado Lavas and Tinombo Lavas, all of Sulawesi (contaminated by overprint); Lower Cretaceous Nakfunu Formation of Timor (internally inconsistent results).

Table A11. *Pre-Late Carboniferous pole positions for the Iapetus blocks and terranes*

Rock unit, location	Age		Pole position				D,	Lat.	Reliability								Reference
	Lo	Hi	Lat.	Long.	k	α_{95}	54°N,	4°W	1	2	3	4	5	6	7	Q	
I. CALEDONIAN OPHIOLITES, NORTHERN BRITAIN – PRE-LATEST SILURIAN (internal rotations possible)																	
Slockenray Fm., Ballantrae, Scotland	468	488	−12,	120	—	4	68,	−29	x	x	x	x				4	Trench *et al.* (1988)
Unst Ophiol. (second.?), E. Shetland	438	517	−47,	64	26	6	135,	−26		x	x			x	x	4	Taylor (1988)
II. LAKE DISTRICT OF BRITAIN, SOUTH OF IAPETUS SUTURE – PRE-LATEST SILURIAN (internal rotations possible)																	
Comb. Siluro-Devonian, Lake District	360	438	22,	149	11	30	25,	−11						x	x	2	Piper (1979a)
Lamprophyre Dike 1, Lake District	360	410	−43,	213	70	9	279,	−63							x	1	Piper (1979a)
Round Knott Dolerite, Lake District	418	468	20,	173	16	10	3,	−16							x	1	Piper (1979a)
Cautley Rhyolite, Lake District	438	446	51,	162	8	13	9,	16	x							1	Piper (1979a)
Carrock Fell Gabbro, Lake District	420	468	−17,	211	14	13	310,	−44		x	x				x	3	Trench & Torsvik (1991)
Stockdale Rhyolite, Lake District	448	458	−82,	123	17	14	168,	−58	x							1	Piper (1979a)
U. Borrowdale Volcs., Lake District	453	461	18,	208	15	11	329,	−13	x	x	x			x	x	5	Piper (1979a)
Borrowdale Volcanics, Lake District	453	461	0,	203	14	7	328,	−32	x	x	x	x			x	5	Faller *et al.* (1977)
High Ireby Fm., Eycott, Lake District	455	469	12,	177	30	8	359,	−24	x			x			x	3	Trench & Torsvik (1991)
Binsey Fm., Eycott, Lake District	461	475	3,	165	17	10	13,	−32	x			x			x	3	Trench & Torsvik (1991)
III. WALES, SE IRELAND, AND MIDLAND CRATON IN BRITAIN – PRE-LATEST SILURIAN (internal rotations possible)																	
Llanddwyn Grp. 2 (Remagn.?), Wales	245?	438	48,	176	461	7	0,	12								0	Piper (1976b)
Remagnetized Ordov. Comb., Ireland	280?	505	36,	168	6	30	7,	0			x					1	Morris (1976)
Caerfai Series (Remagn.?), Wales	300?	570	26,	169	21	9	6,	−10		x		x				2	Briden *et al.* (1971c)
Warren House Group, Welsh Borderlands	270	570	41,	152	11	9	18,	7			x					1	Piper (1982)

Table A11. *Pre-Late Carboniferous pole positions for the Iapetus blocks and terranes – continued*

Rock unit, location	Age		Pole position				D,	Lat.	Reliability								Reference
	Lo	Hi	Lat.	Long.	k	α_{95}	54°N,	4°W	1	2	3	4	5	6	7	Q	
Llanddwyn Dikes Grp. 1, Wales	386	438	7,	130	29	10	54,	−4		x					x	2	Piper (1976b)
Tortworth Inlier Traps, S. England	421	438	−9,	107	12	—	78,	−20	x					x	x	3	Morris *et al.* (1973)
Somerset/Glouc. Lavas, S. England	426	432	−8,	129	29	14	57,	−31	x	x		x		x		4	Piper (1975c)
Breidden Hills, NE Wales	421	458	0,	197	16	12	335,	−33		x				x	x	3	Trench & Torsvik (1991)
Builth Upper Intrusives, C. Wales	438	458	−2,	182	37	11	352,	−38	x	x				x	x	4	Piper & Briden (1973)
Welsh Borderland Tuffs	440	478	23,	178	—	—	358,	−13	x	x		x			x	*	Piper (1978b)
Tramore Volcanics, SE Ireland	454	468	−11,	198	20	9	330,	−44	x	x	x	x		x	x	6	Trench & Torsvik (1991)
Builth Volcanics, C. Wales	464	472	3,	175	22	4	1,	−33	x	x	x	x			x	5	Briden & Mullan (1984)
Builth (Llanelwedd) Volcs., C. Wales	464	472	3,	184	24	7	349,	−33	x	x	x	x			x	5	Trench *et al.* (1991)
Shelve Intrusions, Wales	460	480	−10,	226	15	9	299,	−31	x	x	x	x		x	x	6	Trench & Torsvik (1991)
Stapeley Volcs., Shelve Inlier, Wales	470	478	−27,	216	89	5	297,	−50	x	x	x	x			x	5	McCabe & Channell (1989)
Cadr Idris Basalts, N. Wales	438	505	−73,	229	13	14	209,	−61		x					x	2	Thomas & Briden (1976)
North Wales Intrusives	438	505	−68,	108	17	5	141,	−56		x		x			x	3	Thomas & Briden (1976)
Treffgarne lavas, seds., SW Wales	481	496	−56,	126	—	6	115,	−62	x	x	x	x			x	5	Torsvik & Trench (1991)
Fishguard Volcs. (Secondary?), Wales	478	505	2,	119	5	18	61,	−17								0	Morris *et al.* (1973)
Longmyndian, Welsh Borderland	510	570	−36,	207	20	6	297,	−62		x	x				x	3	Smith & Piper (1984)
Old Radnor Seds., Welsh Borderlands	510	570	−22,	67	40	—	118,	−7		x					x	2	Piper (1982)
Leicester Diorites, S. England	546	552	−63,	321	136	11	198,	−30	x		x				x	3	Duff (1978a)
Batch Volcanics, Welsh Borderland	530	570	−35,	46	10	22	141,	−9			x				x	2	Smith & Piper (1984)
Bangor Red Beds, N. Wales	540	575	−54,	312	13	21	207,	−24	x						x	2	Duff (1978a)
Post-Uriconian Dikes, Welsh Borderl.	540	664	−37,	257	15	12	252,	−14		x					x	2	Piper (1979b)
Uriconian, Welsh Borderland	542	664	−14,	272	17	13	257,	−8		x		x			x	3	Piper (1979b)

Rock unit, location	Age Lo	Age Hi	Pole position Lat.	Pole position Long.	k	α_{95}	D, Lat. 47°N,	0°E	Reliability 1	2	3	4	5	6	7	Q	Reference
IV. ARMORICA (FRANCE, GERMANY, CZECHOSLOVAKIA, POLAND AND IBERIA) – PRE-LATE CARBONIFEROUS (internal rotations possible)																	
Montmartin Remagn., Normandy, France	280?	540	38,	145	27	12	27,	1		x	x					*	Perroud *et al.* (1985)
Trégastel-P. Granite, Trégor, France	295	306	34,	152	121	7	23,	−5	x	x	x				x	4	Duff (1979)
Champ du Feu Massif, Vosges, France	284	350	−16,	109	60	12	87,	−24			x				x	2	Edel *et al.* (1986)
Group B Metam., Massif Central, France	310	320	12,	111	—	6	66,	−5	x	x	x			x	x	5	Edel (1987)
Flamanville Granite, Normandy, France	300	340	30,	151	—	15	25,	−9	x		x				x	3	Van der Voo & Klootwijk (1972)
St. Malo Dikes, N. Brittany, France	313	331	30,	148	110	4	28,	−8	x	x	x	x			x	5	Perroud *et al.* (1986a)
Beja Gabbro, S. Portugal	315	337	7,	149	15	16	36,	−29	x		x			x	x	4	Ruffet (1990)
Group C Metam., Massif Central, France	320	335	−30,	79	—	8	119,	−15	x	x	x			x	x	5	Edel (1987)
Laval Syncl. Volcs., Seds., Brittany	300	360	21,	137	168	19	41,	−12			x				x	2	Edel & Coulon (1984)
Alba Griotte (S. Emiliano), Asturias	315	352	4,	249	53	2	288,	−11	x	x	x	x			x	5	Bonhommet *et al.* (1981)
Alba & Candas Red Beds, Asturias	315	352	35,	175	39	9	4,	−8	x	x	x	x			x	5	Perroud (1983)
Vosges, Frankenwald Sediments	320	360	41,	163	214	14	13,	−1		x	x	x		x		4	Bachtadse *et al.* (1983)
Harz Greywacke, Germany	320	360	42,	200	—	5	345,	1		x	x	x			x	4	Bachtadse *et al.* (1983)
Visean Volcanics, Massif Central, France	333	348	8,	104	19	12	74,	−4	x	x	x	x		x	x	6	Edel *et al.* (1981)
Vosges Visean Volcanics, France	333	352	25,	228	15	9	317,	−6	x	x	x	x		x	x	6	Edel *et al.* (1984)
Group D Metam., Massif Central, France	350	379	−37,	110	—	9	106,	−39	x	x	x			x	x	5	Edel (1987)
Harz, Frankenw. Seds., Volcs, Germany	360	387	30,	189	56	10	352,	−13	x	x	x	x		x	x	6	Bachtadse *et al.* (1983)
Group B Dolerites, Jersey, UK	360	500	1,	159	59	8	27,	−39		x	x					2	Duff (1980)
San Pedro Red Beds, Asturias, Spain	401	421	3,	235	30	10	299,	−21	x	x	x	x				4	Perroud & Bonhommet (1984)

Table A11. *Pre-Late Carboniferous pole positions for the Iapetus blocks and terranes – continued*

Rock unit, location	Age		Pole position				D, Lat.		Reliability								Reference
	Lo	Hi	Lat.	Long.	k	α_{95}	47°N,	0°E	1	2	3	4	5	6	7	Q	
Lamprophyre Dikes, Jersey, UK	400	440	−16,	142	16	24	58,	−46			x				x	2	Duff (1980)
Almadén Volcanics, S. Meseta, Spain	421	438	8,	118	12	14	64,	−12	x	x	x	x		x	x	6	Perroud *et al.* (1991)
Thouars Intr., Vendée/S. Brittany, France	435	453	−34,	185	27	5	342,	−76	x	x	x			x	x	5	Perroud & Van der Voo (1985)
Crozon Dolerites, W. Brittany, Fr.	438	458	−3,	178	34	8	3,	−46	x		x				x	3	Perroud *et al.* (1983)
Cabo de Peñas Volcs., Asturias, Spain	438	476	−22,	165	50	6	31,	−62	x		x				x	3	Perroud (1983)
Vogtland Claystones, Germany	438	505	13,	150	—	—	32,	−24							x	1	Rother (1971)
Pont Réan Fm., C. Brittany, France	478	488	−28,	151	87	7	60,	−60	x	x	x	x			x	5	Cogné (1988)
M. de Châteaupanne Fm., C. Brittany, France	480	486	−34,	163	65	6	51,	−72	x	x	x				x	4	Perroud *et al.* (1986b)
Erquy Volcanics, N. Brittany, France	472	493	−35,	164	47	11	51,	−73	x	x					x	3	Duff (1979)
Arenig Sandstones, Poland	478	488	−22,	189	—	10	341,	−64	x		x				x	3	Lewandowski (1987)
Bohemian Massif Seds., Czechoslovakia	476	505	−48,	92	16	3	126,	−34	x	x	x				x	4	Krs *et al.* (1986)
Buçaco Red Beds, N. Portugal	458	523	−23,	182	83	6	356,	−66			x				x	2	Perroud & Bonhommet (1981)
NW Granite, Jersey, UK	475	505	−73,	173	71	6	175,	−64	x	x					x	3	Duff (1978b)
Paimpol-Bréhec Red Beds, Trégor, France	478	523	−50,	63	13	20	142,	−21			x	x		x	x	*	Jones *et al.* (1979)
Barrandian Andesites, Bohemia, Czechoslovakia	510	517	−29,	131	9	6	82,	−48	x		x				x	3	Krs *et al.* (1987)
SE Granite, Jersey, UK	516	524	−77,	157	54	6	170,	−59	x	x					x	3	Duff (1978b)
Jince Fm., Bohemian Massif, Czechoslovakia	517	536	−11,	154	16	13	40,	−48	x	x	x			x	x	5	Krs *et al.* (1987)
SW Granite, Jersey, UK	475	580	−74,	176	87	5	178,	−63		x					x	2	Duff (1978b)
Jersey Volcanics, UK	521	540	−52,	143	8	19	116,	−66	x		x				x	3	Duff (1980)
NW Diorite, Jersey, UK	505	590	−64,	84	170	9	146,	−39			x				x	2	Duff (1981)
Granite de Porz-Scarff, Trégor, Fr.	544	560	−9,	163	7	30	26,	−49	x		x				x	3	Hagstrum *et al.* (1980)

Loguivy Microgr. Dikes, Trégor, France	544	560	6,	160	29	9	24, −34	x	x	x	x			x	5	Perigo *et al.* (1983)
Loguivy Microgr. Dikes, Trégor, France	544	560	−16,	130	35	11	70, −39	x		x	x			x	4	Hagstrum *et al.* (1980)
Lézardrieux Ignimbr. I, Trégor, France	491	603	16,	163	13	16	18, −25			x					1	Hagstrum *et al.* (1980)
Lézardrieux Ignimbr. II, Trégor, France	491	603	−15,	102	5	25	92, −19			x				x	2	Hagstrum *et al.* (1980)
Carteret Red Beds, Normandy, France	540	590	−59,	151	105	6	136, −69		x	x					2	Perroud *et al.* (1982)
St. Quay Diorite, N. Brittany, Fr.	528	640	−34,	319	226	4	213, 1		x	x				x	3	Hagstrum *et al.* (1980)
Group C Dikes, Jersey, UK	540	630	−26,	68	11	20	123, −5			x				x	2	Duff (1980)

Rock unit, location	Age		Pole position				D, Lat.	Reliability								Reference
	Lo	Hi	Lat.	Long.	k	α_{95}	45°N, 65°W	1	2	3	4	5	6	7	Q	
V. THE AVALON TERRANE IN NORTH AMERICA – PRE-DEVONIAN (internal rotations possible)																
Granites, Metamorph., E. New England	340	490	31,	122	43	9	354, −14		x				x		2	Hurley & Shearer (1981)
Cape Breton Granite, CB-3 (NE-2), NS	380	450	27,	84	27	11	28, −13		x	x					2	Rao *et al.* (1981)
Mascarene Fm., C component, New Brunswick	380?	421	28,	85	666	4	27, −12		x	x					2	Roy & Anderson (1981)
St. George Pluton, I, New Brunswick	390	430	29,	93	43	14	20, −13		x	x				x	3	Roy *et al.* (1979)
St. George Pluton, II, New Brunswick	390	430	−38,	85	66	8	83, −67		x	x			x	x	*	Roy *et al.* (1979)
St. Stephen Pluton, New Brunswick	390	430	−43,	136	26	18	270, −75			x			x	x	*	Roy *et al.* (1979)
Mascarene Fm., A-B compt., New Brunswick	408	421	−5,	87	33	7	40, −43	x	x				x	x	4	Roy & Anderson (1981)
Mascarene Intrusives, New Brunswick	408	430	−2,	122	32	22	350, −47			x			x	x	3	Roy *et al.* (1979)
Cape Breton Granite, CB-1 (SE), NS	380	480	−16,	139	166	4	318, −55		x	x				x	3	Rao *et al.* (1981)
Dunn Point Volcanics, Arisaig, NS	419	449	2,	130	69	4	340, −41	x	x	x	x			x	5	Johnson & Van der Voo (1990)

Table A11. *Pre-Late Carboniferous pole positions for the Iapetus blocks and terranes – continued*

Rock unit, location	Age		Pole position				D, Lat.	Reliability								Reference
	Lo	Hi	Lat.	Long.	k	α_{95}	45°N, 65°W	1	2	3	4	5	6	7	Q	
Dunn Point Volcanics, Arisaig, NS	419	449	0,	118	23	11	356, −45	x	x	x					*	Seguin *et al.* (1987)
Wabana and Bell Island Fms., East NFL	478	505	33,	102	137	6	11, −11	x		x					2	Buchan & Hodych (1982)
Cape Breton Granite, CB-2 (NE-1), NS	400	520	−32,	83	228	4	73, −62			x			x	x	3	Rao *et al.* (1981)
Nahant Gabbro, MA	438	505	−34,	102	—	—	47, −75		x	x			x	x	4	Weisse *et al.* (1985)
Bourinot Volcs., Cape Breton I., NS	530	540	−21,	160	42	8	288, −46	x	x	x	x		x	x	6	Johnson & Van der Voo (1985)
Bourinot Seds., Cape Breton I., NS	530	540	−33,	174	16	9	264, −44	x	x	x			x	x	5	Johnson & Van der Voo (1985)
Brigus Red Shale, East NFL	540	570	−27,	51	47	7	93, −37	x	x	x				x	4	Seguin & Rao (1989)
Morrison R., MacCodrum Fms., CBI, NS	540	570	−20,	34	97	5	99, −20	x	x	x					3	Rao *et al.* (1986)
Cape Breton Granite, CB (NW), NS	490	650	−37,	356	298	3	135, −9		x	x				x	3	Rao *et al.* (1981)
Roxbury Congl. Volcanics, MA	550	600	−13,	87	311	5	46, −50			x	x			x	3	Fang *et al.* (1986)
VI. OTHER APPALACHIAN TERRANES – PRE-DEVONIAN (internal rotations possible)																
Elberton-Sparta Area Granites, GA	281	350	55,	134	49	10	349, 12		x	x			x		3	Ellwood (1982)
Austell Gneiss, GA	325	365	34,	126	27	13	351, −10	x	x	x				x	4	Ellwood & Abrams (1982)
Granites, Gneisses, GA	325	365	39,	130	22	15	348, −5	x	x	x					3	Ellwood (1982)
Lawrenceton Fm., Botwood Gr., C. NFL	421	435	16,	133	54	7	341, −27	x	x	x	x		x	x	6	Gales *et al.* (1989)
Wigwam Red Beds, Central NFL	421	435	25,	100	9	14	14, −19	x		x			x		4	Lapointe (1979)
King George IV Lake, Central NFL	426	436	36,	85	19	11	24, −5	x	x	x	x		x	x	6	Buchan & Hodych (1989)
Mt. Peyton Granitic Batholith, NFL	373	480	−63,	125	24	14	194, −71		x	x					2	Lapointe (1979)
Mt. Peyton Dioritic Batholith, NFL	396	480	−15,	68	61	10	68, −40		x	x				x	3	Lapointe (1979)

Metamorphic Rocks, DE, PA	438	458	−48,	108	15	16	124,	−84	x	x	x				x	4	Brown & Van der Voo (1983)
Arden Pluton, DE	438	458	−16,	123	32	14	344,	−60	x	x	x				x	4	Rao & Van der Voo (1980)
Carolina Slate Belt Metamorphics, NC	450	460	20,	60	10	14	51,	−8	x		x			x	x	4	Noel *et al.* (1988)
Orwell Ls./Hatch Hill Fm., VT	440	476	26,	161	28	8	319,	−8			x				x	2	Tucker & Kent (1988)
Stacyville Volcs., C. Maine	458	480	−14,	188	12	12	272,	−22	x		x			x	x	4	Wellensiek *et al.* (1990)
Robert's Arm Volcanics, Central NFL	455	478	6,	140	14	11	329,	−34	x	x	x	x		x	x	6	Van der Voo *et al.* (1991)
Summerford Group Volcanics, C. NFL	450	485	8,	140	31	9	330,	−32	x	x	x				x	4	Van der Voo *et al.* (1991)
Chanceport group Volcanics, C. NFL	450	485	10,	131	50	7	341,	−33	x	x	x				x	4	Van der Voo *et al.* (1991)
Leading Tickles Dunite, NFL	439	510	19,	98	74	29	18,	−24			x					1	Lapointe (1979)
Moreton's Harbour Volcanics, C. NFL	478	488	29,	135	23	7	342,	−14	x	x	x	x		x	x	6	Johnson *et al.* (1991)
Moreton's Harbour Volcanics, C. NFL	478	488	32,	130	9	16	347,	−12	x						x	2	Deutsch & Rao (1977)
Cid Fm. & Albermarle Grp., NC	505	586	30,	122	119	5	354,	−15		x	x					2	Vick *et al.* (1987)
Comb. Thetford Mines Volcs., Québec	520	580	20,	32	15	33	71,	9						x	x	2	Seguin (1979)

The entries in this Table are for the blocks and terranes thought to have been located in or at the margins of the Iapetus Ocean during Paleozoic times. A '*' is placed in the Q column when the result is not to be used for apparent polar wander path construction or paleolatitude determinations, with reasons given below. For other explanations see Table A1.

*Poles that would have $Q \geq 3$ but are excluded from mean pole determinations and other compilations, are listed below with the reasons in parentheses: *Southern Britain:* Upper Ordovician Welsh Borderland Tuffs (anomalous paleolatitude, remagnetization suspected). *Armorica:* Cambrian Montmartin Formation (now known to be remagnetized); Cambro-Ordovician Paimpol-Bréhec Red Beds (too few samples, poorly determined age). *Avalon:* Siluro-Devonian St. George II and St. Stephen Pluton results (anomalous paleolatitudes, overprint or tilting suspected); Upper Ordovician Dunn Point Volcanics results from Seguin *et al.* (superseded).

References

Abou-deeb, J. M. & Tarling, D. H. (1984). Upper Palaeozoic palaeomagnetic results from Algeria and Tunisia. *Tectonophysics*, **101**, 143–57.

Abdullah, A., Nairn, A. E. M., Saving, D. & Sprague, K. (1984). Paleomagnetic study of some of the dikes from the Esh El Mellaha Range, eastern Desert, Egypt. *J. Afr. Earth Sci.*, 2, 267–75.

Abrahamsen, N. & Schonharting, G. (1987). Palaeomagnetic timing of the rotation and translation of Cyprus. *Earth Planet. Sci. Lett.*, **81**, 409–18.

Abrahamsen, N. & Van der Voo, R. (1987). Palaeomagnetism of middle Proterozoic (c. 1.25 Ga) dykes from central North Greenland. *Geophys. J. R. Astr. Soc.*, **91**, 597–611.

Abrahamsen, N., Wilson, J. R., Thy, P., Olsen, N. O. & Esbensen, K. H. (1979). Palaeomagnetism of the Fongen-Hyllingen gabbro complex, southern Scandinavian Caledonides: plate rotation or polar shift? *Geophys. J. R. Astr. Soc.*, **59**, 231–48.

Achache, J. & Courtillot, V. (1985). An Upper Triassic paleomagnetic pole for the Khorat Plateau (Thailand): consequences for the accretion of Indochina against Eurasia. *Earth Planet. Sci. Lett.*, 73, 147–57.

Achache, J., Courtillot, V. & Xu, Z. Y. (1984). Paleogeographic and tectonic evolution of southern Tibet since Middle Cretaceous-time: new paleomagnetic data and synthesis. *J. Geophys. Res.*, **89**, 10311–39.

Achache, J., Courtillot, V., Zhou, Y. X., Lu, L. Z. & Yuan, X. G. (1982). The paleolatitude and geographical extent of Southern Tibet in the Middle Cretaceous – new paleomagnetic data. In *Palaeomagnetic Research in Southeast and East Asia*, ed. J. C. Briden *et al.*, pp. 47–57. Committee for the Coordination of Offshore Prospecting (CCOP), U.N. Development Project Office.

Ade-Hall, J. M. (1974). The palaeomagnetism of some basic dykes cutting the Loch Ba Felsite ring dyke, Isle of Mull, Scotland. *Geophys. J. R. Astr. Soc.*, **36**, 267–71.

Ade-Hall, J. M., Dagley, P., Wilson, R. L., Evans, A., Riding, A., Smith, P. J., Skelhorne, R. R. & Sloan, T. (1972). A palaeomagnetic study of the Mull regional dyke swarm. *Geophys. J. R. Astr. Soc.*, **27**, 517–45.

Aïfa, T., Feinberg, H. & Pozzi, J.-P. (1990). Devonian–Carboniferous paleopoles for Africa: consequences for Hercynian geodynamics. *Tectonophysics*, **179**, 287–304.

Alberti, A. A., Comin-Chiaramonti, P., Ferriani, A. & Manzoni, M. (1974). Paleomagnetismo di alcune lave del distretto del Savalan (Azerbaidjan, Iran nord-occidentale). *Mem. Soc. Geol. Ital.*, **18**, 69–72.

Alvarez, L. W., Alvarez, W., Asaro, F. & Michel, H. V. (1980b). Extraterrestrial cause for the Cretaceous–Tertiary extinction. *Science*, **208**, 1095–108.

Alvarez, W., Cocozza, T. & Wezel, F. C. (1974). Fragmentation of the Alpine orogenic belt by microplate dispersal. *Nature*, **248**, 309–14.

Alvarez, W., Kent, D. V., Premoli-Silva, I., Schweickert, R. A. & Larson, R. A. (1980a). Franciscan complex limestone deposited at 17 degrees south paleolatitude. *Geol. Soc. Amer. Bull.*, **91**, 476–84.

Alvarez, W. & Lowrie, W. (1978). Upper Cretaceous palaeomagnetic stratigraphy at Moria (Umbrian Apennines, Italy): verification of the Gubbio section. *Geophys. J. R. Astr. Soc.*, **55**, 1–17.

André, L., Hertogen, J. & Deutsch, S. (1986). Ordovician-Silurian magmatic provinces in Belgium and the Caledonian orogeny in middle Europe. *Geology*, **14**, 879–82.

Andrews, J. A. (1985). True polar wander: an analysis of Cenozoic and Mesozoic paleomagnetic results. *J. Geophys. Res.*, **90**, 7737–50.

Andreyeva, O. L., Bucha, V. V. & Petrova, G. N. (1965). Laboratory evaluation of magnetic stability of rocks of the Czech Massif. *Akad. Nauk SSSR, Izvestiya, Earth Phys. Ser.*, 54–64.

Andriamirado, C. R. A. (1971). Recherches paléomagnétiques sur Madagascar: résultats et interprétations dans le cadre de la dislocation de la partie orientale du Gondwana. Ph.D. Thesis, University of Strasbourg.

Appel, E., Müller, R. & Widder, R. W. (1991). Paleomagnetic results from the Tibetan sedimentary series of the Manang Area (north central Nepal). *Geophys. J. Int.*, **104**, 255–66.

Argand, E. (1924). La tectonique de l'Asie. *Proc. Intern. Geol. Congress*, **13**, 171–372.

Arthaud, F. & Matte, P. (1977). Détermination de la position initiale de la Corse et de la Sardaigne à la fin de l'orogénèse hercynienne grâce aux marqueurs géologiques antémésozoïques. *Bull. Soc. Géol. France*, **19**, 833–40.

As, J. A. (1960). Instruments and measuring methods in paleomagnetic research. *Med. Verh. Kon. Nederl. Metereol. Inst.*, **78**, 56 pp.

As, J. A. & Zijderveld, J. D. A. (1958). Magnetic cleaning of rocks in palaeomagnetic research. *Geophys. J. R. Astr. Soc.*, **1**, 308–19.

Athavale, R. N., Hansraj, A. & Verma, R. K. (1972). Palaeomagnetism and age of Bhander and Rewa sandstones from India. *Geophys. J. R. Astr. Soc.*, **28**, 499–509.

Athavale, R. N., Radhakrishnamurty, C. & Sahasrabudhe, P. W. (1963). Palaeomagnetism of some Indian rocks. *Geophys. J. R. Astr. Soc.*, 7, 304–13.

Athavale, R. N., Rao, G. V. S. P. & Rao, M. S. (1980). Palaeomagnetic results from two basic volcanic formations in the western Himalayas and a Phanerozoic polar wandering curve for India. *Geophys. J. R. Astr. Soc.*, **60**, 419–33.

Athavale, R. N. & Sharma, P. V. (1974). Preliminary paleomagnetic results on some Triassic rocks from East Greenland. *Phys. Earth Planet. Int.*, **9**, 51–6.

Athavale, R. N. & Sharma, P. V. (1975). Paleomagnetic results on early Tertiary lava flows from West Greenland and their bearing on the evolution history of the Baffin Bay–Labrador Sea region. *Can. J. Earth Sci.*, **12**, 1–18.

Athavale, R. N. & Verma, R. K. (1970). Palaeomagnetic results on Gondwana dykes from the Domodar coal fields and their bearing on the sequence of Mesozoic igneous activity in India. *Geophys. J. R. Astr. Soc.*, **20**, 303–16.

Athavale, R. N., Verma, R. K., Bhalla, M. S. & Pullaiah, G. (1970). Drift of the Indian sub-continent since Precambrian times. In *Palaeogeophysics*, ed. S. K. Runcorn, pp. 291–305. London: Academic Press.

Aubouin, J. (1965). *Geosynclines*. Amsterdam: Elsevier.

Audley-Charles, M. G. (1988). Evolution of the southern margin of Tethys (North Australian region) from early Permian to late Cretaceous. *Geol. Soc. London Spec. Publ.*, **37**, 79–100.

Baag, C.-G. & Helsley, C. E. (1974). Remanent magnetization of a 50 m core from the Moenkopi Formation. *J. Geophys. Res.*, **79**, 3308–20.

Bachtadse, V. & Briden, J. C. (1989). Palaeomagnetism of the Early to mid-Ordovician Salala igneous ring complex, Red Sea Hills, Sudan. *Geophys. J. Int.*, **99**, 677–85.

Bachtadse, V. & Briden, J. C. (1990). Palaeomagnetic constraints on the position of Gondwana during Ordovician to Devonian times. In *Palaeozoic Palaeogeography and Biogeography*, ed. W. S. McKerrow & C. R. Scotese. *Geol. Soc. London Memoir*, **12**, 43–8.

Bachtadse, V., Heller, F. & Kröner, A. M. (1983). Paleomagnetic investigation in the Hercynian mountain belt of central Europe. *Tectonophysics*, **91**, 285–99.

Bachtadse, V., Soffel, H. C. & Böhm, V. (1991). Paleomagnetism of Ordovician sediments from central and northern Europe (Abstract). *Eos, Trans. Am. Geophys. Union*, **72**, 105.

Bachtadse, V. & Van der Voo, R. (1986). Paleomagnetic evidence for crustal and thin-skinned rotations in the European Hercynides. *Geophys. Res. Lett.*, **13**, 161–4.

Bachtadse, V., Van der Voo, R. & Hälbich, I. W. (1987b). Paleomagnetism of the western Cape fold belt, South Africa, and its bearing on the Paleozoic apparent polar wander path for Gondwana. *Earth Planet. Sci. Lett.*, **84**, 487–99.

Bachtadse, V., Van der Voo, R., Haynes, F. M. & Kesler, S. E. (1987a). Late Paleozoic magnetization of mineralized and unmineralized Ordovician carbonates from east Tennessee: evidence for a post-ore chemical event. *J. Geophys. Res.*, **92**, 14165–76.

Bai, Y., Chen, G., Sun, Q., Li, Y. A., Dong, Y. & Sun, D. (1987). Late Paleozoic polar wander path for the Tarim Platform and its tectonic significance. *Tectonophysics*, **139**, 145–53.

Bailey, R. C. & Halls, H. C. (1984). Estimate of confidence in paleomagnetic directions derived from remagnetization circle and direct observational data. *J. Geophys.*, **54**, 174–82.

Ballard, M. M., Van der Voo, R. & Hälbich, I. W. (1986). Remagnetizations in Late Permian and Early Triassic rocks from southern Africa and their implications for Pangea reconstructions. *Earth Planet. Sci. Lett.*, **79**, 412–18.

Barberi, F., Civetta, L., Gasparini, P., Innocenti, F. & Scandone, R. (1974). Evolution of a section of the Africa-Europe plate boundary: paleomagnetic and volcanological evidence from Sicily. *Earth Planet. Sci. Lett.*, **22**, 123–32.

Bardon, C., Bossert, A., Hamzeh, R., Rolley, J. P. & Westphal, M. (1973). Etude paléomagnétique de formations volcaniques du Crétacé inférieur dans l'Atlas de Beni Mellal (Maroc). *C. R. Acad. Sci. Paris, Ser. D*, **227**, 2141–4.

Barnes, A. E. & Butler, R. F. (1980). A Paleocene paleomagnetic pole from the Gringo Gulch Volcanics. *Geophys. Res. Lett.*, **7**, 545–8.

Barnett, R. L., Arima, M., Blackwell, J. D., Winder, C. G., Palmer, H. C. & Hayatsu, A. (1984). The Picton and Varty Lake ultramafic dikes: Jurassic magmatism in the St Lawrence Platform near Belleville, Ontario. *Can. J. Earth Sci.*, **21**, 1460–72.

Barr, S. M. & MacDonald, A. S. (1979). Paleomagnetism, age and geochemistry of the Denchai basalt, northern Thailand. *Earth Planet. Sci. Lett.*, **46**, 113–24.

Barr, S. M., MacDonald, A. S. & Haile, N. S. (1978). Reconnaissance palaeomagnetic measurements on Triassic and Jurassic sedimentary rocks from Thailand. *Geol. Soc. Malaysia Bull.*, **10**, 53–62.

Barton, C. & Brown, L. (1983). Paleomagnetism of Carboniferous intrusions in North Carolina. *J. Geophys. Res.*, **88**, 2327–35.

Bates, M. P. (1989). Palaeomagnetic evidence for rotations and deformation in the Nogueras Zone, Central Southern Pyrenees, Spain. *J. Geol. Soc. London*, **146**, 459–76.

Bazhenov, M. L. (1987). Paleomagnetism of Cretaceous and Paleogene sedimentary rocks from the Kopet Dagh and its tectonic implications. *Tectonophysics*, **136**, 223–35.

Bazhenov, M. L., Burtman, V. S. & Gurariy, G. Z. (1978). Paleomagnetic results for the Paleogene of the Outer Pamir and Transalai. *Dokl. Akad. Nauk USSR*, **242**, 1137–9. (In Russian.)

Beaubouef, R., Casey, J. F., Hall, S. A. & Evans, I. (1990). A paleomagnetic investigation of the Lower Ordovician St George Group, Port-au-Port Peninsula, Newfoundland: implications for the Iapetus Ocean and evidence for Late Paleozoic remagnetization. *Tectonophysics*, **182**, 337–56.

Beck, M. E. (1972a). Paleomagnetism of Upper Triassic diabase from southeastern Pennsylvania: further results. *J. Geophys. Res.*, **77**, 5673–87.

Beck, M. E. (1972b). Paleomagnetism and magnetic polarity zones in the Jurassic Dufek Intrusion, Pensacola Mountains, Antarctica. *Geophys. J. R. Astr. Soc.*, **28**, 49–63.

Beck, M. E. (1976). Discordant paleomagnetic pole positions as evidence of regional shear in the western Cordillera of North America. *Am. J. Sci.*, **276**, 694–712.

Beck, M. E. (1980). Paleomagnetic record of plate-margin tectonic processes along the western edge of North America. *J. Geophys. Res.*, **85**, 7115–31.

Beckman, G. E. J. (1982). Palaeomagnetism of nine dated Phanerozoic dykes in southeast Greenland. *Geophys. J. R. Astr. Soc.*, **69**, 355–67.

Bellon, H., Coulon, C. & Edel, J. B. (1977). La déplacement de la Sardaigne. Synthèse des données géochronologiques, magmatiques et paléomagnétiques. *Bull. Soc. Géol. France*, **19**, 825–31.

Berberian, F. & Berberian, M. (1981). Tectono-plutonic episodes in Iran. *Geodynamics Ser.*, **3**, 5–32.

Berthold, G., Nairn, A. E. M. & Negendank, J. F. W. (1975). A palaeomagnetic investigation of some of the igneous rocks of the Saar-Nahe Basin. *N. Jb. Geol. Paläont. Mh.*, 134–50.

Besse, J., Courtillot, V., Pozzi, J. P., Westphal, M. & Zhou, Y. X. (1984b). Paleomagnetic estimates of crustal shortening in the Himalayan thrusts and Zangbo suture. *Nature*, **311**, 621–6.

Besse, J., Pozzi, J. P., Mascle, G. & Feinberg, H. (1984a). Paleomagnetic study of Sicily:

consequences for deformation of Italian and African margins over the last 100 million years. *Earth Planet. Sci. Lett.*, **67**, 377–90.

Bhalla, M. S. & Verma, R. K. (1969). Palaeomagnetism of Triassic Parsora sandstones from India. *Phys. Earth Planet. Int.*, **2**, 138–46.

Bina, M. M., Becur, I., Prévot, M., Meyerfeld, Y., Daly, L., Cantagrel, J. M. & Mergoil, J. (1986). Palaeomagnetism, petrology and geochronology of Tertiary magmatic and sedimentary units from Iran. *Tectonophysics*, **121**, 303–29.

Bingham, D. K. & Klootwijk, C. T. (1980). Palaeomagnetic constraints on Greater Indias underthrusting of the Tibetan Plateau. *Nature*, **284**, 336–8.

Biquand, D. (1967). Sur les directions de l'aimantation remanente des couches rouges du Trias suppérieur du bassin de Carentan (Normandie). *C. R. Acad. Sci. Paris Ser. D.*, **264**, 1597–600.

Biquand, D. (1977). Paléomagnétisme de la formation des 'grès-à-Voltzia' (Buntsandstein supérieur) du Massif des Vosges (France). *Can. J. Earth Sci.*, **14**, 1490–514.

Biquand, D. & Prévot, M. (1971). AF demagnetization of viscous remanent magnetization in rocks. *Geophys.*, **37**, 471–85.

Birkenmajer, K., Grocholski, A., Milewicz, J. & Nairn, A. E. M. (1968). Palaeomagnetic studies of Polish rocks, 2. The Upper Carboniferous and Lower Permian of the Sudetes. *Ann. Géol. Pologne*, **38**, 435–74.

Birkenmajer, K. & Nairn, A. E. M. (1964). Palaeomagnetic studies of Polish rocks, 1. The Permian igneous rocks of the Krakow District and some results from the Holy Cross Mountains. *Ann. Géol. Pologne*, **34**, 225–44.

Black, R. F. (1964). Paleomagnetic support of the theory of rotation of the island of Newfoundland. *Nature*, **202**, 945–8.

Blackett, P. M. S. (1961). Comparison of ancient climates with the ancient latitudes deduced from rock magnetic measurements. *Proc. Roy. Soc. London*, **A263**, 1–30.

Bobier, C. (1968). Etude paléomagnétique de quelques formations du complêxe volcanique d'Almopias (Macédoine centrale, Grèce). *C. R. Acad. Sci. Paris*, **267**, 1091–4.

Bobier, C. & Coulon, C. (1970). Résultats préliminaires d'une étude paléomagnétique des formations volcaniques tertiaires et quaternaires du Logudoro (Sardaigne septentrionale). *C. R. Acad. Sci. Paris*, **270**, 1434–7.

Boesen, R., Irving, E. & Robertson, W. A. (1961). The palaeomagnetism of some igneous rock bodies in New South Wales. *J. Proc. R. Soc. New South Wales*, **94**, 227–232.

Bogdanoff, S. & Schott, J. J. (1977). Etude paléomagnétique et analyse tectonique dans les schistes rouges permiens du Sud de l'Argentéra. *Bull. Soc. Géol. France*, **19**, 909–16.

Bond, G. C. & Komintz, M. A. (1988). Evolution of thought on passive continental margins from the origin of geosynclinal theory (c. 1860) to the present. *Geol. Soc. Am. Bull.*, **100**, 1909–33.

Bond, G. C., Nickeson, P. A. & Kominz, M. A. (1984). Break-up of a supercontinent between 625 and 555 Ma: new evidence and implications for continental histories. *Earth Planet. Sci. Lett.*, **70**, 325–45.

Bonhommet, N., Cobbold, P. R., Perroud, H. & Richardson, A. (1981). Paleomagnetism and cross-folding in a key area of the Asturian arc (Spain). *J. Geophys. Res.*, **86**, 1873–87.

Bonhommet, N., Roperch, P. & Calza, F. (1988). Paleomagnetic arguments for block rotations along the Arakapas fault, Cyprus. *Geology*, **16**, 422–5.

Bossart, P., Ottiger, R. & Heller, F. (1989). Paleomagnetism in the Hazara–Kashmir Syntaxis, NE Pakistan. *Eclogae Geol. Helv.*, **82**, 585–601.

Boucot, A. J. (1990). Silurian biogeography. In Palaeozoic palaeogeography and biogeography, ed. W. S. McKerrow & C. R. Scotese. *Geol. Soc. London Memoir*, **12**, 191–6.

Boulin, J. (1991). Structures in southwest Asia and evolution of the eastern Tethys. *Tectonophysics*, **196**, 211–68.

Brailsford, F. (1951). *Magnetic Materials. London: Methuen.*

Briden, J. C. (1966). Variation of intensity of the palaeomagnetic field through geological time. *Nature*, **212**, 246–7.

Briden, J. C. (1967). A new palaeomagnetic result from the Lower Cretaceous of East-Central Africa. *Geophys. J. R. Astr. Soc.*, **12**, 375–80.

Briden, J. C. (1968a). Paleoclimatic evidence of a geocentric axial dipole field. In *History of the Earth's Crust*, ed. R. A. Phinney, pp. 178–94. Princeton: Princeton University Press.

Briden, J. C. (1968b). Paleomagnetism of the Ntonya Ring Complex, Malawi. *J. Geophys. Res.*, **73**, 725–33.

Briden, J. C. (1970). Palaeomagnetic results from the Arrochar and Garabal Hill–Glen Fyne Igneous complexes, Scotland. *Geophys. J. R. Astr. Soc.*, **21**, 457–70.

Briden; J. C., Drewry, G. E. & Smith, A. G. (1974). Phanerozoic equal-area world maps. *J. Geol.*, **82**, 555–74.

Briden, J. C. & Duff, B. A. (1981). Pre-Carboniferous paleomagnetism of Europe north of the Alpine orogenic belt. *Geodynamics Series*, **2**, 137–49.

Briden, J. C., Henthorn, D. I. & Rex, D. C. (1971b). Palaeomagnetic and radiometric evidence for the age of the Freetown igneous complex, Sierra Leone. *Earth Planet. Sci. Lett.*, **12**, 385–91.

Briden, J. C., Irons, J. & Johnson, P. A. (1971c). Palaeomagnetic studies of the Caerfai Series and the Skomer Volcanic Group (Lower Palaeozoic, Wales). *Geophys. J. R. Astr. Soc.*, **22**, 1–16.

Briden, J. C. & Irving, E. (1964). Palaeoclimatic spectra of sedimentary palaeoclimatic indicators. In *Problems in Palaeoclimatology*, ed. A. E. M. Nairn, pp. 199–250. New York: Interscience.

Briden, J. C., Kent, D. V., Lapointe, P. L., Livermore, R. A., Roy, J. L., Seguin, M. K., Smith, A. G., Van der Voo, R. & Watts, D. R. (1988). Palaeomagnetic constraints on the evolution of the Caledonian–Appalachian orogen. *Geol. Soc. London Spec. Publ.*, **38**, 35–48.

Briden, J. C. & Morris, W. A. (1973). Paleomagnetic studies in the British Caledonides, III. Igneous rocks of the northern Lake District, England. *Geophys. J. R. Astr. Soc.*, **34**, 27–46.

Briden, J. C. & Mullan, A. J. (1984). Superimposed recent, Permo-Carboniferous and Ordovician palaeomagnetic remanence in the Builth Volcanic Series, Wales. *Earth Planet. Sci. Lett.*, **69**, 413–21.

Briden, J. C., Smith, A. G. & Sallomy, J. T. (1971a). The geomagnetic field in Permo-Triassic time. *Geophys. J. R. Astr. Soc.*, **23**, 101–17.

Briden, J. C., Turnell, H. B. & Watts, D. R. (1984). British paleomagnetism, Iapetus Ocean and the Great Glen Fault. *Geology*, **12**, 428–31.

Brock, A. (1967). Paleomagnetic results from the Hook Intrusives, Zambia. *Nature*, **216**, 359–60.

Brock, A. (1968). The palaeomagnetism of the Nuanetsi igneous province and its bearing upon the sequence of Karroo igneous activity in southern Africa. *J. Geophys. Res.*, **73**, 1389–97.

Brock, A., Gibson, I. L. & Gacii, P. (1970). The palaeomagnetism of the Ethiopian flood basalt succession near Addis Ababa. *Geophys. J. R. Astr. Soc.*, **19**, 485–97.

Brown, L. & Kelly, W. M. (1980). Paleomagnetic results from northern Maine – reinterpretations. *Geophys. Res. Lett.*, 7, 1109–11.

Brown, P. M. & Van der Voo, R. (1982). Paleomagnetism of the latest Precambrian/Cambrian Unicoi basalts from the Blue Ridge, northeast Tennessee and southwest Virginia: evidence for Taconic deformation. *Earth Planet. Sci. Lett.*, **60**, 407–14.

Brown, P. M. & Van der Voo, R. (1983). A paleomagnetic study of Piedmont metamorphic rocks from N. Delaware. *Geol. Soc. Amer. Bull.*, **94**, 814–22.

Bryan, P. & Gordon, R. G. (1986). Rotation of the Colorado Plateau: an analysis of paleomagnetic data. *Tectonics*, **5**, 661–7.

Bryan, P. & Gordon, R. G. (1988). Rotation of the Colorado Plateau: an updated analysis of paleomagnetic data (abstract). *Geol. Soc. America Abstr. Progr.*, **20**, A63.

Buchan, K. L. (1978). Magnetic overprinting in the Thanet gabbro complex, Ontario. *Can. J. Earth Sci.*, **15**, 1407–21.

Buchan, K. L. & Hodych, J. P. (1982). Paleomagnetic reexamination of the Lower Ordovician Wabana and Bell Island groups of the Avalon Peninsula of Newfoundland. *Can. J. Earth Sci.*, **19**, 1055–69.

Buchan, K. L. & Hodych, J. P. (1989). Early Silurian paleopole for red beds and volcanics of the King George IV Lake area, Newfoundland. *Can. J. Earth Sci.*, **26**, 1904–17.

Bucur, I. (1971). Etude paléomagnétique d'une formation sédimentaire du Sahara algérien, d'âge Cambro-Ordovicien. *Ann. Géophys.*, **27**, 255–261.

Bücker, C., Schult, A., Bloch, W. & Guerreiro, S. D. C. (1986). Rock-magnetism and palaeomagnetism of an Early Cretaceous/Late Jurassic dike swarm in Rio Grande do Norte, Brazil. *J. Geophys.*, **60**, 129–35.

Bullard, E. C., Everett, J. E. & Smith, A. G. (1965). A symposium on continental drift–IV. The fit of the continents around the Atlantic. *Phil. Trans. Roy. Soc.*, **258**, 41–51.

Bunopas, S. (1982). Paleogeographic history of western Thailand and adjacent parts of southeast Asia – a plate tectonics interpretation. *Geol. Surv. Paper* 5. Bangkok: Dept. Mineral Resources.

Bunopas, S., Vella, P., Pitakpaivan, K. & Sukroo, J. (1978). Preliminary paleomagnetic results from Thailand sedimentary rocks. In *Proc. Third Regional conference on Geology and Mineral Resources of Southeast Asia*, ed. P. Nutalaya, pp. 25–32, Bangkok, Thailand.

Burek, P. J. (1969). Device for chemical demagnetization of red beds. *J. Geophys. Res.*, **74**, 6710–12.

Burke, K. & Dewey, J. F. (1974). Two plates in Africa during the Cretaceous? *Nature*, **249**, 313–16.

Burns, K. L., Rickard, M. J., Belbin, L. & Chamalaun, F. (1980). Further paleomagnetic confirmation of the Magellanes orocline. *Tectonophysics*, **63**, 75–90.

Burrett, C., Long, J. & Stait, B. (1990). Early-middle Palaeozoic biogeography of Asian terranes derived from Gondwana. In *Palaeozoic Palaeogeography and Biogeography*, ed. W. S. McKerrow & C. R. Scotese. *Geol. Soc. London Memoir*, **12**, 163–74.

Burrus, J. (1984). Contribution to a geodynamic synthesis of the Provençal Basin (NW Mediterranean). *Marine Geol.*, **55**, 247–69.

Butler, R. F. (1991). *Paleomagnetism: Magnetic Domains to Geologic Terranes*. Cambridge, MA: Blackwell.

Butler, R. F., Hervé, F., Munizaga, F., Beck, M. E., Burmester, R. F. & Oviedo, E. S. (1991). Paleomagnetism of the Patagonian Plateau Basalts, Southern Chile and Argentina. *J. Geophys. Res.*, **96**, 6023–34.

Butler, R. F., Krause, D. W. & Gingerich, P. D. (1987). Magnetic polarity stratigraphy and biostratigraphy of Middle–Late Paleocene continental deposits of south-central Montana. *J. Geol.*, **95**, 647–57.

Butler, R. F. & Taylor, L. H. (1978). A middle Paleocene paleomagnetic pole from the Nacimiento Formation, San Juan Basin, New Mexico. *Geology*, **6**, 495–8.

Bylund, G. (1974). Paleomagnetism of dykes along the southern margin of the Baltic Shield. *Geol. Föreningens i Stockholm Forhandl.*, **96**, 231–5.

Bylund, G. (1981). Paleomagnetism of Jurassic-Cretaceous Basalts from Scania, southern Sweden (abstract). *Eos, Trans. Am. Geophys. Union*, **62**, 231.

Bylund, G. & Zellman, O. (1980). Paleomagnetism of the dolerites of the Saerv Nappe, southern Swedish Caledonides. *Geol. Föreningens i Stockholm Forhandl.*, **102**, 393–402.

Cadet, J. P., et al. (1985). Océanographie dynamique. *C. R. Acad. Sci. Paris, Ser.* **2**(301), 287–296.

Cadet, J. P., et al. (1987). Deep scientific dives in the Japan and Kuril trenches. *Earth Planet. Sci. Lett.*, **83**, 313–28.

Cande, S. C., LaBrecque, J. L. & Haxby, W. F. (1988). Plate kinematics of the South Atlantic: Chron C34 to Present. *J. Geophys. Res.*, **93**, 13479–92.

Caputo, M. V. & Crowell, J. C. (1985). Migration of glacial centers across Gondwana during the Paleozoic. *Geol. Soc. Am. Bull*, **96**, 1020–36.

Carey, S. W. (1958). A tectonic approach to continental drift. In *Continental Drift Symposium*, ed. S. W. Carey, pp. 177–355. Tasmania: University of Hobart.

Carroll, A. R., Liang, Y., Graham, S. A., Xiao, X., Hendrix, M. S., Chu, J. & McKnight, C. L. (1990). Junggar basin, northwest China: trapped Late Paleozoic ocean. *Tectonophysics*, **181**, 1–14.

Catalano, R., Channell, J. E. T., D'Argenio, B. & Napoleone, G. (1976). Palaeogeography of Southern Apennines and Sicily: problems of palaeotectonics and palaeomagnetism. *Mem. Geol. Soc. Ital.*, **15**, 95–118.

Chaimov, T. A., Barazangi, M., Al-Saad, D., Sawaf, T. & Gebran, A. (1990). Crustal shortening in the Palmyride fold belt, Syria, and implications for movement along the Dead Sea Fault System. *Tectonics*, **9**, 1369–86.

Chamalaun, F. (1968). The magnetization of the Dotswood Red Beds (Queensland). *Earth Planet. Sci. Lett.*, **3**, 439–43.

Chamalaun, F. H. (1977a). Paleomagnetic evidence for the relative position of Timor and Australia in the Permian. *Earth Planet. Sci. Lett.*, **34**, 107–12.

Chamalaun, F. H. (1977b). Paleomagnetic reconnaissance results from the Maubisse Formation, East Timor, and its tectonic implications. *Tectonophysics*, **42**, T17–T26.

Chamalaun, F. H. & Creer, K. M. (1964). Thermal demagnetization studies on the Old Red Sandstone of the Anglo-Welsh Cuvette. *J. Geophys. Res.*, **69**, 1607–16.

Chamalaun, F. H. & Sunata, W. (1982). The paleomagnetism of the western Banda Arc system. In *Paleomagnetic Research in Southeast and East Asia*, ed. J. C. Briden et al., pp. 162–94. Committee for the Coordination of Offshore Prospecting (CCOP), U. N. Development Project Office.

Chan, L. S., Wang, C. Y. & Wu, X. Y. (1984). Paleomagnetic results from some Permian–Triassic rocks from southwestern China. *Geophys. Res. Lett.*, **11**, 1157–60.

Channell, J. E. T. (1977). Paleomagnetism of limestones from the Gargano Peninsula (Italy), and the implications of these data. *Geophys. J. R. astr. Soc.*, **51**, 605–16.

Channell, J. E. T., Brandner, R., Spieler, A. & Smathers, N. P. (1990b). Mesozoic paleogeography of the Northern Calcareous Alps – evidence from paleomagnetism and facies analysis. *Geology*, **18**, 828–31.

Channell, J. E. T., Catalano, R. & D'Argenio, B. (1980). Paleomagnetism and deformation of the Mesozoic continental margin in Sicily. *Tectonophysics*, **61**, 391–407.

Channell, J. E. T., D'Argenio, B. & Horvath, F. (1979). Adria, the African promontory in Mesozoic Mediterranean paleogeography. *Earth. Sci. Rev.*, **15**, 213–92.

Channell, J. E. T., Lowrie, W., Medizza, F. & Alvarez, W. (1978). Paleomagnetism and tectonics in Umbria, Italy. *Earth Planet. Sci. Lett.*, **39**, 199–210.

Channell, J. E. T., Lowrie, W., Pialli, P. & Venturi, F. (1984). Jurassic magnetic stratigraphy from Umbrian (Italian) land sections. *Earth Planet. Sci. Lett.*, **68**, 309–25.

Channell, J. E. T., Oldow, J. S., Catalano, R. & D'Argenio, B. (1990a). Paleomagnetically determined rotations in the western Sicilian fold and thrust belt. *Tectonics*, **9**, 641–60.

Channell, J. E. T. & Tarling, D. H. (1975). Palaeomagnetism and the rotation of Italy. *Earth Planet, Sci. Lett.*, **25**, 177–88.

Chen, Y. & Courtillot, V. (1989). Widespread Cenozoic(?) remagnetization in Thailand and its implications for the India–Asia collision. *Earth Planet. Sci. Lett.*, **93**, 113–22.

Chen, Z., Wang, C. & Deng, X. (1965). Some results of paleomagnetic research in China. *Acta Geol. Sin.*, **43**, 241–6.

Cheng, G.-H., Bai, Y.-H. & Sun, Y.-H. (1988). Paleomagnetic study on the tectonic evolution of the Ordos block, North China. *Seism, and Geology*, **10**, 81–7. (In Chinese, with English abstract.)

Cirilli, S., Marton, P. & Vigli, L. (1984). Implications of a combined biostratigraphic and paleomagnetic study of the Umbrian Maiolica Formation. *Earth Planet. Sci. Lett.*, **69**, 203–14.

Claesson, K. C. (1979). Early Paleozoic geomagnetism of Gotland. *Geol. Föreningens i Stockholm Forhandl.*, **101**, 149–55.

Clegg, J. A., Almond, A. & Stubbs, P. H. S. (1954). Remanent magnetism in some sedimentary rocks in Great Britain. *Phil. Mag.*, **45**, 583–98.

Clegg, J. A., Deutsch, E. R., Everitt, C. W. F. & Stubbs, P. H. S. (1957). Some recent palaeomagnetic measurements made at Imperial College, London. *Academic Physics*, **6**, 219–31.

Clegg, J. A., Deutsch, E. R. & Griffiths, D. H. (1956). Rock magnetism in India. *Phil. Mag., Ser. 8*, **1**, 419–531.

Clegg, J. A., Radhakrishnamurty, C. & Sahasrabudhe, P. W. (1958). Remanent magnetism of the Rajmahal traps of northeastern India. *Nature*, **181**, 830–1.

Clube, T. M. M., Creer, K. M. & Robertson, A. H. F. (1985). The palaeorotation of the Troodos microplate. *Nature*, **317**, 522–5.

Cocks, L. R. M. & Fortey, R. A. (1988). Lower Palaeozoic facies and faunas around Gondwana. *Geol. Soc. London Spec. Publ.*, **37**, 183–200.

Cocks, L. R. M. & Fortey, R. A. (1990). Biogeography of Ordovician and Silurian faunas. In *Palaeozoic Palaeogeography and Biogeography*, ed. W. S. McKerrow & C. R. Scotese. *Geol. Soc. London Memoir*, **12**, 97–104.

Cogné, J. P. (1987). Paleomagnetic direction obtained by strain removal in the Pyrenean Permian red beds at the 'Col du Somport' (France). *Earth Planet. Sci. Lett.*, **85**, 162–72.

Cogné, J. P. (1988). Strain, magnetic fabric, and paleomagnetism of the deformed redbeds of the Pont-Réan Formation, Brittany, France. *J. Geophys. Res.*, **93**, 13673–87.

Cogné, J. P., Brun J. P. & Van den Driessche, J. (1990). Paleomagnetic evidence for rotation during Stephano-Permian extension in southern Massif Central (France). *Earth Planet. Sci. Lett.*, **101**, 272–80.

Cogné, J. P. & Perroud, H. (1985). Strain removal applied to paleomagnetic directions in an orogenic belt: the Permian red slates of the Alpes Maritimes, France. *Earth Planet. Sci. Lett.*, 72, 125–40.

Cole, J., McCabe, R., Moriarty, T., Malicse, J. A., Delfin, F. G., Tebar, H. & Ferrer, H. P. (1989). A preliminary Neogene paleomagnetic data set from Leyte and its relation to motion on the Philippine Fault. *Tectonophysics*, **168**, 205–20.

Collinson, D. W. (1983). *Methods in Palaeomagnetism and Rock Magnetism*. London: Chapman and Hall.

Collinson, D. W., Creer, K. M. & Runcorn, S. K. (1967). *Methods in Palaeomagnetism*. Amsterdam: Elsevier.

Coney, P. J., Jones, D. L. & Monger, J. W. H. (1980). Cordilleran suspect terranes. *Nature*, **239**, 329–33.

Conrad, G., Montigny, R., Thuizat, R. & Westphal, M. (1981). Tertiary and Quaternary geodynamics of southern Lut (Iran) as deduced from paleomagnetic, isotopic and structural data. *Tectonophysics*, **75**, T11–T17.

Conrad, J. & Westphal, M. (1975). Paleomagnetic results from the Carboniferous and Jurassic of the Saharan platform. In *Gondwana Geology*, ed. K. S. W. Campbell, pp. 9–13. Canberra: Australian National University Press.

Cornwell, J. D. (1967). The palaeomagnetism of the Exeter Lavas. *Geophys. J. R. Astr. Soc.*, **12**, 181–96.

Coupland, D. H. & Van der Voo, R. (1980). Long-term non-dipole components in the geomagnetic field during the last 130 m.y. *J. Geophys. Res.*, **85**, 3529–48.

Courtillot, V. & Besse, J. (1986). Mesozoic and Cenozoic evolution of the North and South China blocks. *Nature*, **320**, 86–7.

Coward, M. & Dietrich, D. (1989). Alpine tectonics – an overview. In *Alpine Tectonics*, ed. M. P. Coward, D. Dietrich & R. G. Park. *Geol. Soc. London Special Publ.*, **45**, 1–29.

Cowie, J. W. & Johnson, M. R. W. (1985). Late Precambrian and Cambrian geological time scale. In *The Chronology of the Geological Record*, pp. 47–64, Oxford: Memoir 10 of the Geol. Soc. published by Blackwell.

Cox, A. & Doell, R. R. (1960). Review of paleomagnetism. *Geol. Soc. Am. Bull.*, **71**, 645–768.

Creer, K. M. (1957a). Paleomagnetic investigations in Great Britain- IV. The natural remanents magnetization for certain stable rocks from Great Britain. *Phil. Trans. R. Soc. London*, A**250**, 111–29.

Creer, K. M. (1957b). Paleomagnetic investigations in Great Britain, V. The remanent magnetization of unstable Keuper marls. *Phil. Trans. R. Soc. London*, A**250**, 130–43.

Creer, K. M. (1958). Preliminary palaeomagnetic measurements from South America. *Ann. Geophys.*, **15**, 373–90.

Creer, K. M. (1959). AC demagnetization of unstable Triassic Keuper marls from SW England. *Geophys. J. R. Astr. Soc.*, **2**, 261–75.

Creer, K. M. (1965). A symposium on continental drift, III. Palaeomagnetic data from the Gondwanic continents. *Phil. Trans. R. Soc. London*, A**256**, 569–73.

Creer, K. M. (1970). A review of paleomagnetism. *Earth Sci. Rev.*, **6**, 369–466.

Creer, K. M., Embleton, B. J. J. & Valencio, D. A. (1970). Triassic and Permo-Triassic palaeomagnetic data for S. America. *Earth Planet. Sci. Lett.*, **8**, 173–8.

Creer, K. M., Irving, E. & Runcorn, S. K. (1954). The direction of the geomagnetic field in remote epochs in Great Britain. *J. Geomagn. Geoelectr.*, **6**, 163–8.

Creer, K. M., Irving, E. & Runcorn, S. K. (1957). Palaeomagnetic investigations in Great Britain. *Phil. Trans. R. Soc. London*, A**250**, 144–56.

Creer, K. M., Mitchell, J. G. & Valencio, D. A. (1971). Evidence for normal geomagnetic field polarity at 263 ± 5 m.y. B.P. within the Late Palaeozoic reversed interval. *Nature*, **233**, 87–9.

Currie, K. L. & Larochelle, A. (1969). A paleomagnetic study of volcanic rocks from Mistastin Lake, Labrador, Canada. *Earth Planet. Sci. Lett.*, **6**, 309–15.

Dagley, P. (1969). Paleomagnetic results from some British dikes. *Earth Planet. Sci. Lett.*, **6**, 349–54.

Dagley, P. & Mussett, A. E. (1978). Palaeomagnetism of the Fishnish Dykes, Mull, Scotland. *Geophys. J. R. Astr. Soc.*, **53**, 553–8.

Dagley, P. & Mussett, A. E. (1981). Palaeomagnetism of the British Tertiary igneous province: Rhum and Canna. *Geophys. J. R. Astr. Soc.*, **65**, 475–91.

Dagley, P. & Mussett, A. E. (1986). Palaeomagnetism and radiometric dating of the British Tertiary Igneous Province: Muck and Eigg. *Geophys. J. R. Astr. Soc.*, **85**, 221–42.

Dagley, P., Mussett, A. E. & Skelhorn, R. R. (1984). The palaeomagnetism of the Tertiary igneous complex of Ardnamurchan. *Geophys. J. R. Astr. Soc.*, **79**, 911–22.

Dagley, P., Mussett, A. E. & Skelhorn, R. R. (1990). Magnetic polarity stratigraphy of the Tertiary igneous rocks of Skye, Scotland. *Geophys. J. Intern.*, **101**, 395–409.

Dagley, P., Mussett, A. E., Wilson, R. L. & Hall, J. M. (1978). The British Tertiary igneous province: palaeomagnetism of the Arran dykes. *Geophys. J. R. Astr. Soc.*, **54**, 75–91.

D'Agrelho-Filho, M. S. & Pacca, I. G. (1988). Palaeomagnetism of the Itajai, Castro and Bom Jardim groups from southern Brazil. *Geophys. J.*, **93**, 365–76.

Dalrymple, G. B., Grommé, C. S. & White, R. W. (1975). Potassium–Argon age and paleomagnetism of diabase dikes in Liberia: initiation of central Atlantic rifting. *Geol. Soc. Am. Bull.*, **86**, 399–411.

Daly, L. & Irving, E. (1983). Paléomagnétisme des roches carbonifères du Sahara central: analyse des aimantations juxtaposés; configuration de la Pangée. *Ann. Géophys.*, **1**, 207–16.

Daly, L. & Pozzi, J. P. (1976). Résultats paléomagnétiques du Permien inférieur et du Trias Marocain: comparaison avec les données africaines et sud-américaines. *Earth Planet. Sci. Lett.*, **29**, 71–80.

Daly, L. & Pozzi, J. P. (1977a). Conséquences des résultats paléomagnétiques permiens et triasiques du Maroc pour l'histoire du Gondwana. *Bull. Soc. Géol. France*, **19**, 507–12.

Daly, L. & Pozzi, J. P. (1977b). Détermination d'un nouveau pôle paléomagnétique africain sur des formations cambriennes du Maroc. *Earth Planet. Sci. Lett.*, **34**, 264–72.

Dalziel, I. W. D. (1991). Pacific margins of Laurentia and East Antarctica-Australia as a conjugate rift pair: evidence and implications for an Eocambrian supercontinent. *Geology*, **19**, 598–601.

Dalziel, I. W. D. & Forsythe, R. D. (1985). Andean evolution and the terrane concept. In *Tectonostratigraphic Terranes of the Circum-Pacific Ocean*, ed. D. G. Howell, pp. 565–81. Houston, TX: Circum-Pacific Council Energy Mineral Res.

Dalziel, I. W. D., Kligfield, R., Lowrie, W. & Opdyke, N. D. (1973). Palaeomagnetic data from the southernmost Andes and the Antarctandes. In *Implications of Continental Drift to the Earth Sciences*, ed. D. H. Tarling & S. K. Runcorn, vol. 1, pp. 87–101. London: Academic Press.

Dankers, P. (1982). Implications of Early Devonian poles from the Canadian Arctic archipelago for the North American apparent polar wander path. *Can. J. Earth Sci.*, **19**, 1802–9.

Dankers, P. & Lapointe, P. (1981). Paleomagnetism of Lower Cambrian volcanics and a cross-cutting Cambro-Ordovician diabase dyke from Buckingham (Québec). *Can. J. Earth Sci.*, **18**, 1174–86.

Davies, J., Nairn, A. E. M. & Ressetar, R. (1980). The paleomagnetism of certain late Precambrian and early Paleozoic rocks from the Red Sea Hills, eastern Desert, Egypt. *J. Geophys. Res.*, **85**, 3699–710.

Davoudzadeh, M., Soffel, H. & Schmidt, K. (1981). On the rotation of the Central-East Iran microplate. *N. Jb. Geol. Paläontol. Mh.*, 1981, 180–92.

De Boer, J. (1963). The geology of the Vicentinian Alps with special reference to their paleomagnetic history. *Geol. Ultraiectina*, **11**, 178 pp.

De Boer, J. (1965). Paleomagnetic indications of megatectonic movements in the Tethys. *J. Geophys. Res.*, **70**, 931–44.

De Boer, J. (1968). Paleomagnetic differentiation and correlation of the Late Triassic volcanic rocks in the Central Appalachians (with special reference to the Connecticut Valley). *Geol. Soc. Am. Bull.*, 79, 609–26.

De Boer, J. & Brookins, D. G. (1972). Paleomagnetic and radiometric age determination of (Permian) pegmatites in the Middletown district (Connecticut). *Earth Planet. Sci. Lett.*, **15**, 140–4.

De Charpal, O., Guennoc, P., Montadert, L. & Roberts, D. G. (1978). Rifting, crustal attenuation and subsidence in the Bay of Biscay. *Nature*, **275**, 706–11.

De Jong, K. A., Manzoni, M., Stavenga, T., Van Dijk, F., Van der Voo, R. & Zijderveld, J. D. A. (1973). Palaeomagnetic evidence for the rotation of Sardinia during the Early Miocene. *Nature*, **243**, 281–3.

De Jong, K. A., Manzoni, M. & Zijderveld, J. D. A. (1969). Palaeomagnetism of the Alghero trachyandesites. *Nature*, **224**, 67–9.

Dercourt, J. ***et al.*** (1986). Geologic evolution of the Tethys belt from the Atlantic to the Pamirs since the Lias. *Tectonophysics*, **123**, 241–315.

De Ruig, M. J. (1990). Fold trends and stress deviation in the Alicante fold belt, SE Spain. *Tectonophysics*, **184**, 393–403.

Deutsch, E. R. (1969). Paleomagnetism and North Atlantic paleogeography. *Am. Assoc. Petrol. Geol. Memoir*, **12**, 931–54.

Deutsch, E. R. & Kristjansson, L. G. (1974). Palaeomagnetism of Late Cretaceous–Tertiary volcanics from Disko Island, West Greenland. *Geophys. J. R. Astr. Soc.*, **39**, 343–60.

Deutsch, E. R., Kristjansson, L. G. & May, B. T. (1971). Remanent magnetism of Lower Tertiary lavas on Baffin Island. *Can. J. Earth Sci.*, **8**, 1542–52.

Deutsch, E. R. & Prasad, J. N. (1987). Ordovician paleomagnetic results from the St George and Table Head carbonates of western Newfoundland. *Can. J. Earth Sci.*, **24**, 1785–96.

Deutsch, E. R. & Rao, K. V. (1977). New paleomagnetic evidence fails to support rotation of western Newfoundland. *Nature*, **266**, 314–18.

Deutsch, E. R. & Somayajulu, C. (1970). Paleomagnetism of Ordovician ignimbrites from Killary Harbour, Eire. *Earth Planet. Sci. Lett.*, **7**, 337–45.

Deutsch, E. R. & Storetvedt, K. M. (1988). Magnetism of igneous rocks from the Tourmakeady and Glensaul inliers, W. Ireland: mode of emplacement and aspects of the Ordovician field pattern. *Geophys. J.*, **92**, 223–34.

Dewey, J. F., Helman, M. L., Turco, E., Hutton, D. H. W. & Knott, S. D. (1989). Kinematics of the western Mediterranean. In *Alpine Tectonics*, ed. M. P. Coward, D. Dietrich & R. G. Park. *Geol. Soc. London Special Publ.*, **45**, 265–83.

Dewey, J. F., Pitman, W. C. III, Ryan, W. B. F. & Bonnin, J. (1973). Plate tectonics and the evolution of the Alpine System. *Geol. Soc. Am. Bull.*, **84**, 3137–80.

Dewey, J. F., Shackleton, R. M., Chang, C. & Sun, Y. (1988). The tectonic evolution of the Tibetan Plateau. *Phil. Trans. R. Soc. London*, A**327**, 379–413.

Diehl, J. F., Beck, M. E., Beske-Diehl, S., Jacobson, D. & Hearn, B. C. (1983). Paleomagnetism of the Late Cretaceous-early Tertiary north-central Montana alkalic province. *J. Geophys. Res.*, **88**, 10593–609.

Diehl, J. F. & Shive, P. N. (1979). Paleomagnetic studies of the Early Permian Ingelside Formation of northern Colorado. *Geophys. J. R. Astr. Soc.*, **56**, 271–82.

Diehl, J. F. & Shive, P. N. (1981). Paleomagnetic results from the Late Carboniferous/Early Permian Casper Formation: implications for northern Appalachian tectonics. *Earth Planet. Sci. Lett.*, **54**, 281–92.

DiVenere, V. J. & Opdyke, N. D. (1990). Paleomagnetism of the Maringouin and Shepody formations, New Brunswick: a Namurian magnetic stratigraphy. *Can. J. Earth Sci.*, **27**, 803–10.

Dixon, J. E. & Robertson, A. H. F., Editors (1984). *The Geological Evolution of the Eastern Mediterranean*. Oxford: Published for the Geological Society by Blackwell.

Dooley, R. E. (1983). Paleomagnetism of some mafic intrusions in the South Carolina Piedmont. I. Magnetic systems with single characteristic directions. *Phys. Earth Planet. Int.*, **31**, 241–68.

Dooley, R. E. & Smith, W. A. (1982). Paleomagnetism of early Mesozoic diabase dikes in the South Carolina Piedmont. *Tectonophysics*, **90**, 283–307.

Douglass, D. N. (1988). Paleomagnetism of Ringerike Old Red Sandstone and related rocks, southern Norway: implications for pre-Carboniferous separation of Baltica and British terranes. *Tectonophysics*, **148**, 11–27.

Duff, B. A. (1978a). Palaeomagnetism of late Precambrian or Cambrian diorites from Leicestershire, UK. *Geol. Mag.*, **117**, 479–83.

Duff, B. A. (1978b). *Palaeomagnetic and Rock Magnetic Studies of Lower Palaeozoic Rocks on Jersey and the Adjacent Regions of the Armorican Massif.* Ph. D. Thesis, University of Leeds.

Duff, B. A. (1979). The palaeomagnetism of Cambro-ordovician redbeds, the Erquy spillite series and Trégastel-Ploumanac'h granite complex, Armorican Massif (France and the Channel Islands). *Geophys. J. R. Astr. Soc.*, **59**, 345–65.

Duff, B. A. (1980). The palaeomagnetism of Jersey volcanics and dykes, and the Lower Palaeozoic apparent polar wander path for Europe. *Geophys. J. R. Astr. Soc.*, **60**, 355–75.

Dunbar, J. A. & Sawyer, D. S. (1989). Patterns of continental extension along the conjugate margin of the Central and North Atlantic Oceans and Labrador Sea. *Tectonics*, **8**, 1059–77.

Duncan, R. A., Hargraves, R. B. & Brey, G. P. (1978). Age, palaeomagnetism and chemistry of melilite basalts in the southern Cape, South Africa. *Geol. Mag.*, **115**, 317–27.

Dunlop, D. J. (1979). On the use of Zijderveld vector diagrams in multicomponent paleomagnetic studies. *Phys. Earth Planet. Int.*, **20**, 12–24.

Dunlop, D. J. & Stirling, J. M. (1977). 'Hard' viscous remanent magnetization (VRM) in fine-grained hematite. *Geophys. Res. Lett.*, **4**, 163–6.

Dunlop, D. J. & Stirling, J. M. (1985). Post-tectonic magnetizations from the Cordova gabbro, Ontario and Palaeozoic reactivation in the Grenville province. *Geophys. J. R. Astr. Soc.*, **91**, 521–50.

Dunn, W. J. & Elmore, R. D. (1985). Paleomagnetic and petrographic investigation of the Taum Sauk Limestone, southeast Missouri. *J. Geophys. Res.*, **90**, 11469–83.

Edel, J. B. (1979) Paleomagnetic study of the Tertiary volcanics of Sardinia. *J. Geophys.*, **45**, 259–80.

Edel, J. B. (1987). Paleomagnetic evolution of the Central Massif (France) during the Carboniferous. *Earth Planet. Sci. Lett.*, **82**, 180–92.

Edel, J. B. & Coulon, M. (1984). Late Hercynian remagnetization of Tournaisian series from the Laval syncline, Armorican Massif, France. *Earth Planet. Sci. Lett.*, **68**, 343–50.

Edel, J. B., Coulon, M. & Hernot, M. P. (1984). Mise en évidence par le paléomagnétisme d'une importante rotation antihoraire des Vosges méridionales entre le Viséen terminal et le Westphalien supérieur. *Tectonophysics*, **106**, 239–57.

Edel, J. B., Lacaze, M. & Westphal, M. (1981b). Paleomagnetism in the northeastern Central Massif (France): evidence for Carboniferous rotations of the Hercynian orogenic belt. *Earth Planet. Sci. Lett.*, **55**, 48–52.

Edel, J. B., Montigny, R., Royer, J. Y., Thuizat, R. & Trolard, F. (1986). Paleomagnetic investigations and K-Ar dating on the Variscan plutonic massif of the Champ du Feu and its volcanic-sedimentary environment, northern Vosges, France. *Tectonophysics*, **122**, 165–85.

Edel, J. B., Montigny, R. & Thuizat, R. (1981a). Late Paleozoic rotations of Corsica and Sardinia: new evidence from paleomagnetic and K-Ar studies. *Tectonophysics*, **79**, 201–23.

Edwards, J. (1965). Reversals of natural remanent magnetization within the iron grit of Sussex. *Geophys. J. R. Astr. Soc.*, **9**, 389–97.

Ekstrand, E. J. & Butler, R. F. (1989). Paleomagnetism of the Moenave Formation: implications for the Mesozoic North American apparent polar wander path. *Geology*, **17**, 245–8.

Ellwood, B. B. (1982). Paleomagnetic evidence for the continuity and independent movement of a distinct major crustal block in the southern Appalachians. *J. Geophys. Res.*, **87**, 5339–50.

Ellwood, B. B. & Abrams, C. (1982). Magnetization of the Austell gneiss, northwest Georgia Piedmont. *J. Geophys. Res.*, **87**, 3033–43.

Elmore, R. D. & Van der Voo, R. (1982). Origin of hematite and its associated remanence in the Copper Harbor Conglomerate (Keweenawan), Upper Michigan. *J. Geophys. Res.*, **87**, 10918–29.

El Shazly, E. M. & Krs, M. (1973). Paleogeography and paleomagnetism of Nubian sandstone, Eastern Desert of Egypt. *Geol. Rundschau*, **62**, 212–25.

Elston, D. P. & Bressler, S. L. (1977). Paleomagnetic poles and polarity zonation from Cambrian and Devonian strata of Arizona. *Earth Planet. Sci. Lett.*, **36**, 423–33.

Elston, D. P. & Purucker, M. (1979). Detrital magnetization in red beds of the Moenkopi Formation. *J. Geophys. Res.*, **84**, 1653–65.

Embleton, B. J. J. (1970). Palaeomagnetic results for the Permian of South America and a comparison with the African and Australian data. *Geophys. J. R. Astr. Soc.*, **21**, 105–18.

Embleton, B. J. J. (1972). The paleomagnetism of some Proterozoic-Cambrian sediments from the Amadeus Basin, Central Australia. *Earth Planet. Sci. Lett.*, **17**, 217–26.

Embleton, B. J. J. (1977). A Late Devonian palaeomagnetic pole for the Mulga Downs Group, western New South Wales. *J. Proc. R. Soc. New S. Wales*, **110**, 25–7.

Embleton, B. J. J. & Giddings, G. W. (1974). Late Precambrian and Lower Palaeozoic palaeomagnetic results from South Australia and Western Australia. *Earth Planet. Sci. Lett.*, **22**, 355–65.

Embleton, B. J. J. & McDonnell, K. L. (1981). Magnetostratigraphy in the Sydney Basin, southeastern Australia. In *Global Reconstruction and the Geomagnetic Field During the Palaeo-*

zoic, ed. M. W. McElhinny, A. N. Khramov, M. Ozima & D. A. Valencio, *Advances in Earth Planet. Sci.*, vol. 10, pp. 1–10. Dordrecht: Reidel Publ. Co.

Embleton, B. J. J. & McElhinny, M. W. (1975). The paleoposition of Madagascar: Paleomagnetic evidence from the Isalo group. *Earth Planet. Sci. Lett.*, **27**, 329–41.

Embleton, B. J. J., Schmidt, P. W., Hamilton, L. H. & Riley, G. H. (1985). Dating volcanism in the Sydney Basin: evidence from K-Ar ages and palaeomagnetism. *Publ. Geol. Soc. Aust., N.S.W. Div.*, **1**, 59–72.

Embleton, B. J. J. & Shepherd, J. (1977). The Late Devonian paleomagnetic field for southeastern Australia: a new result for the Lochiel Formation and a reassessment of results from the Catombal Group. *J. Geophys. Res.*, **82**, 5423–6.

Embleton, B. J. J. & Williams, G. E. (1986). Low palaeolatitudes of deposition for late Precambrian periglacial varvites in South Australia: implications for palaeoclimatology. *Earth Planet. Sci. Lett.*, **79**, 419–30.

Esang, C. B. & Piper, J. D. A. (1984a). Palaeomagnetism of the Carboniferous E–W dyke swarm in Argyllshire. *Scott. J. Geol.*, **20**, 309–14.

Esang, C. B. & Piper, J. D. A. (1984b). Palaeomagnetism of Caledonian intrusive suites in the northern Highlands of Scotland: Constraints to tectonic movements within the Caledonian orogenic belt. *Tectonophysics*, **104**, 1–34.

Evans, I. & Hall, S. A. (1990). Paleomagnetic constraints on the tectonic evolution of the Sakarya continent, northwestern Anatolia. *Tectonophysics*, **182**, 357–72.

Evans, I., Hall, S. A., Carman, M. F., Senalp, M. & Coskun, S. (1982). A paleomagnetic study of the Bilecik Limestone (Jurassic), NW Anatolia. *Earth Planet. Sci. Lett.*, **61**, 199–208.

Evans, I., Hall, S. A., Saribudak, M. & Aykol, A. (1991). Preliminary paleomagnetic results from Paleozoic rocks of the Istanbul-Zonguldak region, northwestern Turkey. *Bull. Tech. Univ. Istanbul*, **44**, 81–106.

Evans, M. E. (1965). Note on the remanant magnetism of some Upper Jurassic lavas from the Argentine Islands. *Brit. Antarct. Surv. Bull.*, **6**, 49–50.

Evans, M. E. & Maillol, J. M. (1986). A palaeomagnetic investigation of a Permian red bed sequence from a mining drill core. *Geophys. J. R. Astr. Soc.*, **87**, 411–19.

Everitt, C. W. F. & Clegg, J. A. (1962). A field test of palaeomagnetic stability. *Geophys. J. R. Astr. Soc.*, **6**, 312–19.

Facer, R. A. (1977). Palaeomagnetism, radiometric age and geochemistry of an adamellite at Yetholme, N. S. W. – Reply. *J. Geol. Soc. Austr.*, **24**, 122–3.

Fahrig, W. F. & Freda, G. (1975). Paleomagnetism of the Mesozoic coast-parallel dolerite dikes of west Greenland. *Can. J. Earth Sci.*, **12**, 1244–8.

Fairhead, J. D. (1988). Mesozoic plate tectonic reconstructions of the central South Atlantic Ocean: the role of the west and central African rift system. *Tectonophysics*, **155**, 181–92.

Faller, A. M. (1975). Palaeomagnetism of the oldest Tertiary basalts in the Kangerdlugssuaq area of East Greenland. *Bull. Geol. Soc. Denmark*, **24**, 173–8.

Faller, A. M., Briden, J. C. & Morris, W. A. (1977). Palaeomagnetic results from the Borrowdale Volcanic Group, English Lake District. *Geophys. J. R. Astr. Soc.*, **48**, 111–21.

Faller, A. M. & Soper, N. J. (1979). Palaeomagnetic evidence for the origin of the coastal flexure and dyke swarm in central E. Greenland. *J. Geol. Soc.*, **136**, 737–44.

Fang, W. & Van der Voo, R. (1988). Paleomagnetism of Middle–Late Triassic plutons in southern Maine. *Tectonophysics*, **156**, 51–8.

Fang, W., Van der Voo, R. & Johnson, R. J. E. (1986). Eocambrian paleomagnetism of the Boston Basin: evidence for displaced terrane. *Geophys. Res. Lett.*, **13**, 1450–3.

Fang, W., Van der Voo, R. & Liang, Q. (1988/89). Reconnaissance magnetostratigraphy of the Precambrian–Cambrian boundary section at Meishucun, China. *Cuad. Geol. Iber.*, **12**, 205–22.

Fang, W., Van der Voo, R. & Liang, Q. (1989). Devonian paleomagnetism of Yunnan province across the Shan Thai–South China suture. *Tectonics*, **8**, 939–52.

Fang, W., Van der Voo, R. & Liang, Q. (1990a). Paleomagnetism of the Late Permian Emeishan Basalt, Yunnan, China (abstract). *Eos, Trans. Am. Geophys. Union*, **71**, 488.

Fang, W., Van der Voo, R. & Liang, Q. (1990b). Ordovician paleomagnetism of eastern Yunnan, China. *Geophys. Res. Lett.*, **17**, 953–6.

Farrell, W. E. & May, B. T. (1969). Paleomagnetism of Permian red beds from the Colorado Plateau. *J. Geophys. Res.*, **74**, 1495–504.

Faure, M., Marchadier, Y. & Rangin, C. (1989). Pre-Eocene synmetamorphic structure in

the Mindoro-Romblon-Palawan area, west Philippines, and implications for the history of southeast Asia. *Tectonics*, **8**, 963–79.
Finger, F. & Steyrer, H. P. (1990). I-type granitoids as indicators of a late Paleozoic convergent ocean-continent margin along the southern flank of the central European Variscan orogen. *Geology*, **18**, 1207–10.
Fisher, R. A. (1953). Dispersion on a sphere. *Proc. Roy. Soc.*, **A217**, 295–305.
Flynn, J. J. (1986). Correlation and geochronology of Middle Eocene strata from the western United States. *Palaeogeog. Palaeoclim. Palaeoecol.*, **55**, 335–406.
Förster, H., Soffel, H. & Zinsser, H. (1975). Paleomagnetism of rocks from the Eastern Alps from north and south of the Insubric line. *N. Jb. Geol. Paläontol.*, **149**, 112–27.
Foster, J. & Symons, D. T. A. (1979). Defining a paleomagnetic polarity pattern in the Monteregian intrusives. *Can. J. Earth Sci.*, **16**, 1716–25.
Freeman, R., Sabat, F., Lowrie, W. & Fontboté, J.-M. (1989). Paleomagnetic results from Mallorca (Balearic Islands, Spain). *Tectonics*, **8**, 591–608.
Frei, L. S. & Cox, A. V. (1987). Relative displacement between Eurasia and North America prior to the formation of oceanic crust in the North Atlantic. *Tectonophysics*, **142**, 111–36.
French, A. N. (1978). *Paleomagnetism and Rock Magnetism of the Rose Hill Formation*. M.Sc. Thesis, University of Michigan.
French, A. N. & Van der Voo, R. (1979). The magnetization of the Rose Hill Formation at the classical site of Graham's fold test. *J. Geophys. Res.*, **84**, 7688–96.
French, R. B., Alexander, D. H. & Van der Voo, R. (1977). Paleomagnetism of upper Precambrian to lower Paleozoic intrusive rocks from Colorado. *Geol. Soc. Am. Bull.*, **88**, 1785–92.
French, R. B. & Van der Voo, R. (1977). Remagnetization problems with the paleomagnetism of the Middle Silurian Rose Hill Formation of the central Appalachians. *J. Geophys. Res.*, **82**, 5803–6.
Freund, R. & Tarling, D. H. (1979). Preliminary Mesozoic paleomagnetic results from Israel and inferences for a microplate structure in the Lebanon. *Tectonophysics*, **60**, 189–205.
Fuller, M. D. (1985). Paleomagnetism in an accretionary margin, Luzon, northern Philippines. *J. Geodynamics*, **2**, 141–58.
Fuller, M. D., Haston, R. & Almasco, J. (1989). Paleomagnetism of the Zambales ophiolite, Luzon, northern Philippines. *Tectonophysics*, **168**, 171–203.
Fuller, M. D., McCabe, R., Williams, I. S., Almasco, J., Encina, R. Y., Zanoria, A. S. & Wolfe, J. A. (1983). Paleomagnetism of Luzon. *Am. Geophys. Union Geophys. Monograph*, **27**, Pt. 2, 79–94.
Funaki, M. (1984). Paleomagnetic investigations of McMurdo Sound region, South Victoria Land, Antarctica. *Mem. Natl. Inst. Polar Res. Japan. Ser. C, Earth Sci.*, **16**, 1–81.
Galbrun, B. (1985). Magnetostratigraphy of the Berriasian stratotype section (Berrias, France). *Earth Planet. Sci. Lett.*, **74**, 130–6.
Galbrun, B., Gabilly, J. & Rasplus, L. (1988). Magnetostratigraphy of the Toarcian stratotype section at Thouars and Airvault (Deux-Sèvres, France). *Earth Planet. Sci. Lett.*, **87**, 453–62.
Galbrun, B., Rivas, P., Baudin, F., Foucault, A., Fourcade, E. & Vrielynck, B. (1989). Magnétostratigraphie du Toarcien à facies 'ammonitico rosso' de la subbétique (Cordillères bétiques, Espagne). *C. R. Acad. Sci. Paris*, **308**, Serie II, 501–7.
Galdeano, A., Moreau, M. G., Pozzi, J. P., Berthou, P. Y. & Malod, J. A. (1989). New paleomagnetic results from Cretaceous sediments near Lisboa (Portugal) and implications for the rotation of Iberia. *Earth Planet. Sci. Lett.*, **92**, 95–106.
Gale, N. H., Beckinsale, R. D. & Wadge, A. J. (1980). Discussion of a paper by McKerrow, Lambert and Chamberlain on the Ordovician, Silurian and Devonian time scales. *Earth Planet. Sci. Lett.*, **51**, 9–17.
Gales, J. E., van der Pluijm, B. A. & Van der Voo, R. (1989). Paleomagnetism of the Lawrenceton Formation, Silurian Botwood Group, Change Islands, Newfoundland. *Can. J. Earth Sci.*, **26**, 295–304.
Gansser, A. (1981). The geodynamic history of the Himalay. *Geodynamics Series*, **3**, 111–21.
Garfunkel, Z. & Derin, B. (1984). Permian–Early Mesozoic tectonism, and continental margin formation in Israel and its implications for the history of the eastern Mediterranean. In *The Geological Evolution of the Eastern Mediterranean*, ed. J. E. Dixon & A. H. F. Robertson, pp. 187–201. Oxford: published by Blackwell for the Geol. Society.

Gehring, A. U. & Heller, F. (1989). Timing of natural remanent magnetization in ferriferous limestones from the Swiss Jura Mountains. *Earth Planet. Sci. Lett.*, **93**, 261–72.

Geissman, J. W. (1980). Paleomagnetism of ash-flow tuffs: microanalytical recognition of TRM components. *J. Geophys. Res.*, **85**, 1487–99.

Geissman, J. W., Harlan, S. S. & Brearley, A. J. (1988). The physical isolation and identification of carriers of geologically stable remanent magnetization: paleomagnetic and rock magnetic microanalysis and electron microscopy. *Geophys. Res. Lett.*, **15**, 479–82.

Ghorabi, M. (1990). Lower- and Middle-Triassic palaeomagnetic analysis from southern Tunisia. *J. Geodynamics*, **12**, 163–76.

Giang Nguyen (1982). Palaeomagnetic studies of Cenozoic basalts in Vietnam. In *Paleomagnetic Research in Southeast and East Asia*, ed. J. C. Briden *et al.*, pp. 58–63, Committee for the Coordination of Offshore Prospecting (CCOP), UN Development Project Office.

Giddings, J. W. & Embleton, B. J. J. (1974). Large-scale horizontal displacements in southern Australia – Contrary evidence from palaeomagnetism. *J. Geol. Soc. Austr.*, **21**, 431–6.

Giddings, J. W., Tarling, D. H. & Thomas, D. H. (1974). The palaeomagnetism of the Cleveland-Armathwaite Dyke, northern England. *Trans. Nat. Hist. Soc. Northumberland*, **4**, 220–6.

Gidskehaug, A., Creer, K. M. & Mitchell, J. G. (1975). Palaeomagnetism and KAr ages of the southwest African basalts and their bearing on the time of rifting of the South Atlantic Ocean. *Geophys. J. R. Astr. Soc.*, **42**, 1–20.

Girdler, R. W. (1968). A paleomagnetic investigation of some Late Triassic and Early Jurassic volcanic rocks from the-northern Pyrenees. *Ann. Geophys.*, **24**, 1–14.

Glen, W. (1982). *The Road to Jaramillo*. Stanford: Stanford University Press.

Globerman, B. R. & Irving, E. (1988). Mid-Cretaceous paleomagnetic reference field for North America: restudy of 100 Ma intrusive rocks from Arkansas. *J. Geophys. Res.*, **93**, 11,721–33.

Goleby, B. R. (1981). Early Palaeozoic palaeomagnetism in southeast Australia. In *Global Reconstruction and the Geomagnetic Field During the Palaeozoic*, ed. M. W. McElhinny, A. N. Khramov, M. Ozima & D. A. Valencio, pp. 11–21. *Advances in Earth Planet. Sci.*, vol. 10. Dordrecht: Reidel Publ. Co.

Gordon, R. G. (1988). True polar wander and paleomagnetic poles. *Phys. Today*, **41**, S–46.

Gordon, R. G., Cox, A. & O'Hare, S. (1984). Paleomagnetic Euler poles and the apparent polar wander and absolute motion of North America since the Carboniferous. *Tectonics*, **3**, 499–537.

Gose, W. A. & Helsley, C. E. (1972). Paleomagnetic and rock magnetic studies of the Permian Cutler and Elephant Canyon formations in Utah. *J. Geophys. Res.*, 77, 1534–48.

Gough, D. I. & Brock, A. (1964). The palaeomagnetism of the Shawa Ijolite. *J. Geophys. Res.*, **69**, 2489–93.

Gough, D. I., Brock, A., Jones, D. L. & Opdyke, N. D. (1964). The paleomagnetism of the ring complexes at Marangudzi and the Mateke Hills. *J. Geophys. Res.*, **69**, 2499–507.

Gough, D. I. & Opdyke, N. D. (1963). The palaeomagnetism of the Lupata alkaline volcanics. *Geophys. J. R. Astr. Soc.*, 7, 457–68.

Graham, J. W. (1949). The stability and significance of magnetism in sedimentary rocks. *J. Geophys. Res.*, **54**, 131–67.

Graham, K. W. T. & Hales, A. L. (1957). Palaeomagnetic measurements on Karroo dolerites. *Adv. Phys.*, **6**, 149–61.

Graham, K. W. T. & Hales, A. L. (1961). Preliminary palaeomagnetic measurements on Silurian sediments from South Africa. *Geophys. J. R. Astr. Soc.*, **5**, 318–25.

Grasso, M., Lentini, F., Nairn, A. E. M. & Vigliotti, L. (1983). A geological and paleomagnetic study of the Hyblean volcanic rocks, Sicily. *Tectonophysics*, **98**, 271–95.

Grasty, R. L. (1964). *Dating Basic Rocks for use in Paleomagnetism*. Ph.D. Thesis, University of London.

Green, R. (1961). Palaeomagnetism of some Devonian rock formations in Australia. *Tellus*, **13**, 119–34.

Green, R. & Irving, E. (1958). The palaeomagnetism of the Cainozoic basalts from Australia. *Proc. Roy. Soc. Victoria*, **70**, 1–16.

Gregor, C. B., Mertzman, S., Nairn, A. E. M. & Negendank, J. (1974). Paleomagnetism of some Mesozoic and Cenozoic volcanic rocks from the Lebanon. *Tectonophysics*, **21**, 375–95.

Gregor, C. B., Nairn, A. E. M. & Negendank, J. F. W. (1975). Paleomagnetic investigations of the Tertiary and Quaternary rocks; IX. The Pliocene of southeast Sicily and some Cretaceous rocks from Capo Passero. *Geol. Rundschau*, **64**, 948–58.

Gregor, C. B. & Zijderveld, J. D. A. (1964). The magnetism of some Permian red sandstones from northwestern Turkey. *Tectonophysics*, **1**, 289–306.

Grindley, G. W. & Oliver, P. J. (1983). Paleomagnetism of Cretaceous volcanic rocks from Marie Byrd Land. In *Antarctic Earth Science*, ed. R. L. Oliver, P. R. James & J. B. Jago, pp. 573–8. Canberra: Australian Acad. Sci.

Groot, J. J., de Jonge, R. B. G., Langereis, C. G., ten Kate, W. G. H. Z. & Smith, J. (1989). Magnetostratigraphy of the Cretaceous-Tertiary boundary at Agost (Spain). *Earth Planet. Sci. Lett.*, **94**, 385–97.

Grubbs, K. L. & Van der Voo, R. (1976). Structural deformation of the Idaho-Wyoming overthrust belt (USA), as determined by Triassic paleomagnetism. *Tectonophysics*, **33**, 321–36.

Grunow, A. M., Kent, D. V. & Dalziel, I. W. D. (1987). Mesozoic evolution of West Antarctica and the Weddell Sea Basin: new paleomagnetic constraints. *Earth Planet. Sci. Lett.*, **86**, 16–26.

Grunow, A. M., Kent, D. V. & Dalziel, I. W. D. (1991). New paleomagnetic data from Thurston Island: implications for the tectonics of West Antarctica and Weddell Sea opening. *J. Geophys. Res.*, **96**, 17935–54.

Guicherit, R. (1964). Gravity tectonics, gravity field, and palaeomagnetism in northeast Italy. *Geol. Ultraiectina*, **14**, 125 pp.

Guimera, J. & Alvaro, M. (1990). Structure et évolution de la compression alpine dans la chaîne ibérique et la chaîne cotière catalane (Espagne). *Bull. Soc. Géol. France*, Series 8, VI, No. 2, 339–48.

Guner, M. (1982). A palaeomagnetic study of some basaltoids and ores from the Pontic Ranges, northern Turkey: palaeogeographic implications. *Tectonophysics*, **90**, 309–33.

Gurariy, G. Z., Kropotkin, P. N., Pevznev, M. A., Won, P. S. & Trubikhin, V. M. (1966). Laboratory evaluation of the usefulness of North Korean sedimentary rocks for paleomagnetic studies. *Izv. Akad. Sci. USSR Phys. Solid Earth*, **11**, 128–36.

Hagstrum, J. T. (1990). Remagnetization and northward coastwise transport of Franciscan complex rocks, northern California: a reinterpretation of the paleomagnetic data. *Tectonics*, **9**, 1221–33.

Hagstrum, J. T., Van der Voo, R., Auvray, B. & Bonhommet, N. (1980). Eocambrian-Cambrian palaeomagnetism of the Armorican massif. *Geophys. J. R. Astr. Soc.*, **61**, 489–517.

Haile, N. S. (1978). Reconnaissance paleomagnetic results from Sulawesi, Indonesia, and their bearing on paleogeographic reconstruction. *Tectonophysics*, **46**, 77–85.

Haile, N. S. (1979a). Rotation of Borneo microplate completed by Miocene: palaeomagnetic evidence. *Warta Geologi*, **5**, 19–22.

Haile, N. S. (1979b). Paleomagnetic evidence for the clockwise rotation and palaeolatitude of Sumatra. *J. Geol. Soc. London*, **136**, 541–6.

Haile, N. S. (1980). Paleomagnetic evidence from the Ordovician and Silurian of northwest Peninsular Malaysia. *Earth Planet. Sci. Lett.*, **48**, 233–6.

Haile, N. S (1981a). Palaeomagnetic evidence for the rotation of Seram, Indonesia. In *Geodynamics of the West Pacific*, ed. S. Uyeda, *Advances Earth Planet. Sci.*, **6**, 191–8.

Haile, N. S. (1981b). Paleomagnetism of southeast and east Asia. *Geodynamics Series*, **2**, 129–35.

Haile, N. S., Beckinsale, R. D., Chakraborty, K. R., Hussein, A. H. & Hardjono, T. (1983). Paleomagnetism, geochronology and petrology of the dolerite dikes and basaltic lavas from Kuantan, West Malaysia. *Geol. Soc. Malaysia Bull.*, **16**, 71–85.

Haile, N. S. & Khoo Han Peng (1980). Palaeomagnetic measurements on Upper Jurassic to Lower Cretaceous sedimentary rocks from Peninsular Malaysia. *Geol. Soc. Malaysia Bull.*, **12**, 75–8.

Haile, N. S., McElhinny, M. W. & McDougall, I. (1977). Palaeomagnetic data and radiometric ages from the Cretaceous of West Kalimantan (Borneo), and their significance in interpreting regional structure. *J. Geol. Soc. London*, **133**, 133–44.

Haile, N. S. & Tarling, D. H. (1975). Note on reconnaissance palaeomagnetic measurements on Jurassic red beds from Thailand. *Pacific Geology*, **10**, 101–3.

Hailwood, E. A. (1974). Paleomagnetism of the Msissi Norite (Morocco) and the Paleozoic reconstruction of Gondwanaland. *Earth Planet. Sci. Lett.*, **23**, 376–86.

Hailwood, E. A. (1975). The palaeomagnetism of Triassic and Cretaceous rocks from Morocco. *Geophys. J. R. Astr. Soc.*, **41**, 219–35.

Hailwood, E. A. (1977). Configuration of the geomagnetic field in early Tertiary times. *J. Geol. Soc. London*, **133**, 23–36.

Hailwood, E. A. & Mitchell, J. G. (1971). Paleomagnetic and radiometric dating results from Jurassic intrusions in southern Morocco. *Geophys. J. R. Astr. Soc.,* **24**, 351–64.

Hall, J. M., Wilson, R. L. & Dagley, P. (1977). A palaeomagnetic study of the Mull Lava succession. *Geophys. J. R. Astr. Soc*, **49**, 499–514.

Hall, S. A. & Evans, I. (1988). Paleomagnetic study of the Ordovician Table Head Group, Port-au-Port Peninsula, Newfoundland. *Can. J. Earth Sci.*, **25**, 1407–19.

Hallam, A. (1976). Alfred Wegener and the hypothesis of continental drift. In *Continents Adrift and Continents Aground, Readings from Scientific American*, ed. J. T. Wilson, pp. 11–17. San Francisco: Freeman.

Hallam, A. (1983). Supposed Permo-Triassic megashear between Laurasia and Gondwana. *Nature*, **301**, 499–502.

Halls, H. C. (1976). A least-squares method to find a remanence direction from converging remagnetization circles. *Geophys. J. R. Astr. Soc.*, **45**, 297–304.

Halls, H. C. (1978). The use of converging remagnetization circles in paleomagnetism. *Phys. Earth Planet. Int.*, **16**, 1–11.

Halvorsen, E. (1970). Palaeomagnetism and age of the younger diabases in the Ny-Hellesund area, S. Norway. *Norsk Geol. Tidsskr.*, **50**, 157–66.

Halvorsen, E. (1972a). A palaeomagnetic study of two volcanic formations from northern Spitsbergen. *Norsk Polarinst. Årbok 1970, Oslo*, pp. 70–5.

Halvorsen, E. (1972b). On the palaeomagnetism of the Arendal diabases. *Norsk Geol. Tidsskr.*, **52**, 217–28.

Halvorsen, E. (1989). A paleomagnetic pole position of Late Jurassic/Early Cretaceous dolerites from Hinlopenstretet, Svalbard, and its tectonic implications. *Earth Planet. Sci. Lett.*, **94**, 398–408.

Halvorsen, E., Lewandowski, M. & Jelenska, M. (1989). Palaeomagnetism of the Upper Carboniferous Strzegom and Karkonosze granites and the Kudowa granitoid from the Sudet mountains, Poland. *Phys. Earth Planet. Int.*, **55**, 54–64.

Hamilton, W. B. (1970). The Uralides and the motion of the Russian and Siberian platforms. *Geol. Soc. Am. Bull.*, **81**, 2553–76.

Hamilton, W. B. (1979). Tectonics of the Indonesian region. *US Geol. Surv. Prof. Paper 1078*, 345 pp.

Hanna, W. F. (1967). Paleomagnetism of Upper Cretaceous volcanic rocks of southwestern Montana. *J. Geophys. Res.*, **72**, 595–610.

Hanna, W. F. (1973). Paleomagnetism of the Late Cretaceous Boulder Batholith, Montana. *Am. J. Sci.*, **273**, 778–802.

Harbury, N. A., Jones, M. E., Audley–Charles, M. G., Metcalfe, I. & Mohamed, K. R. (1990). Structural evolution of Mesozoic Peninsular Malaysia. *J. Geol. Soc. London,* **147**, 11–26.

Hargraves, R. B. (1975). Problems in palaeomagnetic synthesis illustrated by results from Permo-triassic dolerites in Guyana. *Phys. Earth Planet. Int.*, **16**, 277–84.

Hargraves, R. B. (1989). Paleomagnetism of Mesozoic kimberlites in southern Africa and the Cretaceous apparent polar wander curve for Africa. *J. Geophys. Res.*, **94**, 1851–66.

Hargraves, R. B., Dawson, E. M. & Van Houten, F. B. (1987). Palaeomagnetism and age of mid-Paleozoic ring complexes in Niger, West Africa, and tectonic implications. *Geophys. J. R. Astr. Soc.*, **90**, 705–29.

Hargraves, R. B. & Onstott, T. (1980). Paleomagnetic results from some southern African kimberlites, and their tectonic significance. *J. Geophys. Res.*, **85**, 3587–96.

Harlan, S. S., Geissman, J. W., Lageson, D. R. & Snee, L. W. (1988). Paleomagnetic and isotopic dating of thrust-belt deformation along the eastern edge of the Helena salient, northern Crazy Mountains Basin, Montana. *Geol. Soc. Am. Bull.*, **100**, 492–9.

Harland, W. B., Cox, A., Llewellyn, P. G., Pickton, C. A. G., Smith, A. G. & Walters, R. (1982). *A Geological Time Scale*. London: Academic Press.

Harland, W. B. & Gayer, R. A. (1972). The Arctic Caledonides and earlier oceans. *Geol. Mag.*, **109**, 289–314.

Harris, A. L., Holland, C. H. & Leake, B. E., (Eds) (1979). *The Caledonides of the British Isles Reviewed.* Edinburgh: Published for the Geol. Soc. London by the Scottish Academic Press.

Harrison, C. G. A. & Lindh, T. (1982a). Comparison between the hot spot and geomagnetic field reference frames. *Nature*, **300**, 251–2.

Harrison, C. G. A. & Lindh, T. (1982b). A polar wandering curve for North America during the Mesozoic and Cenozoic. *J. Geophys. Res.*, **87**, 1903–20.

Hasnain, I. & Qureshy, M. N. (1971). Paleomagnetism and geochemistry of some dikes in Mysore State, India. *J. Geophys. Res.*, **76**, 4786–95.

Haston, R., Fuller, M. D. & Schmidtke, E. (1988). Paleomagnetic results from Palau, West Caroline Islands: a constraint on Philippine Sea plate motion. *Geology*, **16**, 654–7.

Hatcher, R. D. Jr. (1988). The third synthesis: Wenlock to mid-Devonian (end of Acadian orogeny). In *The Caledonian–Appalachian Orogen*, ed. A. L Harris & D. J. Fettes, *Geol. Soc. Special Publ.*, vol. 38, pp. 499–504. Oxford: Published by Blackwell for the Geol. Soc.

Heckel, P. H. & Witzke, B. J. (1979). Devonian world palaeogeography determined from distribution of carbonates and related lithic palaeoclimatologic indicators. In *Special Papers in Paleontology*, vol. 23, pp. 99–123, Paleontological Association. Oxford: University Press.

Heiniger, C. (1979). Paleomagnetic and rock magnetic properties of the Permian volcanics in the western Southern Alps. *J. Geophys.*, **46**, 397–411.

Heirtzler, J. R., Dickson, G. O., Herron, E. M., Pitman, W. C. III & Le Pichon, X. (1968). Marine magnetic anomalies, geomagnetic field reversals, and motions of the ocean floor and continents. *J. Geophys. Res.*, **73**, 2119–36.

Heller, F. (1973). Magnetic anisotropy of granitic rocks of the Bergell Massif (Switzerland). *Earth Planet. Sci. Lett.*, **20**, 180–8.

Heller, F. (1977). Palaeomagnetic data from the western Lepontine area. *Schweizer. Mineral. Petrogr. Mitt.*, **57**, 135–43.

Heller, F. (1978). Rock magnetic studies of Upper Jurassic Limestones from southern Germany. *J. Geophys.*, **44**, 525–43.

Heller, F. (1980). Paleomagnetic evidence for late Alpine rotation of the Lepontine area. *Eclogae Helv.*, **73**, 607–18.

Heller, F. & Channell, J. E. T. (1979). Palaeomagnetism of Upper Cretaceous limestones from the Münster Basin, Germany. *J. Geophys.*, **46**, 413–27.

Heller, F., Lowrie, W. & Hirt, A. M. (1989). A review of palaeomagnetic and magnetic anisotropy results from the Alps. In *Alpine Tectonics*, ed. M. P. Coward, D. Dietrich & R. G. Park, *Geol. Soc. London Special Publ.*, vol. 45, pp. 399–420.

Heller, F., Lowrie, W., Huamei, L. & Junda, W. (1988). Magnetostratigraphy of the Permo-Triassic boundary section at Shangsi (Guangyuan, Sichuan Province, China). *Earth Planet. Sci. Lett.*, **88**, 348–56.

Helsley, C. E. (1965). Paleomagnetic results from the Lower Permian Dunkard Series of West Virginia. *J. Geophys. Res.*, **70**, 413–24.

Helsley, C. E. (1969). Magnetic reversal stratigraphy of the Lower Triassic Moenkopi Formation of western Colorado. *Geol. Soc. Am. Bull.*, **80**, 2431–50.

Helsley, C. E. (1971). Remanent magnetization of the Permian Cutler Formation of western Colorado. *J. Geoph. Res.*, **76**, 4842–8.

Helsley, C. E. & Nur, A. (1970). The paleomagnetism of Cretaceous rocks from Israel. *Earth Planet. Sci. Lett.*, **8**, 403–10.

Helsley, C. E & Steiner, M. B. (1974). Paleomagnetism of the Lower Triassic Moenkopi Formation. *Geol. Soc. Am. Bull.*, **85**, 457–64.

Henry, B. (1976). Relations entre contraintes tectoniques et propriétés magnétiques des roches volcaniques permiennes de la vallée du Guil (Briançonnais, Alpes Françaises). *Pure Appl. Geophys.*, **114**, 685–700.

Henry, S. G. (1979). Chemical demagnetization methods, procedures and applications through vector analysis. *Can. J. Earth Sci.*, **16**, 1832–41.

Hernando, S., Schott, J. J., Thuizat, R. & Montigny, R. (1980). Age des andésites et des sédiments interstratifiés de la région d'Atienza (Espagne): étude stratigraphique, géochronologique et paléomagnétique. *Sci. Géol. Bull., (Strasbourg)*, **33**, 119–28.

Herrero-Bervera, E. & Helsley, C. E. (1983). Paleomagnetism of a polarity transition in the Lower(?) Triassic Chugwater Formation, Wyoming. *J. Geophys. Res.*, **88**, 3506–22.

Herron, E. M., Dewey, J. F. & Pitman, W. C. (1974). Plate tectonic model for the evolution of the Arctic. *Geology*, **2**, 377–80.

Hijab, B. R. & Tarling, D. H. (1982). Lower Jurassic palaeomagnetic results from Yorkshire, England, and their implications. *Earth Planet. Sci. Lett.*, **60**, 147–54.

Hillhouse, J. W. (1977). Paleomagnetism of the Triassic Nikolai greenstone, McCarthy Quadrangle. *Can. J. Earth Sci.*, **14**, 2578–92.

Hillhouse, J. W. & Grommé, C. S. (1984). Northward displacement and accretion of Wrangellia; new paleomagnetic evidence from Alaska. *J. Geophys. Res.*, **89**, 4461–77.

Hirt, A. M. & Lowrie, W. (1988). Paleomagnetism of the Umbrian-Marches orogenic belt. *Tectonophysics*, **146**, 91–103.

Hodych, J. P. (1989). Limestones of western Newfoundland that magnetized before Devonian folding but after Middle Ordovician lithification. *Geophys. Res. Lett.*, **16**, 93–6.

Hodych, J. P. & Hayatsu, A. (1988). Paleomagnetism and K/Ar isochron dates of Early Jurassic basalt flows and dikes of Atlantic Canada. *Can. J. Earth Sci.*, **25**, 1972–89.

Hoffman, K. A. & Day, R. (1978). Separation of multi-component NRM: a general method. *Earth Planet. Sci. Lett.*, **40**, 433–8.

Hoffman, P. F. (1988). United Plates of America, the birth of a craton: early Proterozoic assembly and growth of Laurentia. *Ann. Rev. Earth Planet. Sci.*, **16**, 543–603.

Hoffman, P. F. (1991). Did the breakout of Laurentia turn Gondwanaland inside-out? *Science*, **252**, 1409–12.

Horner, F. & Freeman, R. (1983). Paleomagnetic evidence from pelagic limestones for clockwise rotation of the Ionian zone, western Greece. *Tectonophysics*, **98**, 11–27.

Horner, F. & Heller, F. (1983). Lower Jurassic magnetostratigraphy at the Breggia Gorge (Ticino, Switzerland) and Alpe Turati (Como, Italy). *Geophys. J. R. Astr. Soc.*, **73**, 705–18.

Horner, F. & Lowrie, W. (1981). Paleomagnetic evidence from Mesozoic carbonate rocks for the rotation of Sardinia. *J. Geophys.*, **49**, 11–19.

Hospers, J. (1967). Review of palaeomagnetic evidence for the displacement of continents, with particular reference to North America and Europe–northern Asia. In *Mantles of the Earth and Terrestrial Planets*, eds. S. K. Runcorn, pp. 331–349. London: Interscience Publ.

Hospers, J. & Van Andel, S. I. (1968). Palaeomagnetic data from Europe and North American and their bearing on the origin of the North Atlantic Ocean. *Tectonophysics*, **6**, 475–90.

Howell, D. G. (1985). *Tectonostratigraphic Terranes of the Circum-Pacific*. Circum-Pacific Council for Energy and Mineral Resources, Earth Science Series, vol. 1. Tulsa, Oklahoma: Amer. Assoc. Petrol. Geol.

Howell, D. G. (1989). *Tectonics of Suspect Terranes*. London: Chapman and Hall.

Hower, J. C. (1979). A Lower Ordovician geomagnetic pole from Pennsylvania. *Earth Planet. Sci. Lett.*, **44**, 65–72.

Hsü, K. J. (1989). Time and place in Alpine orogenesis – the Fermor Lecture. In *Alpine Tectonics*, ed. M. P. Coward, D. Dietrich & R. G. Park, *Geol. Soc. London Special Publ.*, **45**, pp. 421–43.

Hsu, I-Chi, Kienzle, J., Scharon, L. & Sun, S. S. (1966). Paleomagnetic investigation of Taiwan igneous rocks. *Bull. Geol. Surv. Taiwan*, **16**, 27–81.

Huang, K. & Opdyke, N. D. (1991). Paleomagnetic results from the Upper Carboniferous of the Shan-Thai-Malay block of Western Yunnan, China. *Tectonophysics*, **192**, 333–44.

Huang, K., Opdyke, N. D., Xu, G. Z. & Tang, R. L. (1986). Further paleomagnetic results from the Permian Emeishan Basalt in SW China. *Kexue Tongbao*, **31**, 1195–1201 (English Translation).

Hurley, N. F. & Van der Voo, R. (1987). Paleomagnetism of Upper Devonian reefal limestones, Canning Basin, western Australia. *Geol. Soc. Am. Bull.*, **98**, 138–46.

Hurley, P. M. & Rand, J. R. (1969). Pre-drift continental nuclei. *Science*, **164**, 1229–42.

Hurley, P. M. & Shearer, C. K. (1981). Paleomagnetic investigations in igneous-metamorphic rock units in eastern New England. *Can. J. Earth Sci.*, **18**, 1248–60.

Hussain, A. G. & Aziz, Y. (1983). Paleomagnetism of Mesozoic and Tertiary rocks from East El Weinat area, southwest Egypt. *J. Geophys. Res.*, **88**, 3523–9.

Hussain, A. G., Schult, A. & Soffel, H. C. (1979). Paleomagnetism of basalts of Wadi Abu Tereifiya, Mandisha and dioritic dikes of Wadi Abu Shihat, Egypt. *Geophys. J. R. Astr. Soc.*, **56**, 55–61.

Idnurm, M. & Giddings, J. W. (1988). Australian Precambrian polar wander: a review. *Precambrian Res.*, **40/41**, 61–88.

Idnurm, M. & Senior, B. R. (1978). Palaeomagnetic ages of Late Cretaceous and Tertiary weathered profiles in the Eromange Basin, Queensland. *Palaeogeogr., Palaeoclimatol., Palaeoecol.*, **24**, 263–77.

Incoronato, A., Tarling, D. H. & Nardi, G. (1985). Palaeomagnetic study of an allochthonous terrane: the Scisti Silicei Formation, Lagonegro Basin, southern Italy. *Geophys, J. R. Astr. Soc.*, **83**, 721–9.

Irving, E. (1956). Palaeomagnetic and palaeoclimatological aspects of polar wandering. *Geofis. Pura e Appl.*, **33**, 23–41.

Irving, E. (1963). Palaeomagnetism of the Narrabeen chocolate shale and of the Tasmanian dolerite. *J. Geophys. Res.*, **68**, 2283–7.

Irving, E. (1964). *Paleomagnetism and its Application to Geological and Geophysical Problems.* New York: John Wiley and Sons.

Irving, E. (1966). Paleomagnetism of some Carboniferous rocks from New South Wales and its relation to geological events. *J. Geophys. Res.*, **71**, 6025–51.

Irving, E. (1967). Palaeomagnetic evidence for shear along the Tethys. In *Aspects of Tethyan Biogeography*, ed. C. G. Adams & D. V. Ager, *Systematics Assoc. Publ.*, vol. 7, pp. 59–76.

Irving, E. (1977). Drift of the major continental blocks since the Devonian. *Nature*, **270**, 304–9.

Irving, E. (1979). Paleopoles and paleolatitudes of North America and speculations about displaced terrains. *Can. J. Earth Sci.*, **16**, 669–94.

Irving, E. (1988). The paleomagnetic confirmation of continental drift. *Eos, Trans. Amer. Geophys. Union*, **69**, 994–1014.

Irving, E. & Banks, M. R. (1961). Paleomagnetic results from the Upper Triassic lavas of Massachusetts. *J. Geophys. Res.*, **66**, 1935–9.

Irving, E. & Briden, J. C. (1962). Palaeolatitude of evaporite deposits. *Nature*, **196**, 425–8.

Irving, E. & Irving, G. A. (1982). Apparent polar wander paths Carboniferous through Cenozoic and the assembly of Gondwana. *Geophys. Surv.*, **5**, 141–88.

Irving, E. & Opdyke, N. D. (1965). The palaeomagnetism of the Bloomsburg red beds and its possible application to the tectonic history of the Appalachians. *Geophys. J. R. Astr. Soc.*, **9**, 153–67.

Irving, E. & Parry, L. G. (1963). The magnetism of some Permian rocks from New South Wales. *Geophys. J. R. Astr. Soc.*, 7, 395–411.

Irving, E., Stott, P. M. & Ward, M. A. (1961). Demagnetization of igneous rocks by alternating magnetic fields. *Phil. Mag.*, **6**, 225–41.

Irving, E. & Strong, D. F. (1984). Palaeomagnetism of the early Carboniferous Deer Lake Group, western Newfoundland: no evidence for mid-Carboniferous displacement of Acadia. *Earth Planet. Sci. Lett.*, **69**, 379–90.

Irving, E. & Strong, D. F. (1985). Paleomagnetism of rocks from the Burin Peninsula, Newfoundland: hypothesis of late Paleozoic displacement of Acadia criticized. *J. Geophys. Res.*, **90**, 1949–62.

Irving, E., Tanczyk, J. & Hastie, J. (1976a). *Catalogue of Paleomagnetic Directions and Poles*, 3rd issue, *Paleozoic Results 1949–1975*. Geomagn. Ser., 5, 99 pp. Ottawa: Geomagn. Serv. Canada.

Irving, E., Tanczyk, J. & Hastie, J. (1976b). *Catalogue of Paleomagnetic Directions and Poles*, 4th issue, *Mesozoic results 1954–1975*. Geomagn. Ser., 6, 70 pp. Ottawa: Geomagn. Serv. Canada.

Ito, H. & Tokeida, K. (1980). An interpretation of paleomagnetic results from Cretaceous granites in South Korea. *J. Geomagn. Geoelectr.*, **32**, 275–84.

Jackson, K. C. (1990). A palaeomagnetic study of Apennine thrusts, Italy: Monte Maiella and Monte Raparo. *Tectonophysics*, **178**, 231–40.

Jackson, K. C. & Halls, H. C. (1988). Tectonic implications of paleomagnetic data from sills and dikes in the Sverdrup Basin, Canadian Arctic. *Tectonics*, 7, 463–81.

Jackson, M. & Van der Voo, R. (1985). A Lower Ordovician paleomagnetic pole from the Oneota Dolomite, Upper Mississippi Valley. *J. Geophys. Res.*, **90**, 10449–61.

Jackson, M., Van der Voo, R. & Geissman, J. W. (1988). Paleomagnetism of Ordovician alkalic intrusives and host rocks from the Pedernal Hills, New Mexico: positive contact test in remagnetized rocks? *Tectonophysics*, **147**, 313–23.

Jacobson, D., Beck, M. E., Diehl, J. F. & Hearn, B. C. (1980). A Paleocene paleomagnetic pole for North America from alkalic intrusions, central Montana. *Geophys. Res. Lett.*, 7, 549–52.

Jacobi, R. D. & Wasowski, J. J. (1985). Geochemistry and plate-tectonic significance of the volcanic rocks of the Summerford Group, northcentral Newfoundland. *Geology*, **13**, 126–30.

Jelen, M., Krs, M. & Kubiny, D. (1966). Palaeomagnetism of Triassic rocks from northern Mongolia. *Cas. Mineral. Geol., Prague*, **11**, 445–50.

Jelenska, M. (1987). Aspects of pre-Tertiary paleomagnetism of Spitsbergen and their tectonic implications. *Tectonophysics*, **139**, 99–106.

Jelenska, M. & Lewandowski, M. (1986). A paleomagnetic study of Devonian sandstone from central Spitsbergen. *Geophys. J. R. Astr. Soc.*, **87**, 617–32.

Jesinkey, C., Forsythe, R. D., Mpodozis, C. & Davidson, J. (1987). Concordant late Palaeozoic palaeomagnetizations for the Athacama Desert: implications for tectonic models of the Chilean Andes. *Earth Planet. Sci. Lett.*, **85**, 461–72.

Johnson, A. H. (1976). Palaeomagnetism of the Jurassic Navajo Sandstone from South-Western Utah. *Geophys. J. R. Astr. Soc.*, **44**, 161–75.

Johnson, A. H. & Nairn, A. E. M. (1972). Jurassic paleomagnetism. *Nature*, **240**, 551–2.

Johnson, R. J. E., van der Pluijm, B. A. & Van der Voo, R. (1991). Paleomagnetism of the Moreton's Harbour Group, northeastern Newfoundland Appalachians: evidence for an Early Ordovician island arc near the Laurentian margin of Iapetus. *J. Geophys. Res.*, **96**, 11689–701.

Johnson, R. J. E. & Van der Voo, R. (1985). Middle Cambrian paleomagnetism of the Avalon terrane in Cape Breton Island, Nova Scotia. *Tectonics*, **4**, 629–51.

Johnson, R. J. E. & Van der Voo, R. (1986). Paleomagnetism of the late Precambrian Fourchu Group, Cape Breton Island, Nova Scotia. *Can. J. Earth Sci.*, **23**, 1673–85.

Johnson, R. J. E. & Van der Voo, R. (1989). Dual-polarity Early Carboniferous remagnetization of the Fissett Brook Formation, Cape Breton Island, Nova Scotia. *Geophys. J.*, **97**, 259–73.

Johnson, R. J. E. & Van der Voo, R. (1990). Pre-folding magnetization reconfirmed for the Late Ordovician–Early Silurian Dunn Point volcanics, Nova Scotia. *Tectonophysics*, **178**, 193–205.

Johnson, R. J. E., Van der Voo, R. & Lowrie, W. (1984). Paleomagnetism and late diagenesis of Jurassic carbonates from the Jura Mountains, Switzerland and France. *Geol. Soc. Am. Bull.*, **95**, 478–88.

Jones, D. L., Silberling, N. J., Coney, P. J. & Plafker, G. (1987). *Lithotectonic Terrane Map of Alaska*. US Geol. Survey Map MF-1874-A, scale 1: 2 500 000.

Jones, D. L., Silberling, N. J. & Hillhouse, J. (1977). Wrangellia – a displaced terrane in northwestern North America. *Can. J. Earth Sci.*, **14**, 2565–77.

Jones, M., Van der Voo, R. & Bonhommet, N. (1979). Late Devonian to Early Carboniferous paleomagnetic poles from the Armorican Massif, France. *Geophys. J. R. Astr. Soc.*, **58**, 287–308.

Jurdy, D. M. (1981). True polar wander. *Tectonophysics*, **74**, 1–16.

Jurdy, D. M. & Van der Voo, R. (1975). True polar wander since the Early Cretaceous. *Science*, **187**, 1193–6.

Kadzialko-Hofmokl, M. & Kruczyk, J. (1987). Paleomagnetism of Middle–Late Jurassic sediments from Poland and implications for the polarity of the geomagnetic field. *Tectonophysics*, **139**, 53–66.

Kadzialko-Hofmokl, M., Kruczyk, J. & Westphal, M. (1988). Paleomagnetism of Jurassic sediments from the western border of the Rheingraben, Alsace (France). *J. Geophys.*, **62**, 102–8.

Kang, Y. H. & Li, H. S. (1977). On paleomagnetic characteristics of the formations of the Kyongsang system and block movement in the western part of Korea. *Bull. Acad. Sci., DPRK*, **25**, 289–301.

Kean, B. F. & Strong, D. F. (1975). Geochemical evolution of an Ordovician island arc of the central Newfoundland Appalachians. *Am. J. Sci.*, **275**, 97–118.

Kellogg, K. (1980). Paleomagnetic evidence for oroclinal bending of the southern Antarctic Peninsula. *Geol. Soc. Amer. Bull.*, **91**, 414–20.

Kellogg, K. S. (1988). A paleomagnetic investigation of rocks from the Ohio Range and the Dry Valleys, Transantarctic Mountains. *N.Z. J. Geol. Geophys*, **31**, 77–85.

Kellogg, K. S. & Reynolds, R. L. (1978). Paleomagnetic results from the Lassiter Coast, Antarctica, and a test for oroclinal bending of the Antarctic Peninsula. *J. Geophys. Res.*, **83**, 2293–9.

Kent, D. V. (1982). Paleomagnetic evidence for post-Devonian displacement of the Avalon Platform (Newfoundland). *J. Geophys. Res.*, **87**, 8709–16.

Kent, D. V. (1988). Further paleomagnetic evidence for oroclinal rotation in the central folded Appalachians from the Bloomsburg and Mauch Chunk formations. *Tectonics*, 7, 749–60.

Kent, D. V., Dia, O. & Sougy, J. M. A. (1984). Paleomagnetism of Lower–Middle Devonian and Upper Proterozoic-Cambrian (?) rocks from Mejeria (Mauritania, West Africa). *Geodynamics Series*, **12**, 99–115.

Kent, D. V. & May, S. R. (1987). Polar wander and paleomagnetic reference pole controversies. *Rev. Geophys.*, **25**, 961–70.

Kent, D. V. & Opdyke, N. D. (1978). Paleomagnetism of the Devonian Catskill red beds: evidence for motion of the coastal New England–Canadian maritime region relative to cratonic North America. *J. Geophys. Res.*, **83**, 4441–50.

Kent, D. V. & Opdyke, N. D. (1980). Paleomagnetism of Siluro-Devonian rocks from eastern Maine. *Can. J. Earth Sci.*, **17**, 1653–65.

Kent, D. V. & Opdyke, N. D. (1985). Multicomponent magnetizations from the Mississippian Mauch Chunk Formation of the Central Appalachians and their tectonic implications. *J. Geophys. Res.*, **90**, 5371–83.

Kent, D. V. & Van der Voo, R. (1990). Palaeozoic palaeogeography from palaeomagnetism of the Atlantic-bordering continents. In *Palaeozoic Palaeogeography and Biogeography*, ed. W. S. McKerrow & C. R. Scotese. *Geol. Soc. London Memoir*, **12**, 49–56.

Kent, D. V., Xu, G., Huang, K., Zhang, W. Y. & Opdyke, N. D. (1986). Paleomagnetism of Upper Cretaceous rocks from South China. *Earth Planet. Sci. Lett.*, **79**, 179–84.

Kent, D. V., Zeng, X. S., Zhang, W. Y. & Opdyke, N. D. (1987). Widespread late Mesozoic to Recent remagnetization of Paleozoic and Lower Triassic sedimentary rocks from South China. *Tectonophysics*, **139**, 133–43.

Kerth, M. (1985). A paleomagnetic study of Turonian carbonates from the southeastern Münsterland area, NW Germany. *J. Geophys.*, **57**, 118–24.

Kerth, M. & Hailwood, E. A. (1988). Magnetostratigraphy of the Lower Cretaceous Vectis Formation (Wealden Group) on the Isle of Wight, Southern England. *J. Geol. Soc. London*, **145**, 351–60.

Khan, M. A. (1960). The remanent magnetization of the basic Tertiary igneous rocks of Invernesshire. *Geophys. J. Roy. Ast. Soc.*, **3**, 45–62.

Khattach, D., Perroud, H. & Robardet, M. (1989). Etude paléomagnétique de formations paléozoïques du Maroc. *Sciences Géologiques (Strasbourg) Mem.*, **83**, 97–113.

Khramov, A. N. (1971). *Paleomagnetic Directions and Paleomagnetic Poles*, Issue No. 1. Moscow: Soviet Geophysical Committee of the Acad. Sci. USSR, World Data Center B.

Khramov, A. N. (1975). *Paleomagnetic Directions and Paleomagnetic Poles*, Issue No. 3. Moscow: Soviet Geophysical Committee of the Acad. Sci. USSR, World Data Center B.

Khramov, A. N. (1984). *Paleomagnetic Directions and Pole Positions: Data for the USSR, Summary Catalogue*. Moscow: Soviet Geophys. Comm., USSR. Acad. of Sci.

Khramov, A. N. (1988). Paleomagnetism and the problems of accretional tectonics of the north-west segment of the Pacific belt (in Russian). In *Paleomagnetism and Accretional Tectonics*, pp. 141–53. Leningrad: VNIGRI.

Khramov, A. N. & Ustritsky, V. I. (1990). Paleopositions of some northern Eurasian tectonic blocks: paleomagnetic and paleobiologic constraints. *Tectonophysics*, **184**, 101–9.

Kienzle, J. & Scharon, L. (1966). Paleomagnetic comparison of Cretaceous rocks from South Korea and late Paleozoic and Mesozoic rocks from Japan. *J. Geomagn. Geoelectr.*, **18**, 413–16.

Kilbourne, D. E. (1969). Paleomagnetism of some rocks from the Mesaverde Group, southwestern Wyoming and northeastern Utah. *Geol. Soc. Am. Bull.*, **80**, 2069–74.

Kim, K. H. & Van der Voo, R. (1990). Jurassic and Triassic paleomagnetism of South Korea. *Tectonics*, **9**, 699–719.

King, R. F. (1955). The remanent magnetism of artificially deposited sediments. *Mon. Not. R. Astr. Soc.*, 7, 115–34.

Kipfer, R. & Heller, F. (1988). Paleomagnetism of Permian red beds in the contact aureole of the Tertiary Adamello massif (northern Italy). *Phys. Earth Planet. Int.*, **52**, 365–75.

Kirschvink, J. L. (1978). The Precambrian–Cambrian boundary problem: paleomagnetic directions from the Amadeus Basin, central Australia. *Earth Planet. Sci. Lett.*, **40**, 91–100.

Kirschvink, J. L. (1980). The least squares line and plane and the analysis of paleomagnetic data. *Geophys. J. R. Astr. Soc.*, **62**, 699–718.

Kissel, C. (1989) Unpublishable palaeomagnetic data from Turkey? IAGA Abstract 1.16.45, IAGA Bulletin, 53, part A, p. 78.

Kissel, C., Kondopoulou, D., Laj, C. & Papadopoulou, P. (1986b). New paleomagnetic data from Oligocene formations of northern Aegea. *Geophys. Res. Lett.*, **13**, 1039–42.

Kissel, C. & Laj, C. (1988). The Tertiary geodynamical evolution of the Aegean arc: a paleomagnetic reconstruction. *Tectonophysics*, **146**, 183–201.

Kissel, C., Laj, C. & Mazaud, A. (1986a). First paleomagnetic results from Neogene formations in Evia, Skyros and the Volos region, and the deformation of central Aegea. *Geophys. Res. Lett.*, **13**, 1446–9.

Kissel, C., Laj, C. & Muller, C. (1985). Tertiary geodynamical evolution of northwestern Greece – paleomagnetic results. *Earth Planet. Sci. Lett.*, **72**, 190–204.

Kissel, C., Laj, C., Poisson, A., Savascin, Y., Simeakis, K. & Mercier, J. L. (1986c). Paleomagnetic evidence for Neogene rotational deformation in the Aegean domain. *Tectonics*, **5**, 783–95.

Kissel, C., Laj, C., Şengör, A. M. C. & Poisson, A. (1987). Paleomagnetic evidence for rotation in opposite senses of adjacent blocks in northeastern Aegea and western Anatolia. *Geophys. Res. Lett.*, **14**, 907–10.

Kissel, C. & Poisson, A. (1986). Etude paléomagnétique préliminaire des formations néogènes du bassin d'Antalya (Taurides occidentales, Turquie). *C. R. Acad. Sci. Paris*, **302**, 711–16.

Kissel, C. & Poisson, A. (1987). Etude paléomagnétique préliminaire des formations cénozoïques des Bey Dağlari (Taurides occidentales, Turquie). *C. R. Acad. Sci. Paris*, **304**, 343–8.

Kligfield, R. & Channell, J. E. T. (1979). Paleomagnetic evidence of large fault displacement around the Po Basin – Discussion. *Tectonophysics*, **53**, 139–42.

Klimetz, M. P. (1987). The Mesozoic tectonostratigraphic terranes and accretionary heritage of south-eastern mainland Asia. *Geodynamics Series*, **19**, 221–34.

Klitgord, K. D. & Schouten, H. (1986). Plate kinematics of the Central Atlantic. In *The Geology of North America*, vol. M, *The Western North Atlantic Region*, ed. P. R. Vogt & B. E. Tucholke, pp. 351–78. Boulder: Geol. Soc. America.

Klootwijk, C. T. (1971). Paleomagnetism of the upper Gondwana Rajmahal traps, northeast India. *Tectonophysics*, **12**, 449–67.

Klootwijk, C. T. (1973). Paleomagnetism of Upper Bhander sandstones from central India and implications for a tentative Cambrian Gondwanaland reconstruction. *Tectonophysics*, **18**, 123–45.

Klootwijk, C. T. (1974). Paleomagnetic results from some Panchet clay beds, Karanpura Coalfield, northeastern India. *Tectonophysics*, **21**, 79–92.

Klootwijk, C. T. (1975). Paleomagnetism of Upper Permian red beds in the Wardha Valley, central India. *Tectonophysics*, **25**, 115–37.

Klootwijk, C. T. (1980). Early Paleozoic paleomagnetism in Australia. *Tectonophysics*, **64**, 249–332.

Klootwijk, C. T. (1981). Greater Indias northern extent and its underthrust of the Tibetan Plateau: Paleomagnetic constraints and implications. *Geodynamics Ser.* **3**, 313–23.

Klootwijk, C. T. & Bingham, D. K. (1980). The extent of Greater India, III. Paleomagnetic data from the Tibetan sedimentary series, Thakkola region, Nepal, Himalaya. *Earth Planet. Sci. Lett.*, **51**, 381–405.

Klootwijk, C. T. & Conaghan, P. J. (1979). The extent of Greater India, I. Preliminary paleomagnetic data from the Upper Devonian of the eastern Hindukush, Chitral (Pakistan). *Earth Planet. Sci. Lett.*, **42**, 167–82.

Klootwijk, C. T., Conaghan, P. J. & Powell, C. McA. (1985). The Himalayan arc: large-scale continental subduction, oroclinal bending and back-arc spreading. *Earth Planet. Sci. Lett.*, **75**, 167–83.

Klootwijk, C. T., Jain, A. K. & Khorana, R. (1982). Palaeomagnetic constraints on allochthony and age of the Krol Belt sequence, Garhwal Himalaya, India. *J. Geophys.*, **50**, 127–36.

Klootwijk, C. T., Nazirullah, R. & De Jong, K. A. (1986b). Palaeomagnetic constraints on formation of the Mianwali reentrant, Trans-Indus and western Salt Range, Pakistan. *Earth Planet. Sci. Lett.*, **80**, 394–414.

Klootwijk, C. T., Nazirullah, R., De Jong, K. A. & Ahmed, H. (1981). A palaeomagnetic reconnaissance of northeastern Baluchistan, Pakistan. *J. Geophys. Res.*, **86**, 289–306.

Klootwijk, C. T., Shah, S. K., Gergan, J., Sharma, M. L., Tirkey, B. & Gupta, B. K. (1983). A palaeomagnetic reconnaissance of Kashmir, northwest Himalaya, India. *Earth Planet. Sci. Lett.*, **63**, 305–24.

Klootwijk, C. T., Sharma, M. L., Gergan, J., Shah, S. K. & Gupta, B. K. (1986a). Rotational overthrusting of the northwestern Himalaya: further palaeomagnetic evidence from the Riasi thrust sheet, Jammu foothills, India. *Earth Planet. Sci. Lett.*, **80**, 375–93.

Klootwijk, C. T., Sharma, M. L., Gergan, J., Shah, S. K. & Tirkey, B. (1984). The Indus-Tsangpo suture zone in Ladakh, northwest Himalaya: further paleomagnetic data and implications. *Tectonophysics*, **106**, 215–38.

Klootwijk, C. T., Sharma, M. L., Gergan, J., Tirkey, B., Shah, S. K. & Agarwal, V. (1979). The extent of Greater India, II. Paleomagnetic data from the Ladakh intrusives at Kargil, northwestern Himalayas. *Earth Planet. Sci. Lett.*, **44**, 47–64.

Kluger-Cohen, K., Anderson, T. H. & Schmidt, V. A. (1986). A paleomagnetic test of the proposed Mojave-Sonora megashear in northwestern Mexico. *Tectonophysics*, **131**, 23–51.

Kluth, C. F., Butler, R., Harding, L. E., Shafiqullah, M. & Damon, P. E. (1982). Paleomagnetism of Late Jurassic rocks in the northern Canelo Hills, southeastern Arizona. *J. Geophys. Res.*, **87**, 7079–86.

Kneen, S. J. (1973). The palaeomagnetism of the Foyers Plutonic Complex, Invernesshire. *Geophys. J. R. Astr. Soc.*, **32**, 53–63.

Kodama, K. P. (1988) Remanence rotation due to rock strain during folding and the stepwise application of the fold test. *J. Geophys. Res.*, **93**, 3357–71.

Kondopoulou, D. & Lauer, J. P. (1984). Paleomagnetic data from Tertiary units of the North Aegean zone. In *The Geological Evolution of the Eastern Mediterranean*, ed. J. E. Dixon & A. H. F. Robertson, pp. 681–6. Oxford: published by Blackwell for the Geol. Society.

Kondopoulou, D. & Westphal, M. (1986). Paleomagnetism of the Tertiary intrusives from Chalkidiki (Northern Greece). *J. Geophys.*, **59**, 62–6.

Kono, M., Kinoshita, H. & Aoki, Y. (1972). Paleomagnetism of Deccan trap basalts in India. *J. Geomagn. Geoelectr.*, **24**, 49–67.

Konrad, H. J. & Nairn, A. E. M. (1972). The palaeomagnetism of the Permian rocks of the Black Forest, Germany. *Geophys. J. R. astr. Soc.*, **27**, 369–82.

Kröner, A., McWilliams, M. O., Germs, G. J. B., Reid, A. B. & Schalk, K. E. L. (1980). Paleomagnetism of Late Precambrian to early Paleozoic mixtite-bearing formations in Namibia (Southwest Africa) – the Nama Group and Blaubeker Formation. *Am. J. Sci.*, **280**, 942–68.

Krs, M. (1967). On the palaeomagnetic stability of red sediments. *Geophys. J. R. Astr. Soc.*, 12, 313–317.

Krs, M. (1968). Rheological aspects of paleomagnetism? *13th Intern. Geol. Congress, Prague*, **5**, 87–96.

Krs, M., Krsova, M., Pruner, P., Chvojka, R. & Havlicek, V. (1987). Palaeomagnetism, palaeogeography and the multicomponent analysis of Middle and Upper Cambrian rocks of the Barrandian in the Bohemian Massif. *Tectonophysics*, **139**, 1–20.

Krs, M., Krsova, M., Pruner, P. & Havlicek, V. (1986). Paleomagnetism, paleogeography and multi-component analysis magnetization of Ordovician rocks from the Barrandian area of the Bohemian Massif. *Sbor. geol. Ved. Uzita Geofiz. Praha*, **20**, 9–45.

Kruczyk, J. & Kadzialko-Hofmokl, M. (1988). Paleomagnetism of Oxfordian sediments from the Halokinetic structure in Kujawy, central Poland. *2nd Intl. Symp. Jurassic Stratigraphy (Lisbon)*, pp. 1151–78.

Kruczyk, J., Kadzialko-Hofmokl, M., Nozharov, P., Petkov, N. & Nachev, I. (1990). Paleomagnetic studies on sedimentary Jurassic rocks from southern Bulgaria. *Phys. Earth Planet. Int.*, **62**, 82–96.

Krumsiek, K. (1976). Zur Bewegung der Iranisch-Afghanischen Platte. *Geol. Rundschau*, **65**, 908–29.

Krumsiek, K., Nagel, J. & Nairn, A. E. M. (1968). Record of paleomagnetic measurements on some igneous rocks from the Isfjorden region, Spitsbergen. *Norsk Polarinst., Årbok 1966*, 76–83.

Kruseman, G. P. (1962). Etude paléomagnétique et sédimentologique du bassin Permien de Lodève, Hérault, France. *Geol. Ultraiectina*, **9**, 1–66.

Lackie, M. A. (1988). The palaeomagnetism and magnetic fabric of the Late Permian Dundee rhyodacite, New England. In *New England Orogen: Tectonics and Metallogenesis*, ed. J. D. Kleeman, pp. 157–65. Armidale, NSW, Australia: University of New England.

Lacombe, P. & Roche, A. (1970). Sur l'aimantation de basaltes quaternaires du Cambodge. *C. R. Somm. Séances, Soc. Géol. France*, **8**, 300.

Laj, C., Jamet, M., Sorel, D. & Valente, J. P. (1982). First paleomagnetic results from Mio-Pliocene series of the Hellenic sedimentary arc. *Tectonophysics*, **86**, 45–67.

Lallemand, S., Cullota, R. & Von Huene, R. (1989). Subduction of the Daiichi Kashima Seamount in the Japan trench. *Tectonophysics*, **160**, 231–47.

Lang, J., Yahaya, M., El Hamet, M. O., Besombes, J. C. & Cazoulat, M. (1991). Dépôts glaciaires du Carbonifère inférieur à l'ouest de l'Aïr (Niger). *Geol. Rundschau*, **80**, 611–22.

Langereis, C. G. (1984). *Late Miocene Magnetostratigraphy in the Mediterranean*. Ph. D. Thesis, Univ. of Utrecht.

Lanza, R. (1977). Palaeomagnetic data from the andesitic and lamprophyre dikes of the Sezia-Lanzo zone (western Alps). *Schweizer. Mineral. Petrogr. Mitt.*, **57**, 281–90.

Lapointe, P. L. (1979a). Paleomagnetism and orogenic history of the Botwood Group and Mount Peyton Batholith, Central Mobile Belt, Newfoundland. *Can. J. Earth Sci.*, **16**, 866–76.

Lapointe, P. L. (1979b). Paleomagnetism of the Notre Dame Bay lamprophyre dikes, Newfoundland, and the opening of the North Atlantic Ocean. *Can. J. Earth Sci.*, **16**, 1823–31.

Larochelle, A. (1967). Paleomagnetic directions of a basic sill in Prince Edward Island. *Geol. Surv. Canada Paper*, 67-39, 6 pp.

Larochelle, A. (1971). Note on the paleomagnetism of two diabase dikes, Anticosti Island, Québec. *Geol. Assoc. Canada Proc.*, **23**, 73–6.

Larochelle, A., Black, R. F. & Wanless, R. K. (1965). Paleomagnetism of the Isachsen diabase rocks. *Nature*, **208**, 179–82.

Larson, E. E. & Lafountain, L. (1970). Timing of the breakup of the continents around the Atlantic as determined by paleomagnetism. *Earth Planet. Sci. Lett.*, **8**, 331–51.

Larson, E. E., Mutschler, F. E. & Brinkworth, G. (1969). Paleocene virtual geomagnetic poles determined from volcanic rocks near Golden, Colorado. *Earth Planet. Sci. Lett.*, **7**, 29–32.

Larson, E. E., Patterson, P. E., Curtis, G., Drake, R. & Mutschler, F. E. (1985). Petrologic, paleomagnetic, and structural evidence of a Paleozoic rift system in Oklahoma, New Mexico, Colorado and Utah. *Geol. Soc. Am. Bull.*, **96**, 1364–72.

Larson, E. E. & Strangway, D. W. (1969). Magnetization of the Spanish Dike swarm, Colorado, and Shiprock Dike, New Mexico. *J. Geophys. Res.*, **74**, 1505–14.

Larson, E. E., Walker, T. R., Patterson, P. E., Hoblitt, R. P. & Rosenbaum, J. G. (1982). Paleomagnetism of the Moenkopi Formation, Colorado Plateau: basis for long-term model of acquisition of chemical remanent magnetization in red beds. *J. Geophys. Res.*, **87**, 1081–106.

Latham, A. G. & Briden, J. C. (1975). Palaeomagnetic field directions in Siluro-Devonian lavas of the Lorne Plateau, Scotland, and their regional significance. *Geophys. J. R. Astr. Soc.*, **43**, 243–52.

Laubscher, H. (1988). Material balance in Alpine orogeny. *Geol. Soc. Amer. Bull.*, **100**, 1313–28.

Laubscher, H. & Bernouilli, D. (1977). Mediterranean and Tethys. In *The Ocean Basins and Margins*, vol. 4a, *The Eastern Mediterranean*, ed. A. E. M. Nairn *et al.*, pp. 1–28. New York: Plenum.

Lauer, J. P. (1981a). Origine méridionale des Pontides d'après de nouveaux résultats paléomagnétiques obtenus en Turquie. *Bull. Soc. Géol. France*, **23**, 619–24.

Lauer, J. P. (1981b). *Evolution géodynamique de la Turquie et de Chypre déduite de l'étude paléomagnétique*. Ph. D. Thesis, Univ. Strasbourg.

Lavecchia, G. & Stoppa, F. (1990). The Tyrrhenian zone: a case of lithosphere extension control of intra-continental magmatism. *Earth Planet. Sci. Lett.*, **99**, 336–50.

Lawver, L. & Scotese, C. R. (1988). A revised reconstruction of Gondwanaland. In *Gondwana Six: Structure, Tectonics and Geophysics*, ed. G. D. McKenzie. *Am. Geophys. Union, Geophys. Monogr.*, **40**, 17–23.

Lee, D. S. (Ed.) (1987). *Geology of Korea*. Seoul, Korea: Kyohak-Sa.

Lee, G., Besse, J., Courtillot, V. & Montigny, R. (1987). Eastern Asia in the Cretaceous:

new paleomagnetic data from South Korea and a new look at Chinese and Japanese data. *J. Geophys. Res.*, **92**, 3580–96.

Lee, S. (1983). *A Study of the Time-Averaged Paleomagnetic Field for the Last 195 Million Years.* Ph. D. Thesis, Australian National University, Canberra.

Lefort, J. P., Max, M. D. & Roussel, J. (1988). Geophysical evidence for the location of the NW boundary of Gondwanaland and its relationship with two older satellite sutures. In *The Caledonian–Appalachian Orogen*, ed. A. L. Harris & D. J. Fettes, *Geol. Soc. Special Publ.*, **38**, pp. 49–60. Oxford: published by Blackwell for the Geol. Soc.

Lefort, J. P. & Van der Voo, R. (1981). A kinematic model for the collision and complete suturing between Gondwanaland and Laurasia in the Carboniferous. *J. Geol.*, **89**, 537–50.

LeGrand, H. (1988). *Drifting Continents and Shifting Theories.* Cambridge: Cambridge University Press.

LeGrand, H. (1990). Rise and fall of paleomagnetic research at Carnegie's DTM. *Eos, Trans. Am. Geophys. Union*, **71**, 1043–4.

Le Pichon, X. (1968). Sea-floor spreading and continental drift. *J. Geophys. Res.*, **73**, 3661–97.

Le Pichon, X. & Fox, J. P. (1971). Marginal offsets, fracture zones and the early opening of the North Atlantic. *J. Geophys. Res.*, **76**, 6294–308.

Le Pichon, X. & Sibuet, J. C. (1971). Western extension of boundary between European and Iberian plates during the Pyrenean opening. *Earth Planet. Sci. Lett.*, **12**, 83–8.

Le Pichon, X., Sibuet, J. C. & Francheteau, J. (1977). The fit of the continents around the North Atlantic Ocean. *Tectonophysics,* **38**, 169–209.

Lerbekmo, J. F. & Coulter, K. C. (1985). Late Cretaceous to Early Tertiary magnetostratigraphy of a continental sequence: Red Deer Valley, Alberta, Canada. *Can. J. Earth Sci.*, **22**, 567–83.

Lewandowski, M. (1987). Results of the preliminary paleomagnetic investigation of some lower Paleozoic rocks from the Holy Cross mountains (Poland). *Kwartalnik Geol.*, **17**, 45–54.

Lewandowski, M., Jelenska, M. & Halvorsen, E. (1987). Paleomagnetism of the Lower Devonian sandstones from Holy Cross Mountains Central Poland, part 1. *Tectonophysics,* **139**, 21–9.

Li, C., Wang, Q., Liu, X. & Tang, Y. (1982). Exploratory notes to the tectonic map of Asia, edited by the Research Institute of Geology, Chinese Academy of Geological Sciences, pp. 41–43. Beijing: Cartographic Publishing House. (In Chinese with English abstract.)

Li, Y. (1990). An apparent polar wander path from the Tarim block, China. *Tectonophysics,* **181**, 31–41.

Li, Y., McWilliams, M. O., Cox, A., Sharps, R., Li, Y. A., Gao, Z., Zhang, Z. & Zhai, Y. (1988b). Late Permian paleomagnetic pole from dikes of the Tarim craton, China. *Geology,* **16**, 275–8.

Li, Y., McWilliams, M. O., Sharps, R., Cox, A., Li, Y., Li, Q., Gao, Z., Zhang, Z. & Zhai, Y. (1990). A Devonian paleomagnetic pole from red beds of the Tarim Block, China. *J. Geophys. Res.*, **95**, 19185–98.

Li, Y., Sharps, R., McWilliams, M., Li, Y., Li, Q. & Zhang, W. (1991). Late Paleozoic paleomagnetic poles from the Junggar Block, northwestern China. *J. Geophys. Res.*, **96**, 16047–60.

Li, Y., Sharps, R., McWilliams, M. O., Nur, A., Li, Y. A., Li, Q. & Zhang, W. (1989). Paleomagnetic results from dikes of the Junggar block, northwestern China. *Earth Planet. Sci. Lett.*, **94**, 123–30.

Li, Y., & Zhang, Z. (1986). Preliminary paleomagnetic results of the late Paleozoic from the Qiliangshan terrane. *J. Changchun Geol. Inst.*, **46** (4).

Li, Y., Zhang, Z., McWilliams, M. O., Sharps, R., Zhai, Y., Li, Y. A., Li, Q. & Cox, A. (1988a). Mesozoic paleomagnetic results of the Tarim craton: Tertiary relative motion between China and Siberia? *Geophys. Res. Lett.*, **15**, 217–20.

Li, Z. X., Schmidt, P. W. & Embleton, B. J. J. (1988). Paleomagnetism of the Hervey Group, central New South Wales, and its tectonic implications. *Tectonics,* 7, 351–67.

Liang, Q., Ding, S., Yu, R. & Niu, Z. (1986). A case study of the Eocene paleomagnetic pole sites and characteristics of magnetic strata in eastern Yunnan, China. *Geol. Review,* **32**, 144–9. (In Chinese, with English abstract.)

Lin, J. L. (1984). *The Apparent Polar Wander Paths for the North and South China Blocks.* Ph. D. Thesis, University of California at Santa Barbara.

Lin, J. L. & Fuller, M. D. (1986). Mesozoic and Cenozoic evolution of the North and South China Blocks. *Nature*, **320**, 87.

Lin, J. L., Fuller, M. & Zhang, W. (1985). Preliminary Phanerozoic polar wander paths for the North and South China blocks. *Nature*, **313**, 444–9.

Lin, J. L. & Watts, D. R. (1988). Palaeomagnetic results from the Tibetan Plateau. *Phil. Trans. R. Soc. London*, **A327**, 239–62.

Linares, E. & Valencio, D. A. (1975). Paleomagnetic and K-Ar ages of some trachybasaltic dikes from Rio de los Molinos, Province of Cordoba, Republica Argentina. *J. Geophys. Res.*, **80**, 3315–21.

Linssen, J. (1991). Properties of Pliocene sedimentary geomagnetic reversal records from the Mediterranean. *Geol. Ultraiectina*, **80**, 231 pp.

Liss, M. J., van der Pluijm, B. A. & Van der Voo, R. (1991). Paleogeography of the Middle Ordovician Tetagouche Group, northern New Brunswick: paleomagnetic evidence (abstract). *Eos, Trans. Am. Geophys. Union*, **72**, 106.

Liu Chun, Chen Guo-Liang, Ye Su-juan, Zhu Xiang-Yuan, Lin-Jin-lu & Li Su-ling (1975). The paleomagnetic study on some Cainozoic basalt groups in the vicinity of Nanjing and the preliminary determination of their geological age. *Kexue Tongbao*, **20**, 222–4.

Liu, C. & Liang, Q. (1984). Magnetostratigraphic studies on the boundaries between Yulongshi (Late Silurian) and Cuifengshan (Early Devonian) formations in Qujing, Yunnan. *Kexue Tongbao*, **29**, 232–4. (English Translation.)

Livermore, R. A. & Smith, A. G. (1985). Some boundary conditions for the evolution of the Mediterranean. In *Geological Evolution of the Mediterranean Basin*, ed. D. J. Stanley & F. G. Wezel, pp. 84–98. Berlin: Springer-Verlag.

Livermore, R. A., Vine, F. J. & Smith, A. G. (1984). Plate motions and the geomagnetic field, II. Jurassic to Tertiary. *Geophys. J. R. Astr. Soc.*, **79**, 939–61.

Lock, J. & McElhinny, M. W. (1991). The global paleomagnetic database. *Surv. Geophys.*, **12**, 317–491.

Longshaw, S. K. & Griffiths, D. H. (1983). A paleomagnetic study of Jurassic rocks from the Antarctic Peninsula and its implications. *J. Geol. Soc., London*, **140**, 945–54.

Lottes, A. L. & Rowley, D. B. (1990). Early and Late Permian reconstructions of Pangaea. In *Palaeozoic, Palaeogeography and Biogeography*., ed. W. S. McKerrow & C. R. Scotese. *Geol. Soc. London Memoir*, **12**, 383–95.

Loucks, V. & Elmore, R. D. (1986). Absolute dating of dedolomitization and the origin of a magnetization in the Cambrian Morgan Creek Limestone, central Texas. *Geol. Soc. Am. Bull.*, **97**, 486–96.

Løvlie, R. (1979). Mesozoic paleomagnetism in Vestfjella, Dronning Maud Land, East Antarctica. *Geophys. J. R. Astr. Soc.*, **59**, 529–37.

Løvlie, R. (1981). Palaeomagnetism of coast-parallel alkaline dykes from western Norway: ages of magmatism and evidence for crustal uplift and collapse. *Geophys. J. R. Astr. Soc.*, **66**, 417–26.

Løvlie, R., Gidskehaug, A. & Storetvedt, K. M. (1972). On the magnetization history of the northern Irish Basalts. *Geophys. J. R. Astr. Soc.*, **27**, 487–98.

Løvlie, R. & Kvingedal, M. (1975). A palaeomagnetic discordance between a lava sequence and an associated interbasaltic horizon from the Faeroe Islands. *Geophys. J. R. Astr. Soc.*, **40**, 45–54.

Løvlie, R. & Opdyke, N. D. (1974). Rock magnetism and paleomagnetism of some intrusions from Virginia. *J. Geophys. Res.*, **79**, 343–9.

Lowrie, W. (1986). Paleomagnetism and the Adriatic promontory: A reappraisal. *Tectonics*, **5**, 797–807.

Lowrie, W. & Alvarez, W. (1975). Paleomagnetic evidence for rotation of the Italian Peninsula. *J. Geophys. Res.*, **80**, 1579–92.

Lowrie, W. & Alvarez, W. (1977). Late Cretaceous geomagnetic polarity sequence: detailed rock and paleomagnetic studies of the Scaglia Rossa limestone at Gubbio, Italy. *Geophys. J. R. Astr. Soc.*, **51**, 561–81.

Lowrie, W., Alvarez, W., Napoleone, G., Perch-Nielsen, K., Premoli-Silva, I. & Tourmarkine, M. (1982). Paleogene magnetic stratigraphy in Umbrian pelagic rocks: the Contessa sections, Gubbio. *Geol. Soc. Am. Bull.*, **93**, 414–32.

Lowrie, W., Alvarez, W., Premoli-Silva, I. & Monechi, S. (1980). Lower Cretaceous magnetic stratigraphy in Umbrian pelagic carbonate rocks. *Geophys. J. R. Astr. Soc.*, **60**, 263–81.

Lowrie, W. & Heller, F. (1982). Magnetic properties of limestones. *Rev. Geophys. Space Phys.*, **20**, 171–92.

Lowrie, W. & Hirt, A. M. (1986). Paleomagnetism in arcuate mountain belts. In *The origin of Arcs*, ed. F. C. Wezel, pp. 141–58. Amsterdam: Elsevier.

Lowrie, W. & Ogg, J. G. (1985/86). A magnetic polarity time scale for the early Cretaceous and Late Jurassic. *Earth Planet. Sci. Lett.*, **76**, 341–9.

Lu, G., Marshak, S. & Kent, D. V. (1990) Characteristics of magnetic carriers responsible for Late Paleozoic remagnetization in carbonate strata of the mid-continent, USA *Earth Planet. Sci. Lett.*, **99**, 351–61.

Luck, G. R. (1972). Paleomagnetic results from Paleozoic sediments of northern Australia. *Geophys. J. R. Astr. Soc.*, **28**, 475–87.

Luck, G. R. (1973). Palaeomagnetic results from Palaeozoic rocks of southeast Australia. *Geophys. J. R. Astr. Soc.*, **32**, 35–52.

Luo Huilin, et al. (1984). *Sinian-Cambrian Boundary Stratotype Section at Meishucun, Jinning, Yunnan, China.* Yunnan, China: People's Publishing House.

Luyendijk, B. P., Laws, B. R., Day, R. & Collinson, T. B. (1982). Paleomagnetism of the Samail Ophiolite, Oman, I. The sheeted dike complex at Ibra. *J. Geophys. Res.*, **87**, 10883–902.

Lynnes, C. S. & Van der Voo, R. (1984). Paleomagnetism of the Cambro-Ordovician McClure Mountain alkalic complex, Colorado. *Earth Planet. Sci. Lett.*, **71**, 163–72.

MacDonald, W. D. (1980). Net rotation, apparent tectonic rotation, and the structural correction in paleomagnetic studies. *J. Geophys. Res.*, **85**, 3659–69.

MacDonald, W. D. & Opdyke, N. D. (1974). Triassic paleomagnetism of northern South America. *Am. Assoc. Petrol. Geol. Bull.*, **58**, 208–15.

Mäkel, G. H., Rondeel, H. E. & Vandenberg, J. (1984). Triassic paleomagnetic data from the Subbetic and the Malaguide complex of the Betic Cordilleras (Southeast Spain). *Tectonophysics*, **101**, 131–41.

Malod, J. A. (1989). Ibérides et plaque ibérique. *Bull. Soc. Géol. France*, Series 8, 5, 927–34.

Malod, J. A. & Mauffret, A. (1990). Iberian plate motions during the Mesozoic. *Tectonophysics*, **184**, 261–78.

Manten, A. A. (1971). *Silurian Reefs of Gotland.* Amsterdam: Elsevier, 539 pp.

Manwaring, E. A. (1963). The palaeomagnetism of some igneous rocks of the Sydney Basin, NSW. *J. & Proc. Roy. Soc. NSW*, **96**, 141–51.

Manwaring, E. A. (1971). Palaeomagnetism of some Recent basalts from New Guinea. *Bur. Miner. Resour. Austr., Rec.*, 1971/45.

Manwaring, E. A. (1974). A palaeomagnetic reconnaissance of Papua New Guinea. *Bur. Miner. Resour. Austr., Rec.*, 1974/92.

Manzoni, M. (1970). Paleomagnetic data of Middle and Upper Triassic age from the Dolomites (eastern Alps, Italy). *Tectonophysics*, **10**, 411–24.

Manzoni, M. (1974). A review of paleomagnetic data from Italy and their interpretations. *Giorn. Geol.* **39**, 513–50.

Manzoni, M. (1975). Paleomagnetic evidence for rotation in northern Calabria. *Geophys. Res. Lett.*, **2**, 427–9.

Manzoni, M. (1979). Paleomagnetic evidence for non-Apenninic origin of the Sila Nappes (Calabria). *Tectonophysics*, **60**, 169–88.

Manzoni, M. & Nanni, T. (1977). Paleomagnetism of Ordovician lamprophyres from Taylor Valley, Victoria Land, Antarctica. *Pure Appl. Geoph.*, **115**, 961–77.

Manzoni, M., Venturini, C. & Vigliotti, L. (1989). Paleomagnetism of Upper Carboniferous Limestone from the Carnic Alps. *Tectonophysics*, **165**, 73–80.

Manzoni, M. & Vigliotti, L. (1983). Further paleomagnetic data from northern Calabria: their bearing on directions of emplacement of the Calabrian nappes. *Boll. Geofis. Teor. Appl.*, **25**, 27–43.

Maranate, S. & Vella, P. (1986). Paleomagnetism of the Khorat Group, Mesozoic, northern Thailand. *J. Southeast Asian Earth Sci.*, **1**, 23–31.

Martin, D. L. (1975). A paleomagnetic polarity transition in the Devonian Columbus limestone of Ohio: a possible stratigraphic tool. *Tectonophysics*, **28**, 125–34.

Martin, D. L., Nairn, A. E. M., Noltimier, H. C., Petty, M. H. & Schmitt, T. J. (1978). Paleozoic and Mesozoic paleomagnetic results from Morocco. *Tectonophysics*, **44**, 91–114.

Marton, E. (1984). Tectonic implications of paleomagnetic results for the Carpatho-Balkan and adjacent areas. In *The Geological Evolution of the Eastern Mediterranean*, ed. J. E. Dixon & A. H. F. Robertson. *Geol. Society, London, Spec. Publ.*, **17**, pp. 645–54.

Marton, E. & Marton P. (1976). A palaeomagnetic study of the Nigerian volcanic provinces. *Pageoph*, **114**, 61–9.

Marton, E. & Mauritsch, H. J. (1990). Structural applications and discussion of a paleomagnetic post-Paleozoic data base for the Central Mediterranean. *Phys. Earth Planet. Int.*, **62**, 46–59.

Marton, E., Milicevic, V. & Veljovic, D. (1990a). Paleomagnetism of the Kvarner islands, Yugoslavia. *Phys. Earth Planet. Int.*, **62**, 70–81.

Marton, E., Papanikolaou & E. Lekkas (1990b). Palaeomagnetic results from the Pindos, Paxos and Ionian zones of Greece. *Phys. Earth Planet. Int.*, **62**, 60–9.

Marton, E. & Veljovich, D. (1983). Paleomagnetism of the Istria Peninsula, Yugoslavia. *Tectonophysics*, **91**, 73–87.

Mattauer, M. & Seguret, M. (1971). Les relations entre la chaîne des Pyrénées et le Golfe de Gascogne. In *Histoire Structurale du Golfe de Gascogne*, ed. J. Debyser, X. Le Pichon & L. Montadert, pp. IV.4.1–24. Paris: Editions Technip.

Matte, P. (1991). Accretionary history and crustal evolution of the Variscan belt in western Europe. *Tectonophysics*, **196**, 309–37.

Mauritsch, H. J. & Frisch, W. (1978). Paleomagnetic data from the central part of the Northern Calcareous Alps, Austria. *J. Geophys.*, **44**, 623–37.

Mauritsch, H. J. & Rother, K. (1983). Paleomagnetic investigations in the Thüringer Forest (GDR). *Tectonophysics*, **99**, 63–72.

May, S. R. & Butler, R. F. (1986). North American Jurassic apparent polar wander: implications for plate motion, paleogeography and Cordilleran tectonics. *J. Geophys. Res.*, **91**, 11519–44.

May, S. R., Butler, R. F., Shafiqullah, M. & Damon, P. E. (1986). Paleomagnetism of Jurassic volcanic rocks in the Patagonia mountains, southeastern Arizona: implications for the North American 170 Ma reference pole. *J. Geophys. Res.*, **91**, 11545–55.

Mayhew, M. A. (1986). Approximate paleomagnetic poles for some of the New England Seamounts. *Earth Planet. Sci. Lett.*, **79**, 185–94.

Mazaud, A., Galbrun, B., Azema, J., Enay, R., Fourcade, E. & Rasplus, L. (1986). Données magnétostratigraphiques sur le Jurassique supérieur et le Berriasien du NE des Cordillères bétiques. *C. R. Acad. Sci. Paris*, **302**, Série II, 1165–70.

McCabe, C. & Channell, J. E. T. (1989). Paleomagnetic results from volcanic rocks of the Shelve inlier, Wales: evidence for a wide Late Ordovician Iapetus Ocean in Britain. *Earth Planet. Sci. Lett.*, **96**, 458–68.

McCabe, C., Channell, J. E. T. & Woodcock, N. H. (1991). Paleomagnetism of Middle Ordovician rocks from the Builth inlier, Wales (abstract). *Eos, Trans. Am. Geophys. Union*, **72**, 105.

McCabe, C. & Elmore, R. D. (1989). The occurrence and origin of late Paleozoic remagnetization in the sedimentary rocks of North America. *Rev. Geophys.*, **27**, 471–94.

McCabe, C., Van der Voo, R. & Ballard, M. M. (1984). Late Paleozoic remagnetization of the Trenton limestone. *Geophys. Res. Lett.*, **11**, 979–82.

McCabe, C., Van der Voo, R. & Wilkinson, B. H. (1982). Paleomagnetic and rock magnetic results from the Twin Creek Formation (Middle Jurassic), Wyoming. *Earth Planet. Sci. Lett.*, **60**, 140–6.

McCabe, C., Van der Voo, R., Wilkinson, B. H. & Devaney, K. (1985). A Middle–Late Silurian paleomagnetic pole from limestone reefs of the Wabash Formation (Indiana, USA). *J. Geophys. Res.*, **90**, 2959–65.

McCabe, R., Almasco, J. & Diegor, W. (1982b). Geologic and paleomagnetic evidence for a possible Miocene collision in western Panay, central Philippines. *Geology*, **10**, 325–9.

McCabe, R., Almasco, J., Shibuya, H., Zanoria, W., Pagauitan, H. & Torii, M. (1982a). Evidence against late Neogene rotations of the island of Luzon. *Paleomagnetic Research in Southeast and East Asia*, ed. J. C. Briden *et al.*, pp. 117–29, Committee for the Coordination of Offshore Prospecting (CCOP), U. N. Development Project Office.

McCabe, R., Celaya, M., Cole, J., Han, H. C. & Ohnstad, T. (1988). Extension tectonics:

the Neogene opening of the north-south trending basins of central Thailand. *J. Geophys. Res.*, **93**, 11899–910.

McCabe, R. & Cole, J. (1986). Speculations on the late Mesozoic and Cenozoic evolution of the southeast Asian margin. In *Transactions of the Fourth Circum-Pacific Energy and Mineral Resources Conference*, ed. M. K. Horn, Ch. 30, pp. 375–94.

McCabe, R., Kikawa, E., Cole, J. T., Malicse, A. J., Baldauf, P. E., Yumul, J. & Almasco, J. (1987). Paleomagnetic results from Luzon and the central Philippines. *J. Geophys. Res.*, **92**, 555–80.

McClelland-Brown, E. A. (1981). Paleomagnetic estimates of temperatures reached in contact metamorphism. *Geology*, **9**, 112–16.

McElhinny, M. W. (1964). Statistical significance of the fold test in palaeomagnetism. *Geophys. J. R. Astr. Soc.*, **8**, 338–40.

McElhinny, M. W. (1970). The paleomagnetism of the Cambrian purple sandstone from the Salt Range, West Pakistan, *Earth Planet. Sci. Lett.*, **8**, 149–56.

McElhinny, M. W. (1973). *Paleomagnetism and plate tectonics.* Cambridge: Cambridge University Press.

McElhinny, M. W., Briden, J. C., Jones, D. L. & Brock, A. (1968). Geological and geophysical implications of paleomagnetic results from Africa. *Rev. Geophys.*, **6**, 201–38.

McElhinny, M. W. & Brock, A. (1975). A new paleomagnetic result from East Africa and estimates of the Mesozoic paleoradius. *Earth Planet. Sci. Lett.*, **27**, 321–8.

McElhinny, M. W. & Cowley, J. A. (1980). Palaeomagnetic directions and pole positions, XVI. *Geophys. J. R. Astr. Soc.*, **61**, 549–71.

McElhinny, M. W., Cowley, J. A. & Edwards, D. J. (1978). Palaeomagnetism of some rocks from Peninsular India and Kashmir. *Tectonophysics*, **50**, 41–54.

McElhinny, M. W., Embleton, B. J. J., Daly, L. & Pozzi, J. P. (1976). Paleomagnetic evidence for the location of Madagascar in Gondwanaland. *Geology*, **4**, 455–7.

McElhinny, M. W., Embleton, B. J., Ma, X. & Zhang, Z. (1981). Fragmentation of Asia in the Permian. *Nature*, **293**, 212–16.

McElhinny, M. W., Embleton, B. J. J. & Wellman, P. (1974a). A synthesis of Australian Cenozoic palaeomagnetic data. *Geophys. J. R. Astr. Soc.*, **36**, 141–51.

McElhinny, M. W., Haile, N. S. & Crawford, A. R. (1974b). Palaeomagnetic evidence shows Malay Peninsula was not part of Gondwanaland. *Nature*, **252**, 641–5.

McElhinny, M. W. & Jones, D. L. (1965). Palaeomagnetic measurements on some Karroo dolerites from Rhodesia. *Nature*, **206**, 921–2.

McElhinny, M. W. & Lock, J. (1990). IAGA global palaeomagnetic database. *Geophys. J. Intern.*, **101**, 763–6.

McElhinny, M. W. & Luck, G. R. (1970). The palaeomagnetism of the Antrim Plateau Volcanics of Northern Australia. *Geophys. J. R. Astr. Soc.*, **20**, 191–205.

McElhinny, M. W. & Opdyke, N. D. (1968). Paleomagnetism of some Carboniferous glacial varves from central Africa. *J. Geophys. Res.*, 73, 689–96.

McFadden, P. L. (1990). A new fold test for palaeomagnetic studies. *Geophys. J. Int.*, **103**, 163–9.

McFadden, P. L. & Jones, D. L. (1981). The fold test in palaeomagnetism. *Geophys. J. R. Astr. Soc.*, **67**, 53–8.

McFadden, P. L., Ma, X., McElhinny, M. W. & Zhang, Z. (1988). Permotriassic magnetostratigraphy in China: northern Tarim. *Earth Planet. Sci. Lett.*, **87**, 152–60.

McFadden, P. L. & McElhinny, M. W. (1990). Classification of the reversal test in palaeomagnetism. *Geophys. J. Int.*, **103**, 725–9.

McIntosh, W. C., Hargraves, R. B. & West, C. L. (1985). Paleomagnetism and oxide mineralogy of Upper Triassic to Lower Jurassic red beds and basalts in the Newark Basin. *Geol. Soc. Am. Bull.*, **96**, 463–80.

McKenzie, D. (1972). Active tectonics of the Mediterranean region. *Geophys. J. R. Astr. Soc.*, **30**, 109–85.

McKenzie, D. (1978). Active tectonics of the Alpine–Himalayan belt: the Aegean Sea and surrounding regions. *Geophys. J. R. Astr. Soc.*, **55**, 217–54.

McKerrow, W. S., St Lambert, R. & Chamberlain, V. W. (1980). The Ordovician, Silurian and Devonian time scales. *Earth Planet. Sci. Lett.*, **51**, 1–8.

McKerrow, W. S. & Ziegler, A. M. (1972). Paleozoic oceans. *Nature*, **240**, 92–4.

McMahon, B. E. & Strangway, D. W. (1968). Stratigraphic implications of paleomagnetic data from Upper Paleozoic–Lower Triassic red beds of Colorado. *Geol. Soc. Am. Bull.*, **79**, 417–28.

McQueen, D. M., Scharnberger, C. K., Scharon, L. & Halpern, M. (1972). Cambro-Ordovician paleomagnetic pole position and rubidium-strontium total rock isochron for charnockitic rocks from Mirnyy Station, Antarctica. *Earth Planet. Sci. Lett.*, **16**, 433–8.

McWhinnie, S. T., van der Pluijm, B. A. & Van der Voo, R. (1990). Remagnetizations and thrusting in the Idaho-Wyoming Overthrust Belt. *J. Geophys. Res.*, **95**, 4551–9.

McWilliams, M. O. (1981). Paleomagnetism and Precambrian tectonic evolution of Gondwana. In *Precambrian Plate Tectonics*, ed. A. Kröner, pp. 647–687. Amsterdam: Elsevier.

McWilliams, M. O. & Kröner, A. (1981). Paleomagnetism and tectonic evolution of the Pan-African Damara Belt, southern Africa. *J. Geophys. Res.*, **86**, 5147–62.

Mendia, J. E. (1978). Palaeomagnetism of alkaline lava flows from El Salto-Almafuerte, Cordoba Province, Argentina. *Geophys. J. R. Astr. Soc.*, **54**, 539–46.

Menning, M. (1989). A synopsis of numerical time scales 1917–1986. *Episodes*, **12**, 3–5.

Merabet, N. & Daly, L. (1986). Détermination d'un pôle paléomagnétique et mise en évidence d'aimantations à polarité normale sur les formations du Permien supérieur du Massif des Maures (France). *Earth Planet. Sci. Lett.*, **80**, 156–66.

Merabet, N. & Guillaume, A. (1988). Palaeomagnetism of the Permian rocks of Lodève (Hérault, France). *Tectonophysics*, **145**, 21–9.

Merrill, R. T. & McElhinny, M. W. (1983). *The Earth's Magnetic Field*. London: Academic Press.

Metcalfe, J. (1988). Origin and assembly of south-east Asian continental terranes. *Geol. Soc. London Spec. Publ.*, **37**, 101–18.

Miller, J. D. & Kent, D. V. (1986a). Paleomagnetism of the Upper Devonian Catskill Formation from the southern limb of the Pennsylvania salient: possible evidence of oroclinal rotation. *Geophys. Res. Lett.*, **13**, 1173–6.

Miller, J. D. & Kent, D. V. (1986b). Synfolding and prefolding magnetizations in the Upper Devonian Catskill Formation of eastern Pennsylvania: implications for the tectonic history of Acadia. *J. Geophys. Res.*, **91**, 12791–803.

Miller, J. D. & Kent, D. V. (1988). Paleomagnetism of the Siluro-Devonian Andreas red beds: evidence for an Early Devonian supercontinent? *Geology*, **116**, 195–8.

Miller, J. D. & Kent, D. V. (1989a). Paleomagnetism of the Upper Ordovician Juniata Formation of the central Appalachians revisited again. *J. Geophys. Res.*, **94**, 1843–9.

Miller, J. D. & Kent, D. V. (1989b). Paleomagnetism of selected Devonian age plutons from Maine, Vermont and New York. *Northeastern Geol.*, **11**, 66–76.

Miller, J. D. & Opdyke, N. D. (1985). Magnetostratigraphy of the Red Sandstone Creek section – Vail, Colorado. *Geophys. Res. Lett.*, **12**, 133–6.

Mital, G. S., Verma, R. K. & Pullaiah, G. (1970). Paleomagnetic study of Satyavedu sandstones of Cretaceous age from Andrah Pradesh, India. *Pure Appl. Geophys.*, **81**, 177–91.

Mitchell, A. H. G. (1981). Phanerozoic plate boundaries in mainland SE Asia, the Himalayas and Tibet. *J. Geol. Soc. London*, **138**, 109–22.

Mizutani, S. (1987). Mesozoic terranes in the Japanese Islands and neighbouring East Asia. *Geodynamics Series*, **19**, 263–73.

Molina-Garza, R. S., Geissman, J. W. & Van der Voo, R. (1989). Paleomagnetism of the Dewey Lake Formation (Late Permian), northwest Texas: end of the Kiaman Superchron in North America. *J. Geophys. Res.*, **94**, 17881–8.

Molina-Garza, R. S., Geissman, J. W., Van der Voo, R., Lucas, S. G. & Hayden, S. N. (1991). Paleomagnetism of the Moenkopi and Chinle formations in central New Mexico: implications for the North American apparent polar wander path and Triassic magnetostratigraphy. *J. Geophys. Res.*, **96**, 14239–62.

Molnar, P. & Tapponnier, P. (1978). Active tectonics of Tibet. *J. Geophys. Res.*, **83**, 5361–75.

Montigny, R., Edel, J. B. & Thuizat, R. (1981). Oligo-Miocene rotation of Sardinia: K-Ar ages and paleomagnetic data of Tertiary volcanics. *Earth Planet. Sci. Lett.*, **54**, 261–71.

Moon, T. & Merrill, R. (1985). Nucleation theory and domain states in multidomain magnetic material. *Phys. Earth Planet. Int.*, **37**, 214–22.

Moores, E. M. (1991). Southwest US – East Antarctic (SWEAT) connection: a hypothesis. *Geology*, **19**, 425–8.

Morel, P. & Irving, E. (1978). Tentative paleocontinental maps for the Early Phanerozoic and Proterozoic. *J. Geol.*, **86**, 535–61.

Morel, P. & Irving, E. (1981). Paleomagnetism and the evolution of Pangea. *J. Geophys. Res.*, **86**, 1858–72.

Morel, P., Irving, E., Daly, L. & Moussine-Pouchkine, A. (1981). Paleomagnetic results from Permian rocks of the northern Saharan craton and motions of the Moroccan Meseta and Pangea. *Earth Planet. Sci. Lett.*, **55**, 65–74.

Morgan, W. J. (1968). Rises, trenches, great faults and crustal blocks. *J. Geophys. Res.*, **73**, 1959–82.

Morgan, W. J. (1972). Plate motions and deep mantle convection. *Geol. Soc. America Memoir*, **132**, 7–22.

Morgan, W. J. (1983). Hotspot tracks and the early rifting of the Atlantic. *Tectonophysics*, **94**, 123–39.

Morley, L. W. (1991). The role of magnetic surveying in developing the concept of plate tectonics. *Tectonophysics*, **187**, 23–5.

Morris, A., Creer, K. M. & Robertson, A. H. F. (1990). Paleomagnetic evidence for clockwise rotations related to dextral shear along the southern Troodos transform fault. *Earth Planet. Sci. Lett.*, **99**, 250–62.

Morris, W. A. (1976). Paleomagnetic results from the lower Paleozoic of Ireland. *Can. J. Earth Sci.*, **13**, 294–304.

Morris, W. A., Briden, J. C., Piper, J. D. A. & Sallomy, J. T. (1973). Paleomagnetic studies in the British Caledonides, 5. Miscellaneous new data. *Geophys. J. R. Astr. Soc.*, **34**, 69–105.

Morrison, J. & Ellwood, B. B. (1986). Paleomagnetism of Silurian-Ordovician sediments from the Valley and Ridge province, northwest Georgia. *Geophys. Res. Lett.*, **13**, 189–92.

Mulder, F. G. (1971). Paleomagnetic research in some parts of central and southern Sweden. *Sveriges Geol. Undersökn. Ser. C. 653, Avhandl. Uppsätser*, 64.

Mulder, F. G. (1972). Paleomagnetic results from Irish Carboniferous and Triassic rocks. *Scient. Proc. Roy. Dublin Soc.*, **A4**, 343–9.

Mumme, W. G. (1963). Thermal and alternating field demagnetization experiments on Cainozoic basalts of Victoria, Australia. *Geophys. J. R. Astr. Soc.*, 7, 314–27.

Murphy, J. B. & Nance, R. D. (1991). Supercontinent model for the contrasting character of late Proterozoic orogenic belts. *Geology*, **19**, 469–72.

Murthy, G. S. (1983a). Paleomagnetism of the Deadman's Bay diabase dikes from northeastern central Newfoundland. *Can. J. Earth Sci.*, **20**, 195–205.

Murthy, G. S. (1983b). Paleomagnetism of diabase dikes from the Bonavista Bay area of northeastern central Newfoundland. *Can. J. Earth Sci.*, **20**, 206–16.

Murthy, G. S. (1985). Paleomagnetism of certain constituents of the Bay St George subbasin, western Newfoundland. *Phys. Earth Planet. Int.*, **39**, 89–107.

Murthy, G. S. & Rao, K. V. (1976). Paleomagnetism of Steel Mountain and Indian Head anorthosites from western Newfoundland. *Can. J. Earth Sci.*, **13**, 75–83.

Mussett, A. E., Dagley, P. & Eckford, M. (1976). The British Tertiary igneous province: Palaeomagnetism and age of dykes, Lundy Island, Bristol Channel. *Geophys. J. R. Astr. Soc.*, **46**, 595–603.

Nagata, T. (1961). *Rock Magnetism*. 2nd edn. Tokyo: Maruzan.

Nairn, A. E. M. & Negendank, J. (1973). Paleomagnetic investigations of the Tertiary and Quaternary igneous rocks, VII. The Tertiary rocks of southwest Germany. *Geol. Rundschau*, **62**, 126–37.

Nairn, A. E. M., Schmitt, T. J. & Smithwick, M. E. (1981). A palaeomagnetic study of the Upper Mesozoic succession in northern Tunisia. *Geophys. J. R. Astr. Soc.*, **65**, 1–18.

Nairn, A. E. M. & Westphal, M. (1967). A second virtual pole from Corsica, the Ota Gabbrodiorite. *Palaeogeogr., Palaeoclimatol., Palaeoecol.*, **3**, 277–86.

Nairn, A. E. M. & Westphal, M. (1968). Possible implications of the paleomagnetic study of late Paleozoic rocks of northwestern Corsica. *Palaeogeogr., Palaeoclimatol., Palaeoecol.*, **5**, 179–204.

Napoleone, G., Premoli-Silva, I., Heller, F., Cheli, P., Çorezzi, S. & Fischer, A. G. (1982). Eocene magnetic stratigraphy at Gubbio, Italy, and its implications for Paleogene geochronology. *Geol. Soc. Am. Bull.*, **94**, 181–91.

Nestor, H. (1990). Biogeography of Silurian stromatoporoids. In *Palaeozoic Palaeogeography* *and Biogeography*, ed. W. S. McKerrow & C. R. Scotese. *Geol. Soc. London Memoir*, **12**, 215–21.

Neuman, R. B. (1984). Geology and paleobiology of islands in the Ordovician Iapetus Ocean: review and implications. *Geol. Soc. Am. Bull.*, **95**, 1188–201.

Nick, K. E. & Elmore, R. D. (1990). Paleomagnetism of the Cambrian Royer Dolomite and Pennsylvanian Collings Ranch Conglomerate, southern Oklahoma: an early Paleozoic magnetization and nonpervasive remagnetization by weathering. *Geol. Soc. Am. Bull.*, **102**, 1517–25.

Nie, S. (1991). Paleoclimatic and paleomagnetic constraints on the Paleozoic reconstructions of South China, North China and Tarim. *Tectonophysics*, **196**, 279–308.

Nie, S., Rowley, D. B. & Ziegler, A. M. (1990). Constraints on the location of the Asian microcontinents in Palaeo-Tethys during the late Palaeozoic. In *Palaeozoic Palaeogeography and Biogeography*, ed. W. S. McKerrow & C. R. Scotese. *Geol. Soc. London Memoir*, **12**, 397–409.

Nie, S., Van der Voo, R., Li, M. & Rowley, D. B. (1992). Paleomagnetism of late Paleozoic rocks of Tianshan, northwestern China. Tectonics. (In press.)

Nijenhuis, G. H. W. (1961). A palaeomagnetic study of the Permian volcanics in the Nahe region, southwestern Germany. *Geol. Mijnbouw*, **40**, 26–38.

Nishimura, S. & Suparka, S. (1990). Tectonics of East Indonesia. *Tectonophysics*, **181**, 257–66.

Noel, J. R., Spariosu, D. J. & Dallmeyer, R. D. (1988). Paleomagnetism and 40Ar/39Ar ages from the Carolina Slate Belt, Albemarle, North Carolina: implications for terrane amalgamation with North America. *Geology*, **16**, 64–8.

Nozharov, P. B. & Velevich, D. (1974). Paleomagnetism of some Upper Cretaceous vulcanites in the Timok eruptive region and Srednogoriye. *C. R. Acad. Bulg. Sci.*, **27**, 199–202.

Nur, A. & Helsley, C. E. (1971). Paleomagnetism of Tertiary and Recent lavas of Israel. *Earth Planet. Sci. Lett.*, **10**, 375–9.

Odin, G. S., Gale, N. H., Auvray, B., Bielski, M., Doré, F., Lancelot, J.-R. & Pasteels, P. (1983). Numerical dating of the Precambrian-Cambrian boundary. *Nature*, **301**, 21–3.

Ogg, J. G., Steiner, M. B., Company, M. & Tavera, J. M. (1988). Magnetostratigraphy across the Berriasian-Valanginian stage boundary (Early Cretaceous) at Cehegin (Murcia Province, southern Spain). *Earth Planet. Sci. Lett.*, **87**, 205–15.

Ogg, J. G., Steiner, M. B., Oloriz, F. & Tavera, J. M. (1984). Jurassic magnetostratigraphy, 1. Kimmeridgian-Tithonian of Sierra Gorda and Carcabuey, southern Spain. *Earth Planet. Sci. Lett.*, **71**, 147–62.

Ogg, J. G., Steiner, M. B., Wieczorek, J. & Hoffmann, M. (1991). Jurassic magnetostratigraphy: 4. Early Callovian through Middle Oxfordian of the Krakow Uplands (Poland). *Earth Planet. Sci. Lett.*, **104**, 488–504.

Olivet, J. L., Bonnin, J., Beuzart, P. & Auzende, J. M. (1984). Cinématique de l'Atlantique nord et central. *Rapports Scientifiques*. Paris: CNEXO.

Opdyke, N. D. (1961). The paleomagnetism of the New Jersey Triassic: a field study of the inclination error in red sediments. *J. Geophys. Res.*, **66**, 1941–9.

Opdyke, N. D. (1964). The paleomagnetism of the Permian red beds of southwest Tanganyika. *J. Geophys. Res.*, **69**, 2477–87.

Opdyke, N. D. & Henry, K. W. (1969). A test of the dipole hypothesis. *Earth Planet. Sci. Lett.*, **6**, 139–51.

Opdyke, N. D., Huang, K., Xu, G., Zhang, W. Y. & Kent, D. V. (1986). Paleomagnetic results from the Triassic of the Yangtze Platform. *J. Geophys. Res.*, **91**, 9553–68.

Opdyke, N. D., Huang, K., Xu, G., Zhang, W. Y. & Kent, D. V. (1987a). Paleomagnetic results from the Silurian of the Yangtze Paraplatform. *Tectonophysics*, **139**, 123–32.

Opdyke, N. D., Jones, D. S., MacFadden, B. J., Smith, D. L., Mueller, P. A. & Shuster, R. D. (1987b). Florida as an exotic terrane: paleomagnetic and geochronologic investigations of lower Paleozoic rocks from the subsurface of Florida. *Geology*, **15**, 900–3.

Opdyke, N. D. & MacDonald, W. D. (1973). Paleomagnetism of Late Cretaceous Poços de Caldas Alkaline Complex, southern Brazil. *Earth Planet. Sci. Lett.*, **18**, 37–44.

Opdyke, N. D. & Runcorn, S. K. (1960). Wind direction in the western United States in the late Paleozoic. *Geol. Soc. Am. Bull.*, **71**, 959–72.

Opdyke, N. D. & Wensink, H. (1966). Paleomagnetism of rocks from the White Mountain plutonic-volcanic series in New Hampshire and Vermont. *J. Geophys. Res.*, **71**, 3045–51.

Orbay, N. & Bayburdi, A. (1979). Palaeomagnetism of dykes and tuffs from the Mesudiye region and rotation of Turkey. *Geophys. J. R. Astr. Soc.*, **59**, 437–44.

O'Reilly, W. (1984). *Rock and Mineral Magnetism*. Glasgow: Blackie.

Osete, M. L., Freeman, R. & Vegas, R. (1988). Preliminary paleomagnetic results from the Subbetic zone (Betic Cordillera, southern Spain): kinematic and structural implications. *Phys. Earth Planet. Int.*, **52**, 283–300.

Ostrander, J. H. (1971). Paleomagnetic investigations of the Queen Alexandra Range, Antarctica. *Antarctic J. US*, **6**, 183–5.

Otofuji, Y., Funahara, S., Matsuo, J., Murata, G., Nishiyama, T., Zhang, X. & Yaskawa, K. (1989). Paleomagnetic study of western Tibet: deformation of a narrow zone along the Indus Zangbo suture between India and Asia. *Earth Planet. Sci. Lett.*, **92**, 307–16.

Otofuji, Y., Ho, K. K., Inokuchi, H., Morinaga, H., Murata, F., Katao, H. & Yaskawa, K. (1986). A paleomagnetic reconnaissance of Permian to Cretaceous sedimentary rocks in southern part of Korean Peninsula. *J. Geomagn. Geoelectr.*, **38**, 387–402.

Otofuji, Y., Inoue, Y., Funahara, S., Murata F. & Zheng, X. (1990). Palaeomagnetic study of eastern Tibet – deformation of the Three Rivers region. *Geophys. J. Int.*, **103**, 85–94.

Otofuji, Y., Oh, J. Y., Hirajima, T., Min, K. D. & Sasajima, S. (1983). Paleomagnetism and age determination of Cretaceous rocks from Geyongsang Basin, Korean Peninsula. *Am. Geophys. Union, Geophys. Monogr.*, **27**, 388–96.

Otofuji, Y., Sasajima, S., Nishimura, S., Dharma, A. & Hehuwat, F. (1981b). Paleomagnetic evidence for clockwise rotation of the northern arm of Sulawesi, Indonesia. *Earth Planet. Sci. Lett.*, **54**, 272–80.

Otofuji, Y., Sasajima, S., Nishimura, S., Yokoyama, T., Hadiwastra, S. & Hehuwat, F. (1981a). Paleomagnetic evidence for the paleoposition of Sumba Islands, Indonesia. *Earth Planet. Sci. Lett.*, **52**, 93–100.

Owen, T. R. (1976). *The Geological Evolution of the British Isles*. Oxford: Pergamon Press.

Pacca, I. G. & Hiodo, F. Y. (1976). Paleomagnetic analysis of Mesozoic Serra Geral basaltic lava flows in southern Brazil. *An. Acad. Bras. Cienc.* (supl.), **48**, 207–14.

Pacca, I. G. & Valencio, D. A. (1972). Preliminary paleomagnetic study of igneous rocks from the Abrolhos Islands. *Nature Phys. Sci.*, **240**, 163–4.

Palmer, A. R. (1983). The decade of North American geology (DNAG) 1983 geologic time scale. *Geology*, **11**, 503–4.

Palmer, H. C., Hayatsu, A. & MacDonald, W. D. (1980). Palaeomagnetic and K-Ar age studies of a 6 km thick Cretaceous section from the Chilean Andes. *Geophys. J. R. Astr. Soc.*, **62**, 133–53.

Palmer, H. C., Merz, B. A. & Hayatsu, A. (1977). The Sudbury dikes of the Grenville Front region: paleomagnetism, petrochemistry, and K-Ar age studies. *Can. J. Earth Sci.*, **14**, 1867–87.

Panuska, B. C. & Stone, D. B. (1985). Latitudinal motion of the Wrangellia and Alexander terranes and the southern Alaska superterrane. In *Tectonostratigraphic Terranes of the Circum-Pacific*, ed. D. G. Howell. *Circum-Pacific Council for Energy and Mineral Resources, Earth Science Series*, vol. 1, pp. 109–20. Tulsa, Oklahoma: Amer. Assoc. Petrol. Geol.

Parès, J. M. (1988). Estúdio paleomagnético de las rocas tardihercinianas de la Cadena Costera Catalana: primeros resultados. *Cuadern. Geol. Iberica*, **12**, 171–9.

Parès, J. M., Banda, E. & Santanach, P. (1988a). Paleomagnetic results from the southeastern margin of the Ebro Basin (NE Spain): evidence for a Tertiary clockwise rotation. *Phys. Earth Planet. Int.*, **52**, 267–82.

Parès, J. M., Freeman, R. & Sabat, F. (1988b). Sintesis de los resultados paleomagnéticos de los bordes de la Cuenca Catalano-Baleár. *Cuadern. Geol. Ibérica*, **12**, 59–74.

Paris, F. & Robardet, M. (1990). Early Paleozoic paleobiogeography of the Variscan region. *Tectonophysics*, **177**, 193–213.

Park, J. K. & Irving, E. (1972). Magnetism of dikes of the Frontenac axis. *Can. J. Earth Sci.*, **9**, 763–5.

Parrish, J. T., Ziegler, A. M. & Scotese, C. R. (1982). Rainfall patterns and the distribution of coals and evaporites in the Mesozoic and Cenozoic. *Palaeogeogr., Palaeoclimatol., Palaeoecol.*, **40**, 67–101.

Pascholati, E. M. & Pacca, I. G. (1976). Estudo paleomagnético de seçoes do Subgrupo Itarare. *November 1976, 29th Congress Brasileiro de Geologia, Belo Horizonte, Brazil.*

Pascholati, E. M., Pacca, I. G. & Vilas, J. F. A. (1976). Paleomagnetism of sedimentary rocks from the Permian Irati Formation, southern Brazil. *Rev. Bras. Geoci.*, **6**, 156–63.

Pascual, J. O. & Parès, J. M. (1990). Paleomagnetismo de las calizas y areniscas del tránsito Eoceno-Oligoceno en el borde SE de la cuenca del Ebro. *Rev. Soc. Geol. Espana*, **3**, 323–33.

Pavlides, S. B., Kondopoulou, D. P., Kilias, A. A. & Westphal, M. (1988). Complex rotational deformations in the Serbo-Macedonian Massif (North Greece): structural and paleomagnetic evidence. *Tectonophysics,* **145**, 329–35.

Payne, M. A., Shulik, S. J., Donahue, J., Rollins, H. B. & Schmidt, V. A. (1981). Paleomagnetic poles for the Carboniferous Brush Creek limestone and Buffalo siltstone of Southwestern Pennsylvania. *Phys. Earth Planet. Int.*, **25**, 113–18.

Pearce, J. A. & Deng, W. (1988). The ophiolites of the Tibet Geotraverse, Lhasa to Golmud (1985) and Lhasa to Kathmandu (1986). *Phil. Trans. R. Soc. London*, **A327**, 215–38.

Pe-Piper, G. & Piper, D. J. W. (1984). Tectonic setting of the Mesozoic Pindos Basin of the Peloponnese, Greece. In *The Geological Evolution of the Eastern Mediterranean*, ed. J. E. Dixon & A. H. F. Robertson, pp. 563–7. Oxford: published by Blackwell for the Geol. Society.

Perigo, R., Van der Voo, R., Auvray, B. & Bonhommet, N. (1983). Paleomagnetism of Late Precambrian-Cambrian volcanics and intrusives from the Armorican Massif, France. *Geophys. J. R. Astr. Soc.*, **75**, 236–60.

Perroud, H. (1982). The change of paleomagnetic vector orientation induced by Eulerian rotations: applications for the relative rotations of Spain and Europe. *Tectonophysics*, **81**, T15–23.

Perroud, H. (1983). Paleomagnetism of Paleozoic rocks from the Cabo de Peñas, Astúrias, Spain. *Geophys. J. R. Astr. Soc.*, **75**, 201–15.

Perroud, H. (1986). Paleomagnetic evidence for tectonic rotations in the Variscan Mountain Belt. *Tectonics*, **5**, 205–14.

Perroud, H., Auvray, B., Bonhommet, N., Macé, J. & Van der Voo, R. (1986). Palaeomagnetism and K-Ar dating of Lower Carboniferous dolerite dikes from northern Brittany. *Geophys. J. R. Astr. Soc.*, **87**, 143–54.

Perroud, H. & Bonhommet, N. (1981). Palaeomagnetism of the Ibero-Armorican arc and the Hercynian orogeny in western Europe. *Nature*, **292**, 445–8.

Perroud, H. & Bonhommet, N. (1984). A Devonian palaeomagnetic pole for Armorica. *Geophys. J. R. Astr. Soc.*, 77, 839–45.

Perroud, H., Bonhommet, N. & Robardet, M. (1982). Comment on 'A paleomagnetic study of Cambrian red beds from Carteret, Normandy, France' by Morris. *Geophys. J. R. Astr. Soc.*, **69**, 573–8.

Perroud, H., Bonhommet, N. & Thébault, J. P. (1987). Palaeomagnetism of the Ordovician Moulin de Châteaupanne formation, Vendée, western France. *Geophys. J. R. Astr. Soc.*, **85**, 573–82.

Perroud, H., Bonhommet, N. & Van der Voo, R. (1983). Paleomagnetism of the Ordovician dolerites of the Crozon peninsula (France). *Geophys. J. R. Astr. Soc.*, **72**, 307–19.

Perroud, H., Calza, F. & Khattach, D. (1991). Paleomagnetism of the Silurian volcanism at Almadén, southern Spain. *J. Geophys. Res.*, **96**, 1949–62.

Perroud, H., Robardet, M. & Bruton, D. L. (1992). Paleomagnetic constraints upon the paleogeographic position of the Baltic Shield in the Ordovician. *Tectonophysics*, **201**, 97–120.

Perroud, H., Robardet, M., Van der Voo, R., Bonhommet, N. & Paris, F. (1985). Revision of the age of magnetization of the Montmartin red beds, Normandy, France. *Geophys. J. R. Astr. Soc.*, **80**, 541–9.

Perroud, H. & Van der Voo, R. (1985). Paleomagnetism of the Late Ordovician Thouars Massif, Vendée Province, France. *J. Geophys. Res.*, **90**, 4611–25.

Perroud, H., Van der Voo, R. & Bonhommet, N. (1984). Paleozoic evolution of the Armorica plate on the basis of paleomagnetic data. *Geology*, **12**, 579–82.

Pesonen, L. J., Torsvik, T. H., Elming, S.-A. & Bylund, G. (1989). Crustal evolution of Fennoscandia – palaeomagnetic constraints. *Tectonophysics*, **162**, 27–49.

Peterson, D. N. & Nairn, A. E. M. (1971). Paleomagnetism of Permian red beds from the southwestern United States. *Geophys. J. R. Astr. Soc.*, **23**, 191–205.

Peybernes, B., Tchoumatchenco, P., Dercourt, J., Ivanov, Z., Lachkar, G., Rolando, J.-P., Surmont, J. & Thierry, J. (1989). Données nouvelles sur les flyschs jurassiques de la zone de Luda Kamcija (Balkanides externes, Bulgarie): conséquences paléogéographiques. *C. R. Acad. Sci. Paris*, **309**, 115–24.

Phillips, J. D. & Forsyth, D. (1972). Plate tectonics, paleomagnetism and the opening of the Atlantic. *Geol. Soc. Am. Bull.*, **83**, 1579–600.

Phillips, W. E. A., Stillman, C. J. & Murphy, T. (1976). A Caledonian plate tectonic model. *J. Geol. Soc. London*, **132**, 579–609.

Pigram, C. J. & Panggabean, H. (1984). Rifting of the northern margin of the Australian continent and the origin of some microcontinents in eastern Indonesia. *Tectonophysics*, **107**, 331–53.

Pindell, J. L. & Dewey, J. F. (1982). Permo-Triassic reconstruction of western Pangea and the evolution of the Gulf of Mexico/Caribbean region. *Tectonics*, **1**, 179–211.

Piper, J. D. A. (1972). A palaeomagnetic study of the Bukoban System, Tanzania. *Geophys. J. R. Astr. Soc.*, **28**, 111–27.

Piper, J. D. A. (1974). Sulitjelma gabbro, N. Norway: a paleomagnetic result. *Earth Planet. Sci. Lett.*, **21**, 383–8.

Piper, J. D. A. (1975a). 'Sulitjelma gabbro, N. Norway: a paleomagnetic result'. Reply to comment by R. Mason. *Earth Planet. Sci. Lett.*, **25**, 90.

Piper, J. D. A. (1975b). A palaeomagnetic study of the coast-parallel Jurassic dyke swarm in southern Greenland. *Phys. Earth Planet. Int.*, **11**, 36–42.

Piper, J. D. A. (1975c). Palaeomagnetism of Silurian lavas of Somerset and Gloucestershire, England. *Earth Planet. Sci. Lett.*, **25**, 355–60.

Piper, J. D. A. (1976a). Palaeomagnetic evidence for a Proterozoic supercontinent. *Phil. Trans. R. Soc. London*, **A280**, 469–90.

Piper, J. D. A. (1976b). Magnetic properties of Precambrian pillow lavas of the Mona complex and a related dyke swarm, Anglesey, Wales. *Geol. J.*, **11**, 189–201.

Piper, J. D. A. (1978a). Palaeomagnetism and palaeogeography of the Southern Uplands block in Ordovician times. *Scott. J. Geol.*, **14**, 93–107.

Piper, J. D. A. (1978b). Palaeomagnetic survey of the (Palaeozoic) Shelve inlier and Berwyn Hills, Welsh Borderlands. *Geophy. J. R. Astr. Soc.*, **53**, 355–71.

Piper, J. D. A. (1979a). Aspects of Caledonian paleomagnetism and their tectonic implications. *Earth Planet. Sci. Lett.*, **44**, 176–92.

Piper, J. D. A. (1979b). Palaeomagnetic study of late Precambrian rocks of the Midland craton of England and Wales. *Phys. Earth Planet. Int.*, **19**, 59–72.

Piper, J. D. A. (1981a). Magnetic properties of the Alnön Complex. *Geol. Föreningens i Stockholm Forhandl.*, **103**, 9–15.

Piper, J. D. A. (1981b). Palaeomagnetic study of the (Late Precambrian) West Greenland kimberlite-lamprophyre suite: Definition of the Hadrynian track. *Phys. Earth Planet. Int.*, **27**, 164–86.

Piper, J. D. A. (1982). A palaeomagnetic investigation of the Malvernian and Old Radnor Precambrian, Welsh Borderlands. *Geol. J.*, **17**, 69–88.

Piper, J. D. A. (1987). *Palaeomagnetism and the Continental Crust*. New York: Halsted Press.

Piper, J. D. A. (1988a). *Paleomagnetic database*. Milton Keynes: Open University Press.

Piper, J. D. A. (1988b). Palaeomagnetism of (late Vendian-earliest Cambrian) minor alkaline intrusions, Fen Complex, southeast Norway. *Earth Planet. Sci. Lett.*, **90**, 422–30.

Piper, J. D. A. & Briden, J. C. (1973). Palaeomagnetic studies in the British Caledonides, I. Igneous rocks of the Builth Wells Ordovician inlier, Radnorshire, Wales. *Geophys. J. R. Astr. Soc.*, **34**, 1–12.

Pitman, W. C., III & Talwani, M. (1972). Sea-floor spreading in the North Atlantic. *Geol. Soc. Am. Bull.*, **83**, 619–46.

Pohl, J. & Soffel, H. (1971). Paleomagnetic age determination of the Rochechouart Impact Structure (France). *Zeitschr. Geophys.*, **37**, 857–66.

Poorter, R. P. E. (1972). Preliminary results from the Fen Carbonatite Complex, S. Norway. *Earth Planet. Sci. Lett.*, **17**, 194–8.

Potts, S., Van der Voo, R. & van der Pluijm, B. A. (1991): Paleomagnetic results from the Munsungun Anticlinorium of northern Maine: paleogeographic implications for a northern Appalachian volcanic terrane (Abstract). *Eos, Trans. Am. Geophys. Union*, **72**, 106.

Pouchan, P. & Roche, A. (1971). Etude paléomagnétique de formations volcaniques du Territoire des Afars et Issas. *C. R. Acad. Sci. Paris*, **272**, 531–4.

Pozzi, J. P., Westphal, M., Girardeau, J., Besse, J. P. & Yaoxin, Z. (1984). Paleomagnetism of the Xigaze Ophiolite and flysch: Latitude and direction of spreading. *Earth Planet. Sci. Lett.*, **70**, 383–94.

Pozzi, J. P., Westphal, M., Zhou, Y. X., Xing, L. S. & Chen, X. Y. (1982). Position of the Lhasa block, south Tibet, during the Late Cretaceous. *Nature*, **297**, 319–21.

Prasad, S. N. & Sharma, P. V. (1978). Palaeomagnetism of the Nexø Sandstone from Bornholm Island, Denmark. *Geophys. J. R. Astr. Soc.*, **54**, 669–80.

Prévot, M. & McWilliams, M. O. (1989). Paleomagnetic correlation of Newark Supergroup volcanics. *Geology*, **17**, 1007–10.

Proko, M. S. & Hargraves, R. B. (1973). Paleomagnetism of the Beemerville (New Jersey) alkaline complex. *Geology*, **1**, 185–6.

Pruner, P. (1987). Paleomagnetism and paleogeography of Mongolia in the Cretaceous, Permian and Carboniferous – preliminary data. *Tectonophysics*, **139**, 155–67.

Pucher, R., Bannert, D. & Fromm, K. (1974). Paleomagnetism in Greece: indications for relative block movement. *Tectonophysics*, **22**, 31–9.

Puigdefabregas, C. & Souquet, P. (1986). Tecto-sedimentary cycles and depositional sequences of the Mesozoic and Tertiary from the Pyrenees. *Tectonophysics*, 129, 173–203.

Purucker, M. E., Elston, D. P. & Shoemaker, E. M. (1980). Early acquisition of characteristic magnetization in red beds of the Moenkopi Formation (Triassic), Gray Mountain, Arizona. *J. Geophys. Res.*, **85**, 997–1012.

Quennell, A. M. (1984). The western Arabian rift system. *Geol. Soc. London Spec. Publ.*, **17**, 775–88.

Rabinowitz, P. D. & LaBrecque, J. L. (1979). The Mesozoic South Atlantic Ocean and evolution of its continental margins. *J. Geophys. Res.*, **84**, 5973–6002.

Rage, J.-C. (1988). Gondwana, Tethys, and terrestrial vertebrates during the Mesozoic and Cainozoic. *Geol. Soc. London Spec. Publ.*, **37**, 255–73.

Raja, P. K. S. & Vise, J. B. (1973). Paleomagnetism of the Tororo Ring Complex, SE Uganda. *Earth Planet. Sci. Lett.*, **19**, 438–42.

Rao, K. V. & Deutsch, E. R. (1976). Paleomagnetism of the Lower Cambrian Bradore Sandstones, and the rotation of Newfoundland. *Tectonophysics*, **33**, 337–57.

Rao, K. V., Seguin, M. K. & Deutsch, E. R. (1981). Paleomagnetism of Siluro-Devonian and Cambrian granitic rocks from the Avalon Zone in Cape Breton Island, Nova Scotia. *Can. J. Earth Sci.*, **18**, 1187–210.

Rao, K. V., Seguin, M. K. & Deutsch, E. R. (1986). Paleomagnetism of Early Cambrian redbeds on Cape Breton Island, Nova Scotia. *Can. J. Earth Sci.*, **23**, 1233–42.

Rao, K. V. & Van der Voo, R. (1980). Paleomagnetism of a Paleozoic anorthosite from the Appalachian Piedmont, northern Delaware: possible tectonic implications. *Earth Planet. Sci. Lett.*, **47**, 113–20.

Rapalini, A. E. & Vilas, J. F. (1991). Preliminary paleomagnetic data from the Sierra Grande Formation: tectonic consequences of the first mid-Paleozoic paleopoles from Patagonia. *J. South Am. Earth Sci.*, **4**, 25–41.

Rapp, S. D., MacFadden, B. J. & Schiebout, J. A. (1983). Magnetic polarity stratigraphy of the Early Tertiary Black Peaks Formation, Big Bend National Park, Texas. *J. Geol.*, **91**, 555–72.

Rast, N., Sturt, B. A. & Harris, A. L. (1988). Early deformation in the Caledonian–Appalachian orogen. In *The Caledonian–Appalachian Orogen*, ed. A. L. Harris & D. J. Fettes, *Geol. Soc. Special Publ.*, **38**, pp. 111–22. Oxford: published by Blackwell for the Geol. Soc.

Rayner, D. (1967). *The Stratigraphy of the British Isles*. Cambridge: Cambridge University Press.

Reeve, S. C. (1975). *Paleomagnetic studies of sedimentary rocks of Cambrian and Triassic age*. Ph.D. Thesis, Dallas, University of Texas.

Reeve, S. C. & Helsley, C. E. (1972). Magnetic reversal sequence in the upper portion of the Chinle Formation, Montoya, New Mexico. *Geol. Soc. America Bull.*, **83**, 3795–812.

Reeve, S. C., Leythaeuser, D., Helsley, C. E. & Bay, K. W. (1974). Paleomagnetic results from the Upper Triassic of East Greenland. *J. Geophys. Res.*, **79**, 3302–7.

Rehault, J.-P., Boillot, G. & Mauffret, A. (1985). The Western Mediterranean Basin. In *Geological Evolution of the Mediterranean Basin*, ed. D. J. Stanley & F.-C. Wezel, pp. 101–29. Heidelberg: Springer-Verlag.

Reid, A. B. (1968). *A Palaeomagnetic Study of the Sijarira Group, Rhodesia*. M. Phil. Thesis, University of London.

Reinhardt, B. M. (1969). On the genesis and emplacement of ophiolites in the Oman mountains geosyncline. *Schweiz. Min. Petr. Mitt.*, 49/1.

Ressetar, R. & Martin, D. L. (1980). Paleomagnetism of Eocene igneous intrusives in the Valley and Ridge Province, Virginia and West Virginia. *Can. J. Earth Sci.*, **17**, 1583–7.

Ressetar, R., Nairn, A. E. M. & Monrad, J. R. (1981). Two phases of Cretaceous–Tertiary magmatism in the Eastern Desert of Egypt: paleomagnetic, chemical and K-Ar evidence. *Tectonophysics*, **73**, 169–93.

Reynolds, R. L., Hudson, M. R., Fishman, N. S. & Campbell, J. A. (1985). Paleomagnetic and petrologic evidence bearing on the age and origin of uranium deposits in the Permian Cutler Formation, Lisbon Valley, Utah. *Geol. Soc. America Bull.*, **96**, 719–30.

Ricou, L. E. (1971). Le croissant ophiolitique péri-arabe, une ceinture de nappes mises en place au Crétacé supérieur. *Rev. Géogr. Phys. Géol. Dynamique*, **13**, 327–49.

Ricou, L. E., Marcoux, J. & Whitechurch, H. (1984). The Mesozoic organization of the Taurides: one or several ocean basins? In *The Geological Evolution of the Eastern Mediterranean*, ed. J. E. Dixon & A. H. F. Robertson, pp. 349–59. Oxford: published by Blackwell for the Geol. Society.

Ridd, M. F. (1971). Southeast Asia as part of Gondwanaland. *Nature*, **234**, 531–3.

Robardet, M. & Paris, F. (1985). Paleomagnetism of some lithological units from the lower Paleozoic of the Vendée, France: Discussion. *Tectonophysics*, **118**, 85–7.

Roberts, D. (1988). Timing of Silurian to Middle Devonian deformation in the Caledonides of Scandinavia, Svalbard and E. Greenland. In *The Caledonian–Appalachian Orogen*, ed. A. L. Harris & D. J. Fettes, *Geol. Soc. Special Publ.*, **38**, pp. 429–35. Oxford: published by Blackwell for the Geol. Soc.

Robertson, A. H. F. & Dixon, J. E. (1984). Introduction: Aspects of the geological evolution of the eastern Mediterranean. *Geol. Soc. London Spec. Publ.*, **17**, 1–74.

Robertson, D. J. (1988). Palaeomagnetism of the Connemara Gabbro, Western Ireland. *Geophys. J. R. Astr. Soc.*, **94**, 51–64.

Robertson, W. A. (1963). The paleomagnetism of some Mesozoic intrusives and tuffs from eastern Australia. *J. Geophys. Res.*, **68**, 2299–312.

Robertson, W. A. (1964). Paleomagnetism of the monzonite porphyry from Milton, New South Wales. *Pure Appl. Geophys.*, **59**, 93–9.

Robertson, W. A. (1979). Palaeomagnetic results from some Sydney Basin igneous rock deposits. *J. Proc. Roy. Soc. N. S. W.*, **112**, 31–5.

Robertson, W. A. & Hastie, L. (1962). A palaeomagnetic study of the Cygnet alkaline complex of Tasmania. *J. Geol. Soc. Austr.*, **8**, 259–68.

Robinson, M. A. (1985). Palaeomagnetism of volcanics and sediments of the Eday Group, S. Orkney. *Scott. J. Geol.*, **21**, 285–300.

Roche, A., Saucier, H. & Lacaze, J. (1962). Etude paléomagnétique des roches volcaniques Permiennes de la region Nideck-Donon. *Bull. Serv. Carte Géol. Alsace-Lorraine*, **15**, 59–68.

Rogerson, R. & Moaina, R. B. (1987). Concerning palaeomagnetic research in Papua New Guinea. In *Palaeomagnetic Research in Southeast and East Asia*, ed. J. C. Briden *et al.*, pp. 212–15. Committee for the Coordination of Offshore Prospecting (CCOP), U.N. Development Project Office.

Roggenthen, W. M. & Napoleone, G. (1977). Upper Cretaceous-Paleocene magnetic stratigraphy at Gubbio, Italy: IV. Upper Maastrichtian-Paleocene magnetic stratigraphy. *Geol. Soc. Am. Bull.*, **88**, 378–82.

Ron, H. (1987). Deformation along the Yammuneh, the restraining bend of the Dead Sea Transform: Paleomagnetic and kinematic implications. *Tectonics*, **6**, 653–66.

Ron, H., Nur, A. & Eyal, Y. (1990). Multiple strike-slip fault sets: A case study from the Dead Sea Transform. *Tectonics*, **9**, 1421–31.

Roperch, P. & Bonhommet, N. (1986). Paleomagnetism of Miocene volcanics from South Syria. *J. Geophys.*, **59**, 98–102.

Rother, K. (1971). Gesteins- und paläomagnetische Untersuchungen an Gesteinsproben vom Territorium der DDR aus dem Präkambrium bis zum Tertiär und Fölgerungen für die Veränderungen des geomagnetischen Hauptfeldes sowie für geologisch-geotektonische Interpretationsmöglichkeiten. *Veröffentl*, **5**, 92 pp. Potsdam: Zentralinstitut Physik der Erde.

Rowley, D. B. & Lottes, A. L. (1988). Plate-kinematic reconstructions of the North Atlantic and Arctic: Late Jurassic to Present. *Tectonophysics*, **155**, 73–120.

Roy, J. L. (1966). Désaimantation thermique et analyse statistique de sédiments Carbonifères et Permiens de l'est du Canada. *Can. J. Earth Sci.*, **3**, 139–61.

Roy, J. L. (1969). Paleomagnetism of the Cumberland Group and other Paleozoic formations. *Can. J. Earth Sci.*, **6**, 663–9.

Roy, J. L. (1972). A pattern of rupture of the eastern North American-western European paleoblock. *Earth Planet. Sci. Lett.*, **14**, 103–14.

Roy, J. L. (1977). La position stratigraphique déterminée paléomagnétiquement de sédiments carbonifères de Minudie Point, Nouvelle Ecosse: à propos de l'horizon répère magnétique du Carbonifère. *Can. J. Earth Sci.*, **14**, 1116–27.

Roy, J. L. & Anderson, P. (1981). An investigation of the remanence characteristics of three sedimentary units of the Silurian Mascarene Group of New Brunswick, Canada. *J. Geophys. Res.*, **86**, 6351–68.

Roy, J. L., Anderson, P. & Lapointe, P. L. (1979). Paleomagnetic results from three rock units of New Brunswick and their bearing on the lower Paleozoic tectonics of North America. *Can. J. Earth Sci.*, **16**, 1210–27.

Roy, J. L., Opdyke, N. D. & Irving, E. (1967). Further paleomagnetic results from the Bloomsburg Formation. *J. Geophys. Res.*, **72**, 5075–86.

Roy, J. L. & Park, J. K. (1969). Paleomagnetism of the Hopewell Group, New Brunswick. *J. Geophys. Res.*, **74**, 594–604.

Roy, J. L. & Park, J. K. (1974). The magnetization process of certain red beds: vector analysis of chemical and thermal results. *Can. J. Earth Sci.*, **11**, 437–71.

Roy, J. L., Robertson, W. A. & Park, J. K. (1968). Stability of the magnetization of the Hurley Creek Formation. *J. Geophys. Res.*, **73**, 697–702.

Ruffet, G. (1990). 40Ar-39Ar dating of the Beja Gabbro: timing of the accretion of southern Portugal. *Geophys. Res. Lett.*, **17**, 2121–4.

Ruffet, G., Perroud, H. & Feraud, G. (1990). 40Ar/39Ar dating of a late Proterozoic palaeomagnetic pole for the Armorican Massif (France). *Geophys. J. Int.*, **102**, 397–409.

Runcorn, S. K. (1955). Palaeomagnetism of sediments from the Colorado Plateau. *Nature*, **176**, 505.

Runcorn, S. K. (1956). Paleomagnetic comparisons between Europe and North America. *Geol. Assoc. Canada Proc.*, **9**, 77–85.

Runcorn, S. K. (1961). Climatic change through geological time in the light of the palaeomagnetic evidence for polar wandering and continental drift. *Quart. J. Roy. Met. Soc.*, **87**, 282–313.

Rutten, M. G. (1969). *The Geology of Western Europe*. Amsterdam: Elsevier.

Sacks, P. E. & Secor, D. T. Jr. (1990). Kinematics of late Paleozoic continental collision between Laurentia and Gondwana. *Science*, **250**, 1702–5.

Sallomy, J. T. & Briden, J. C. (1975). Palaeomagnetic studies of Lower Jurassic rocks in England and Wales. *Earth Planet. Sci. Lett.*, **24**, 369–76.

Sallomy, J. T. & Krs, M. (1980). A palaeomagnetic study of some igneous rocks from Jordan. *Inst. Appl. Geol. (Jeddah) Bull.*, **3**, 155–64.

Sallomy, J. T. & Piper, J. D.A. (1973). Palaeomagnetic studies in the British Caledonides – 4. Lower Devonian lavas of the Strathmore region, Scotland. *Geophys. J. R. Astr. Soc.*, **34**, 47–68.

Salmon, E., Montigny, R., Edel, J. B., Pique, A., Thuizat, R. & Westphal, M. (1986). The Msissi norite revisited: K/Ar dating, petrography, and paleomagnetism. *Geophys. Res. Lett.*, **13**, 741–3.

Samson, S., Palmer, A. R., Robison, R. A. & Secor, D. T. (1990). Biogeographical significance of Cambrian trilobites from the Carolina Slate Belt. *Geol. Soc. Am. Bull.*, **102**, 1459–70

Sandal, S. T. & Halvorsen, E. (1973). Late Mesozoic paleomagnetism from Spitsbergen: implications for continental drift in the Arctic. *Phys. Earth Planet. Int.*, 7, 125–32.

Saradeth, S., Soffel, H. C., Horn, P., Müller-Söhnius, D. & Schult, A. (1989). Upper Proterozoic and Phanerozoic pole positions and potassium-argon (K-Ar) ages from the East Sahara craton. *Geophys. J.*, **97**, 209–21.

Saribudak, M. (1989). New results and a paleomagnetic overview of the Pontides in northern Turkey. *Geophys. J. Int.*, **99**, 521–31.

Saribudak, M. (1991). Paleomagnetic constraints on the tectonic evolution of the Sakarya continent, northwestern Anatolia – a comment. *Tectonophysics*, **200**, 317–19.

Saribudak, M., Sanver, M. & Ponat, E. (1989). Location of the western Pontides during Triassic time: preliminary palaeomagnetic results. *Geophys. J.*, **96**, 43–50.

Saribudak, M., Sanver, M., Şengör, A. M. C. & Gorur, N. (1990). Palaeomagnetic evidence for substantial rotation of the Almacik flake within the North Anatolian Fault Zone. *Geophys. J. Intern.*, **102**, 563–8.

Sasajima, S., Nishimura, S., Hirooka, K., Otofuji, Y., Van Leeuwen, T. & Hehuwat, F. (1980). Paleomagnetic studies combined with fission-track datings on the western arc of Sulawesi, East Indonesia. *Tectonophysics*, **64**, 163–72.

Sasajima, S., Otofuji, Y., Hirooka, K., Suparka, Suwijanto & Hehuwat, F. (1978). Paleomagnetic studies on Sumatra Island: on the possibility of Sumatra being part of Gondwanaland. *Rockmagnetism and Paleogeophys.*, **5**, 104–10.

Savostin, L. A., Sibuet, J. C., Zonenshain, L. P., Le Pichon, X. & Roulet, M. J. (1986). Kinematic evolution of the Tethys Belt from the Atlantic Ocean to the Pamirs since the Triassic. *Tectonophysics*, **123**, 1–35.

Scharnberger, C. H. & Scharon, L. (1972). Paleomagnetism and plate tectonics of Antarctica. In *Antarctic Geology and Geophysics*, ed. R. J. Adie, pp. 843–7. Oslo: Universitetsforlaget.

Schmidt, P. W. (1976). A new paleomagnetic investigation of Mesozoic igneous rocks in Australia. *Tectonophysics*, **33**, 1–13.

Schmidt, P. W., Currey, D. J. & Ollier, C. D. (1976). Sub-basaltic weathering, damsites, palaeomagnetism, and the age of lateritization. *J. Geol. Soc. Austr.*, **23**, 367–70.

Schmidt, P. W. & Embleton, B. J. J. (1981). Magnetic overprinting in southeastern Australia and the thermal history of its rifted margin. *J. Geophys. Res.*, **86**, 3998–4008.

Schmidt, P. W., Embleton, B. J. J., Cudahy, T. J. & Powell, C. McA. (1986). Prefolding and pre-megakinking magnetization from the Devonian Comerong Volcanics, New South Wales, Australia, and their bearing on the Gondwana pole path. *Tectonics*, **5**, 135–50.

Schmidt, P. W., Embleton, B. J. J. & Palmer, H. C. (1987). Pre- and post-folding magnetizations from the Early Devonian Snowy River Volcanics and Buchan Caves limestone, Victoria. *Geophys. J. R. Astr. Soc.*, **91**, 155–77.

Schmidt, P. W. & McDougall, I. (1977). Palaeomagnetic and potassium-argon dating studies of the Tasmanian dolerites. *J. Geol. Soc. Austr.*, **25**, 321–8.

Schmidt, P. W., Powell, C. McA., Li, Z. X. & Thrupp, G. A. (1990). Reliability of Palaeozoic palaeomagnetic poles and APWP of Gondwanaland. *Tectonophysics*, **184**, 87–100.

Schmidtke, E. A., Fuller, M. D. & Haston, R. (1990). Paleomagnetic data from Sarawak, Malaysian Borneo, and the late Mesozoic and Cenozoic tectonics of Sundaland. *Tectonics*, **9**, 123–40.

Schneider, D. A. & Kent, D. V. (1990a). The time-averaged paleomagnetic field. *Rev. Geophys.*, **28**, 71–96.

Schneider, D. A. & Kent, D. V. (1990b). Testing models of the Tertiary paleomagnetic field. *Earth Planet. Sci. Lett.*, **101**, 260–271.

Schott, J. J., Montigny, R. & Thuizat, R. (1981). Paleomagnetism and potassium-argon age of the Messejana dike (Portugal and Spain): angular limitation to the rotation of the Iberian Peninsula since the Middle Jurassic. *Earth Planet. Sci. Lett.*, **53**, 457–70.

Schott, J. J. & Peres, A. (1987a). Paleomagnetism of Lower Cretaceous red beds from northern Spain: evidence for a multistage acquisition of magnetization. *Tectonophysics*, **139**, 239–53.

Schott, J. J. & Peres, A. (1987b). Paleomagnetism of Permo-triassic red beds from the Asturias and Cantabric Chain (northern Spain): evidence for strong lower Tertiary remagnetizations. *Tectonophysics*, **149**, 179–191.

Schott, J. J. & Peres, A. (1988). Paleomagnetism of Permo-Triassic red beds in the western Pyrenees: evidence for strong clockwise rotations of the Paleozoic units. *Tectonophysics*, **156**, 75–88.

Schouten, H., Srivastava, S. P. & Klitgord, K. D. (1984). Iberian plate kinematics: jumping plate boundaries, an alternative to ball-bearing plate tectonics (Abstract). *Eos, Trans. Am. Geophys. Union*, **65**, 190.

Schult, A. (1973). Paleomagnetism of Upper Cretaceous volcanic rocks in Sicily. *Earth Planet. Sci. Lett.*, **19**, 97–100.

Schult, A. (1976). Paleomagnetism of Cretaceous and Jurassic volcanic rocks in West Sicily. *Earth Planet. Sci. Lett.*, **31**, 454–7.

Schult, A. & Guerreiro, S. D. C. (1979). Paleomagnetism of Mesozoic igneous rocks from the Maranhao Basin, Brazil, and the time of opening of the South Atlantic. *Earth Planet. Sci. Lett.*, **42**, 427–36.

Schult, A. & Guerreiro, S. D. C. (1980). Palaeomagnetism of Upper Cretaceous volcanic rocks from Cabo de Sto. Agostinho, Brazil. *Earth Planet. Sci. Lett.*, **50**, 311–15.

Schult, A., Hussain, A. G. & Soffel, H. C. (1981). Palaeomagnetism of Upper Cretaceous volcanics and Nubian sandstones of Wadi Natash, SE Egypt and implications for the polar wander path for Africa in the Mesozoic. *J. Geophys.*, **50**, 16–22.

Schult, A. & Soffel, H. C. (1973). Palaeomagnetism of Tertiary basalts from Lybia. *Geophys. J. R. Astr. Soc.*, **32**, 373–80.

Schult, A., Soffel, H. C. & Hussain, A. G. (1978). Paleomagnetism of Cretaceous Nubian Sandstones, Egypt. *J. Geophys.*, **44**, 333–40.

Schutts, L. D., Brecher, A., Hurley, P. M., Montgomery, C. W. & Krueger, H. W. (1976). A case study of the time and nature of paleomagnetic resetting in a mafic complex in New England. *Can. J. Earth Sci.*, **13**, 898–907.

Schwab, F. L., Nystuen, J. P. & Gunderson, L. (1988). Pre-Arenig evolution of the Appalachian-Caledonide orogen: sedimentation and stratigraphy. In *The Caledonian–Appalachian Orogen*, ed. A. L. Harris & D. J. Fettes, *Geol. Soc. Special Publ.*, **38**, pp. 75–91. Oxford: published by Blackwell for the Geol. Soc.

Schwartz, S. Y. & Van der Voo, R. (1984). Paleomagnetic study of thrust sheet rotation during foreland impingement in the Wyoming-Idaho overthrust belt. *J. Geophys. Res.*, **89**, 10077–86.

Schwarz, E. J. (1963). A paleomagnetic investigation of Permo-Triassic red beds and andesites from the Spanish Pyrenees. *J. Geophys. Res.*, **68**, 3265–71.

Schwarz, E. J., Coleman, L. C. & Cattroll, H. M. (1979). Paleomagnetic results from the Skaergaard Intrusion, East Greenland. *Earth Planet. Sci. Lett.*, **42**, 437–43.

Sclater, J. G., Hellinger, S. & Tapscott, C. (1977). The paleobathymetry of the Atlantic Ocean from the Jurassic to present. *J. Geol.*, **85**, 509–52.

Scotese, C. R. (1985). *The assembly of Pangea: Middle and Late Paleozoic paleomagnetic results from North America*. Ph.D. Thesis University of Chicago.

Scotese, C. R. (1986). *Phanerozoic Reconstructions: A New Look at the Assembly of Asia*. Tech. Report, 66, 54 pp., Austin: Institute for Geophysics, University of Texas.

Scotese, C. R. & Barrett, S. F. (1990). Gondwana's movement over the south pole during the Palaeozoic: evidence from lithological indicators of climate. In *Palaeozoic palaeogeography and biogeography*, ed. W. S. McKerrow & C. R. Scotese. *Geol. Soc. London Memoir*, **12**, 75–85.

Scotese, C. R., Barton, C., Van der Voo, R. & Ziegler, A. M. (1979). Paleozoic base maps. *J. Geol.*, **87**, 217–77.

Scotese, C. R. & McKerrow, W. S. (1991). Ordovician plate tectonic reconstructions. In *Advances in Ordovician Geology*, ed. C. R. Barnes & S. H. Williams, Geol. Survey Canada, Paper 90–9, pp. 271–282.

Scotese, C. R., Van der Voo, R., Johnson, R. J. E. & Giles, P. S. (1984). Paleomagnetic results from the Carboniferous of Nova Scotia. In *Plate Reconstructions from Paleozoic Paleomagnetism*, ed. R. Van der Voo, C. R. Scotese & N. Bonhommet. *Am. Geophy. Union Geodynamics Ser.*, **12**, 63–81.

Seguin, M.-K. (1977). Paleomagnetism of Middle Ordovician volcanic rocks from Québec. *Phys. Earth Planet. Int.*, **15**, 363–73.

Seguin, M.-K. (1979.) Paleomagnetism of the Thetford Mines Ophiolites, Québec. *J. Geomag. Geoelect.*, **31**, 103–13.

Seguin, M.-K. (1983). Paleomagnetism of some lithological units from the lower Paleozoic of the Vendée, France. *Tectonophysics*, **96**, 257–79.

Seguin, M.-K. (1986). Palaeomagnetism of Lower Devonian units from Gaspé, Québec. *Earth Planet. Sci. Lett.*, **78**, 129–38.

Seguin, M.-K. (1987). Paleomagnetism of Carboniferous diabase dykes from Gaspé, Québec. *Can. J. Earth Sci.*, **24**, 1705–14.

Seguin, M.-K. & Fyffe, L. (1986). Paleomagnetism of Lower Carboniferous (Tournaisian to Namurian) volcanic rocks of New Brunswick, Canada, and its tectonic implications. *Tectonophysics*, **122**, 357–79.

Seguin, M.-K. & Gahé, E. (1985). Paleomagnetism of Lower Devonian volcanics and

Devonian dykes from northcentral New Brunswick, Canada. *Phys. Earth Planet. Int.*, **38**, 262–76.

Seguin, M.-K. & Petryk, A. A. (1986). Paleomagnetic study of the Late Ordovician to Early Silurian platform sequence of Anticosti Island, Quebec. *Can. J. Earth Sci.*, **23**, 1880–90.

Seguin, M.-K. & Rao, K. V. (1989). Paleomagnetism of Early Cambrian Brigus red shale, eastern Newfoundland. *Phys. Earth Planet. Int.*, **56**, 242–53.

Seguin, M.-K., Rao, K. V. & Arnal, P. (1981). Paleomagnetic study of Cambrian Potsdam Group sandstones, St Lawrence Lowlands, Québec. *Earth Planet. Sci. Lett.*, **55**, 433–49.

Seguin, M.-K., Rao, K. V. & Deutsch, E. R. (1987). Paleomagnetism and rock magnetism of Early Silurian Dunn Point Volcanics, Avalon Zone, Nova Scotia. *Phys. Earth Planet. Int.*, **46**, 369–80.

Seguin, M.-K., Rao, K. V. & Pineault, R. (1982). Paleomagnetic study of Devonian rocks from the Ste. Cécile-St Sébastien region, Québec Appalachians. *J. Geophys. Res.*, **87**, 7853–64.

Seguin, M.-K., Singh, A. & Fyffe, L. (1985). New paleomagnetic data from Carboniferous volcanics and red beds from central New Brunswick. *Geophys. Res. Lett.*, **12**, 81–4.

Şengör, A. M. C. (1979). Mid-Mesozoic closure of the Permo-Triassic Tethys and its implications. *Nature*, **279**, 590–3.

Şengör, A. M. C. (1987). Tectonics of the Tethysides: Orogenic collage development in a collisional setting. *Ann. Rev. Earth Planet. Sci.*, **15**, 213–44.

Şengör, A. M. C., Altiner, D., Cin, A., Ustaomer, T. & Hsü, K. J. (1988). Origin and assembly of the Tethyside orogenic collage at the expense of Gondwanaland. *Geol. Soc. London Spec. Publ.*, **37**, 119–81.

Şengör, A. M. C., Yilmaz, Y. & Sungurlu, O. (1984). Tectonics of the Mediterranean Cimmerides: Nature and evolution of the western termination of Paleo-Tethys. *Geol. Soc. London Spec. Publ.*, **17**, 77–112.

Sharps, R., McWilliams, M. O., Li, Y., Cox, A., Zhang, Z., Zhai, Y., Gao, Z., Li, Y. A. & Li, Q. (1989). Lower Permian paleomagnetism of the Tarim block, northwestern China. *Earth Planet. Sci. Lett.*, **92**, 275–91.

Sheriff, S. D. & Shive, P. N. (1980). The Rattlesnake Hills of central Wyoming revisited: further paleomagnetic results. *Geophys. Res. Lett.*, 7, 589–92.

Shibuya, H., Min, K. D., Lee, Y. S., Sasajima, S. & Nishimura, S. (1988). Paleomagnetism of Cambrian to Jurassic sedimentary rocks from the Ogcheon Zone, southern part of Korean Peninsula. *J. Geomagn. Geoelectr.*, **40**, 1469–80.

Shive, P. N. & Frerichs, W. E. (1974). Paleomagnetism of the Niobrara Formation in Wyoming, Colorado, and Kansas. *J. Geophys. Res.*, **79**, 3001–9.

Shive, P. N. & Pruss, E. F. (1977). A paleomagnetic study of basalt flows from the Absaroka Mountains, Wyoming. *J. Geophys. Res.*, **82**, 3039–48.

Shive, P. N., Steiner, M. B. & Huycke, D. T. (1984). Magnetostratigraphy, paleomagnetism, and remanence acquisition in the Triassic Chugwater Formation of Wyoming. *J. Geophys. Res.*, **89**, 1801–15.

Sichler, B., Oliver, J. L., Auzende, J. M., Jonquet, H., Bonin, J. & Bonifay, A. (1980). Mobility of Morocco. *Can. J. Earth Sci.*, **17**, 1546–58.

Sinito, A. M., Valencio, D. A. & Vilas, J. F. (1979). Paleomagnetism of a sequence of Upper Palaeozoic-Lower Mesozoic red beds from Argentina. *Geophys. J. R. Astr. Soc.*, **58**, 237–47.

Skehan, J. W. (1988). Evolution of the Iapetus Ocean and its borders in pre-Arenig times: a synthesis. In *The Caledonian–Appalachian Orogen*, ed. A. L. Harris & D. J. Fettes, *Geol. Soc. Special Publ.*, **38**, pp. 185–229. Oxford: published by Blackwell for the Geol. Soc.

Smethurst, M. A. & Briden, J. C. (1988). Palaeomagnetism of Silurian sediments in W. Ireland: evidence for block rotation in the Caledonides. *Geophys. J.*, **95**, 327–46.

Smith, A. B. & Xu, J. (1988). Palaeontology of the 1985 Tibet Geotraverse, Lhasa to Golmud. *Phil. Trans. R. Soc. London*, **A327**, 53–105.

Smith, A. G. (1971). Alpine deformation and the oceanic areas of the Tethys, Mediterranean, and Atlantic. *Geol. Soc. Am. Bull.*, **82**, 2039–70.

Smith, A. G. & Hallam, A. (1970). The fit of the southern continents. *Nature*, **225**, 139–49.

Smith, A. G., Hurley, A. M. & Briden, J. C. (1980). *Phanerozoic Palaeocontinental World Maps*. Cambridge: Cambridge University Press.

Smith, A. G. & Livermore, R. A. (1991). Pangea in Permian to Jurassic time. *Tectonophysics*, **187**, 135–79.

Smith, P. J. (1966). Tertiary geomagnetic field reversals in Scotland. *Earth Planet. Sci. Lett.*, **1**, 341–7.

Smith, R. L. & Piper, J. D. A. (1979). Palaeomagnetism of the Sarna alkaline body. *Geol. Föreningens i Stockholm Forhandl.*, **101**, 167–8.

Smith, R. L. & Piper, J. D. A. (1984). Palaeomagnetic study of the (Lower Cambrian) Longmyndian sediments and tuffs, Welsh Borderlands. *Geophys. J. R. Astr. Soc.*, **79**, 875–92.

Smith, R. L., Stearn, J. E. F. & Piper, J. D. A. (1983). Palaeomagnetic studies of the Torridonian sediments, NW Scotland. *Scott. J. Geol.*, **19**, 29–45.

Smith, T. E. & Noltimier, H. C. (1979). Paleomagnetism of the Newark-trend igneous rocks of the north-central Appalachians and the opening of the central Atlantic Ocean. *Am. J. Sci.*, **279**, 778–807.

Smith, W. A. (1987). Paleomagnetic results from a crosscutting system of northwest and north–south trending diabase dikes in the North Carolina Piedmont. *Tectonophysics*, **136**, 137–50.

Snider-Pelligrini, A. (1858). *La création et ses mystères dévoilés*. Paris: Franck et Dentu.

Soffel, H. C. (1975). The paleomagnetism of the Permian effusives near St Anton, Vorarlberg (Austria) and the anticlockwise rotation of the Northern Calcareous Alps through 60 degrees. *N. Jhrb. Geol. Paläontol. Monatsh.*, **6**, 375–82.

Soffel, H. C. (1978). Reinterpretation of paleomagnetism of the Colli Eugani and Monte Lessini (Italy). *J. Geophys.*, **45**, 35–9.

Soffel, H. C. (1979). Paleomagnetism of Permo-triassic red sandstones from the Northern Calcareous Alps. *J. Geophys.*, **45**, 447–50.

Soffel, H. C. (1981). Palaeomagnetism of a Jurassic ophiolite series in east Elba (Italy). *J. Geophys.*, **49**, 1–10.

Soffel, H. C. & Förster, H. G. (1981). Apparent polar wander path of Central Iran and its geotectonic implications. In *Global Reconstruction and the Geomagnetic Field during the Palaeozoic*, ed. M. W. McElhinny, A. N. Khramov, M. Ozima & D. A. Valencio, *Advances in Earth Planet. Sci.*, vol. 10, pp. 117–35. Dordrecht: Reidel Publ. Co.

Soffel, H. C., Pohl, W. & Buser, S. (1983). Paleomagnetism of Permo-Triassic rocks from northern Slovenia, Yugoslavia, and the eastern margin of the Adriatic plate. *Tectonophysics*, **91**, 301–20.

Soffel, H. C., Saradeth, S., Briden, J. C., Bachtadse, V. & Rolf, C. (1990). The Sabaloka Ring Complex revisited: palaeomagnetism and rock magnetism,. *Geophys. J. Int.*, **102**, 411–20.

Sopeña, A., Lopez, J., Arche, A., Perez-Arlucea, M., Ramos, A., Virgili, C. & Hernando, S. (1988). Permian and Triassic rift basins of the Iberian Peninsula. In *Triassic-Jurassic Rifting*, ed. W. Manspeizer, pp. 757–86. Amsterdam: Elsevier.

Sopeña, A., Virgili, C., Arche, A., Ramos, A. & Hernando, S. (1983). El Triásico. In *Libro Jubilar* ed J. M. Rios, *Geologia de España*, vol. 2, pp. 47–62. Madrid: Inst. Geol. Min. España.

Soper, N. J. & Woodcock, N. H. (1990). Silurian collision and sediment dispersal patterns in Southern Britain. *Geol. Mag.*, **127**, 527–42.

Spall, H. (1968). Paleomagnetism of basement granites of southern Oklahoma and its implications: progress report. *Okla. Geol. Surv. Notes,* **28**, 65–80.

Spall, H. (1970). Paleomagnetism of basement granites in southern Oklahoma: final report. *Okla. Geol. Surv. Notes*, **30**, 136–50.

Spariosu, D. J. & Kent, D. V. (1983). Paleomagnetism of the Lower Devonian Traveler Felsite and the Acadian orogeny in the New England Appalachians. *Geol. Soc. Am. Bull.*, **94**, 1319–28.

Spariosu, D. J., Kent, D. V. & Keppie, J. D. (1984). Late Paleozoic motions of the Meguma terrane, Nova Scotia: new paleomagnetic evidence. In *Plate Reconstructions from Paleozoic Paleomagnetism*, ed. R. Van der Voo, C. R. Scotese & N. Bonhommet, *Am. Geophys. Union Geodynamics Ser.*, **12**, 82–98.

Spray, J. G., Bébien, J., Rex, D. C. & Roddick, J. C. (1984). Age constraints on the igneous and metamorphic evolution of the Hellenic-Dinaride ophiolites. In *The Geological Evolution of the Eastern Mediterranean*, ed. J. E. Dixon & A. H. F. Robertson, pp. 619–27. Oxford: published by Blackwell for the Geol. Society.

Srivastava, S. P., Roest, W. R., Kovacs, L. C., Oakey, G., Levesque, S., Verhoef, J. & Macnab, R. (1990). Motion of Iberia since the Late Jurassic: results from detailed aeromagnetic measurements in the Newfoundland Basin. *Tectonophysics*, **184**, 229–60.

Srivastava, S. P. & Tapscott, C. (1986). Plate kinematics of the North Atlantic. In *The Geology of North America*, ed. B. E. Tucholke & P. R. Vogt, vol. M, *The Western Atlantic Region*, pp. 379–404. Boulder: Geol. Soc. America.

Stacey, F. D. & Banerjee, S. K. (1974). *The Physical Principles of Rock Magnetism*. Amsterdam: Elsevier, Amsterdam.

Stauffer, P. H. (1983). Unraveling the mosaic of Paleozoic crustal blocks in southeast Asia. *Geol. Rundschau*, 72, 1061–80.

Stearns, C., Mauk, F. J. & Van der Voo, R. (1982). Late Cretaceous to Early Tertiary paleomagnetism of Aruba and Bonaire (Netherlands Leeward Antilles). *J. Geophys. Res.*, **87**, 1127–41.

Stearns, C. & Van der Voo, R. (1988). Dual-polarity magnetizations from the Upper Devonian McAras Brook Formation, Nova Scotia, and their implications for the North American apparent polar wander path. *Tectonophysics*, **156**, 179–91.

Stearns, C., Van der Voo, R. & Abrahamsen, N. (1989). A new Siluro-Devonian paleopole from early Paleozoic rocks of the Franklinian Basin, North Greenland fold belt. *J. Geophys. Res.*, **94**, 10669–83.

Steiner, M. B. (1978). Magnetic polarity during the Middle Jurassic as recorded in the Summerville and Curtis Formations. *Earth Planet. Sci. Lett.*, **38**, 311–45.

Steiner, M. B. (1988). Paleomagnetism of the Late Pennsylvanian and Permian: A test of the rotation of the Colorado Plateau. *J. Geophys. Res.*, **93**, 2201–15.

Steiner, M. B. & Helsley, C. E. (1974). Magnetic polarity sequence of the Upper Triassic Kayenta Formation. *Geology*, **2**, 191–4.

Steiner, M. B. & Helsley, C. E. (1975). Reversal patterns and apparent polar wander for the Late Jurassic. *Geol. Soc. Am. Bull.*, **86**, 1537–43.

Steiner, M. B., Ogg, J. G., Melendez, G. & Sequeiros, L. (1985). Jurassic magnetostratigraphy, 2. Middle-Late Oxfordian of Aguilón, Iberian Cordillera, northern Spain. *Earth Planet. Sci. Lett.*, 76, 151–66.

Steiner, M. B., Ogg, J. G. & Sandoval, J. (1987). Jurassic magnetostratigraphy, 3. Bathonian-Bajocian of Carcabuey, Sierra Harana and Campillo de Arenas (Subbetic Cordillera, southern Spain). *Earth Planet. Sci. Lett.*, **82**, 357–72.

Steiner, M. B., Ogg, J., Zhang, Z. & Sun, S. (1989). The Late Permian/Early Triassic magnetic polarity time scale and plate motions of South China. *J. Geophys. Res.*, **94**, 7343–63.

Steinmann, G. (1905). Geologische Beobachtungen in den Alpen. II. Die Schärdtsche Uberfaltungstheorie und die geologische Bedeutung der Tiefseeabsätze und der ophiolitischen Massengesteine. *Ber. Natürf. Ges. Freiburg*, **16**, 435–68.

Stöcklin, J. (1977). Structural correlation of the Alpine ranges between Iran and Central Asia. *Mém. hors sér. Soc. Géol. France*, **8**, 333–53.

Stöcklin, J. (1984). The Tethys paradox in plate tectonics. *Geodynamics Ser.*, **12**, 27–8.

Storetvedt, K. M. (1968). The permanent magnetism of some basic intrusions in the Kragerø archipelago, Southern Norway, and its geologic implications. *Norsk Geol. Tidsskr.*, **48**, 153–63.

Storetvedt, K. M. (1990). The Tethys Sea and the Alpine–Himalayan orogenic belt: mega-elements in a new global tectonic system. *Phys. Earth Planet. Int.*, **62**, 141–84.

Storetvedt, K. M., Carmichael, C. M., Hayatsu, A. & Palmer, H. C. (1978b). Palaeomagnetism and K/Ar results from the Duncansby volcanic neck, NE Scotland: Superimposed magnetizations, age of igneous activity, and tectonic implications. *Phys. Earth Planet. Int.*, **16**, 379–92.

Storetvedt, K. M. & Gidskehaug, A. (1969). The magnetization of the Great Whin Sill, northern England. *Phys. Earth Planet. Int.*, **2**, 105–11.

Storetvedt, K. M. & Løvlie, R. (1983). Magnetization properties of intrusive/extrusive rocks from East Maio (Republic of Cape Verde) and their geologic implications. *Geophys. J. R. Astr. Soc.*, 73, 197–212.

Storetvedt, K. M. & Markhus, L. A. (1978). Multivectorial magnetization in late Palaeozoic volcanics from North Sardinia: partial remagnetization and rotation. *Geophys. J. R. Astr. Soc.*, **53**, 245–57.

Storetvedt, K. M. & Meland, A. H. (1985). Geological interpretation of palaeomagnetic results from Devonian rocks of Hoy, Orkney. *Scott. J. Geol.*, **21**, 337–52.

Storetvedt, K. M., Mitchell, J. G., Abranches, M. C. & Oftedahl, S. (1990). A new kinematic model for Iberia; further palaeomagnetic and isotopic age evidence. *Phys. Earth Planet. Int.*, **62**, 109–25.

Storetvedt, K. M., Mogstad, H., Abranches, M. C., Mitchell, J. G. & Serralheiro, A. (1987). Palaeomagnetism and isotopic data from Upper Cretaceous igneous rocks of West Portugal; geological correlation and plate tectonic aspects. *Geophys. J. R. Astr. Soc.* **88**, 241–63.

Storetvedt, K. M., Pedersen, S., Løvlie, R. & Halvorsen, E. (1978a). Palaeomagnetism in the Oslo Rift Zone. In *Tectonics and Geophysics of Continental Rifts*, ed. I. B. Ramberg & E. R. Neumann, p. 289–96. Dordrecht: Reidel Publ. Co.

Storetvedt, K. M. & Petersen, N. (1972). Palaeomagnetic properties of the Middle-Upper Devonian volcanics of the Orkney Islands. *Earth Planet. Sci. Lett.*, **14**, 269–78.

Storetvedt, K. M. & Petersen, N. (1976). Postulated rotation of Corsica not confirmed by new palaeomagnetic data. *J. Geophys.*, **42**, 59–71.

Storetvedt, K. M., & Steel, R. J. (1977). Palaeomagnetic evidence for the age of the Stornoway Formation. *Scott. J. Geol.*, **13**, 263–9.

Storetvedt, K. M., Svalestad, S., Thomassen, K., Langlie, A., Nergard, A. & Gidskehaug, A. (1978c). Magnetic discordance in Gran Canaria/Tenerife and its possible relevance to the formation of the NW African continental margin. *J. Geophys.*, **44**, 317–32.

Storetvedt, K. M. & Torsvik, T. H. (1983). Palaeomagnetic re-examination of the basal Caithness Old Red Sandstone: aspects of local and regional tectonics. *Tectonophysics*, **98**, 151–64.

Storetvedt, K. M. & Torsvik, T. H. (1985). Palaeomagnetism of the Middle-Upper Devonian Esha Ness ignimbrite, W. Shetland. *Phys. Earth Planet Int.*, **37**, 169–73.

Storetvedt, K. M., Vaga, A. M., Aase, S. & Løvlie, R. (1979). Paleomagnetism and the early magmatic history of Fuerteventura (Canary Islands). *J. Geophys.*, **46**, 319–34.

Storhaug, K. & Storetvedt, K. M. (1985). Palaeomagnetism of the Sarclet Sandstone (Orcadian Basin): age perspectives. *Scott. J. Geol.*, **21**, 275–84.

Strangway, D. W. & McMahon, B. E. (1973). Paleomagnetism of annually banded Eocene Green River sediments. *J. Geophys. Res.*, **78**, 5237–45.

Strong, D. F. & Payne, J. G. (1973). Early Paleozoic volcanism and metamorphism of the Moreton's Harbour – Twillingate area, Newfoundland. *Can. J. Earth Sci.*, **10**, 1363–79.

Stump, E. (1992). The Ross Orogen of the Transantarctic Mountains in light of the Laurentia – Gondwana split. *GSA Today*, **2**, 25–31.

Stupavsky, M., Symons, D. T.A. & Gravenor, C. P. (1982). Evidence for metamorphic remagnetisation of upper Precambrian tillite in the Dalradian Supergroup of Scotland. *Trans. Roy. Soc. Edinburgh, Earth Sci.*, **73**, 59–65.

Sturt, B. A. & Torsvik, T. H. (1987). A Late Carboniferous palaeomagnetic pole recorded from a syenite sill, Stabben, central Norway. *Phys. Earth Planet. Int.*, **49**, 350–9.

Suess, E. (1893). Are great ocean depths permanent? *Nat. Sci.*, **2**, 180–7.

Suess, E. (1901). *The Face of the Earth*, vol. 3/1. Oxford: Clarendon. (English edition published in 1908.)

Suk, D., Peacor, D. R. & Van der Voo, R: (1990b). Replacement of pyrite framboids by magnetite in limestone and implications for paleomagnetism. *Nature*, **345**, 611–13.

Suk, D., Van der Voo, R. & Peacor, D. R. (1990a). Scanning and transmission electron microscope observations of magnetite and other iron-phases in Ordovician carbonates from east Tennessee. *J. Geophys. Res.*, **95**, 12327–36.

Surmont, J. (1989). Paléomagnétisme dans les Hellénides internes: analyse des aimantations superposées par la méthode des cercles de réaimantation. *Can. J. Earth Sci.*, **26**, 2479–94.

Surmont, J., Nikolov, T., Thierry, J., Peybernes, B. & Sapunov, I. (1991). Paléomagnétisme des formations sédimentaires jurassiques et éocretacées des zones de Stara Planina-Prebalkan et de Luda Kamcija (Balkanides externes, Bulgarie). *Bull. Soc. Géol. France*, **162**, 57–68.

Swenson, P. & McWilliams, M. O. (1989). Paleomagnetic results from the Upper Cretaceous Maudlow and Livingston Formations, Southwest Montana. *Geophys. Res. Lett.*, **16**, 669–72.

Swinden, H. S. (1987). Geology and mineral occurrences in the central and northern parts of the Robert's Arm Group, central Newfoundland. *Current Research*, part A, *Geol. Surv. Canada, Paper* 87–1A, pp. 381–390.

Symons, D. T. A. (1990). Early Permian pole: evidence from the Pictou red beds, Prince Edward Island, Canada. *Geology*, **18**, 234–7.

Symons, D. T. A., Bormann, R. E. & Jans, R. P. (1989). Paleomagnetism of the Triassic red beds of the lower Fundy Group and Mesozoic tectonism of the Nova Scotia platform, Canada. *Tectonophysics*, **164**, 13–24.

Symons, D. T. A. & Chiasson, A. D. (1991). Paleomagnetism of the Callander Complex and the Cambrian apparent polar wander path for North America. *Can. J. Earth Sci.*, **28**, 355–63.

Tanaka, H., Tsunakawa, H. & Amano, K. (1988). Palaeomagnetism of the Cretaceous El Way and Coloso formations from the northern Chilean Andes. *Geophys. J.*, **95**, 195–203.

Tanczyk, E. I., Lapointe, P., Morris, W. A. & Schmidt, P. W. (1987). A paleomagnetic study of the layered mafic intrusion at Sept-Iles, Québec. *Can. J. Earth Sci.*, **24**, 1431–8.

Tapponnier, P. & Molnar, P. (1977). Active faulting and Cenozoic tectonics of China. *J. Geophys. Res.*, **82**, 2905–30.

Tapponnier, P., Peltzer, G. & Armijo, R. (1986). On the mechanics of the collision between India and Asia. In *Collision Tectonics*, ed. M. P. Coward & A. C. Ries. *Geol. Soc. London Spec. Publ.*, **19**, 115–57.

Tapponnier, P., Peltzer, G., LeDain, A. Y., Armijo, R. & Cobbold, P. (1982). Propagating extrusion tectonics in Asia: new insights from simple experiments with plasticine. *Geology*, **10**, 611–16.

Tarduno, J. A., McWilliams, M. O. & Sleep, N. (1990). Fast instantaneous oceanic plate velocities recorded by the Cretaceous Laytonville limestone: paleomagnetic analysis and kinematic implications. *J. Geophys. Res.*, **95**, 15503–27.

Tarduno, J. A., McWilliams, M. O., Sliter, W. V., Cook, H. E., Blake, M. C. Jr. & Premoli-Silva, I. (1986). Southern hemisphere origin of the Cretaceous Laytonville limestone of California. *Science*, **231**, 1425–8.

Tarling, D. H. (1967). The palaeomagnetic properties of some Tertiary lavas from East Greenland. *Earth Planet. Sci. Lett.*, **3**, 81–8.

Tarling, D. H. (1970). Paleomagnetic results from the Faeroe Islands. In *Palaeogeophysics*, ed. S. K. Runcorn, ch. 23, pp. 193–208. London: Academic Press.

Tarling, D. H. (1983). *Palaeomagnetism, Principles and Applications in Geology, Geophysics and Archaeology*. London: Chapman and Hall.

Tarling, D. H., Mitchell, J. G. & Spall, H. (1973). A palaeomagnetic and isotopic age for the Wackerfield dyke of northern England. *Earth Planet. Sci. Lett.*, **18**, 427–32.

Tarling, D. H. & Otulana, H. I. (1972). The palaeomagnetism of some Tertiary igneous rocks from Ubekendt Ejland, West Greenland. *Bull. Geol. Soc. Denmark*, **21**, 395–406.

Tauxe, L. & Clark, D. R. (1987). New paleomagnetic results from the Eureka Sound Group: Implications for the age of early Tertiary Arctic biota. *Geol. Soc. Am. Bull.*, **99**, 739–47.

Taylor, G. K. (1988). A palaeomagnetic study of a Caledonian ophiolite. *Geophys. J.*, **94**, 157–66.

Ten Haaf, E., Van der Voo, R. & Wensink, H. (1971). The geology of the external zone of the Spanish Pyrenees in the province of Huesca. *Geol. Rundschau*, **60**, 996–1009.

Thio, H. K. (1988). Magnetotectonics in the Piedmont Tertiary Basin. *Phys. Earth Planet. Int.*, **52**, 308–19.

Thomas, C. & Briden, J. C. (1976). Anomalous geomagnetic field during the Late Ordovician. *Nature*, **259**, 380–2.

Thompson, J. F. C., Guillaume, A. & Daly, L. (1986). Palaeomagnetism of the Permian volcanic rocks of Moissey (French Jura): implications for the palaeofield and tectonic evolution. *Geophys. J. R. Astr. Soc.*, **86**, 103–17.

Thompson, R. (1972). Paleomagnetic results from the Paganzo Basin of northwest Argentina. *Earth Planet. Sci. Lett.*, **15**, 145–56.

Thompson, R. (1973). South American Palaeozoic palaeomagnetic results and the welding of Pangaea. *Earth Planet. Sci. Lett.*, **18**, 266–78.

Thompson, R. & Clark, R. M. (1981). Fitting polar wander paths. *Phys. Earth Planet. Int.*, **27**, 1–7.

Thorning, L. (1974). Palaeomagnetic results from Lower Devonian rocks of the Cheviot Hills, northern England. *Geophys. J. R. Astr. Soc.*, **36**, 487–96.

Thorning, L. & Abrahamsen, N. (1980). Palaeomagnetism of Permian multiple intrusion dykes in Bohuslän, SW Sweden. *Geophys. J. R. Astr. Soc.*, **60**, 163–85.

Thrupp, G. A., Kent, D. V., Schmidt, P. W. & Powelll, C. McA. (1991). Paleomagnetism of red beds of the Late Devonian Worange Point Formation, SE Australia. *Geophys. J. Int.*, **104**, 179–201.

Thuizat, R., Whitechurch, H., Montigny, R. & Juteau, T. (1981). K-Ar dating of some infra-ophiolitic metamorphic soles from the Eastern Mediterranean: new evidence for oceanic thrusting before subduction. *Earth Planet. Sci. Lett.*, **52**, 302–10.

Torsvik, T. H. (1984). Palaeomagnetism of the Foyers and Strontian granites, Scotland. *Phys. Earth Planet. Int.*, **36**, 163–77.

Torsvik, T. H. (1985a). Palaeomagnetic results from the Peterhead granite, Scotland: implication for regional late Caledonian magnetic overprinting. *Phys. Earth Planet. Int.*, **39**, 108–16.

Torsvik, T. H. (1985b). Magnetic properties of the Lower Old Red Sandstone lavas in the Midland Valley, Scotland; palaeomagnetic and tectonic considerations. *Phys. Earth Planet. Int.*, **39**, 194–207.

Torsvik, T. H., Løvlie, R. & Storetvedt, K. M. (1983). Multicomponent magnetization in the Helmsdale granite, North Scotland; geotectonic implications. *Tectonophysics*, **98**, 111–29.

Torsvik, T. H., Lyse, O., Atteras, G. & Bluck, B. J. (1989). Palaeozoic palaeomagnetic results from Scotland and their bearing on the British apparent polar wander path. *Phys. Earth Planet. Int.*, **55**, 93–105.

Torsvik, T. H., Olesen, O., Ryan, P. D. & Trench, A. (1990b). On the palaeogeography of Baltica during the Palaeozoic: new palaeomagnetic data from the Scandinavian Caledonides. *Geophys. J. Int.*, **103**, 261–79.

Torsvik, T. H., Smethurst, M. A., Briden, J. C. & Sturt, B. A. (1990a). A review of Palaeozoic palaeomagnetic data from Europe and their palaeogeographical implications. In *Palaeozoic Palaeogeography and Biogeography*, ed. W. S. McKerrow & C. R. Scotese. *Geol. Soc. London Memoir*, **12**, 25–41.

Torsvik, T. H. & Trench, A. (1991a). The Ordovician history of the Iapetus Ocean in Britain: new palaeomagnetic constraints. *J. Geol. Soc., London*, **148**, 423–5.

Torsvik, T. H. & Trench, A. (1991b). The Lower–Middle Ordovician paleofield of Scandinavia: southern Sweden 'revisited.' *Phys. Earth Planet. Int.*, **65**, 283–91.

Tozzi, M., Kissel, C., Funiciello, R., Laj, C. & Parotto, M. (1988). A clockwise rotation of southern Apulia? *Geophys. Res. Lett.*, **15**, 681–4.

Trench, A., Bluck, B. J. & Watts, D. R. (1988). Palaeomagnetic studies within the Ballantrae Ophiolite, SW Scotland: magnetotectonic and regional tectonic implications. *Earth Planet. Sci. Lett.*, **90**, 431–48.

Trench, A. & Torsvik, T. H. (1991). A revised Palaeozoic apparent polar wander path for Southern Britain (Eastern Avalonia). *Geophys. J. Int.*, **104**, 227–33.

Trench, A., Torsvik, T. H., Smethurst, M. A., Woodcock, N. H. & Metcalfe, R. (1991). A palaeomagnetic study of the Builth Wells-Llandridnod Wells Ordovician inlier: palaeogeographic and structural implications. *Geophys. J. Int.*, **105**, 477–89.

Tucker, R. D., Krogh, T. E., Ross, R. J. & Williams, S. H. (1990). Time-scale calibration by high-precision U-Pb zircon dating of interstratified volcanic ashes in the Ordovician and Lower Silurian stratotypes of Britain. *Earth Planet. Sci. Lett.*, **100**, 51–8.

Tucker, S. & Kent, D. V. (1988). Multiple remagnetizations of lower Paleozoic limestones from the Taconics of Vermont. *Geophys. Res. Lett.*, **15**, 1251–4.

Tuckey, M. E. (1990). Distributions and extinctions of Silurian Bryozoa. In *Palaeozoic Palaeogeography and Biogeography*, ed. W. S. McKerrow & C. R. Scotese. *Geol. Soc. London Memoir*, **12**, 197–206.

Turnell, H. B. (1985). Palaeomagnetism and Rb-Sr ages of the Ratagan and Comrie intrusions. *Geophys. J. R. Astr. Soc.*, **83**, 363–78.

Turnell, H. B. (1988). Mesozoic evolution of Greek microplates from palaeomagnetic measurements. *Tectonophysics*, **155**, 307–16.

Turnell, H. B. & Briden, J. C. (1983). Palaeomagnetism of NW Scotland syenites in relation to local and regional tectonics. *Geophys. J. R. Astr. Soc.*, 75, 217–34.

Turner, P. (1981). Relationship between magnetic components and diagenetic features in reddened Triassic alluvium (St Bees Sandstone, Cumbria, UK). *Geophys. J. R. Astr. Soc.*, 67, 395–413.

Turner, P. & Ixer, R. A. (1977). Diagenetic development of unstable and stable magnetization in the St Bees Sandstone (Triassic) of northern England. *Earth Planet. Sci. Lett.*, **34**, 113–24.

Turner, P., Turner, A., Ramos, A. & Sopeña, A. (1989). Diagenetic processes and remagnetization in Permo-triassic red beds. *Cuadern. Geol. Iberica*, **12**, 131–146.

Untung, M., Sickarnto, R., Sunata, W. & Wahyano, H. (1987). *Paleomagnetism along transect VII, Geologic Report, Jawa-Kalimantan SEATAR Transect VII*, pp. 73–85. Bandung, Indonesia: Geological Research and Development Centre.

Valencio, D. A. (1972). Paleomagnetism of the Lower Cretaceous vulcanites of the Cerro Colorado Formation of the Sierra de los Condores Group, Province of Cordoba, Argentina. *Earth Planet. Sci. Lett.*, **16**, 370–8.

Valencio, D. A. (1980). *El Magnetismo de las Rocas*. Buenos Aires, Argentina: Editorial Univ. Buenos Aires.

Valencio, D. A., Mendia, J. E., Giudici, A. & Gascon, J. O. (1977a). Paleomagnetism of the Cretaceous Pirgua Subgroup (Argentina) and the age of the opening of the South Atlantic. *Geophys. J. R. Astr. Soc.*, **51**, 47–58.

Valencio, D. A., Mendia, J. E. & Vilas, J. F. (1979). Palaeomagnetism and K-Ar age of Mesozoic and Cenozoic igneous rocks from Antarctica. *Earth Planet. Sci. Lett.*, **45**, 61–8.

Valencio, D. A., Mendia, J. E. & Vilas, J.F. (1975a). Palaeomagnetism and K-Ar ages of Triassic igneous rocks from the Ischigualasto-Ischichuca Basin and Puesto Viejo formation, Argentina. *Earth Planet. Sci. Lett.*, **26**, 319–30.

Valencio, D. A., Rocha-Campos, A. C. & Pacca, I. G. (1975b). Paleomagnetism of some sedimentary rocks of the late Paleozoic Tubarao and Passa Dois Groups, from the Paraná Basin, Brazil. *Rev. Brasil. Geociencias*, **5**, 186–97.

Valencio, D. A., Rocha-Campos, A. C. & Pacca, I. G. (1978). Paleomagnetism of the Cassanje 'Series' (Karroo System), Angola. *An. Acad. Bras. Cienc.*, **50**, 353–64.

Valencio, D. A. & Vilas, J. F. (1976). Sequence of the continental movements occurring prior to and after the formation of the South Atlantic. *An. Acad. Bras. Cienc.* (Supl.), **48**, 377–86.

Valencio, D. A. & Vilas, J. F. (1985). Evidence of a microplate in the southern Andes? *J. Geodynamics*, **2**, 183–92.

Valencio, D. A., Vilas, J. F. & Mendia, J. E. (1977b). Palaeomagnetism of a sequence of red beds of the middle and upper sections of Paganzo Group (Argentina) and the correlation of upper Palaeozoic-lower Mesozoic rocks. *Geophys. J. R. Astr. Soc.*, **51**, 59–74.

Valencio, D. A., Vilas, J. F. & Mendia, J. E. (1980). Palaeomagnetism and K- Ar ages of Lower Ordovician and Upper Silurian-Lower Devonian rocks from northwest Argentina. *Geophys. J. R. Astr. Soc.*, **62**, 27–39.

Valencio, D. A., Vilas, J. F., Sola, P. & Lopez, M. G. (1983). Palaeomagnetism of Upper Cretaceous to Lower Tertiary rocks from central Argentina. *Geophys. J. R. Astr. Soc.*, **73**, 129–34.

Van Alstine, D. R. (1979). *Apparent Polar Wander with Respect to North America Since the Late Precambrian*. Ph.D. Thesis, Pasadena, Cal. Inst. Techn.

Van Bemmelen, R. W. (1949). *The Geology of Indonesia*. The Hague: Martinus Nijhoff.

Vandamme, D., Courtillot, V., Besse, J. & Montigny, R. (1991). Paleomagnetism and age determinations of the Deccan traps (India): results of a Nagpur–Bombay traverse and review of earlier work. *Rev. Geophys.*, **29**, 159–90.

VandenBerg, J. (1979a). Implications of new paleomagnetic data from the Verrucano (Tuscany, Siena) for its age and tectonic position. *Geol. Ultraiectina*, **20**, 137–45.

VandenBerg, J. (1979b). Preliminary results of paleomagnetic research on Eocene to Miocene rocks of the Piedmont Basin (NW Apennines, Italy). *Geol. Ultraiectina*, **20**, 147–54.

VandenBerg, J. (1980). New paleomagnetic data from the Iberian Peninsula. *Geol. Mijnbouw*, **59**, 49–60.

VandenBerg, J. (1983). Reappraisal of paleomagnetic data from Gargano (south Italy). *Tectonophysics*, **98**, 29–41.

VandenBerg, J., Klootwijk, C. T. & Wonders, A. A. H. (1978). The late Mesozoic and Cenozoic movements of the Italian Peninsula: further paleomagnetic data from the Umbrian sequence. *Geol. Soc. Am. Bull.*, **89**, 133–50.

VandenBerg, J. & Wonders, A. A. H. (1976). Paleomagnetic evidence of large fault displacements around the Po-basin. *Tectonophysics*, **33**, 301–20.

VandenBerg, J. & Wonders, A. A. H. (1980). Paleomagnetism of late Mesozoic pelagic limestones from the Southern Alps. *J. Geophys. Res.*, **85**, 3623–7.

Van den Ende, C. (1977). *Palaeomagnetism of Permian Red Beds of the Dôme of Barrot (S. France)*. Ph.D. Thesis, Univ. of Utrecht.

Van der Lingen, G. J. (1960). Geology of the Spanish Pyrenees, north of Canfranc, Huesca Province. *Estúd. Geol. Inst. Invest. Geol. 'Lucas Mallada,' Madrid*, **16**, 205–42.

van der Pluijm, B. A. & van Staal, C. R. (1988). Characteristics and evolution of the Central Mobile Belt, Canadian Appalachians. *J. Geol.*, **96**, 535–47.

Van der Voo, R. (1967). The rotation of Spain: paleomagnetic evidence from the Spanish Meseta. *Palaeogeogr., Palaeoclimatol., Palaeoecol.*, **3**, 393–416.

Van der Voo, R. (1968). Paleomagnetism and the Alpine tectonics of Eurasia, IV. Jurassic, Cretaceous and Eocene pole positions from northeastern Turkey. *Tectonophysics*, **6**, 251–69.

Van der Voo, R. (1969). Paleomagnetic evidence for the rotation of the Iberian Peninsula. *Tectonophysics*, 7, 5–56.

Van der Voo, R. (1982). Pre-Mesozoic paleomagnetism and plate tectonics. *Ann. Rev. Earth Planet. Sci.*, **10**, 191–220.

Van der Voo, R. (1988). Paleozoic paleogeography of North America, Gondwana, and intervening displaced terranes: comparison of paleomagnetism with paleoclimatology and biogeographical patterns. *Geol. Soc. Amer. Bull.*, **100**, 311–24.

Van der Voo, R. (1989). Paleomagnetism of North America, its margins and the Appalachian Belt. In *Geophysical Framework of the Continental United States*, ed. L. Pakiser & W. D. Mooney, *Geol. Soc. America Memoir*, **172**, 447–70.

Van der Voo, R. (1990a). Phanerozoic paleomagnetic poles from Europe and North America and comparisons with continental reconstructions. *Rev. Geophys.*, **28**, 167–206.

Van der Voo, R. (1990b). The reliability of paleomagnetic data. *Tectonophysics*, **184**, 1–9.

Van der Voo, R. & Boessenkool, A. (1973). A Permian paleomagnetic result from the western Pyrenees delineating the plate boundary between the Iberian Peninsula and stable Europe. *J. Geophys. Res.*, **78**, 5118–27.

Van der Voo, R. & Channell, J. E. T. (1980). Paleomagnetism in orogenic belts. *Rev. Geophys. Space Phys.*, **18**, 455–81.

Van der Voo, R., French, A. N. & French, R. B. (1979). A paleomagnetic pole position from the folded Upper Devonian Catskill red beds, and its tectonic implications. *Geology*, 7, 345–8.

Van der Voo, R. & French, R. B. (1974). Apparent polar wander for the Atlantic-bordering continents: Late Carboniferous to Eocene. *Earth Science Rev.*, **10**, 99–119.

Van der Voo, R. & French, R. B. (1977). Paleomagnetism of the Late Ordovician Juniata Formation and the remagnetization hypothesis. *J. Geophys. Res.*, **82**, 5796–802.

Van der Voo, R., French, R. B. & Williams, D. W. (1976b). Paleomagnetism of the Wilberns Formation (Texas) and the Late Cambrian paleomagnetic field for North America. *J. Geophys. Res.*, **81**, 5633–8.

Van der Voo, R. & Grubbs, K. L. (1977). Paleomagnetism of the Triassic Chugwater red beds revisited (Wyoming, USA). *Tectonophysics*, **41**, T27–T33.

Van der Voo, R., Johnson, R. J. E., van der Pluijm, B. A. & Knutson, L. C. (1991). Paleogeography of some vestiges of Iapetus: paleomagnetism of the Ordovician Robert's Arm, Summerford and Chanceport groups, central Newfoundland. *Geol. Soc. Am. Bull.*, **103**, 1564–75.

Van der Voo, R. & Klootwijk, C. T. (1972). Paleomagnetic reconnaissance study of the Flamanville granite, with special reference to the anisotropy of its susceptibility. *Geol. Mijnbouw*, **51**, 609–17.

Van der Voo, R., Mauk, F. J. & French, R. B. (1976a). Permian-Triassic continental configurations and the origin of the Gulf of Mexico. *Geology*, **4**, 177–80.

Van der Voo, R. & Meert, J. G. (1991). Late Proterozoic paleomagnetism and tectonic models: a critical appraisal. *Precambrian Res.*, **53**, 149–63.

Van der Voo, R., Peinado, J. & Scotese, C. R. (1984). A paleomagnetic re-evaluation of Pangea reconstructions. *Geodynamics Ser.*, **12**, 11–26.

Van der Voo, R. & Scotese, C. R. (1981). Paleomagnetic evidence for a large (c. 2000 km) sinistral offset along the Great Glen Fault during the Carboniferous. *Geology*, **9**, 583–9.

Van der Voo, R. & Van der Kleijn, P. H. (1970). The complex NRM of the Permo-carboniferous Bademli red beds (Tauride Chains, southern Turkey). *Geol. Mijnbouw*, **49**, 391–5.

Van der Voo, R. & Zijderveld, J. D. A. (1971). A renewed paleomagnetic study of the Lisbon volcanics. *J. Geophys. Res.*, **76**, 3913–21.

Van Dongen, P. G. (1967). The rotation of Spain: paleomagnetic evidence from the eastern Pyrenees. *Palaeogeogr., Palaeoclimatol., Palaeoecol.*, **3**, 417–32.

Van Dongen, P. G., Van der Voo, R. & Raven, T. (1967). Paleomagnetism and the alpine tectonics of Eurasia, III, Paleomagnetic research in the Central Lebanon mountains and in the Tartous area (Syria). *Tectonophysics*, **4**, 35–53.

Van Everdingen, R. O. (1960). Studies on the igneous rock complex of the Oslo region, 17. Palaeomagnetic analysis of Permian extrusives in the Oslo region, Norway. *Skrifter Norske Vidensk. Akad. Oslo, Mat. Naturv. Kl.*, p. 1–80.

Van Eysinga, F. W. B. (1975). *Geological Time Scale Table.* 1st edn. Amsterdam: Elsevier.

Van Fossen, M. C., Flynn, J. J. & Forsythe, R. D. (1986). Paleomagnetism of Early Jurassic rocks, Watchung Mountains, Newark Basin: evidence for complex rotations along the border fault. *Geophys. Res. Lett.*, **13**, 185–8.

Van Fossen, M. C. & Kent, D. V. (1990). High-latitude paleomagnetic poles from Middle Jurassic plutons and Moat Volcanics in New England and the controversy regarding Jurassic apparent polar wander for North America. *J. Geophys. Res.*, **95**, 17503–16.

Van Hilten, D. (1960). Geology and Permian paleomagnetism of the Val-di-Non area. *Geol. Ultraiectina*, **5**, 95 pp.

Van Hilten, D. (1962). Presentation of paleomagnetic data, polar wandering and continental drift. *Amer. J. Sci.*, **260**, 401–26.

Van Hilten, D. (1964). Evaluation of some geotectonic hypotheses by paleomagnetism. *Tectonophysics*, **1**, 3–71.

Van Zijl, J. S., Graham, K. W. T. & Hales, A. L. (1962a). The palaeomagnetism of the Stormberg Lavas of South Africa, I. Evidence for a genuine reversal of the Earth's field in Triassic-Jurassic times. *Geophys. J. R. Astr. Soc.*, 7, 23–9.

Van Zijl, J. S., Graham, K. W. T. & Hales, A. L. (1962b). The palaeomagnetism of the Stormberg Lavas. II. The behaviour of the magnetic field during a reversal. *Geophys. J. R. Astr. Soc.*, 7, 169–82.

Veldkamp, J., Mulder, F. G. & Zijderveld, J. D. A. (1971). Palaeomagnetism of Surinam dolerites. *Phys. Earth Planet. Int.*, **4**, 370–80.

Verma, R. K. & Mital, G. S. (1974). Paleomagnetic study of a vertical sequence of traps from Mount Pavagarh, Gujrat, India. *Phys. Earth Planet. Interiors*, **8**, 63–74.

Verma, R. K. & Pullaiah, G. (1967). Paleomagnetism of Tirupati sandstones from Godavary Valley, India. *Earth Planet. Sci. Lett.*, **2**, 310–16.

Vialon, P., Rochette, P. & Menard, G. (1989). Indentation and rotation in the western Alpine Arc. In *Alpine Tectonics*, ed. M. P. Coward, D. Dietrich & R. G. Park, *Geol. Soc. Special Publ.*, **45**, 329–38.

Vick, H. K., Channell, J. E. T. & Opdyke, N. D. (1987). Ordovician docking of the North Carolina Slate Belt: Paleomagnetic data. *Tectonics*, **6**, 573–83.

Vigliotti, L., Alvarez, W. & McWilliams, M. O. (1990). No relative rotation detected between Corsica and Sardinia. *Earth Planet. Sci. Lett.*, **98**, 313–18.

Vigliotti, L. & Kent, D. V. (1990). Paleomagnetic results of Tertiary sediments from Corsica: evidence of post-Eocene rotation. *Phys. Earth Planet. Int.*, **62**, 97–108.

Vilas, J. F. (1969). Resultados preliminares del estúdio paleomagnetico de algunas formaciones triásicas del sudoeste de Mendoza. *Cuartas J. Geol. Argent.*, III.

Vilas, J. F. (1974). Paleomagnetism of some igneous rocks of the Middle Jurassic Chon-Aike Formation from Estáncia La Reconquista, Province of Santa Cruz, Argentina. *Geophys. J. R. Astr. Soc.*, **39**, 511–22.

Vilas, J. F. (1976). Palaeomagnetism of the Lower Cretaceous Sierra de los Condores Group, Cordoba Province, Argentina. *Geophys. J. R. Astr. Soc.*, **46**, 295–305.

Vilas, J. F. & Valencio, D. A. (1978a). Palaeomagnetism and K-Ar dating of the Carboniferous Andacollo Series (Argentina) and the age of its hydrothermal overprinting. *Earth Planet. Sci. Lett.*, **40**, 101–6.

Vilas, J. F. & Valencio, D. A. (1978b). Palaeomagnetism and K-Ar age of the Upper Ordovician Alcaparrosa Formation, Argentina. *Geophys. J. R. Astr. Soc.*, **55**, 143–54.

Vincenz, S. A. & Jelenska, M. (1985). Paleomagnetic investigations of Mesozoic and Paleozoic rocks from Svalbard. *Tectonophysics*, **114**, 163–80.

Vitorello, I. & Van der Voo, R. (1977). Late Hadrynian and Helikian pole positions from the Spokane Formation, Montana. *Can. J. Earth Sci.*, **14**, 67–73.

Vogt, P. R., Anderson, C. N. & Bracey, D. R. (1971). Mesozoic magnetic anomalies, sea-floor spreading, and geomagnetic reversals in the southwestern North Atlantic. *J. Geophys. Res.*, **76**, 4796–823.

Vollstadt, H., Rother, K. & Nozharov, P. (1967). The palaeomagnetic stability and petrology of some Caenozoic and Cretaceous andesites from Bulgaria. *Earth Planet. Sci. Lett.*, **3**, 399–408.

Von Hoegen, J., Kramm, U. & Walter, R. (1991). The Brabant Massif as part of Armorica/Gondwana: U-Pb isotopic evidence from detrital zircons. *Tectonophysics*, **185**, 37–50.

Vugteveen, R. W., Barnes, A. E. & Butler, R. F. (1981). Paleomagnetism of the Roskruge and Gringo Gulch Volcanics, southeast Arizona. *J. Geophys. Res.*, **86**, 4021–8.

Walker, T. R., Larson, E. E. & Hoblitt, R. P. (1981). Nature and origin of hematite in the Moenkopi Formation (Triassic), Colorado Plateau: a contribution to the origin of magnetism in red beds. *J. Geophys. Res.*, **86**, 317–33.

Wang, C. (1984). The past and present situation of plate tectonic theory and contributions to its development by Chinese geologists. *Bull. Chinese Acad. Geol. Sci.*, **10**, 35–45.

Watts, D. R. (1982). A multicomponent, dual-polarity palaeomagnetic regional overprint from the Moine of northwest Scotland. *Earth Planet. Sci. Lett.*, **61**, 190–8.

Watts, D. R. (1985a). Palaeomagnetism of the Lower Carboniferous Billefjorden Group, Spitsbergen. *Geol. Mag.*, **122**, 383–8.

Watts, D. R. (1985b). Palaeomagnetic resetting in the Barrovian zones of Scotland and its relationship to the late structural history. *Earth Planet. Sci. Lett.*, **75**, 258–64.

Watts, D. R. & Bramall, A. M. (1981). Paleomagnetic evidence for a displaced terrain in West Antarctica. *Nature*, **293**, 638–41.

Watts, D. R. & Briden, J. C. (1984). Palaeomagnetic signature of slow postorogenic cooling of the northeast Highlands of Scotland recorded in the Newer Gabbros of Aberdeenshire. *Geophys. J. R. Astr. Soc.*, **77**, 775–88.

Watts, D. R. & Van der Voo, R. (1979). Paleomagnetic results from the Ordovician Moccasin, Bays, and Chapman Ridge Formations of the Valley and Ridge Province, eastern Tennessee. *J. Geophys. Res.*, **84**, 645–55.

Watts, D. R., Van der Voo, R. & French, R. B. (1980a). Paleomagnetic investigation of the Cambrian Waynesboro and Rome formations in the Valley and Ridge Province of the Appalachian mountains. *J. Geophys. Res.*, **85**, 5331–43.

Watts, D. R., Van der Voo, R. & Reeve, S. C. (1980b). Cambrian paleomagnetism of the Llano uplift, Texas. *J. Geophys. Res.*, **85**, 5316–30.

Watts, D. R., Watts, G. C. & Bramall, A. M. (1984). Cretaceous and Early Tertiary paleomagnetic results from the Antarctic Peninsula. *Tectonics*, **3**, 333–46.

Wegener, A. (1915). *Die Entstehung der Kontinente und Ozeane.* Braunschweig: Friedr. Vieweg u. Sohn.

Weisse, P. A., Haggerty, S. E. & Brown, L. L. (1985). Paleomagnetism and magnetic mineralogy of the Nahant Gabbro and Tonalite, eastern Massachusetts. *Can. J. Earth Sci.*, **22**, 1425–35.

Wellensiek, M. R., van der Pluijm, B. A., Van der Voo, R. & Johnson, R. J. E. (1990). Tectonic history of the Lunksoos composite terrane in the Maine Appalachians. *Tectonics*, **9**, 719–34.

Wellman, P., McElhinny, M. W. & McDougall, I. (1969). On the polar wander path for Australia during the Cenozoic. *Geophys. J. R. Astr. Soc.*, **18**, 371–95.

Wells, J. M. & Verhoogen, J. (1967). Late Paleozoic paleomagnetic poles and the opening of the Atlantic Ocean. *J. Geophys. Res.*, **72**, 1777–81.

Wendt, J. (1985). Distribution of the continental margin of northwestern Gondwana: Late Devonian of the eastern Anti-Atlas (Morocco). *Geology*, **13**, 815–18.

Wensink, H. (1968). Paleomagnetism of some Gondwana red beds from central India. *Palaeogeogr., Palaeoclimatol., Palaeoecol.*, **5**, 323–43.

Wensink, H. (1972). The paleomagnetism of the Salt Pseudomorph beds of Middle Cambrian age from the Salt Range, West Pakistan. *Earth Planet. Sci. Lett.*, **16**, 189–94.

Wensink, H. (1973). Newer paleomagnetic results of the Deccan Traps, India. *Tectonophysics*, **17**, 41–59.

Wensink, H. (1975). The paleomagnetism of the speckled sandstones of Early Permian age from the Salt Range, Pakistan. *Tectonophysics*, **26**, 281–92.

Wensink, H. (1979). The implications of some paleomagnetic data from Iran for its structural history. *Geol. Mijnbouw*, **58**, 175–85.
Wensink, H. (1981). Le contact Gondwana-Eurasie en Iran d'après les recherches paléomagnétiques. *Bull. Soc. géol. France*, **23**, 547–52.
Wensink, H. (1982). Tectonic inferences of paleomagnetic data from some Mesozoic formations in Central Iran. *J. Geophys.*, **51**, 12–23.
Wensink, H. (1983). Paleomagnetism of red beds of Early Devonian age from Central Iran. *Earth Planet. Sci. Lett.*, **63**, 325–34
Wensink, H. & Hartosukohardjo, S. (1990a). Paleomagnetism of younger volcanics from western Timor, Indonesia. *Earth Planet. Sci. Lett.*, **100**, 94–107.
Wensink, H. & Hartosukohardjo, S. (1990b). The palaeomagnetism of Late Permian to Early Triassic and Late Triassic deposits on Timor: an Australian origin? *Geophys. J. Int.*, **101**, 315–28.
Wensink, H., Hartosukohardjo, S. & Kool, K. (1987). Paleomagnetism of the Nakfunu Formation of Early Cretaceous age, western Timor, Indonesia. *Geol. Mijnbouw*, **66**, 89–99.
Wensink, H., Hartosukohardjo, S. & Suryana, Y. (1989). Paleomagnetism of Cretaceous sediments from Misool, NE Indonesia. *Netherl. J. Sea Res.*, **24**, 287–301.
Wensink, H. & Klootwijk, C. T. (1968). The paleomagnetism of the Talchir Series of the Lower Gondwana System, central India. *Earth Planet. Sci. Lett.*, **4**, 191–6.
Wensink, H. & Klootwijk, C. T. (1971). Paleomagnetism of the Deccan Traps in the western Ghats near Poona (India). *Tectonophysics*, **11**, 175–90.
Wensink, H. & Varekamp, J. C. (1980). Paleomagnetism of basalts from Alborz: Iran part of Asia in the Cretaceous. *Tectonophysics*, **68**, 113–29.
Wensink, H., Zijderveld, J. D. A. & Varekamp, J. C. (1978). Paleomagnetism and ore mineralogy of some basalts of the Geirud Formation of Late Devonian to Early Carboniferous age from the southern Alborz, Iran. *Earth Planet. Sci. Lett.*, **41**, 441–50.
Westphal, M. (1972). Etude paléomagnétique de certaines formations du Paléozoïque supérieur. *Mem. Bur. Recherches Géol. Minières*, 77, 857–60.
Westphal, M. (1973). Etudes paléomagnétiques de quelques formations permiennes et triasiques dans les Alpes Occidentales (France). *Tectonophysics*, **17**, 323–35.
Westphal, M. (1986). *Paléomagnétisme et magnétisme des roches*. Paris: Doin-Editeurs.
Westphal, M., Bazhenov, M. J., Lauer, J. P., Pechersky, D. M. & Sibuet, J. C. (1986). Paleomagnetic implications on the evolution of the Tethys belt from the Atlantic Ocean to the Pamirs since the Triassic. *Tectonophysics*, **123**, 37–82.
Westphal, M. & Durand, J. P. (1989). An Upper Cretaceous paleomagnetic pole for stable Europe from Aix-en-Provence (France). *Earth Planet. Sci. Lett.*, **94**, 143–50.
Westphal, M., Edel, J. B., Kadzialko-Hofmokl, M., Jelenska, M. & Grocholski, A. (1987). Paleomagnetic study of Upper Carboniferous volcanics from Sudetes (Poland). *J. Geophys.*, **61**, 90–6.
Westphal, M., Montigny, R., Thuizat, R., Bardon, C., Bossert, A., Hamzeh, R. & Rolley, J. P. (1979). Paléomagnétisme et datation du volcanisme permien, triasique et crétacé du Maroc. *Can. J. Earth Sci.*, **16**, 2150–64.
Westphal, M., Orsini, J. & Vellutini, P. (1976). Le microcontinent Corso-Sarde, sa position initiale: données paléomagnétiques et records géologiques. *Tectonophysics*, **30**, 141–57.
Westphal, M., Pozzi, J.-P., Zhou, Y. X., Xing, L. S. & Chen, X. Y. (1983). Palaeomagnetic data about southern Tibet (Xizang), I. The Cretaceous formations of the Lhasa Block. *Geophys. J. R. Astr. Soc.*, 73, 507–21.
Whitechurch, H., Juteau, T. & Montigny, R. (1984). Role of the Eastern Mediterranean ophiolites (Turkey, Syria, Cyprus) in the history of the Neo-Tethys. In *The Geological Evolution of the Eastern Mediterranean*, ed. J. E. Dixon & A. H. F. Robertson, pp. 301–17. Oxford: published by Blackwell for the Geol. Society.
Williams, C. A. (1975). Sea-floor spreading in the Bay of Biscay and its relationship to the North Atlantic. *Earth Planet. Sci. Lett.*, **24**, 440–56.
Williams, G. E. (1975). Late Precambrian glacial climate and the Earth's obliquity. *Geol. Mag.*, **112**, 441–65.
Williams, H. (1978). *Tectonic Lithofacies map of the Appalachian Orogen. Map 1.* St John's, Newfoundland: Memorial University of Newfoundland.
Williams, H. (1979). Appalachian orogen in Canada: *Can. J. Earth Sci.*, **16**, 792–807.
Williams, H., Colman-Sadd, S. P. & Swinden, H. S. (1988). *Tectonostratigraphic Sub-*

divisions of Central Newfoundland. Current Research, part B, Geol. Survey Canada, Paper 88–1B, pp. 91–8.

Williams, H. & Payne, J. G. (1975). The Twillingate Granite and nearby volcanic groups, an island arc complex in northeast Newfoundland. *Can. J. Earth Sci.*, **12**, 982–95.

Williamson, P. & Robertson, W. A. (1976). Iterative method of isolating primary and secondary components of remanent magnetization illustrated by using the Upper Devonian Catombal Group of Australia. *J. Geophys. Res.*, **81**, 2531–8.

Wilson, J. T. (1966). Did the Atlantic close and then re-open? *Nature*, **211**, 676–81.

Wilson, R. L. (1970a). Permanent aspects of the Earth's non-dipole magnetic field over the past 25 million years. *Geophys. J. R. Astr. Soc.*, **19**, 417–37.

Wilson, R. L. (1970b). Palaeomagnetic stratigraphy of Tertiary lavas from Northern Ireland. *Geophys. J. R. astr. Soc.*, **20**, 1–9.

Wilson, R. L. (1971). Dipole offset – the time-averaged palaeomagnetic field over the past 25 million years. *Geophys. J. R. Astr. Soc.*, **22**, 491–504.

Wilson, R. L., Ade-Hall, J. M., Skelhorn, R. R., Speicht, J. M. & Dagley, P. (1974). The British Tertiary igneous province: palaeomagnetism of the Vaternish dyke swarm on North Skye, Scotland. *Geophys. J. R. astr. Soc.*, **37**, 23–30.

Wilson, R. L., Dagley, P. & Ade-Hall, J. M. (1972). Palaeomagnetism of the British Tertiary province: the Skye Lavas. *Geophys. J. R. astr. Soc.*, **28**, 285–93.

Wilson, R. L. & Everitt, C. W. F. (1963). Thermal demagnetization of some Carboniferous lavas for palaeomagnetic purposes. *Geophys. J. R. astr. Soc.*, **8**, 149–64.

Wilson, R. L., Hall, J. M. & Dagley, P. (1982). The British Tertiary igneous province: Palaeomagnetism of the dyke swarm along the Sleat coast of Skye. *Geophys. J. R. Astr. Soc.*, **68**, 317–23.

Windley, B. F. (1977). *The Evolving Continents.* New York: John Wiley and Sons.

Wisniowiecki, M., Van der Voo, R., McCabe, C. & Kelly, W. C. (1983). Pennsylvanian paleomagnetic pole from the mineralized Late Cambrian Bonneterre Formation, SE Missouri. *J. Geophys. Res.*, **88**, 6540–8.

Wissman, G. & Roeser, H. A. (1982). A magnetic and halokinetic structural Pangaea fit of northwest Africa and North America. *Geol. Jhrb., Ser. E., Geophysics*, **23**, 43–61.

Witte, W. K. & Kent, D. V. (1989). A middle Carnian to early Norian (c. 225 Ma) paleopole from sediments of the Newark Basin, Pennsylvania. *Geol. Soc. Am. Bull.*, **101**, 1118–26.

Witte, W. K. & Kent, D. V. (1990). The paleomagnetism of red beds and basalts of the Hettangian extrusive zone, Newark Basin, New Jersey. *J. Geophys. Res.*, **95**, 17533–45.

Witzke, B. J. (1990). Palaeoclimatic constraints for Palaeozoic palaeolatitudes of Laurentia and Euramerica. In *Palaeozoic Palaeogeography and Biogeography*, ed. W. S. McKerrow & C. R. Scotese. *Geol. Soc. London Memoir*, **12**, 57–73.

Wyllie, P. J. (1976). *The Way the Earth Works.* New York: John Wiley and Sons.

Wynne, P. J., Irving, E. & Osadetz, K. (1983). Paleomagnetism of the Esayoo Formation (Permian) of northern Ellesmere Island: possible clue to the solution of the Nares Strait dilemma. *Tectonophysics*, **100**, 241–56.

Wynne, P. J., Irving, E. & Osadetz, K. G. (1988). Paleomagnetism of Cretaceous volcanic rocks of the Sverdrup Basin – magnetostratigraphy, paleolatitudes, and rotations. *Can. J. Earth Sci.*, **25**, 1220–39.

Yang, K.-M., Wang, Y., Tsai, Y.-B. & Hsu, V. (1982). Paleomagnetic studies of the coastal range, Lutao and Lanhsu in eastern Taiwan and their tectonic implications. *Bull. Inst. Earth Sci. Acad. Sinica*, 3, 173–188.

Yilmaz, P. O. (1984). Fossil and K-Ar data for the age of the Antalya complex. In *The Geological Evolution of the Eastern Mediterranean*, ed. J. E. Dixon & A. H. F. Robertson, pp. 335–47. Oxford: published by Blackwell for the Geol. Society.

Yokoyama, T. & Dharma, A. (1982). Magnetostratigraphy of the Toba Tuffs in Sumatra, Indonesia. In *Palaeomagnetic Research in Southeast and East Asia*, ed. J. C. Briden *et al.*, pp. 201–11. Committee for the Coordination of Offshore Prospecting (CCOP), U.N. Development Project Office.

Yole, R. W. & Irving, E. (1980). Displacement of Vancouver Island: paleomagnetic evidence from the Karmutsen Formation. *Can. J. Earth Sci.*, **17**, 1210–28.

Young, G. C. (1990). Devonian vertebrate distribution patterns and cladistic analysis of palaeogeographic hypotheses. In *Palaeozoic Palaeogeography and Biogeography*, ed. W. S. McKerrow & C. R. Scotese. *Geol. Soc. London Memoir*, **12**, 243–55.

Yousif, I. A. & Beckman, G. E. (1981). A palaeomagnetic study of some Tertiary and

Cretaceous rocks in Western Saudi Arabia: evidence for the movement of the Arabian plate. *Bull. Fac. Earth Sci. King Abdullaziz Univ.*, **4**, 89–106.

Zhang, H. & Zhang, W. (1985). Palaeomagnetic data, Late Precambrian magnetostratigraphy and tectonic evolution of eastern China. *Precambrian Res.*, **29**, 65–75.

Zhang, Z. K. (1984). Sino-Korean blocks and Yangtze block as part of the Pacifica continent in the late Paleozoic. *Bull. Chinese Acad. Sci.*, **9**, 45–54. (In Chinese with English Abstract.)

Zhao, X. & Coe, R. S. (1987). Paleomagnetic constraints on the collision and rotation of North and South China. *Nature*, **327**, 141–4.

Zhao, X., Coe, R. S., Zhou, Y., Wu, H. & Wang, J. (1990). New paleomagnetic results from northern China: collision and suturing with Siberia and Kazakhstan. *Tectonophysics*, **181**, 43–81.

Zheng, Z., Kono, M., Tsunakawa, H., Kimura, G., Wei, Q., Zhu, X. & Hao, T. (1991). The apparent polar wander path for the North China Block since the Jurassic. *Geophys. J. Int.*, **104**, 29–40.

Zhou, Y. X., Lu, L. Z. & Zheng, B. M. (1986). Paleomagnetic polarity of the Permian Emeishan Basalt in Sichuan. *Geol. Review*, **32**, 465–9. (In Chinese with English Abstract.)

Zhu, X., Liu, C., Ye, S. & Lin, J. (1977). Remanence of red beds from Linzhou, Xizang and the northward movement of the Indian plate. *Scientia Geol. Sin.*, pp. 44–51.

Zhu, Z., Zhu, X. & Zhang, Y. (1981). Palaeomagnetic observation in Xizang and continental drift. *Acta Geophys. Sin.*, **24**, 40–9.

Ziegler, A. M. (1990). Phytogeographic patterns and continental configurations during the Permian period. In *Palaeozoic Palaeogeography and Biogeography*, ed. W. S. McKerrow & C. R. Scotese. *Geol. Soc. London Memoir*, **12**, 363–79.

Ziegler, A. M., Bambach, R. K., Parrish, J. T., Barrett, S. F., Gierlowski, E. H., Raymond, A. & Sepkoski, J. J. Jr. (1981). Palaeozoic biogeography and climatology. In *Palaeobotany, Palaeoecology and Evolution*, ed. K. J. Niklas, vol. 2, pp. 231–66. New York: Praeger.

Ziegler, A. M., Hansen, K. S., Johnson, M. E., Kelly, M. A., Scotese, C. R. & Van der Voo, R. (1977). Silurian continental distributions, paleogeography, climatology, and biogeography. *Tectonophysics*, **40**, 13–51.

Ziegler, A. M., Raymond, A. L., Gierlowski, T. C., Horrell, M. A., Rowley, D. B. & Lottes, A. L. (1987). Coal, climate, and terrestrial productivity: the present and Early Cretaceous compared. In *Coal and Coal-Bearing Strata: Recent Advances*, ed. A. C. Scott. *Geol. Soc. Special Publ.*, **32**, 25–49.

Ziegler, P. (1988). Evolution of the Arctic-North Atlantic and the western Tethys. *Memoir*, **43**, 198 pp. Tulsa, Oklahoma: Am. Assoc. Petrol. Geol.

Zijderveld, J. D. A. (1967a). AC demagnetization of rocks: analysis of results. In *Methods in Palaeomagnetism*, ed. S. K. Runcorn, K. M. Creer & D. W. Collinson, pp. 254–86. Amsterdam: Elsevier.

Zijderveld, J. D. A. (1967b). The natural remanent magnetization of the Exeter volcanic traps (Permian, Europe). *Tectonophysics*, **4**, 121–53.

Zijderveld, J. D. A. (1968). Natural remanent magnetizations of some intrusive rocks from the Sør Rondane mountains, Queen Maud Land, Antarctica. *J. Geophys. Res.*, **73**, 3773–85.

Zijderveld, J. D. A. (1975). *Paleomagnetism of the Estérel rocks*. Ph.D. Thesis, Univ. of Utrecht.

Zijderveld, J. D. A. & De Jong, K. A. (1969). Paleomagnetism of some late Paleozoic and Triassic rocks from the eastern Lombardic Alps. *Geol. Mijnbouw*, **48**, 559–64.

Zijderveld, J. D. A., De Jong, K. A. & Van der Voo, R. (1970b). Rotation of Sardinia: palaeomagnetic evidence from Permian rocks. *Nature*, **226**, 933–4.

Zijderveld, J. D. A., Hazeu, G. J. A., Nardin, M. & Van der Voo, R. (1970a). Shear in the Tethys and the Permian paleomagnetism in the southern Alps, including new results. *Tectonophysics*, **10**, 639–61.

Zonenshain, L. P., Korinevsky, V. G., Kazmin, V. G. & Pechersky, D. M. (1984). Plate tectonic model of the south Urals development. *Tectonophysics*, **109**, 95–135.

Zonenshain, L. P., Verhoef, J., Macnab, R. & Meyers, H. (1991). Magnetic imprints of continental accretion in the USSR. *Eos, Trans. Am. Geophys. Union*, 72, 305–310.

Zwart, H. J. & Dornsiepen, V. F. (1978). The tectonic framework of central and western Europe. *Geol. Mijnbouw*, **57**, 627–54.

Index